Springer Series on
SIGNALS AND COMMUNICATION TECHNOLOGY

Signals and Communication Technology

Riad I. Hammoud (Ed.)

Passive Eye Monitoring

Algorithms, Applications and Experiments

With 180 Figures and 35 Tables

 Springer

Dr. Riad Ibrahim Hammoud (Ed.)
Delphi Electronics & Safety
Adv. Product & Business Development
Adv. Controls & Security
M.S. E110
1800 E. Lincoln Road
Kokomo, IN 46902
USA
riad.hammoud@delphi.com

ISBN 978-3-540-75411-4   e-ISBN 978-3-540-75412-1

DOI 10.1007/978-3-540-75412-1

Springer Series on Signals and Communication Technology ISSN 1860-4862

Library of Congress Control Number: 2007937297

Typesetting and production: LE-TEX Jelonek, Schmidt & Vöckler GbR, Leipzig, Germany
Cover design: WMXDesign GmbH, Heidelberg

Printed on acid-free paper

9 8 7 6 5 4 3 2 1

springer.com

To
Amine, Ghada & my parents,
J. Eitha & his brothers,
and to my dear Hiba.

Preface

The advent of digital technology and image processing in the 1970s has marked a new era of human machine interaction. Since then there has been a rapid development in the eye monitoring field – a sub-area of intelligent human computer interaction – that brought up efficient ways for computers to support and assist humans while engaging in non-trivial tasks. The purpose of this book is to provide a comprehensive in-depth coverage of both theory and practice of *Passive Eye Monitoring*.

Passive Eye Monitoring is the science of teaching computers to automatically determine, in digital video streams, the spatial location, pupil's activity, 3D gaze direction, movements and closure state of an individual's eye, in real-time. It gained a lot of attention as opposed to invasive eye monitoring techniques like *Electro Oculography* and *Magnetic Search Coil*. The theory and practice of passive eye monitoring have now reached a level of maturity where good results can be achieved for applications – in automotive, military, medicine, information security and retrieval, typing and reading, online search, marketing, augmented and virtual reality, and video games – that were certainly non realistic a few decades ago. For instance, this technology has the potential to allow a disabled person to compose electronic messages, browse and navigate through web pages, interact with consumer – electronics devices, and while steering an operating machine, this technology monitors his/her fatigue and distraction state and issues appropriate vigilance warnings. Excitement about the apparent magic of Passive Eye Monitoring comes along from several other ends including that this technology is passive and transparent to users, as well as a valuable factor in bringing down the financial burden caused by fatigue-related vehicle accidents, design and advertisement, and care-giving, to mention a few.

This book offers students and software engineers a detailed description of vision algorithms and core building-blocks of a Passive Eye Monitoring system (often referred to as Eye Tracking or Eye-Gaze Tracking system). It also offers a detailed coverage of recent successful practice of eye monitoring systems in advanced applications and Human Factors research studies. The book contains six distinctive parts and a brief introductory chapter. This latter places the reader into the context of the book's terminologies, motivations, system's building-blocks, challenges and summary of inventions and innovations portrayed in the remaining seventeen chapters. All six chapters of Part I and Part II are reserved for the description of the fundamental algorithmic building-blocks. These chapters emphasis major issues pertaining to some (commercially) available eye tracking systems and state-of-the-art video-based approaches. Furthermore, they present theoretical and experimental details of effective methodologies for building a solid eye monitoring system, including: automatic detection of human eyes in videos, modeling the eyes' structure and appearance among diverse set of human subjects, face poses and lighting conditions, tracking of spatial eye coordinates, identifying pupil, iris, eyelids, and eye-corners locations, estimating, calibrating and tracking a person's gaze direction, as well as interpreting and reconstructing displayed worlds' images on the eye's cornea.

Parts III through VI of this book fulfill the purpose for which eye monitoring systems are built to a great extent. They present advanced applications for which (commercially) available eye monitoring systems are either used as interaction input devices and/or reliable source of data. Part III is on *Gaze-Based Interactions* and contains three chapters. The first one provides a general introduction to the exciting topic of user's interaction with diverse interfaces through eye-gaze as opposed to using hands to control input computers devices. Chapters 9 and 10 highlight the emerging topic of Gaze-Contingent Display – a smart display that enriches the area surrounding the user's gaze fixation point by means of higher resolution rendering and virtual data overlaying as in medical and geophysical imaging visualization, reading and searching.

Part IV includes two chapters on applications of eye monitoring in Military and Security Realms. Chapter 11 reviews the use of eye monitoring in military for interface design and evaluation, performance appraisal, and workload assessment, and gives specific examples of eye tracking contributions in each application area. Chapter 12 proposes to use eye monitoring technology as a "hands-free mouse" without on-screen feedback to allow users to enter information (not limited to passwords) without disclosing it to shoulder surfers.

Part V presents recent deployment of passive eye monitoring technology in automotive and medicine in order to reduce driving accidents and surgeons errors caused by fatigue and distraction. Chapter 13 presents a novel data mining and fusion framework to infer a driver's cognitive distraction state, where various eye movement measures like fixation duration and saccade distance are used as input. Chapter 14 reviews various state-of-the-art approaches and research studies for detecting and mitigating driver drowsiness and fatigue. The last chapter of this part (Chap. 15) provides a general overview of eye tracking technologies in medical surgery and radiology with particular emphasis on the role of eye trackers as tools to enhance visual search, and understand the nature of expertise in radiology.

The last part on *Eye Monitoring in Information Retrieval and Interface Design Assessment* answers questions about how a user scans and searches for online web content and how net surfers respond to banner ads and other advertisements (Chap. 16), as well as how the visual attention of a viewer is distributed when the monitor presents several adjacent windows or screens (Chaps. 17 and 18).

This practical reference offers a thorough understanding of how the core low-level building blocks of Passive Eye Monitoring systems are implemented, and further it describes in details where and how this technology is employed successfully either as an *interaction input device* or a *reliable source of data*. This book contains enough material to fill a two-semester upper-division or advanced graduate course in the fields of eye tracking, computer vision, Human Factors research, computer graphics and interactive visualization, controls and conmmunications, cognitive and behavioral understanding, and intelligent human computer interaction in general. Scientists, teachers, interface designers, software developers and futurists will find in-depth coverage of recent state-of-the-art eye monitoring theory and practice. Moreover, the book helps readers of all levels understand the motivations, activities, trends and directions of developers and researchers in the passive eye monitoring field in today's market, and offers them a view of the future of this rapidly evolving inter-disciplinary technological area. However, deep diving into the first two parts on building-blocks requires a basic knowledge in image processing, pattern recognition and computer vision.

This effort couldn't have been achieved without the valuable contributions made by a number of my colleagues working in an inter-disciplinary fields in both academia and industry. Their biographies are listed at the end of this book along with their acknowledgments. I'm so grateful for their collaborations; their expertise, contributions, feedbacks and reviewing added significant value to this groundbreaking resource.

I would like to extend thanks to all folks at Springer-Verlag, and in particular to Dr. Christoph Baumann for his warm support.

Kokomo, Indiana, USA   *Dr. Riad Ibrahim Hammoud*
August 14, 2007

Contents

3 Advanced Off-Line Statistical Modeling of Eye and Non-Eye Patterns

4 Automatic Eye Position Detection and Tracking Under Natural Facial Movement

Part III Gaze-based Interactions

8 Gaze-based Interaction

9 Gaze-Contingent Volume Rendering

Part V Eye Monitoring in Automotive and Medicine

1

Introduction to Eye Monitoring

Riad I. Hammoud and Jeffrey B. Mulligan

1.1 A Century of Eye Monitoring

Scientific study of human eye movements began in the late 19th century, and employed a variety of measurement techniques. Many systems, such as that of Delabarre [122] were mechanical, while others like Dodge and Cline [132] used photography. Since their inception, eye monitoring systems have not been considered important innovations in-and-of themselves, but rather tools that allow scientists to study the gaze behavior of experimental subjects. While much information concerning basic mechanisms can be gleaned from responses to controlled laboratory studies, there is also much interest in behaviors occurring in the "real world". Over time, technological advances have allowed eye monitoring systems to emerge from the laboratory, and today the use of eye tracking in the study of the performance of everyday tasks is routine in a number of applied disciplines [522].

In the late 1940s, researchers used cameras to record the eye movements of pilots in the cockpit [11]. Eye tracking systems such as these were refined in the following decades, and a host of techniques appeared, including *electro-oculography (EOG)* and *magnetic search coil* [178,527]. But it was the advent of digital technology and image processing in the 1970s which marked the opening of a new era of *video-based eye monitoring*. Research continued in the 1970s under sponsorship from the U.S. Air Force to improve cockpit usability [11]. Today, many companies offer video-based eye monitoring systems at affordable prices.

1.2 From Active to Passive Eye Monitoring

Historically, eye monitoring systems were developed in support of physiological research on the oculomotor system. Eye monitoring systems could be classified into two categories: invasive and active vs. non-invasive and passive. In this latter category the experimental subject is often not aware of the presence of the eye monitoring system as no device is being actively attached to the physical body.

In an early study of fixational eye movements, Horace Barlow placed a drop of mercury in his eye, while an iron bar pressed his head firmly against a granite slab [21]. Fortunately, it is unnecessary to go to such heroic lengths today! Nevertheless, it is still difficult to make measurements having a precision comparable to the physiological noise level: subjects can generally maintain fixation to within a few minutes of arc, while a typical video-based system will produce errors as large as a degree unless special care is taken. To make ultra-precise measurements, a number of approaches have been developed over the years. For example, in the magnetic search coil system [178, 527], a small loop of wire is placed in the eye; in humans this is usually done with a special contact lens, while for animal research the coils are usually surgically placed under the conjunctiva. The position and orientation of the coil (and hence the eye) is determined by measuring the currents induced by three, mutually orthogonal external magnetic fields. The search coil has the advantage of not requiring that the head be fixed, although it must remain in the volume enclosed by the field coils. It is fairly complex and expensive, but is unique in enabling very accurate determination of gaze while still allowing free head movement. But the placement of the coil in the eye is intrusive and can cause discomfort to the subject.

Another high-precision eye tracker is the Dual Purkinje Image (DPI) tracker [109]. It uses fast optical servos to track the first and fourth Purkinje images (reflections of the illuminator from the refracting surfaces of the eye). The first Purkinje image is a virtual image formed by the front surface of the cornea, while the fourth Purkinje image is a real image formed by the (concave) rear surface of the crystalline lens. These two images fortuitously both fall more-or-less in the plane of focus of the pupil, and can sometimes be observed in video images of the eye, although good sharp focus is required to see the dim fourth image. The design of the DPI tracker is such that there is no relative motion between the two images if the eye translates without rotating, while the relative positions encode the rotational state. While the DPI tracker provides excellent performance in terms of sensitivity and temporal bandwidth, it requires stabilization of the head (usually with a dental impression or "bite-bar"), and is thus unsuitable for measurements in "natural" conditions.

The electro-oculogram (EOG) is a measurement made using electrodes attached to the skin around the eye region (see Fig. 1.1). The accumulation of electrical charges in the retina gives the eye a dipole moment, and motion of the eye causes the electrical potential to vary in the surrounding region. After calibration, readings of these voltages can be used to infer into eye gaze direction. Unfortunately, for our purposes, the dipole moment changes as the visual stimulation impinging on the retina changes, which limits the accuracy

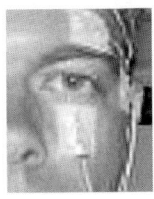 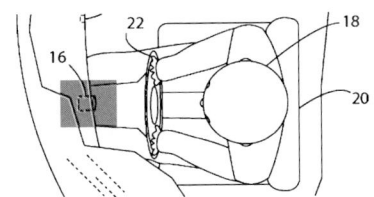 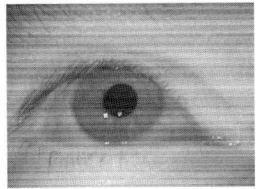

Fig. 1.1. Active electro-oculography eye recordings (*left*) vs. passive video-based eye monitoring (*middle*); the EOG instrument places several electrodes on the skin surrounding the eye (*left image*), while the passive system is image based that uses remote camera (16, *middle image*) to capture images of the subject's eye (*right image*) and compute the direction of gaze

in practical situations. The EOG is simple to implement, and relatively low in cost, but the electrodes are somewhat intrusive, and require a bit of setup time.

The aforementioned methods are relatively invasive and can cause discomfort to the subject. Moreover, they provide indirect measures of what is falling on the retina, and are subject to mechanical artifacts. For instance, the DPI tracker generates spurious transients caused by "wobble" of the eye's lens following rapid saccadic movements, while users of search coil systems must worry about slippage of the coil relative to the eye. Retinal imaging, on the other hand, provides a direct measure of what is on the retina. In the scanning laser ophthalmoscope (SLO) [668–670], a laser beam provides illumination for imaging retinal structures, while simultaneously allowing modulation of the beam intensity to deliver patterned stimulation. Recent SLO designs have incorporated adaptive optics to correct the eye's aberrations [536], enabling the acquisition of diffraction-limited images of retinal structures, and images of the foveal cone mosaic. Using an adaptive-optics-enhanced SLO (AOSLO), it has been shown that the functional foveal "center" (defining the line-of-sight) has a slightly different position for steady fixation target versus smooth pursuit of a moving target [587].

Today, the majority of eye monitoring systems in general use are based on digital images of the front of the eye, captured with a remote video camera and coupled with image processing and machine vision hardware and software. Such systems are called *passive eye monitors*, or video-based eye trackers. When these systems first appeared, they generally required special-purpose image-processing hardware to allow real-time measurement of gaze. Today, thanks to the steady increase of microprocessor power, it is possible to do a decent job entirely with software.

Passive eye monitoring technology became popular by being completely remote and non-intrusive, while offering reasonable accuracy at an affordable cost. It can be slower (sampling at about 30 to 60 Hz) and less accurate than invasive eye monitoring techniques, but provides a more natural experience for the subjects. In many applications, the available accuracy may be adequate to answer the questions of interest. This technology, for example, allows a disabled person to compose electronic messages, browse and navigate through web pages, turn on-off monitors, or call an assistant. Installed in a motor vehicle, a passive eye monitor can continuously evaluate the driver's fatigue and distraction [1,158,226], and generate appropriate vigilance warnings. Numerous other application areas have benefitted from the recent advances in the theory and practice of video based approaches: military, medicine, information security and retrieval, typing and reading, online search, marketing, augmented and virtual reality, and video games, to name a few.

1.3 Terminologies and Definitions

1.3.1 Eye Monitoring Terms

In the past few decades, there has been a rapid development in the passive eye monitoring field [221,358,522], a sub-area of intelligent human computer interaction [220,286,630]. Passive Eye Monitoring is the science of teaching computers to automatically determine, in digital video streams, the spatial location, 3D gaze direction, pupil activity, and closure level of an individual's eye, in real-time. Thus, eye monitoring encompasses a number of component technologies, including: *spatial eye position tracking, eye-gaze tracking, eye closure state tracking, eye movement tracking,* and *pupil-size monitoring.* Perhaps the most widely-known of these is *gaze tracking,* which means the determination of where a person is looking. The first step in analyzing the eye is finding it. *Spatial eye position tracking* refers to finding the eye coordinates in a sequence of face images [215,217,220,229,230]. Sometimes we are interested in monitoring the eye without reference to the gaze direction. For instance, in an eye-based biometric system [10,219] we might want to find the eye and the iris, without being particularly concerned with what the subject is looking at. *Eye closure state tracking* refers to determining the degree of closure of the eyelids. For example, in a fatigue monitoring system, we might be primarily interested in whether the eyes are open or closed [222,226], assuming that if the eyes are open then the subject is awake, and, hopefully, looking where he or she ought to be. *Pupil-size tracking* monitors variations in pupil's size including dilation. *Eye movement tracking* refers to interpreting different eye movements, captured by an eye tracker, as indicators of ocular behaviors, namely fixations, saccades, pupil dilation, and scanpaths. These terms are defined below.

1.3.2 Human Eyeball Model

The eyes are approximately spherical globes having a diameter of around 2 centimeters. Light enters the front of the eye through the *cornea,* a transparent membrane which bulges out from the spherical eyeball. As can be

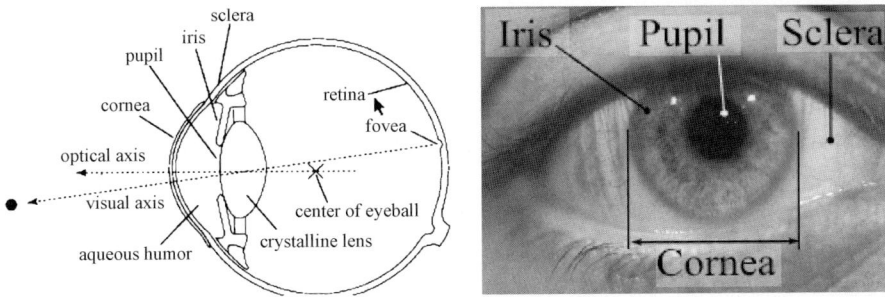

Fig. 1.2. Cross section and frontal view of a human eyeball. The frontal eye view is captured in the visible spectrum

seen in Fig. 1.2(right), the most distinct visual features of the eye are the cornea and the sclera. Figure 1.2(left) shows a schematic of a horizontal cross-section of a human eyeball. The cornea consists of a lamellar structure of submicroscopic collagen fibrils arranged in a manner that makes it transparent [119, 315, 675]. The external surface of the cornea is very smooth. In addition, it has a thin film of tear fluid on it. As a result, the surface of the cornea behaves like a mirror. The *sclera* is the tough, white tissue that makes up most of the ball; the cornea joins the sclera at an interface known as the *limbus*. Suspended in the circular aperture bounded by the limbus is an annular piece of tissue known as the *iris*. The iris is the pigmented part of the eye that determines a person's eye color. The central hole in the iris forms the *pupil*, through which light enters the eye. Small muscles enable the pupil to change size in response to changing illumination conditions.

Seen from the front, with a remote digital infrared camera, the eyes generally appear as a dark disk (the pupil), surrounded by a colored annulus (the iris), usually partially occluded top and bottom by the eyelids, with a triangular patch of white sclera visible on one or both sides (see Fig. 1.3). These are the anterior structures of the eye which are most commonly used for gaze tracking. Behind the iris lies the *crystalline lens*, the structure which allows the eye to focus on objects at different depths, a process known as *accomodation*. The lens is difficult to observe without special equipment, but small dim reflections from its front and rear surfaces can sometimes be observed in images of the anterior structures of the eye.

The eyes rotate within cavities in the skull known as the *orbits*. The movements are accomplished through the action of 3 pairs of antagonistic muscles, which individually produce horizontal rotation (pan), vertical rotation (tilt), and torsion about the line of sight (roll). Horizontal and vertical rotations are generally estimated from the movement of an anterior structure such as the pupil or limbus, while torsion is generally ignored in most commercial gaze-tracking systems. In some circumstance, however, it may be important to measure torsion, which is often done by tracking spatial structure in the iris.

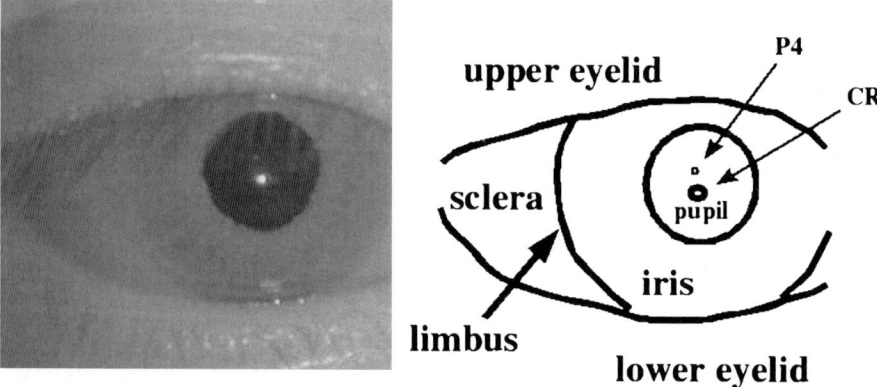

Fig. 1.3. Image of an eye captured under near-infrared (NIR) illumination, with a segmentation of the major parts indicated on the right. CR refers to the corneal reflex (the reflection of the illuminator from the front surface of the cornea); P4 refers to the fourth Purkinje image, a dim reflection from the rear surface of the crystalline lens, which sits behind the iris. See text for explanation of other terms

When the eye is observed by a head mounted camera, motion of the pupil will be directly related to eye-in-head rotation. When the camera is located off the head, however, even small translations of the head induce motions of the eye which can be confounded with eye rotations if corrective measures are not taken. A common approach is to utilize reflections from the front surface of the cornea. These may be controlled by illuminating the eye with near-infrared (NIR) light, and equipping the camera with an infrared filter to block reflections of visible light from the environment. Under these conditions, the only reflection will be from the illuminator; this feature is variously referred to as the *corneal reflex* (CR) or the *glint* (see Fig. 1.3). When the camera and illuminator are sufficiently distant from the eye, the glint will translate with the eye, and eye rotation can be estimated from the differential motion between the glint and the pupil. When either the camera or illuminator is near enough to the eye that the small-angle approximation no longer holds, then the simple difference is no longer the ideal estimator, a poorly-appreciated fact discussed in a recent paper [332]. Other features may also be tracked to provide an estimate of head movement, such as the angles formed by the juncture of the upper and lower eyelid, known as the *nasal canthus* and *temporal canthus*; artificial marks may also be made on the skin or sclera.

1.3.3 Types of Eye Movement

The eyes move in a number of different ways, simultaneously responding to commands from a number of different brain areas. One of the most important types of eye movement is not really a movement at all, but rather the ability to keep the eye trained on a fixed spot in the world. This is known as *fixation*. The ability to maintain steady fixation on an object in the world in the presence of head movement is aided by the *vestibulo-ocular reflex* (VOR). The VOR is effected by a short and fast neural circuit between the vestibular organs in the inner ear and the eye muscles; rotations of the head are detected by the *semi-circular canals*, while linear acceleration (including gravity) is detected by the *otoliths*.

Vestibular signals are not the only cues available to indicate that the head has moved: motions of the head which carry along the eyes also generate full field motion, and the eyes reflexively move to stabilize this motion. This type of eye movement is known as *ocular following*. Ocular following is generally thought to be distinct from *smooth pursuit* eye movements, in which the eyes track the movement of a small target on a stationary (or differentially moving) background. Both smooth pursuit and ocular following have a long latency compared to the VOR, a difference which can be observed easily with simple equipment: if you hold your finger up in front of your face, and try to maintain fixation on the finger while shaking the head back and forth at 2–3 Hz (a small amplitude of a few degrees is sufficient), you will find that it is not hard to keep the image of the finger stable. However, if the finger itself is moved back and forth at the same rate, the eyes are incapable of tracking the motion and the finger appears as a blurred streak, even though in this case the brain potentially has access to the the finger motor control signals which might be used to predict the motion.

Our visual experience is generally made up of a series of fixations on different objects. To get from one fixation to the next, the eyes make rapid, ballistic movements known as *saccades*. Saccadic movements are the fastest movements made by any part of the human body, with rotational velocities as high as 500 deg/sec for large saccades. During saccades, events are relatively invisible, not just because of blur but due to a presumed neural process known as *saccadic suppression*.

Because the two eyes are separated in space, they must point in slightly different directions in order to fixate a near object (such as the hand). The angle between the two lines-of-sight is known as the *convergence angle*, and movements of the eyes which change this angle are known as *vergence movements*. Horizontal vergence movements are an example of *disjunctive* eye movements, in which the two eyes move oppositely. *Conjunctive* eye movements, on the other hand, refer to events in which the two eyes make the same movement; in the case of horizontal and vertical eye rotations, this is referred to as *version*. Horizontal vergence, used to binocularly fixate objects at different depths, can be initiated voluntarily, as when a person "crosses their eyes". Vertical vergence movements are also made in response to vertical image mis-registrations, but these movements cannot be made voluntarily, but only occur in response to a visual stimulus.

Roll of the eye about the line of sight, or *torsion*, is another example of a movement that cannot be generated voluntarily. Torsion can be either conjunctive (*cycloversion*) or disjunctive (*cyclovergence*). Cycloversion can be induced by visual rotation in the image plane, or by rolling the head itself, while cyclovergence is generally associated with other vergence movements made in response to depth changes.

1.4 Building Blocks

The minimal hardware requirements for an eye monitoring system are a digital camera and a personal computer with a suitable interface. The feasibility of creating a low-cost eye monitoring device for personal use has been examined in [11]. Their study shown that the hardware components of their low-cost system can be assembled from readily available consumer electronics and off-the-shelf parts for under 30 dollars with an existing personal computer. On an Apple PowerBook laptop, their proof of concept eye monitoring device operates in near real-time using prototype Jitter software at over 9 frames per second, and within 1 degree of gaze error.

Figure 1.4 illustrates a block diagram of a typical passive eye monitoring system including hardware and software. This system includes a video imaging camera and an infrared light illuminator, coupled to a vision processor which, in turn, is coupled to a host processor. The video imaging cam-

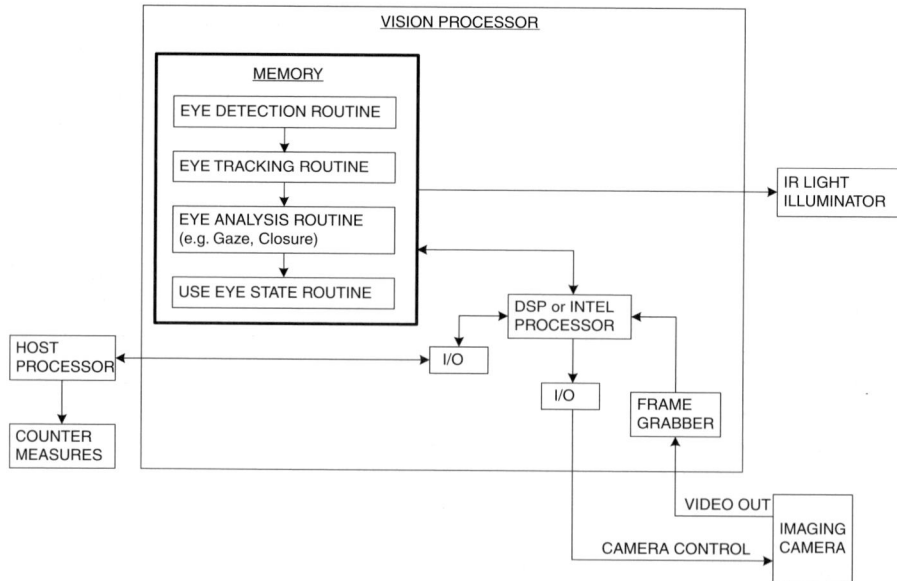

Fig. 1.4. Block diagram illustrating a passive eye monitoring system

era may include a CCD/CMOS active-pixel digital image sensor mounted as a chip onto a circuit board. One example of a CMOS active-pixel digital image sensor is Model No. PB-0330, commercially available from Photobit, which has a resolution of $640\,\mathrm{H} \times 480\,\mathrm{V}$. Other imaging cameras, like Webcams, may be employed depending on the application field. To achieve robustness to environmental variations in the ambient illumination, it is common to image the eye in the near-infrared (NIR), with a filter on the camera to block visible light, and controlled NIR illumination provided by light-emitting diodes (LEDs). In many situations useful images may be obtained with available natural light, but the use of controlled NIR illumination assures consistent lighting conditions across a range of environmental conditions. In Fig. 1.1(right), the eye was illuminated by NIR LED's with a peak wavelength around 870 nanometers, while in Fig. 1.3 the illuminator was a broadband source (a 50 W quartz-halogen lamp, filtered with a Wratten 87 C filter). Note that the contrast between the colored iris and the white sclera varies significantly, showing high contrast under the narrow band LED illumination but much less contrast under the broad-band illumination from the incandescent lamp. Lighting conditions are one of the most important problems to be addressed when the eye monitoring system is employed in real operating scenarios like driving a vehicle. In order to minimize the interference from light sources beyond the IR light emitted by the LEDs, a narrow bandpass filter centered at the LED wavelength could be attached between the CCD camera and the lens [43]. Bergasa et al [43] reported that when such a filter is employed in their real-time video-based pupil tracking system for monitoring driver vigilance, the problem of artificial lights and vehicle light has been solved almost completely. Further they made two important points: (1) this filter added a new drawback for it reduces the intensity of the image and the noise is considerably amplified by the automatic gain controller (AGC) integrated in the camera, and (2) this filter does not eliminate the sunlight interference in real-driving scenarios, except for cases when the light intensity is very low.

Once an image has been acquired by the camera, it is generally passed to a processor using a frame grabber or digital interface. In some systems, a "smart camera" may be employed which performs significant computations on board using dedicated hardware such as a fully-programmable gate array (FPGA). Regardless of the specific implementation, the vision processor is generally responsible for controlling activation of the IR light illuminator(s), controlling the camera, and processing the acquired video images. Control of the video imaging camera may include automatic adjustment of the pointing orientation, focus, exposure, and magnification. Each video frame image is processed to detect and track the spatial location of one or both eyes of the subject (eye finding and eye tracking routines). The detected eye(s) may be analyzed (eye analysis routine) to determine eye gaze vector(s) and eye closure state of one or both eye(s). After determining the gaze direction and/or the eye closure state, the eye monitoring system can then detect inattention and fatigue. The host processor may also interface to control devices that employ the determined eye information. For example, eye closure state may be used in a drowsy driver application to initiate a countermeasure such as an audible warning.

The robustness and accuracy of the software building blocks is crucial in practice. False eye position detection or mis-tracking of the pupil will produce large errors in gaze vector estimation and noisy eye monitoring data. Thus, much effort has been expended in the computer vision community to develop effective low-level building blocks, including automatic detection of human eyes, modeling of the eyes' structure, appearance changes across subjects and lighting conditions, tracking of spatial eye coordinates, identifying pupil, iris, eyelids, and eye-corners locations, estimating, calibrating and tracking a person's gaze direction. These building blocks are described in detail in parts I and II of this book, and could be employed with any eye monitoring hardware.

The accuracy of gaze tracking, pupillometry, or other measures, will generally scale linearly with the number of pixels per eye. Thus we are faced with a trade-off: high accuracy demands high magnification (narrow field of view), while the ability to track moderate head movements requires a wide field of view. One approach is to satisfy both desires by having two or more cameras: a wide-field camera finds the face and the eyes within it, while a steerable, narrow-field camera provides a magnified image of the eye. Moderately-priced pan/tilt/zoom cameras are attractive, but do not move fast enough to follow rapid head movements. Mirror galvonometers allow redirection of the narrow-field camera's line-of-sight during vertical blanking, but add significantly to system cost. Thanks to the introduction of CMOS camera chips supporting area-of-interest readout, these mechanical solutions can now be superseded by a single high-resolution sensor with a wide angle view of the working volume. In this case, the entire image is searched to find the eyes, while the programmable area-of-interest provides a "digital pan".

In a pupil-imaging system, accuracy can be increased by increasing the optical magnification, up to the point at which the pupil fills the entire image – or slightly less than the entire image, to allow some range of movement. Under these conditions, a typical eye with a 7 mm pupil, imaged by a 640×480 sensor, will produce 1 pixel of image motion for a rotational movement of around 5–10 minutes of arc. To obtain a higher optical gain, it is necessary to image a structure smaller than the pupil. Retinal imaging offers the possibility of imaging extremely small structures; even a crude imaging setup is capable of achieving an optical gain of 1 pixel of image motion per arc minute [426]. The AOSLO described above produces 1 pixel of image motion for a few arc seconds. It has also been suggested that blood vessels in the sclera might be imaged at high magnification for high-precision gaze tracking; while this approach deserves to be investigated, potential problems include shallow depth-of-field at high magnification (necessitating dynamic focusing), and the fact that the blood vessels are not rigidly embedded in the sclera, but are supported above it by a thin clear membrane known as the *conjunctiva*, and so may themselves move with respect to the eye as the eye moves.

The capability to actively control multiple illuminators can enhance the performance of an eye monitoring system. The pupil finding task can be simplified by exploiting the fact that the eye is a natural retroreflector. This approach is well known by dark-bright pupil technique. Light entering the eye is reflected by the retina, and passes back out towards the illuminator. When the eye is viewed from the direction of the illuminator the pupil appears filled with (red) light. This is the origin of the "red-eye" effect in flash photographs, which can be eliminated by moving the flash away from the camera lens. The pupil reflex can be isolated by subtracting an image collected with off-axis illumination from one collected with on-axis illumination [155, 422]. The two images may be collected using a single camera and temporal multiplexing, or multiple cameras using wavelength multiplexing. In an interesting twist on the idea of wavelength multiplexing, a sensor chip has been demonstrated which incorporates a checkerboard NIR filter array, providing a single-camera solution with wavelength multiplexing [240]. Wavelength multiplexing is generally superior to temporal multiplexing, which suffers from motion and interlace artifacts. The subtraction of dark-pupil image from bright-pupil image results into a very short list of potential eye/pupil candidates which are further filtered using machine learning techniques. Besides the pupil detection algorithm, the system includes pupil tracking and glint localization in order to estimate the gaze vector. While clean images of the eye, generated by such an imaging system, are relatively easy to analyze, it is nevertheless difficult to build a passive eye monitoring system that works reliably with all members of a large population in all operating scenarios. Eyewear such as prescription glasses and sunglasses are particularly common challenges, introducing clutter around the eye and occlusion of some key eye features. Changes in magnification of the eye caused by the power of the eyeglass lens can generally be calibrated out, but bright specular highlights from metal frames can confound simple searches for the glint which are based on finding pixels whose values exceed a fixed threshold. Similarly, threshold-based approaches to finding the (dark) pupil can be foiled by mascara and cosmetic products applied to the eyelashes. These types of problems arise indoors under the best of conditions; outdoors, the problems are multiplied. For example, many people wear sunglasses while driving which make the eyes nearly invisible; while sunglasses are generally designed to block visible and ultraviolet (UV) light, little can be safely assumed about their NIR transmission. In [43] it is been reported that when the sunlight or the lights of moving vehicles directly illuminate the driver's face, an increase of the pixel levels is noticed, causing the pupil effect to disappear.

A number of these aforementioned issues will be discussed further in the next two parts of this book along with some methodologies to overcome them.

1.5 Applications

Passive eye monitoring is used today in a broad range of applications. Advances in hardware and software have led to robust passive eye monitoring systems that can now be used in operational settings, as well as in conventional laboratory studies. Scientists and engineers from diverse fields have come together to build products which provide better support to humans engaged in complex tasks, and research programs which are uncovering new aspects of human perception, attention, and mental effort during the performance of real-life jobs.

Many of the application areas of passive eye monitoring fall under basic or applied research, in which the eye measurements are studied, but do not affect the subject's interaction with the environment. In another class of applications, the measurements are an integral part of the system. For example, "eye typing" allows disabled users to interact with a computer, and research is underway to determine how gaze might be exploited to increase the productivity of normal users. Another area in which gaze and other measures are used is known as "affective computing". In this case, the subject's gaze is not used for direct control, but rather to infer the subject's "affect", i.e. are they happy or sad, interested or disinterested, etc.

Gaze tracking plays a significant role in Human–Computer Interaction [287, 656, 717] where it may be used as an alternative input device for computers. This holds particular promise for those engaged in multiple tasks as well as for those with reduced physical mobility. Gaze information is useful in information security like hands-free password entry. Gaze tracking can also be used as a data collection tool as well as in evaluation of the usability of interfaces in Human Computer Interaction studies [287], and further for studying the human factors aspects of almost any sort of interface.

Because fixation is generally linked with focal attention and information gathering, the sequence of fixations (sometimes referred to as the "scan path") can provide detailed information about how information is acquired and used. In an early study of fixational eye movements during naturalistic tasks [20], it was discovered that extra fixations are often preferred in lieu of memory: in a pattern-copying task, subjects would often take a second look at the pattern after picking up a piece, suggesting that the first glance served merely to determine the next color, but that no position information was stored. Scan path analysis of complex interfaces can reveal how users get into trouble, in ways that are not always available to introspection.

Eye movements can be affected by a variety of neurological diseases, and their measurement is thus an important diagnostic tool. Additionally, oculomotor measures such as peak saccadic velocity have proved useful in detection of fatigue and intoxication [542]. In recent years, there has been a real push to employ eye monitoring systems in vehicles in order to reduce crashes through estimating, in real time, the level of driver cognitive distraction and driver fatigue [157, 226]. Differences between the eye movements of individuals can reveal differences in aptitude, expertise [502, 711], and even pathology. Eye monitoring systems have been used for medical image perception in radiology since the 1960s [348, 457]. For instance, eye monitoring research has shown important behavioral differences between novices and experts performing minimally invasive surgery (MIS).

Gaze tracking is crucial for disabled user who, due to accident or disease, suffer from near-total paralysis, with the exception of the eyes. For these people, the ability to engage in "eye typing" provides them with the ability to communicate. Early eye typing systems often were based on an image of a traditional keyboard; in order to prevent random eye movements from generating spurious inputs, a relatively long duration fixation may be required to input a character, with the result that input speeds are rather low. A recent innovation in eye typing is the "Dasher" program, in which the user selects characters or word fragments which slowly drift in from the right-hand edge of the screen. A statistical model based on both the language and the user's own typing history is used to allot more screen area to the more probable continuations.

There are some applications where we are interested in monitoring gaze, but are unable to exploit traditional gaze-tracking methods. For example, if we wish to track the gaze of a working air-traffic controller, we are unlikely to be permitted to clutter his or her workspace with our equipment. Under these circumstances, we may still be able to obtain useful data by estimating the pose of the head, and relying on the empirical observation that under natural conditions the eyes rarely deviate by more than 10–15 degrees away from primary position (straight ahead with respect to the head). When an observer

Fig. 1.5. A frame from a video collected in an air-traffic control simulator, with an approximate gaze vector (computed from head pose) indicated

wishes to fixate an object requiring a greater deviation, it is generally done by a combined eye-head movement, with final position of the eyes deviating from primary position by no more than an idiosyncratic threshold, constant for a given subject but varying somewhat between individuals. Figure 1.5 shows an example where such a "head gaze" vector has been computed, and used to infer the gaze target in the environment [70]. It should be noted that in order to identify the target in the scene, it was necessary to use a 3-D model of the environment, and to determine a depth value for the observer's head.

1.6 Overview of the Book's Contents

The remainder of this book consists of seventeen chapters, organized into six distinctive parts. We begin with an in-depth description of the components needed to build a solid eye monitoring system from scratch, and continue with detailed coverage of the growing range of applied studies built upon this technology.

Part I and Part II are reserved for the description of the fundamental building-blocks that constitute an eye monitoring system. The emphasis is on the critical issues pertinant to commercially available systems as well as state-of-the-art video-based approaches. We examine theoretical and experimental details of image-based algorithms for detecting human eyes, and modeling their structure and appearance across diverse subjects and lighting conditions. Other problems considered include identification and tracking of the major features of the eye (the pupil, iris, eyelids, and eye-corners or canthi), as well as calibration and estimation of gaze direction.

In Chap. 2 we tackle the problem of detailed analysis of eye region images in terms of the position of the iris, degree of eyelid opening, and the shape, complexity, and texture of the eyelids. The challenge comes from the diversity of appearance of the eyes due to both individual differences in structure and motion of eyes. We define a rectangular region around the eye as the region for analysis, and exploit a generative model of the two-dimensional appearance. The model consists of multiple components corresponding to the anatomical features of the eye: the iris, upper and lower eyelids, the white scleral region around the iris, dark regions near the inner and outer corners of the white region, and other features observed in images which vary with gaze direction. The model of each component is rendered in a separate rectangular layer. When overlaid, these layers represent the eye region. The structure parameters represent structural individuality of the eye, including the size and color of the iris, the width, boldness, and complexity of the eyelids, the width of the bulge below the eye, and the width of the illumination reflection on the bulge. The motion parameters represent movement of the eye, including the up-down position of the upper and lower eyelids and the 2D position of the iris. The system first registers the eye model to the input in a particular frame and individualizes it by adjusting the structure parameters. The system then tracks motion of the eye by estimating the motion parameters across the entire image sequence. Individualization of the structure parameters is automated by using an Active Appearance Model (AAM). The system performance is investigated with respect to the quality of the input images, initialization and complexity of the eye model, and robustness to ethnic and cosmetic variations in eyelid structure.

In Chap. 3, statistical models have been deployed to represent the individual differences for the eye region as a whole. Generic models for eye and non-eye are built off-line and used online to locate eyes in input test images. The chapter presents first an overview of some advanced statistical approaches for eye-pattern modeling such as Principal Components Analysis and Linear Discriminant Analysis. The focus then shifts to a novel statistical framework, based on generalized Dirichlet mixture to model low-resolution infrared eye images. The chapter presents compelling experimental evidence for the robustness, efficiency and accuracy of the generalized Dirichlet mixture for modeling and classification of eye vs. non-eye.

The last chapter of Part I (Chap. 4) presents a hierarchical multi-state pose-dependent approach for spatial eye position detection and tracking under varying facial expression and pose. For effective and efficient representation of facial feature points, the chapter describes a hybrid representation that integrates Gabor wavelets and gray-level profiles. It proposes a hierarchical statistical face shape model to characterize both the global shape of the human face and the local structural details of each facial component. Furthermore, the chapter introduces multi-state local shape models to deal with shape variations of some facial components under changes in expression. During detection and tracking, both facial component states and feature point positions are dynamically estimated, constrained by the hierarchical face shape model. Experimental results demonstrate that the method presented in this chapter accurately and robustly detects and tracks eyes in real-time, across variations of expression and pose.

Part II starts with Chap. 5 that introduces the reader to personal calibration for eye gaze tracking. Personal calibration is necessary because individual differences constitute some of the most important factors contributing to gaze tracking errors. This task is a troublesome process for users of passive eye monitoring systems, because it requires the user to cooperate by looking at a series of reference points (around twenty markers). Chapter 5 starts by describing the background of personal calibration, and then reviews prior research done to reduce the calibration cost. Finally, algorithms are described which require only one and two calibration points.

Another major problem that limits the use of passive eye-gaze monitoring is that the user's head must often remain within a limited volume. This problem is tackled in Chap. 6. First, this chapter describes several camera based techniques that allow free head motion, and then focuses on a recently developed single-camera technique based on the cross-ratio invariance property of projective transformations. This method uses an array of lights placed around a computer screen, that is then observed via corneal reflection. The position of the center of the pupil relative to the projected pattern is used to compute the point of regard using the cross-ratio invariance property. Extensive analysis of similar methods is performed, using a realistic eye model. Based on these results, the β calibration cross-ratio technique is introduced. It compensates for the angular difference between the visual axis (line-of-sight) and the optical axis (axis of symmetry), as well as for the curvature of the cornea. This chapter also presents a real time (30 fps) prototype using a single camera and 5 light sources to generate the light pattern. Experimental results show that the accuracy of the method in Chap. 6 is about 1° of visual angle, and that this accuracy is maintained even for large head displacements.

The last chapter of Part II (Chap. 7) provides a comprehensive analysis of exactly the visual information about the world which can be recovered from a single image of the eye, using the corneal reflections of external objects. It presents a detailed analysis of the characteristics of the corneal imaging system, including field of view, resolution, locus of viewpoints, and how to compute a 3D structure of the scene.

Parts III through VI of this book fulfill the purpose for which eye monitoring systems are built to a great extent. They present advanced applications for which commercial eye monitoring systems are either used as *interaction input devices* and/or reliable *source of data*. The fundamental idea of gaze-based interaction is to measure the user's gaze position on the computer screen with the eye tracking system, and the application interface reacts to the user's gaze.

Part III is on *Gaze-Based Interactions* and contains three chapters. The first one (Chap. 8) provides a general introduction to the exciting topic of users' interactions with diverse interfaces through eye-gaze, as opposed to using hands to control input computers devices. It reviews the two categories of gaze-based user's interfaces: "Command-Based" and "non-command" interfaces. In the first category, users are required to pay close attention to the selection and control of their computers, as in eye-typing, menu selection and other gaze-controllable Graphical User Interface (GUI). In the second category, the gaze is used as a control medium, or gaze-contingent; that is, the system is aware of the user's gaze and may adapt its behavior based on the visual attention of the user. The chapter walks the readers through numerous gaze-based applications of both categories and further discusses the problem of unintended consequences, or the *Midas Touch*.

In Chap. 9 we present a Gaze-Contingent Volume Rendering approach. A gaze-contingent display (GCD) is one whose image is modified in response to where a user looks. A volume rendered form of such a display is presented built around a mirror stereoscope and integrated video-based gaze-tracking system. A calibration phase is used to configure the system so that the fixation depth accuracy necessary for 3-D GCD is achieved. In comparison with conventional volume rendering, volumetric GCD provides both computational savings and increased clarity for regions buried within the volume. Applications for the volumetric GCD system include those requiring human scrutiny of complex structures such as diagnosis and screening. Subtle prompting of an observer is also possible, with automatic algorithms being applied to identify features of interest within the volume and appropriately enhance them when the observer fixates within their locality.

The last chapter of Part III (Chap. 10) presents a new type of gaze-contingent display that manipulates the temporal resolution of an image sequence. It demonstrates that the temporal filtering effect can remain undetected by the observer if the cutoff frequency lies above an eccentricity-dependent threshold. Evidences indicating that gaze-contingent temporal filtering can reduce the number of saccades to the periphery of the visual field are also presented. This type of display is seen as a valuable tool for psychophysical research on the spatio-temporal characteristics of the human visual system when presented with natural scenes. Gaze-contingent spatial filtering (foveation of video) can be applied to a number of applications, such as video compression.

Part IV includes two chapters on applications of eye monitoring in Military and Security Realms. Chapter 11 reviews the use of eye monitoring for interface design and evaluation, performance appraisal, and workload assessment in a number of military applications. Chapter 12 proposes to use eye monitoring technology as a "hands-free mouse" without on-screen feedback to allow users to enter information (not limited to passwords) without disclosing it to illicit onlookers, or "shoulder surfers." This chapter reviews the usability of eye monitoring authentication systems, given the current development stage of eye monitoring technologies. The lack of on-screen feedback makes it more critical to tolerate calibration errors, as well as users' uncertainty in remembering the exact password. The chapter includes techniques for handling both random errors (due to limitations in sampling the gaze direction for eye monitoring) and systematic errors (due to improper calibration or because the user moved his head).

Part V presents recent deployments of passive eye monitoring technology in automotive and medical applications, all having the goal of reducing errors caused by fatigue and distraction. Chapter 13 presents a novel data mining and fusion framework to infer a driver's cognitive distraction state, where various eye movement measures such as fixation duration and saccade amplitude are used as inputs. Driver distraction is an important and growing safety concern as information technologies (such as navigation systems, cell phones and internet-content services) have become increasingly common in vehicles. To allow people to benefit from these technologies without compromising safety, an adaptive In-Vehicle Information System (IVIS) is needed.

Such systems can help drivers manage their workload and mitigate distraction by monitoring and responding to driver states and roadway conditions. For the system to be effective, however, it is critical that driver distraction can be identified accurately in real time. This chapter will discuss approaches to identify driver cognitive distraction. Eye movements and driving performance were chosen as promising indicators. A robust data fusion system using data mining techniques was proposed to integrate these indicators to detect when a driver was distracted. This chapter presents comparative analysis of two data mining methods for identifying driver cognitive distraction using eye movements and driving performance: Support Vector Machines (SVMs) and Bayesian Networks (BNs). In Chap. 14 we review various state-of-the-art approaches and research studies for detecting and mitigating driver drowsiness and fatigue. The last chapter of this part (Chap. 15) provides a general overview of eye monitoring technologies in surgery and radiology, with particular emphasis on the role of eye trackers as tools to enhance visual search, and understand the nature of expertise in radiology.

The last part of this book (Part VI) in on *Eye Monitoring in Information Retrieval and Interface Design Assessment.* It answers questions about how a user scans and searches for online web content, and how net surfers respond to banner ads and other advertisements (Chap. 16), as well as how the visual attention of a viewer is distributed when the monitor presents several adjacent windows (Chaps. 17 and 18).

Chapter 16 emphasizes the use of eye monitoring technology in online search. The world-wide-web introduces both new opportunities and challenges for eye tracking research. While eye monitoring is still a relatively new analytical tool for studying online human–computer interaction, a number of effective analysis methods have already emerged. This chapter reviews how researchers have used eye monitoring to investigate a number of interesting topics, including online viewing of ads, web homepages, and search results. In web search, eye monitoring analysis has illuminated what people decide to look at, how they navigate search results, and what aspects of search pages are the most important for finding information online. Techniques are now available to quantify, compare, and aggregate eye movements relative to these online environments. Unfortunately, while readily available, general eye monitoring analysis software has not kept pace with all of the analysis techniques used by researchers. There are certainly obstacles to using eye monitoring in online contexts, but tangible results have demonstrated its value and the need for further research. Eye monitoring technologies have the power to reveal how a user scans and searches for online content, and this behavioral data can be combined with implicit forms of feedback, such as server log data. This chapter discusses how passive eye monitoring is an effective tool for augmenting standard analysis methods for studying information retrieval. Chapter 17 presents a method for statistically identifying trends in eye monitoring datasets that are sparse.

The last chapter of this book (Chap. 18) summarizes two research reports – one on television news and another on Web portal pages. These research studies, grounded in visual perception and information-processing theories, are adding to the understanding of how information acquisition takes place in today's living rooms and offices. Together, they are beginning to help us see how information will be acquired in tomorrow's new media environments of interactive television and Web 2.0 sites. This chapter shows results indicating the complexity of the screen design influenced the distribution of fixation time across the different areas of interest. The theoretical approaches, methodology and results presented in this chapter will serve to inform future research on the media forms emerging from the convergence of television and the Internet.

Part I

Low-Level Building-Blocks – Eye Region
Modeling, Automatic Eye Position Detection,
and Spatial Eye Position Tracking

Meticulously Detailed Eye Region Model

Tsuyoshi Moriyama, Takeo Kanade, Jing Xiao, and Jeffrey F. Cohn

2.1 Position of the Problem

Automated analysis of facial images has found eyes still to be a difficult target [90, 96, 97, 125, 215, 221, 229, 230, 248, 360, 509, 692, 709]. The difficulty comes from the diversities in the appearance of eyes due to both structural individuality and motion of eyes, as shown in Fig. 2.1. Past studies have failed to represent these diversities adequately. For example, Tian et al. [616] used a pair of parabolic curves and a circle as a generic eye model, but parabolic curves have too few parameters to represent the complexity of eyelid shape and motion. Statistical models have been deployed to represent such individual differences for the whole eye region [322, 569, 635], but not for subregions, such as the eyelids, due in part to limited variation in training samples.

This chapter presents and evaluate a generative eye region model that can meticulously represent the detailed appearance of the eye region for eye motion tracking. The model parameterizes both the structural individualities and the motions of eyes. Structural individualities include the size and the color of the iris, the width and the boldness of the eyelid, which may have a single or double fold, the width of the bulge below the eye, the furrow below it, and the width of illumination reflection on the bulge. Eye motion includes the up-down positions of upper and lower eyelids and the 2D position of the iris. The input image sequence first is stabilized to compensate for appearance change due to head motion. The system then registers the eye region model to the input eye region and individualizes it by adjusting the structure parameters and accurately tracks the motion of the eye.

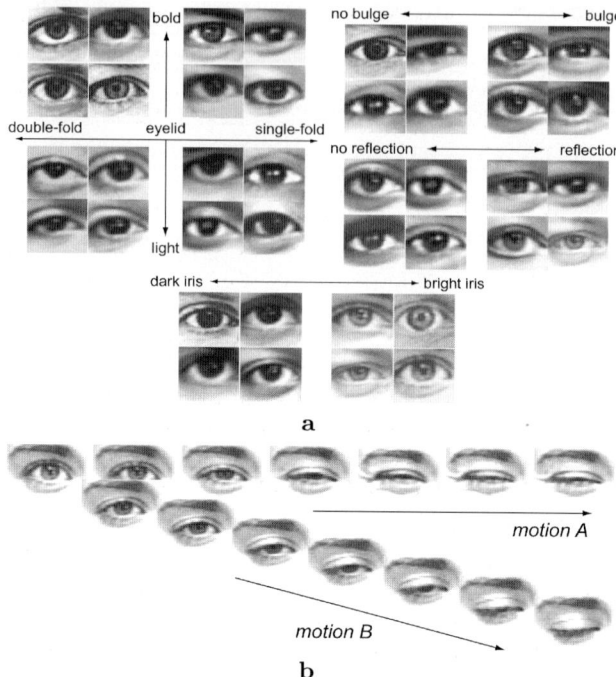

Fig. 2.1a,b. Diversity in the appearance of eye images. **a** Variance from structural individuality. **b** Variance from motion of a particular eye

2.2 Eye Region Model

We define a rectangular region around the eye as an eye region for analysis. We exploit a 2D, parameterized, generative model that consists of multiple components corresponding to the anatomy of an eye. These components include the iris, upper and lower eyelids, a white region around the iris (*sclera*), dark regions near the inner and outer corners of the white region, a bulge below the lower eyelid, a bright region on the bulge, and a furrow below the bulge (the *infraorbital furrow*). The model for each component is rendered in a separate rectangular layer. When overlaid, these layers represent the eye region as illustrated in Fig. 2.2. Within each layer, pixels that render a component are assigned color intensities or transparency so that the color in a lower layer appears in the final eye region model if all the upper layers above it have transparent pixels at the same locations. For example, the iris layer (the third layer from the bottom) has a circular region to represent the iris. The eyelid layer (the fourth layer, one above the iris layer) has two curves to represent upper and lower eyelids, in which the region between those curves (*palpebral fissure*) is transparent while the region above the upper curve and

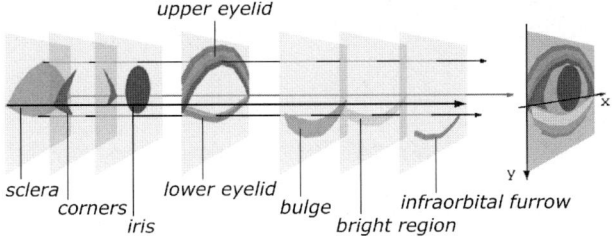

upper eyelid

sclera

corners

iris

lower eyelid

bulge

bright region

infraorbital furrow

Fig. 2.2. Multilayered 2D eye region model

the region below the lower curve are filled with skin color. When the eyelid layer is superimposed over the iris layer, only the portion of the circular region between the eyelid curves appears in the final eye region image while the rest is occluded by the skin pixels in the eyelid layer. When the upper curve in the eyelid layer is lowered, corresponding to eyelid closure, a greater portion of the circular region in the iris layer is occluded.

Table 2.1 shows the eye components represented in the multilayered eye region model along with their control parameters. We call parameters d_u, f, d_b, d_r, r_i, and I_{r7} the *structure parameters* (denoted by $\mathbf{s}$) that define the static and structural detail of an eye region model, while we call parameters ν_{height}, ν_{skew}, λ_{height}, η_x, and η_y the time-dependent *motion parameters* (denoted by $\mathbf{m}_t$, t: time) that define the dynamic detail of the model. The eye region model defined and constructed by the structure parameters $\mathbf{s}$ and the motion parameters $\mathbf{m}_t$ is denoted by $T(\mathbf{x}; \mathbf{s}, \mathbf{m}_t)$, where $\mathbf{x}$ denotes pixel positions in the model coordinates. Table 2.2 and Table 2.3 show examples of the appearance changes due to the different values of $\mathbf{s}$ and $\mathbf{m}_t$ in the eye region model $T(\mathbf{x}; \mathbf{s}, \mathbf{m}_t)$.

2.2.1 Upper Eyelid

The upper eyelid is a skin region that covers the upper area of the *palpebral fissure* (the eye aperture). It has two descriptive features: 1) a boundary between the upper eyelid and the *palpebral fissure* and 2) a furrow running nearly in parallel to the boundary directly above the upper eyelid.

The model represents these features by two polygonal curves (curve1 and curve2) and the region (region1) surrounded by them. Both curve1 and curve2 consist of N_u vertices denoted by $\mathbf{u}_1$ and $\mathbf{u}_2$, respectively (Table 2.1).

2.2.1.1 Structure of Upper Eyelid

To represent the distance between the boundary and the furrow, parameter d_u [01] gives the ratio to the predefined maximum distance between curve1 and curve2. When curve1 and curve2 coincide ($d_u = 0$), the upper eyelid

Table 2.1. Detailed description of the eye region model

parts	models	parameters	
		structure **s**	motion **m**$_t$
Upper eyelid	region1, curve2 ($\mathbf{u}_2$), curve1 ($\mathbf{u}_1$), l_u, d_u, w_u, v_{skew}, v_{height}	d_u: distance between curve1 and 2 I_{r1}: intensity of region1 w_{c2}: line width of curve2 (I_{r1}, w_{c2}: controlled by "boldness" parameter f)	v_{height}: degree of curve1 raising v_{skew}: degree of horizontal skew of curve1
Lower eyelid	curve3 ($\mathbf{l}_1$), region3, curve4 ($\mathbf{l}_2$) region2, curve5 ($\mathbf{l}_3$), curve6 ($\mathbf{l}_4$), l_{c3}, λ_{height}, l_d, l_{c5}, d_b	d_b: distance between curve3 and 4 d_r: distance between curve3 and 6	λ_{height}: degree of curve3 raising I_{c3}: intensity of curve3 I_{c5}: intensity of curve5 (I_{c3}, I_{c5}: subject to λ_{height})
Sclera	curve1, v_{height}, v_{skew}, l_{r4}, region4, curve3, λ_{height}		(upper boundary) ≡ curve1 (lower boundary) ≡ curve3 I_{r4}: intensity of region4 (I_{r4}: subject to v_{height})
Corners	curve1, v_{height}, v_{skew}, curve7 curve8, region6, region5, curve3, λ_{height}		(upper boundary) ∈ curve1 (lower boundary) ∈ curve3
Iris	r_i, l_{r7}, (η_x, η_y), region7	r_i: radius of region7 I_{r7}: intensity of region7	η_x, η_y: position of the center of region7

Table 2.2. Appearance changes controlled by structure parameters

parameters	0.0	0.5	1.0
d_u			
f			
d_b			
d_r			
r_i			
I_{r7}			

Table 2.3. Appearance changes controlled by motion parameters

parameters	0.0	0.5	1.0
ν_{height}			
ν_{skew}			
λ_{height}			
η_x			
η_y			

appears to be a uniform region, which we refer to as a single eyelid fold. Single eyelid folds are common in East Asians. "Boldness" parameter f [01] controls both the intensity I_{r1} of region1 and the line width w_{c2} of curve2, simultaneously by $I_{r1} = I_{r1}^{\text{brightest}} - \beta_1 \cdot f$ and $w_{c2} = \beta_2 \cdot f + w_{c2}^{\text{thickest}}$ (β_1, β_2: constant). The appearance changes controlled by d_u and f are shown in Table 2.2.

2.2.1.2 Motion of Upper Eyelid

When an upper eyelid moves up and down in its motion (e.g., blinking), the boundary between the upper eyelid and the *palpebral fissure* moves up and down. The model represents this motion by moving the vertices of curve1 ($\mathbf{u}_1$). They move between the predefined curve for a completely open eye ($\mathbf{u}_1^{\text{top}}$) and that for a closed eye ($\mathbf{u}_1^{\text{bottom}}$), as shown in Fig. 2.3. Parameter ν_{height} [01] specifies the position of curve1 within this range and, thus, the ith vertex position of curve1 ($\mathbf{u}_{1i}$) is defined by parameter ν_{height} as,

$$\mathbf{u}_{1i} = \sin\left(\frac{\pi}{2} \cdot \nu_{\text{height}}\right) \cdot \mathbf{u}_{1i}^{\text{top}} + \left(1 - \sin\left(\frac{\pi}{2} \cdot \nu_{\text{height}}\right)\right) \cdot \mathbf{u}_{1i}^{\text{bottom}} , \qquad (2.1)$$

where $\mathbf{u}_{1i}^{\text{top}}$ and $\mathbf{u}_{1i}^{\text{bottom}}$ are the positions of the ith vertices of $\mathbf{u}_1^{\text{top}}$ and $\mathbf{u}_1^{\text{bottom}}$, respectively. The sinusoidal term in (2.1) moves the vertices rapidly when ν_{height} is small and slowly when ν_{height} is large with respect to the linear change of ν_{height}. This corresponds to the possible rapid movement of the upper eyelid when it lowers in motion such as blinking.

The furrow on the upper eyelid also moves together with the boundary. The model represents this motion by moving the vertices of curve2 ($\mathbf{u}_2$). The positions of the vertices of curve2 ($\mathbf{u}_2$) are defined by using parameters ν_{height} and d_u such that they move in parallel to curve1 ($\mathbf{u}_1$) when ν_{height} is larger than a preset threshold ν_{height}^T or move slowly keeping the distance between curve1 and curve2 wide otherwise.

If ν_{height} is larger than ν_{height}^T, then

$$u_{2i}^x = u_{1i}^x , \qquad (2.2)$$

$$u_{2i}^y = u_{1i}^y - \left(\alpha_1 \cdot \frac{|u_{1i}^x - \bar{u}_1^x|}{|u_{11}^x - \bar{u}_1^x|} + \alpha_2\right) \cdot d_u , \qquad (2.3)$$

$$\bar{u}_1^x = \frac{u_{11}^x + u_{1N_u}^x}{2}$$

else

$$u_{2i}^x = \left(1 - \gamma_{d_u, \nu_{\text{height}}}\right) \cdot u_{1i}^{x, \nu_{\text{height}} = \nu_{\text{height}}^T} \qquad (2.4)$$
$$+ \gamma_{d_u, \nu_{\text{height}}} \cdot u_{1i}^{x, \text{bottom}} ,$$

$$u_{2i}^y = \left(1 - \gamma_{d_u, \nu_{\text{height}}}\right) \cdot \tilde{u}_{1i}^{y, \nu_{\text{height}} = \nu_{\text{height}}^T} \qquad (2.5)$$
$$+ \gamma_{d_u, \nu_{\text{height}}} \cdot \tilde{u}_{1i}^{y, \text{bottom}} ,$$

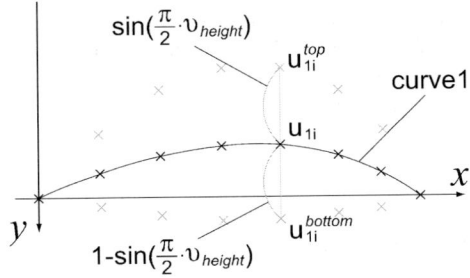

Fig. 2.3. The up-down position of curve1

$$\gamma_{d_u,\nu_{\text{height}}} = \left(1 - \frac{\nu_{\text{height}}}{\nu_{\text{height}}^T}\right) \cdot (1 - d_u) \,,$$

$$\tilde{u}_{1i}^y = u_{1i}^y - \left(\alpha_1 \cdot \frac{|u_{1i}^x - \bar{u}_1^x|}{|u_{11}^x - \bar{u}_1^x|} + \alpha_2\right) \cdot d_u \,,$$

end if, where α_1 and α_2 are constant.

The boundary also appears skewed horizontally when the eye is not straight to the camera because it is on a spherical eyeball. The model represents it by horizontally skewing curve1 by using parameter ν_{skew} [01]. As shown in Fig. 2.4, the vertices of curve1 ($\mathbf{u}_1$) defined by (2.1) are transformed into the skewed positions ($\mathbf{u}_1^{\text{skewed}}$) under orthographic projection, where C denotes the center of the eyeball and θ defines the opening of the eye. The coordinate of C in the x_{eye}–z_{eye} plane is

$$C_{x_{\text{eye}}} = \left(u_{1N_u}^{x_{\text{eye}}} + u_{11}^{x_{\text{eye}}}\right)/2 \,, \tag{2.6}$$

$$C_{z_{\text{eye}}} = C_{x_{\text{eye}}} \cdot \tan\theta \,. \tag{2.7}$$

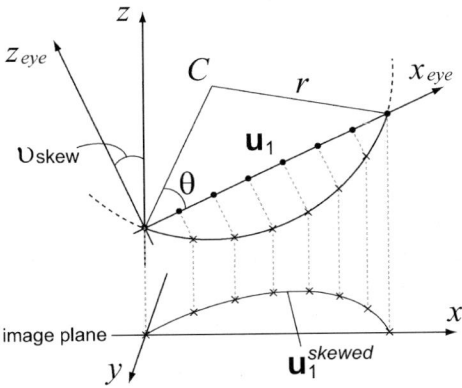

Fig. 2.4. The horizontal skew of curve1

The coordinates of $\mathbf{u}_{1i}$ projected onto the spherical surface, $(u_{1i}^{x_{\mathrm{eye}}}, u_{1i}^{z_{\mathrm{eye}}})$, should satisfy (2.8), with r being the radius of the sphere:

$$\left(u_{1i}^{x_{\mathrm{eye}}} - C_{x_{\mathrm{eye}}}\right)^2 + \left(u_{1i}^{z_{\mathrm{eye}}} - C_{z_{\mathrm{eye}}}\right)^2 = r^2 . \tag{2.8}$$

The x coordinate of horizontally skewed positions of $\mathbf{u}_{1i}$ ($u_{1i}^{\mathrm{skewed},x}$) in the x–z plane is obtained as

$$u_{1i}^{\mathrm{skewed},x} = u_{1i}^{x_{\mathrm{eye}}} \cdot \cos(\nu_{\mathrm{skew}}) + |u_{1i}^{z_{\mathrm{eye}}}| \cdot \sin(\nu_{\mathrm{skew}}) . \tag{2.9}$$

The first two rows of Table 2.3 shows examples of the appearance changes due to parameters ν_{height} and ν_{skew}.

2.2.2 Lower Eyelid

A lower eyelid is a skin region that covers the lower area of the *palpebral fissure*. It has four descriptive features:

1. a boundary between the lower eyelid and the *palpebral fissure*,
2. a bulge below the boundary, which results from the shape of the covered portion of the eye, shortening of the inferior portion of the *orbicularis oculi* muscle (a sphincter muscle around the eye) on its length, and the effects of gravity and aging,
3. an *infraorbital furrow* parallel to and below the lower eyelid, running from near the inner corner of the eye and following the cheek bone laterally [160], and
4. a brighter region on the bulge, which is mainly caused by the reflection of illumination.

As shown in Table 2.1, the model represents these features by four polygonal curves (curve3, curve4, curve5, and curve6) and two regions (region2 surrounded by curve3 and curve4 and region3 surrounded by curve3 and curve6). Curve3, curve4, and curve6 consist of N_l vertices and are denoted by $\mathbf{l}_1$, $\mathbf{l}_2$, and $\mathbf{l}_4$, respectively. Curve5 is the middle portion of curve4, consisting of N_f vertices denoted by $\mathbf{l}_3$.

2.2.2.1 Structure of Lower Eyelid

Distance ratio parameter d_b [0 1] controls the distance between curve3 and curve4. The vertices of curve4 ($\mathbf{l}_2$) have the predefined positions for both the thinnest bulge ($\mathbf{l}_2^{\mathrm{top}}$) and the thickest bulge ($\mathbf{l}_2^{\mathrm{bottom}}$), as shown in Fig. 2.5. The positions of jth vertex of $\mathbf{l}_2$ are defined by using parameter d_b as

$$\mathbf{l}_{2j} = d_b \cdot \mathbf{l}_{2j}^{\mathrm{bottom}} + (1 - d_b) \cdot \mathbf{l}_{2j}^{\mathrm{top}} \tag{2.10}$$

where $\mathbf{l}_{2j}^{\mathrm{top}}$ and $\mathbf{l}_{2j}^{\mathrm{bottom}}$ are the positions of the jth vertices of $\mathbf{l}_2^{\mathrm{top}}$ and $\mathbf{l}_2^{\mathrm{bottom}}$, respectively.

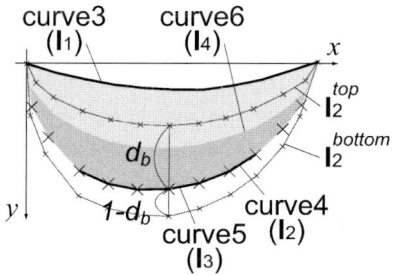

Fig. 2.5. The model for a lower eyelid

Distance ratio parameter d_r [01] controls the distance between curve3 and curve6. The position of the jth vertex of l_4 is defined by using l_1, l_2, and parameter d_r as

$$l_{4j} = d_r \cdot l_{2j} + (1 - d_r) \cdot l_{1j} \ . \tag{2.11}$$

2.2.2.2 Motion of Lower Eyelid

When the lower eyelid moves up or down (e.g., eyelid tightening), the boundary between the lower eyelid and the *palpebral fissure* moves, correspondingly changing in area. The bulge, the *infraorbital furrow*, and the brighter region on the bulge also move together with the boundary.

Our model represents this motion by moving the vertices of curve3, curve5, and curve6. The vertices of curve3 have predefined positions for both the highest (l_1^{top}) and the lowest (l_1^{bottom}). Parameter λ_{height} [01] gives the position within this range. The position of the jth vertex of l_1 is obtained using parameter λ_{height} as

$$l_{1j} = \lambda_{\text{height}} \cdot l_{1j}^{\text{top}} + (1 - \lambda_{\text{height}}) \cdot l_{1j}^{\text{bottom}} \ , \tag{2.12}$$

where l_{1j}^{top} and l_{1j}^{bottom} are the positions of the jth vertices of l_1^{top} and l_1^{bottom}, respectively. Likewise, parameter λ_{height} controls the positions of l_2, l_2^{top} and l_2^{bottom} in (2.10).

$$l_{2j}^{\text{top}} = \lambda_{\text{height}} \cdot l_{2j}^{\text{top},t} + (1 - \lambda_{\text{height}}) \cdot l_{2j}^{\text{top},b} \ , \tag{2.13}$$

$$l_{2j}^{\text{bottom}} = \lambda_{\text{height}} \cdot l_{2j}^{\text{bottom},t} + (1 - \lambda_{\text{height}}) \cdot l_{2j}^{\text{bottom},b} \ , \tag{2.14}$$

where $(l_{2j}^{\text{top},t}, l_{2j}^{\text{top},b})$ and $(l_{2j}^{\text{bottom},t}, l_{2j}^{\text{bottom},b})$ are the preset dynamic ranges for l_{2j}^{top} and l_{2j}^{bottom}.

Parameter λ_{height} also controls both the intensity of curve3 and that of curve5 (I_{c3} and I_{c5}) by $I_{c3} = I_{c3}^{\text{brightest}} - \beta_3 \cdot \lambda_{\text{height}}$ and $I_{c5} = I_{c5}^{\text{brightest}} - \beta_4 \cdot \lambda_{\text{height}}$ (β_3, β_4 : constant).

Table 2.3 shows examples of the appearance changes controlled by parameter λ_{height}.

2.2.3 Sclera

The sclera is the white portion of the eyeball. We limit it to the region that can be seen in the *palpebral fissure*, which is surrounded by the upper eyelid and the lower eyelid. Our model represents the sclera by a region (region4) surrounded by curve1 and curve3, which are defined to represent upper and lower eyelids, as shown in Table 2.1.

When the upper eyelid and/or the lower eyelid move, the sclera changes its shape. Our model controls the change indirectly by parameters ν_{height}, ν_{skew}, and λ_{height}. These primarily control the appearance changes of the upper eyelid and the lower eyelid due to the motions. Parameter ν_{height} also controls the intensity of region4 by $I_{r4} = \beta_5 \cdot \nu_{\text{height}} + I_{r4}^{\text{darkest}}$ (β_5: constant).

2.2.4 Corners

Corners are regions at the medial (close to the midline) and lateral regions of the sclera. They are usually darker than other parts of the sclera due to shadow and color of the *caruncle* (a small, red portion of the corner of the eye that contains sebaceous and sweat glands). As shown in Table 2.1, our model represents the outer corner by a region surrounded by three polygonal curves (curve1, curve3, and curve7) and the inner corner by curve1, curve3, and curve8. Both curve7 and curve8 consist of N_c vertices, denoted by $\mathbf{c}_1$ and $\mathbf{c}_2$, respectively. Figure 2.6 depicts the details of the outer corner model.

When the upper eyelid and/or the lower eyelid move, the shape of the eye corners changes. Our model controls the motion of the upper and the lower boundaries by parameters ν_{height}, ν_{skew}, and λ_{height} as mentioned. The x coordinates of $\mathbf{c}_{12}$ and $\mathbf{c}_{13}$ are moved from predefined neutral positions based on parameter ν_{skew} according to the horizontal proportion $\overline{P\mathbf{c}_{12}}/\overline{PQ}$ and $\overline{P\mathbf{c}_{13}}/\overline{PQ}$, respectively and their y coordinates are so determined as to keep the vertical proportions same.

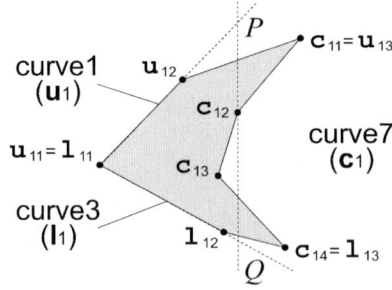

Fig. 2.6. The model for the outer corner

2.2.5 Iris

The iris is a circular and colored region on the eyeball. The apparent color of the iris is mainly determined by reflection of environmental illumination and the iris' texture and patterns including the *pupil* (an aperture in the center of the iris). Our model represents the iris by a circular region, region7, as shown in Table 2.1. Parameter r_i and parameter I_{r7} control the radius and the variable single color of region7, respectively. The color of the iris is represented as the average gray level inside the iris.

The position of the iris center moves when gaze direction moves. Our model represents the motion by moving the vertex of the center coordinate (i_x, i_y) of region7. It has predefined positions for gaze left (i_x^l), gaze right (i_x^r), gaze up (i_y^u), and gaze down (i_y^d), respectively. Parameters η_x [0 1] and η_y [0 1] give the position within these ranges as

$$i_x = \eta_x \cdot i_x^r + (1 - \eta_x) \cdot i_x^l, \qquad (2.15)$$

$$i_y = \eta_y \cdot i_y^u + (1 - \eta_y) \cdot i_y^d. \qquad (2.16)$$

Table 2.3 includes examples of the appearance changes due to parameters η_x and η_y.

2.3 Model-based Eye Image Analysis

Figure 2.7 shows a schematic overview of the whole process of a model-based eye region image analysis system. An input image sequence contains facial behaviors of a subject. Facial behaviors usually accompany spontaneous head motions. The appearance changes of facial images thus comprise both rigid 3D head motions and nonrigid facial actions. Decoupling these two components is realized by recovering the 3D head pose across the image sequence and by accordingly warping the faces to a canonical head pose (frontal and upright), which we refer to as the stabilized images. Stabilized images are intended to include appearance changes due to facial expression only. Eye image analysis proceeds on these stabilized images. For a given stabilized image sequence, the system registers the eye region model to the input in the initial frame and individualizes the model by adjusting the structure parameters **s** (Table 2.1). Motion of the eye is then tracked by estimating the motion parameters $\mathbf{m}_t$ across the entire image sequence. If the tracking results at any time t are off the right positions, the model is manually readjusted, otherwise we get the estimated motion together with the structure of the eye.

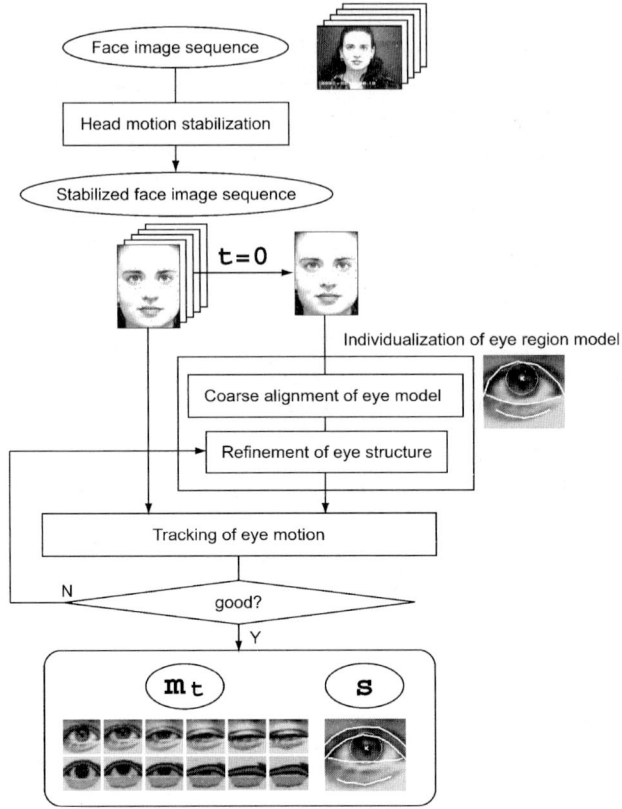

Fig. 2.7. A schematic overview of the model-based eye image analysis system

2.3.1 Head Motion Stabilization

We use a head tracker that is based on a 3D cylindrical head model [690]. Manually given the head region with the pose and feature point locations (e.g., eye corners) in an initial frame, the tracker automatically builds the cylindrical model and recovers 3D head poses and feature point locations across the rest of the sequence. The initial frame is selected such that it has the most frontal and upright face in it. The tracker recovers full 3D rigid motions (three rotations and three translations) of the head. The performance evaluation on both synthetic and real images has demonstrated that it can track as large as 40 degrees and 75 degrees of yaw and pitch, respectively, within 3.86 degrees of average error.

As shown in Fig. 2.8, the stabilized face images cancel out most of the effect of 3D head pose, and contain only the remaining nonrigid facial expression.

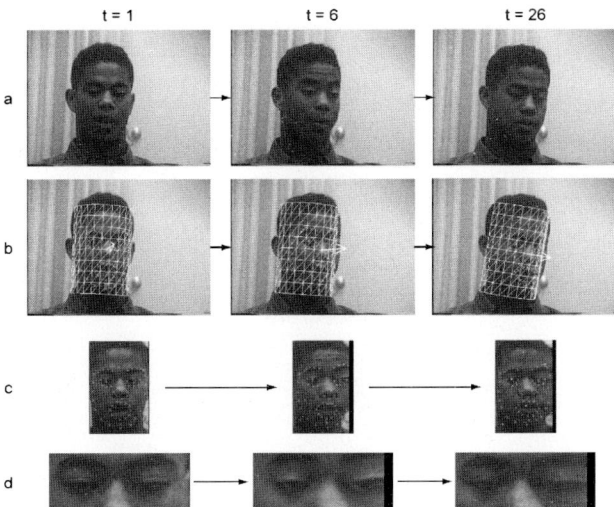

Fig. 2.8a–d. Automatic recovery of 3D head motion and image stabilization [690]. **a** Frames 1, 10, and 26 from original image sequence. **b** Face tracking in corresponding frames. **c** Stabilized face images. **d** Localized face regions

2.3.2 Individualization of Eye Region Model

The system first registers the eye region model to a stabilized face in an initial frame $t = 0$ by scaling and rotating the model so that both ends of curve1 ($\mathbf{u}_1$) of the upper eyelid coincide with the eye corner points in the image. The initial frame is such a frame that contains a neutral eye (an open eye with the iris at the center), which may be different from the initial frame used in head tracking. The individualized structure parameters $\mathbf{s}$ are obtained either manually or automatically and fixed across the entire sequence. Manual individualization was used for the experiments reported in Sect. 2.5.1 and 2.5.2 and automatic individualization was used for those reported in Sect. 2.5.3.

2.3.2.1 Manual Individualization

The first implementation manually individualizes the structure parameters $\mathbf{s}$ by using a graphical user interface. Example results of manual individualization with respect to each factor of the appearance diversities in Fig. 2.1 are shown in Table 2.4.

2.3.2.2 Automated Individualization

Individualization of the structure parameters $\mathbf{s}$ is automated by using AAM (Active Appearance Models) [104] in the second implementation. We define

Table 2.4. Example results of manual structure individualization

	input	normalized	model
(a1) Single-fold eyelid			
(a2) Double-fold eyelid			
(a3) Thick eyelid			
(a4) Revealing eyelid			
(b1) Bright iris			
(b2) Dark iris			
(c1) Bulge			
(c2) Reflection			

three subregions in the eye region image, as shown in Fig. 2.9. Manually labeled contour points of each subregion for all the training samples are used for generating AAM. AAM represents both the shape and the texture of each subregion of an eye region image in the low dimensional subspace by using only the significant principal components.

For each subregion of the input eye region image, the system first searches an optimal AAM that represents the appearance of the input in the subspace while leaving the positions of the corner points of both the eye and the eyebrow fixed (constrained) [104]. It then finds the nearest training sample to the input in the subspace and initializes the structure parameters of the input with those of the selected training sample.

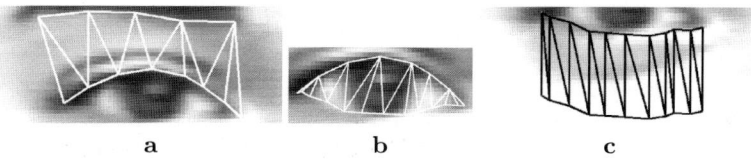

Fig. 2.9a–c. Subregions in an example eye region image and their triangulation. **a** The upper eyelid region has 5 points on the eyebrow and 8 on the upper eyelid. **b** The *palpebral fissure* region has 8 points on the upper eyelid and 11 on the lower eyelid. **c** The lower eyelid region has 11 points on the lower eyelid and the same number parallel to and at half the width of the eye region below it

2.3.3 Tracking of Eye Motion

The pixel intensity values of both the input eye region and the eye region model are normalized prior to eye motion tracking so that they have the same average and standard deviation. The motion parameters in the initial frame $\mathbf{m}_0$ are manually adjusted when the eye region model is individualized.

With the initial motion parameters $\mathbf{m}_0$ and the structure parameters $\mathbf{s}$, the system tracks the motion of the eye across the rest of the sequence starting from $t = 0$ to obtain $\mathbf{m}_t$ at all t. The system tracks the motion parameters by an extended version of the Lucas-Kanade gradient descent algorithm [384], which allows the template searched (the eye region model here) to deform while tracking. Starting with the values in the previous frame, the motion parameters $\mathbf{m}_t$ at the current frame t are estimated by minimizing the following objective function D:

$$D = \sum \left[T\left(\mathbf{x}; \mathbf{m}_t + \delta\mathbf{m}_t\right) - I\left(W\left(\mathbf{x}; \mathbf{p}_t + \delta\mathbf{p}_t\right)\right)\right]^2 , \qquad (2.17)$$

where I is the input eye region image, W is a warp from the coordinate system of the eye region model to that of the eye region image, and $\mathbf{p}_t$ is a vector of the warp parameters that includes only translation in this implementation. Structure parameters $\mathbf{s}$ do not show up in T because they are fixed while tracking.

$\delta\mathbf{m}_t$ and $\delta\mathbf{p}_t$ are obtained by solving the simultaneous equations obtained from the first-order Taylor expansion of (2.17) as explained in detail in the Appendix which can be viewed for free at http://computer.org/tpami/archives.htm. $\mathbf{m}_t$ and $\mathbf{p}_t$ are updated:

$$\mathbf{m}_t \leftarrow \mathbf{m}_t + \delta\mathbf{m}_t , \qquad \mathbf{p}_t \leftarrow \mathbf{p}_t + \delta\mathbf{p}_t . \qquad (2.18)$$

The iteration process at a particular frame t converges when the absolute values of $\delta\mathbf{m}_t$ and $\delta\mathbf{p}_t$ become less than the preset thresholds or the number of iterations reaches the maximum. The region surrounded by curve1

($\mathbf{u}_1$) and curve3 ($\mathbf{l}_1$) of the eyelids is used for the calculation process so that more weight is placed on the structure inside the eye (the *palpebral fissure*) and other facial components (such as an eyebrow) that may appear in the eye region will not interfere. When parameter ν_{height} is less than a preset threshold, the position of region7 (η_x and η_y) is not updated because the iris is so occluded that its position estimation is unreliable. Also, the warp parameters $\mathbf{p}_t$ are not updated when ν_{height} is less than a preset threshold because a closed (or an almost closed) eye appears to have only horizontal structure that gives only the vertical position of the eye region reliably.

2.4 Experiments

We applied the proposed system to 577 image sequences from two independently collected databases: the Cohn–Kanade AU-coded Facial Expression Image Database [310] and the Ekman–Hagar Facial Action Exemplars [161]. The subjects in these databases are young adults and include both men and women of varied ethnic backgrounds. They wear no glasses or other accessories that could occlude their faces. With few exceptions, head motion ranges from none (Ekman-Hager) to small (Cohn–Kanade), and head pose is frontal. Image sequences were recorded using VHS or S-VHS video and digitized into 640 by 480 grayscale or 16-bit color pixel arrays. Image sequences begin with a neutral or near-neutral expression and end with a target expression (e.g., lower eyelids tightened). In Cohn–Kanade, image sequences are continuous (30 frames per second). In Ekman-Hager, they are discontinuous and include the initial neutral or near-neutral expression and two each of low, medium, and high intensity facial action sampled from a longer image sequence.

In the experiments reported here, we empirically chose the following parameter values for the eye model: $N_u = 8$, $N_l = 11$, $N_f = 8$, $\alpha_1 = 30$, $\alpha_2 = 40$, $\beta_1 = 20$, $\beta_2 = 10$, $\beta_3 = 80$, $\beta_4 = 30$, $\beta_5 = 70$, $I_{r1}^{\text{brightest}} = 160$, $I_{c3}^{\text{brightest}} = 160$, $I_{c5}^{\text{brightest}} = 130$, $I_{r4}^{\text{darkest}} = 120$, $w_{c2}^{\text{thickest}} = 5$, and $\theta = \pi/6$. The initialization for tracking was done to the first neutral or near-neutral expression frame in each sequence. The system generates the eye region model as a graphic image with a particular resolution. Because the size and positions of the graphics objects (e.g., lines) are specified in integers, resolution and sharpness of the graphic images must be high enough for the model to represent the fine structures of an eye region. In our implementation, resolution was set at 350 by 250 pixels. The system then registered the model to the input eye region by scaling and rotating it as explained in Sect. 2.3.2. We examined the results for diverse static eye structures and for the whole range of appearance changes from the neutral to the utmost intensities in dynamic motion.

2.4.1 Cohn–Kanade AU-coded Facial Expression Image Database

This database was collected by the Carnegie Mellon and University of Pittsburgh group. A large part of this database has been publicly released. For this experiment, we used 490 image sequences of facial behaviors from 101 subjects, all but one of which were from the publicly released subset of the database. The subjects are adults that range from 18 to 50 years old with both genders (66 females and 35 males) and a variety of ethnicities (86 Caucasians, 12 African Americans, 1 East Asian, and two from other groups). Subjects were instructed by an experimenter to perform single AUs and their combinations in an observation room. Their facial behavior was then manually FACS labeled [160]. Image sequences that we used in this experiment began with a neutral face and had out-of-plane motion as large as 19 degrees.

2.4.2 Ekman–Hager Facial Action Exemplars

This database was provided by Ekman at the Human Interaction Laboratory, University of California San Francisco, whose images were collected by Hager, Methvin, and Irwin. For this experiment, we used 87 image sequences from 18 Caucasian subjects (11 females and 7 males). Some sequences have large lighting changes between frames. For these, we normalized the intensity so as to keep the average intensity constant throughout the image sequence. Each image sequence in this database consists of six to eight frames that were sampled from a longer sequence. Image sequences begin with neutral expression (or a weak facial action) and end with stronger facial actions.

2.5 Results and Evaluation

We used both qualitative and quantitative approaches to evaluate system performance. Qualitatively, we evaluated the system's ability to represent the upper eyelids, localize and track the iris, represent the *infraorbital furrow*, and track widening and closing of the eyelids. Successful performance ensures that the system is robust to ethnic and cosmetic differences in eyelid structure (e.g., single versus double fold) and features that would be necessary for accurate action unit recognition (direction of gaze, *infraorbital furrow* motion, and eyelid widening and closing). In quantitative evaluation, we investigated system performance with respect to resolution and sharpness of input eye region images, initialization, and complexity of the eye model.

Table 2.5. Example results for a variety of upper eyelids

(a) Single-fold	(b) Double-fold	(c) Thick	(d) Revealing

2.5.1 Examples

Of the total 577 image sequences with 9,530 frames, the eye region model failed to match well in only 5 image sequences (92 frames total duration) from two subjects. One of the sequences contained relatively large and rapid head motion (approximately 20 degrees within 0.3 seconds) not otherwise present in either database. This motion caused interlacing distortion in the stabilized image that was not parameterized in the model. The other four error cases from a second subject were due to limitations in individualization as discussed below.

2.5.1.1 Upper Eyelids

A most likely failure would be that a curve of the upper eyelid model matches with the second (upper) curve of a double-fold eyelid in the input when they have similar appearance. As shown in Table 2.5(b), our system was not compromised by such double-fold eyelids. Note that these eye region images shown in the table are after the image stabilization. The face itself moves in the original image sequence. (This is true with the subsequent tables through Table 2.12.)

When an upper eyelid appears thick due to cosmetics, eyelashes, or shadow, a model with a single thin line could match mistakenly at many locations within the area of thickness. Such errors did not occur; by considering boldness of the upper eyelids as a variable, our system was able to track the correct positions of upper eyelids, as shown in Table 2.5(c).

Some subjects had double-fold eyelids that appeared single-folded when the face was at rest (i.e., neutral expression). In these cases, the second (hidden) curves were revealed when the eyelids began to widen or narrow, which

Table 2.6. Example results for irises of different colors

(a) Bright iris	(b) Dark iris

unfolded the double-fold. The boldness parameter absorbed this "revealing effect" and the system was able to track correctly the upper eyelid contour, as shown in Table 2.5(d).

2.5.1.2 Irises

A most likely failure in tracking irises would be for an iris model to match another dark portion in the eye region, such as shadow around the hollow between the inner corner of the eye and the root of the nose. An especially bright iris could contribute to this type of error. This situation could happen if one were to try to find the location of the iris by finding only a circular region with a fixed dark color (e.g., Tian et al. [616]). Because our method uses a whole eye region as a pattern in matching and includes color and size of the irises as variables, the system was able to track the positions of irises accurately over a wide range of brightness as shown in Table 2.6(a).

2.5.1.3 Bulge with Reflection Below the Eye

A most likely failure would be that a curve of the lower eyelid model matches with the lower edge of the bulge or the *infraorbital furrow*. This could occur when the appearance of a bright bulge and the furrow below it are similar to that of the sclera with a lower eyelid curve below it. By considering the bulge, the illumination reflection on the bulge, and the *infraorbital furrow* in modeling the appearance below the eye, our system tracked lower eyelids accurately, as shown in Table 2.7.

2.5.1.4 Motion

Of 44 AUs defined in FACS [160], six single AUs are defined in the eye region. These include AU 5 (upper lid raiser), AU 6 (cheek raiser and lid

Table 2.7. Example results for differences in appearance below the eye

(a) Bulge	(b) Bulge with reflection

compressor), AU 7 (lid tightener, which encompasses AU 44 in the 2002 edition of FACS), AU 43 (eye closure), AU 45 (blink), and AU 46 (wink). Gaze directions are also defined as AU 61 (turn left), AU 62 (turn right), AU 63 (up), and AU 64 (down). Tables 2.8(a), (b), (c), (d), and (e) are correspondent with AU 5, AU 6+62, AU 6, AU 45 & AU 7, and AU 6+7, respectively, which cover the AUs related to the eye region. The frames shown range from neutral to maximum intensity of the AUs. A most likely failure due to appearance changes by the motion of an eye would be that tracking of the upper eyelid and the lower eyelid fails when the distance between them closes, such as in blinking (AU45). Our system tracked blinking well, as shown in Table 2.8. Tracking eye motion by matching an eye region increased system robustness relative to individually tracking feature points (such as in [617], [616], [482], [375]) or using a generic eye model. Studies [616], [707] that have used parabolic curves to represent eye shape have been less able to represent skewed eyelid shapes. Our model explicitly parameterizes skewing in the upper eyelid model; accordingly, the system was able to track such skewing upper eyelids in their motions as shown in Table 2.8(b) and (d).

2.5.1.5 Failure

Iris localization failed in a Caucasian female who had a bright iris with strong specular reflection and a thick and bold outer eye corner. Fig. 2.10 shows the error. While the eyelids were correctly tracked, the iris model mistakenly located the iris at the dark eye corner. Failure to correctly model the texture inside the iris appeared to be the source of this error. To solve this problem in future work, we anticipate that recurrently incorporating the appearance of the target eye region into the model during tracking would be effective and, more generally, would improve ability to accommodate unexpected appearance variation.

Table 2.8. Example results for motions

(a) Upper eyelid raising	(b) Gaze change and cheek raising	(c) Cheek raising	(d) Blinking and eyelid tightening	(e) Cheek raising and eyelid tightening

2.5.2 Quantitative Evaluation

To quantitatively evaluate the system's accuracy, we compared the positions of the model points for the upper and lower eyelids and the iris center ($\mathbf{u}_1$, $\mathbf{l}_1$, η_x, and η_y in Table 2.1, respectively) with ground truth. Ground truth was determined by manually labeling the same number of points around the upper- and lower eyelids and the iris center using a computer mouse. These points then were connected using polygonal curves. We then computed the Euclidean distance from each of the model points to the closest line segment between manually labeled points. If model points were located horizontally outside of the eye, the line segment from the closest manually labeled end-

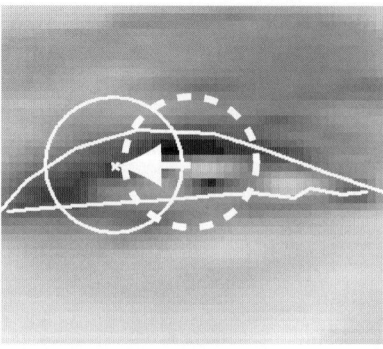

Fig. 2.10. A failure case with a bright and specular iris. The dashed circle indicates the correct position manually labeled, and the solid circle system's output. The eyelids were correctly tracked, whereas the iris model mistakenly located at the dark eye corner

point was used. For the iris center, the Euclidean distance to the manually labeled iris center was computed. The Euclidean distances were normalized by dividing them by the width of the eye region. The vector of tracking errors is denoted as vector ϵ.

2.5.2.1 Sensitivity to Input Image Size and Sharpness

When the size of the input eye region is small relative to the actual size of the eye or the input image is not sufficiently sharp, fine structure of the eye may not be sufficiently visible. Image sharpness refers to large gain in the high-frequency components of an image. To evaluate system robustness to input image size and sharpness, we compared tracking error with respect to multiple sizes and sharpness of input eye region images. Sharpness of the input images was sampled by applying a high pass filter to the image sequences. We selected for analysis nine sequences based on the response: three sequences that had the strongest response, three the weakest, and three in halfway between these. To vary image size, we resampled the images into make three levels: the original scale, 50 percent scale (0.5×0.5), and quarter scale (0.25×0.25). Eye motion tracking in the smaller scales used the same structure parameters as those used in the original scale. Table 2.9 shows an example of multiple scales of a particular eye region image. The table also shows the computation time for updating the model parameters (Pentium M, 1.6 GHz, 768 MB RAM, Windows XP, the average over 10 time trials). Figure 2.11 shows the tracking error plotted against the widths of the image sequences. Tracking error up to about 10 percent of eye region width may have resulted from error in manual labeling. The most likely cause of small error in manual labeling was ambiguity of the boundary around the *palpebral fissure* (Fig. 2.12). We found

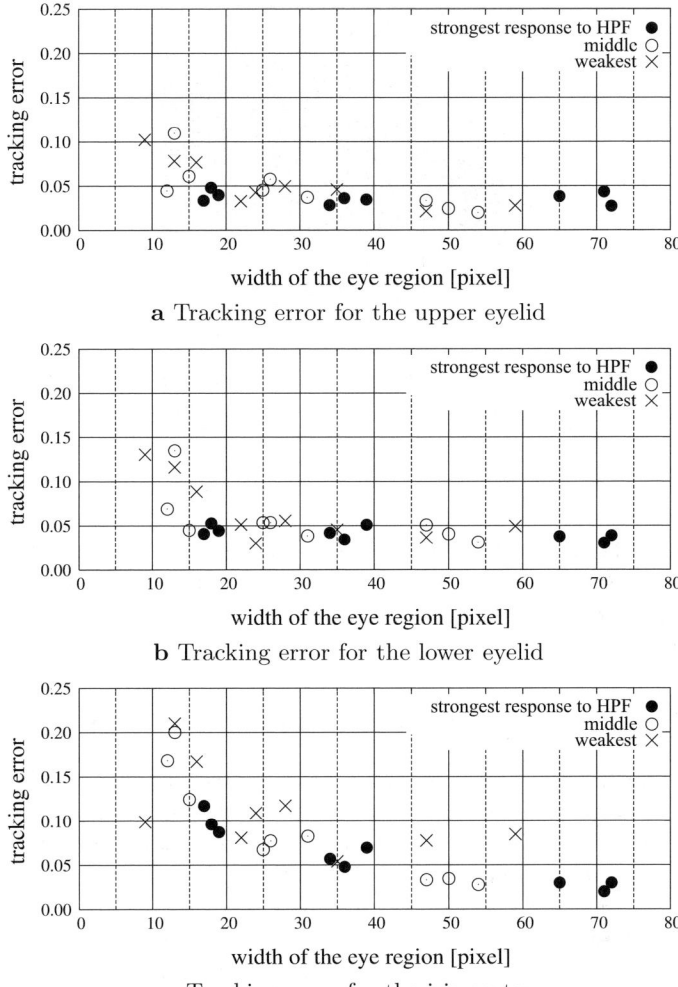

a Tracking error for the upper eyelid

b Tracking error for the lower eyelid

c Tracking error for the iris center

Fig. 2.11a–c. Sensitivity to image resolution

that an eye region width of about 15 pixels was the margin under which tracking became impaired for the upper eyelid, lower eyelid, and the iris position. Above this value, performance was relatively robust with respect to both size and sharpness of the input eye region.

2.5.2.2 Effect of Eye Model Details

The eye region model defines many structural components to represent the diversities of eye structure and motion. To investigate whether all are necessary, we systematically omitted each component and examined the resulting

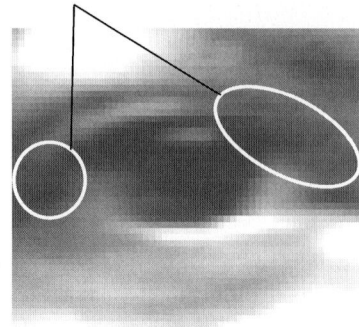

ambiguous boundary

Fig. 2.12. An example of ambiguous boundaries around the *palpebral fissure*

Table 2.9. Computation time for multiple image resolution

		original (59 pixels)	half (28 pixels)	quarter (16 pixels)
multiple scales of a particular eye region image				
the model individualized in the original scale				
computation time	generating the model	13.1 ms	6.6 ms	5.6 ms
	warping input image to the model	2.2 ms	0.4 ms	0.1 ms
	updating the model (using extended LK)	134.0 ms	77.9 ms	61.3 ms
	total	149.3 ms	84.9 ms	67.0 ms

change in tracking error. Table 2.10 shows the results of this comparison. When the model for double eyelid folds was omitted, tracking of the upper eyelid (Table 2.10(a)) was compromised. Omitting components for the appearance below the eye (Table 2.10(b)) and only the brightness region on the bulge (Table 2.10(c)) had similar effects. To achieve accurate and robust eye motion tracking for diverse eye appearances and motion, all the detailed components of the eye region model proven necessary.

In Table 2.10(a), the tracking error ϵ shows that tracking of the other parts of the eye model was also compromised without the model for double

Table 2.10. Different levels of detail of the model and their effects

detailed structure	results				
(a) Double eyelid folds		the first frame		the last frame	
	input				
		individualization		motion tracking	
		eye region model	superposition to the input	eye region model	superposition to the input
	w/o				
					$\epsilon = (0.06, 0.06, 0.09)$
	w/				
					$\epsilon = (0.01, 0.03, 0.05)$
(b) Bulge, *infraorbital furrow*, and reflection on the bulge		the first frame		the last frame	
	input				
		individualization		motion tracking	
		eye region model	superposition to the input	eye region model	superposition to the input
	w/o				
					$\epsilon = (0.02, 0.08, 0.13)$
	w/				
					$\epsilon = (0.03, 0.01, 0.09)$

Table 2.10. Continued

detailed structure	results			
(c) Reflection on the bulge		the first frame		the last frame
	input			

		individualization		motion tracking	
		eye region model	superposition to the input	eye region model	superposition to the input
w/o					$\epsilon = (0.05, 0.09, 0.04)$
w/					$\epsilon = (0.04, 0.06, 0.04)$

eyelid folds (the error for the upper eyelid curve $\mathbf{u}_1$, the lower eyelid curve $\mathbf{l}_1$, and the iris center η_x and η_y are shown in parentheses). This indicates that the model components support tracking accuracy as a whole and erroneous individualization of one component affected tracking accuracy of the other parts.

2.5.2.3 Sensitivity to Model Initialization

The eye region model is manually initialized in the first frame with respect to both the structure parameters $\mathbf{s}$ and the motion parameters $\mathbf{m}_t$. We observed that the initialization of the structure parameters (individualization of the model) dominantly affected the tracking results. To evaluate sensitivity of the system to initialization, we individually manipulated each structure parameter in turn while leaving the others fixed. Figure 2.13b is the eye region model individualized to an example of input eye region images shown in Fig. 2.13a and Fig. 2.13c shows changes in tracking error when parameter d_u was varied from 0.0 to 1.0 while leaving other parameters fixed to $f = 0.7$, $d_b = 0.71$, $d_r = 0.7$, $r_i = 0.5$, and $I_{r7} = 0.7$. Figures 2.13d–h were similarly obtained. Each of the individualized structure parameters that provided stable tracking also locally minimized the tracking error. Only the parameter

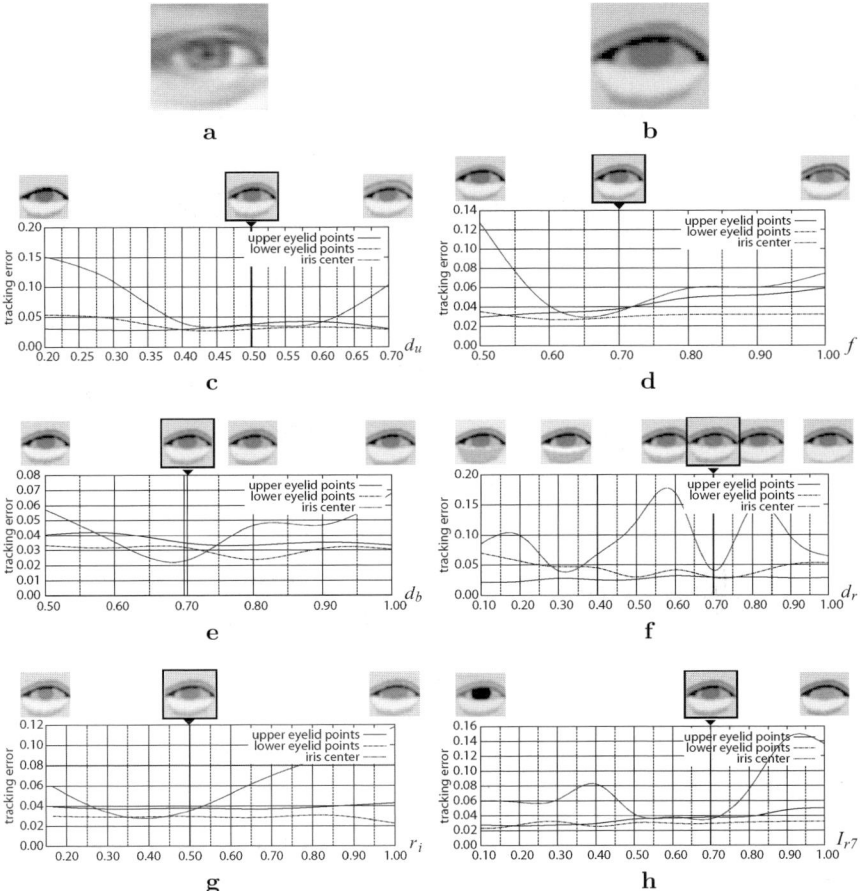

Fig. 2.13a–h. An example of individualization of the model and the sensitivity to the parameter changes. **a** Input eye region image. **b** Individualized eye model, $s = \{d_u, f, d_b, d_r, r_i, I_{r7}\} = 0.5, 0.7, 0.71, 0.7, 0.5, 0.7\}$. **c** Sensitivity to parameter d_u. **d** Sensitivity to parameter f. **e** Sensitivity to parameter d_b. **f** Sensitivity to parameter d_r. **g** Sensitivity to parameter r_i. **h** Sensitivity to parameter I_{r7}

d_r was sensitive to the initialization in this particular example (the tracking error rapidly increased for the slight change of d_r). We also observed that parameters were intercorrelated. Figure 2.14 shows a contour plot of tracking error against the changes of an example pair of structure parameters for the same image sequence used in Fig. 2.13. Nonlinearity is obvious, yet with weak linearity.

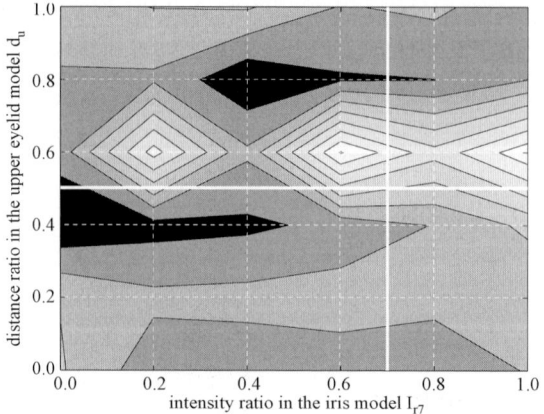

Fig. 2.14. A contour plot of tracking error against an example pair of the structure parameters: the intensity ratio of the iris model I_{r7} versus the distance ratio between eyelid folds d_u. The other parameters were left fixed. The brighter region indicates larger error. White lines are the values individualized in Fig. 2.13

2.5.3 Tracking Results by Automated Individualization

Initialization of the structure parameters of the eye region model is automatically done by the method explained in Sect. 2.3.2. The same image frames that were used for manual initialization were collected (one for each subject) as training samples. In selecting the most similar training sample to an input eye region image, all the subjects but the input's were used for training AAM. The dimensionalities of the trained AAMs were 19, 23, and 23 for the upper eyelid region, the *palpebral fissure* region, and the lower eyelid region, respectively. Example results of automated initialization of the eye region model are shown in Table 2.11. The third column of the table shows the training samples that were found nearest in the subspace of AAM (from the top, the upper eyelid region, the *palpebral fissure* region, and the lower eyelid region). Those that have similar appearances were automatically selected for the subregions in each example except Table 2.11(d) where the training sample with double eyelid fold was selected for the input with single eyelid fold. This occurred because there was no other example of single eyelid fold in the training samples than the input (The database used contains only 1 East Asian with single eyelid fold.). Including more subjects with single eyelid fold would solve this problem.

Tracking results for these examples are shown in Table 2.12. In 2.12(a) and (b) performance for automated and manual initialization were comparable. Tracking in 2.12(c) was unstable because the upper eyelid became brighter in the last frame and illumination reflection on the upper eyelid was not pa-

Table 2.11. Example results of automated structure individualization

	input eye region images	selected training samples	individualized structure parameters
(a) Revealing double-fold, dark iris, and no bulge below the eye			$d_u = 0.3$ $f = 1.0$ $I_{r7} = 86$ $r_i = 10$ $d_b = 0.0$ $d_r = 0.0$
(b) Double-fold, dark iris, and no bulge below the eye			$d_u = 0.5$ $f = 1.0$ $I_{r7} = 50$ $r_i = 10$ $d_b = 0.68$ $d_r = 0.0$
(c) Double-fold, bright iris, and bulge below the eye			$d_u = 0.3$ $f = 0.7$ $I_{r7} = 50$ $r_i = 9$ $d_b = 0.05$ $d_r = 0.48$
(d) Single-fold, dark iris, bulge and reflection on it below the eye			$d_u = 0.6$ $f = 0.75$ $I_{r7} = 50$ $r_i = 9$ $d_b = 0.56$ $d_r = 0.15$

rameterized in the model. In 2.12(d) tracking error occurred in the last frame because the model for the double eyelid fold matched with the single eyelid fold in the input. Other examples of tracking with automated initialization for the same set of sequences used in Table 2.8 are shown in Table 2.13. Automated initialization again led to good tracking. This means that neutral eyes with similar appearances similarly change the appearance in their mo-

Table 2.12. Example comparison of tracking results between manual and automated individualization

	first frame	last frame	manual	automated
(a)				
(b)				
(c)				
(d)				

tions, which suggests that the appearance changes caused from motion can be predicated from the structure in the neutral expression.

2.6 Conclusion

The appearance of the eyes varies markedly due to both individual differences in structure and the motion of the eyelids and iris. Structural individuality includes the size and color of the iris, the width, boldness, and number of eyelid folds, the width of the bulge below the eye, and the width of the illumination reflection on the bulge. Eye motion includes the up-down action of the upper and lower eyelids and the 2D movement of the iris. This variation together with self-occlusion and change of reflection and shape of furrows and bulges has made robust and precise analysis of the eye region a challenging problem. To meticulously represent detailed appearance variation in both structural individuality and eye motion, we developed a generative eye region model and evaluated its effectiveness. We tested the model in two large, independent face image databases. The detailed eye region model led to substantially better results than those previously reported in the literature. The system achieved precise tracking of the eyes over a variety of eye appearances and motions.

Table 2.13. Example results of tracking eye motions with automated structure individualization

	initial frame	selected training samples	individualized structure parameters	tracking results	
				manual	automated
(a) Upper eyelid raising			$d_u = 0.5$ $f = 0.7$ $I_{r7} = 50$ $r_i = 10$ $d_b = 0.0$ $d_r = 0.0$		
(b) Gaze change and cheek raising			$d_u = 0.6$ $f = 0.7$ $I_{r7} = 95$ $r_i = 10$ $d_b = 0.54$ $d_r = 0.0$		
(c) Cheek raising			$d_u = 0.4$ $f = 0.7$ $I_{r7} = 50$ $r_i = 10$ $d_b = 0.1$ $d_r = 0.66$		
(d) Blinking and eyelid tightening			$d_u = 0.7$ $f = 0.8$ $I_{r7} = 77$ $r_i = 10$ $d_b = 0.05$ $d_r = 0.09$		
(e) Cheek raising and eyelid tightening			$d_u = 0.3$ $f = 1.0$ $I_{r7} = 50$ $r_i = 10$ $d_b = 0.05$ $d_r = 0.48$		

3

Advanced Off-Line Statistical Modeling of Eye and Non-Eye Patterns

Nizar Bouguila, Riad I. Hammoud, and Djemel Ziou

3.1 Position of the Problem

Modeling of eye patterns is a fundamental process in a vision-based eye monitoring system. It helps identifying the unique structure and appearance of the eye across as many subjects and operating scenarios.

From an application stand point, constructing accurate eye models is a challenging problem which provide excellent cues for different applications such as eyes tracking [229] for driver fatigue and behavior analysis [170, 215, 217, 220, 222, 226], face detection and recognition [697], eye typing [231], communication via eye blinks [208], drowsiness detection [158], and human computer interaction [284]. Another important application is facial expression understanding, since eyes play a crucial role in interpreting a person's emotional and affective states [134]. This problem has been presenting several challenges to existing computer vision, pattern recognition and machine learning algorithms because of many facts such as the difference of the eye shape from one subject to another (Asian, European, African, etc), the drastic occlusions, changes in brightness-level and lighting conditions (indoor, outdoor, direct sun-light, etc), and eye closure due to blinking [215]. In addition, the eye appearance changes significantly with glasses [686]. Dealing with glasses is an important issue in eyes modeling if we note that 34% of the total US population wore prescription glasses in 2000 [10]. The glares on the glasses caused by light reflections represent a real challenge to modeling.

Two categories of eye modeling have appeared in the vision community: image-based passive approaches [470, 718] and active IR-based approaches [236, 420]. Image-based passive methods fall themselves broadly within three categories: deformable templates, appearance-based and features-based methods [230]. While deformable templates and appearance-based methods build models directly on the appearance of the eye region, feature-based methods extract particular local features of the eye region [230]. In this chapter, we are interested in IR eyes modeling. Indeed, IR-based approaches permit simple and effective modeling [719] and may offer better performance because it provides a capability for modeling under different lighting conditions and it produces the bright/dark pupil effect in contrast of visible imagery [93,94,719].

The focus of this chapter is on low-resolution infrared eye modeling using advanced statistical approaches. Indeed, an important step, after developing relevant features describing a given object (the eye in our case), is the selection of a statistical generative model to capture the intra-class and inter-class variations in the feature space. Machine learning techniques are widely used for this goal. Such techniques provides a natural and successful framework for bringing eye modeling to practical use. Processing brightness, shape and texture-based eye features, a number of learning approaches have shown success in modeling eyes images and discriminating them from non-eye objects. Eyes modeling can be viewed as a two-class (eye vs non-eye) classification problem. Eye and non-eye models are learned off-line using a set of training images and then a new patch (test image) is assigned online, using a classification decision rule, to one of these two classes (see Fig. 3.1).

Haro et al. [236] use probabilistic principal component analysis (PPCA) to model off-line the inter-class variability within the eye space and the non-eye distribution where the probability is used as a measure of confidence about the classification distribution. Osuna et al. [476] trained support vector machines (SVM) [641] on a set of eye samples captured under different poses, rotation and scale changes. Hammoud [215] used relevance vector machines (RVM) [619] instead of SVM in order to find the sparse vectors representing the optimal space boundary between two classes of eye and non-eye patches. Hansen and Hammoud [230] used Haar wavelet classifiers for eyes modeling and detection.

In this chapter, we review and compare several pattern recognition techniques in the case of statistical eyes modeling. Moreover, we propose the use of finite mixture models for this task. In Sect. 3.2 different pattern recognition techniques will be briefly reviewed. These techniques will be employed to model eye and non-eye patterns offline as well as to classify un-seen patterns on-line. In order to have a self-contained chapter, we will recall some important properties of the generalized Dirichlet mixture in Sect. 3.3. Section 3.4 presents generalized Dirichlet estimation and selection using both deterministic and pure Bayesian approaches. In Sect. 3.5, we show some experimental results. Finally the conclusion of this work is drawn in Sect. 3.5.

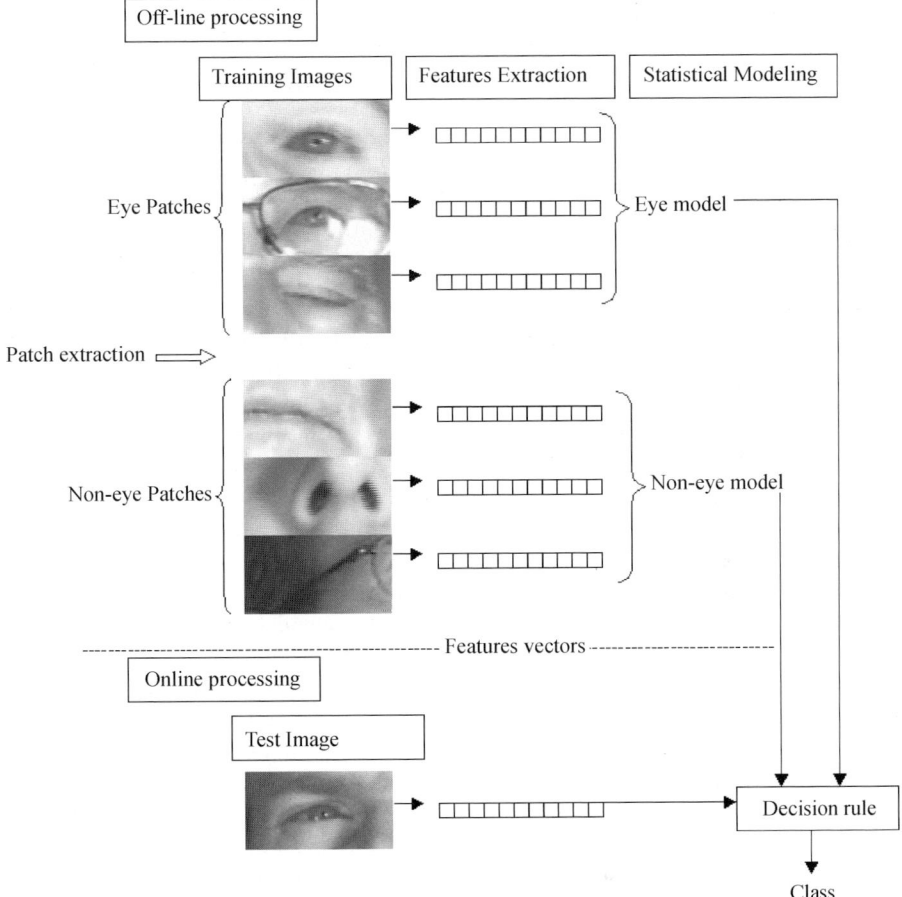

Fig. 3.1. Eye modeling flowchart

3.2 Review of Pattern Classification Techniques

As Rosenfeld said *the general goal of image pattern recognition is to generate descriptions of images, and relates those descriptions to models characterizing classes of images* [539]. Many algorithms for learning and pattern recognition have appeared and applied with success for different computer vision tasks. There are several reference books about image pattern recognition techniques that can be recommended such as [47, 48, 89, 150, 187, 203, 524, 612, 667]. In this section, we will focus on four important techniques which are: Principal components analysis (PCA), linear discriminant analysis (LDA), support vector machine (SVM) and finite mixture models. An example of finite mixture models based on generalized Dirichlet distributions will be developed in details in the next two sections.

3.2.1 Principal Components Analysis

PCA and LDA have been used especially for dimensionality reduction and feature extraction [398]. PCA is one of the most used algorithms in computer vision. Its application to face recognition is perhaps the most popular [634]. Given a training set of N feature vectors representing N images $\mathcal{X} = (\mathbf{X}_1, \ldots, \mathbf{X}_N)$, where $\mathbf{X}_i \in \mathbb{R}^d$ $i = 1, \ldots, N$, PCA finds a linear transformation W^t which maps the d-dimensional features vectors space into a new feature space with lower dimension $d^{\text{new}} < d$. The d^{new}-dimensional features vectors $\mathbf{X}_i^{\text{new}}$ are given by:

$$\mathbf{X}_i^{\text{new}} = W^T \mathbf{X}_i \tag{3.1}$$

With PCA we try to find the optimal projection E which maximizes the determinant of the scatter matrix $W^T SW$ of the new projected samples $\mathcal{X}^{\text{new}} = (\mathbf{X}_1^{\text{new}}, \ldots, \mathbf{X}_N^{\text{new}})$

$$E = \arg \max_W |W^T SW| \tag{3.2}$$

where S is the scatter matrix of the original data

$$S = \sum_{i=1}^{N} (\mathbf{X}_i - \bar{\mathbf{X}})(\mathbf{X}_i - \bar{\mathbf{X}})^T \tag{3.3}$$

$\bar{\mathbf{X}}$ is the mean vector of $\mathcal{X}$

$$\bar{\mathbf{X}} = \frac{1}{N} \sum_{i=1}^{N} \mathbf{X}_i \tag{3.4}$$

and $E = [E_1, \ldots, E_{d^{\text{new}}}]$ is composed of the d-dimensional eigenvectors of S correspoding to the d^{new} largest eigenvalues [150].

3.2.2 Linear Discriminant Analysis

The goal of LDA [524], also known as Fisher Discriminant Analysis (FDA), is to find the set of vectors in the underlying space that best discriminate the different classes describing the data. LDA has been applied to different problems such as face recognition [249] and image retrieval [599]. LDA assumes that the classes are linearly separable and follow homoscedastic gaussian distributions. Under this assumption, one can show that the optimal subspace where we can perform the classification is given by the vectors W which are the solution of the following generalized eigenvalue problem

$$S_b W = \lambda S_w W \tag{3.5}$$

where S_w is the within-class scatter matrix and given by

$$S_w = \sum_{j=1}^{M} \sum_{i=1}^{N_j} (\mathbf{X}_i - \bar{\mathbf{X}}_j)(\mathbf{X}_i - \bar{\mathbf{X}}_j)^T \tag{3.6}$$

where N_j is the number of vectors in class j and $\bar{\mathbf{X}}_j$ is the mean of class j. S_b is the between-class scatter matrix and given by

$$S_b = \sum_{j=1}^{M} (\bar{\mathbf{X}}_j - \bar{\mathbf{X}})(\bar{\mathbf{X}}_j - \bar{\mathbf{X}})^T \qquad (3.7)$$

Despite its effectiveness, a major inconvenient of LDA is the linearity of the classification surface. To overcome this problem, SVM can be used to offer both linear and non-linear flexible classification surfaces.

3.2.3 Support Vector Machine

Support Vector Machine (SVM) is a two-class classification method that have been used successfully in many applications dealing with data and images classification [641]. In the following, we briefly summarize the theory of SVM. For two-class pattern recognition, we try to estimate a function $f : \mathbb{R}^d \rightarrow \{\pm 1\}$ using l training d-dimensional vectors $\mathbf{X}_i$ and class labels y_i,

$$(\mathbf{X}_1, y_1), \ldots, (\mathbf{X}_l, y_l) \in \mathbb{R}^d \times \{\pm 1\} \qquad (3.8)$$

after the training the function f should be able to correctly classify new test vectors $\mathbf{X}$ into one of the two classes. Suppose that we have a hyperplane separating the first class (positive class) for the the second class (negative class). The idea behind SVM is to find the optimal hyperplane permitting a maximal margin of separation between the two classes and defined by

$$\mathbf{w}.\mathbf{X} + b = 0 \quad \mathbf{w} \in \mathbb{R}^d, b \in \mathbb{R} \qquad (3.9)$$

corresponding to decision function

$$f(\mathbf{X}) = \text{sign}(\mathbf{w}.\mathbf{X} + b) \qquad (3.10)$$

where b is the distance to the hyperplane from the origin and $\mathbf{w}$ is the normal of the hyperplane which can be estimated through the use of training data by solving a quadratic optimization problem [641]. $\mathbf{w}$ can be estimated by

$$\mathbf{w} = \sum_{i=1}^{l} v_i \mathbf{X}_i \qquad (3.11)$$

where v_i are coefficient weights.

In general, classes are not linearly separable. In order to overcome this problem, SVM can be extended by introducing a kernel K to map the data into another dot product space F using a nonlinear map

$$\Phi : \mathbb{R}^d \rightarrow F \qquad (3.12)$$

In this new space F, the classes will be linearly separable. The kernel K is given by

$$K(\mathbf{X}, \mathbf{X}_i) = (\Phi(\mathbf{X}).\Phi(\mathbf{X}_i)) \tag{3.13}$$

and measures the similarity between data vectors $\mathbf{X}$ and $\mathbf{X}_i$. Then, the decision rule is

$$f(\mathbf{X}) = \text{sign}\left(\sum_{i=1}^{l} v_i K(\mathbf{X}_i, \mathbf{X}) + b\right) \tag{3.14}$$

An important issue here is the choice of the kernel function. However, in most of the applications, the intrinsic structure of the data has been ignored by standard kernels such as Gaussian, Radial Basis Function (RBF), polynomial and Fisher kernels. An interesting approach to this problem is the generation of Mercer Kernels directly from data, using finite mixture models [577] [578], as follows:

$$K(\mathbf{X}_l, \mathbf{X}_m) = \Phi(\mathbf{X}_l)\Phi^T(\mathbf{X}_m) = \frac{1}{N(\mathbf{X}_l, \mathbf{X}_m)} \sum_{j=1}^{M} p(j|\mathbf{X}_l)p(j|\mathbf{X}_m)$$

$$= \frac{1}{f(\mathbf{X}_l, \mathbf{X}_m)} \sum_{j=1}^{M} \hat{Z}_{lj}\hat{Z}_{mj} \tag{3.15}$$

where M is the number of clusters defining the mixture model, $p(j|\mathbf{X}_l) = \hat{Z}_{lj}$ represents the posterior probability, i.e the conditional probability that observation l belongs to class j, $f(\mathbf{X}_l, \mathbf{X}_m)$ is a normalization factor so that $f(\mathbf{X}_l, \mathbf{X}_l) = 1$. With this approach, we have $\Phi(\mathbf{X}_l) \propto (\hat{Z}_{l1}, \hat{Z}_{l2}, \ldots, \hat{Z}_{lM})$. Note that this Kernel verify Mercer's condition as shown in [577] and then can be applied for SVM. Intuitively, two vectors are similar if they have similar posterior probabilities. Indeed, from a probabilistic point of view two data vectors will have a larger similarity if they are placed in the same cluster of a mixture distribution.

3.2.4 Finite Mixture Models

Another powerful statistical pattern recognition approach is finite mixture models. Indeed, finite mixture models have been a useful tool for different challenging computer vision and pattern recognition problems [173,410,621]. In statistical pattern recognition, finite mixtures permit a formal approach to unsupervised learning. Indeed, finite mixtures can be viewed as a superimposition of a finite number of component densities and thus adequately model situations in which each data element is assumed to have been generated by one (unknown) component. More formally, a finite mixture model with M components is defined as

$$p(\mathbf{X}|\Theta) = \sum_{j=1}^{M} p(\mathbf{X}|\theta_j)p_j \tag{3.16}$$

The parameters of a mixture for M clusters are denoted by $\Theta = (\theta_1, \ldots, \theta_M, \mathbf{P})$, where $\mathbf{P} = (p_1, \cdots, p_M)^T$ is the mixing parameter vector. Of course, being probabilities, the p_j must satisfy

$$0 < p_j \leq 1, \quad j = 1, \ldots, M \tag{3.17}$$

$$\sum_{j=1}^{M} p_j = 1 \tag{3.18}$$

The choice of the component model $p(\mathbf{X}|\theta_j)$ is very critical in mixture decomposition. The number of components required to model the mixture and the modeling capabilities are directly related to the component model used [180]. In the past two decades, much effort has been devoted to Gaussian mixture models estimation, selection and application. However, several studies have stressed the need of alternative approaches in the case of applications involving non-Gaussian data. In recent development, finite Dirichlet mixture models are considered a robust alternative way to deal with non-Gaussian data [61,63,64]. Dirichlet mixture have the flexibility to represent very irregular forms of densities which makes it very useful in different applications [64]. Despite its flexibility the Dirichlet distribution has a restrictive negative covariance structure as shown in [57]. Bouguila and Ziou proposed then the Generalized Dirichlet mixture as an efficient solution to overcome this problem and to model high-dimensional data [57]. In [57] a hybrid stochastic expectation maximization algorithm (HSEM) was developed to estimate the parameters of the generalized Dirichlet mixture. In this chapter, we propose also a Bayesian algorithm for generalized Dirichlet mixture learning. The estimation of the parameters is based on the Monte Carlo simulation technique of Gibbs sampling mixed with a Metroplis-Hastings step. For the selection of the number of clusters, we use Bayes factors. In the next two sections, we sill discuss in details finite mixture models based on generalized Dirichlet distributions.

3.3 The Generalized Dirichlet Mixture

In dimension d, the generalized Dirichlet pdf is defined by [551]:

$$p(X_1, \ldots, X_d) = \prod_{i=1}^{d} \frac{\Gamma(\alpha_i + \beta_i)}{\Gamma(\alpha_i)\Gamma(\beta_i)} X_i^{\alpha_i - 1} \left(1 - \sum_{j=1}^{i} X_j\right)^{\gamma_i} \tag{3.19}$$

for $\sum_{i=1}^{d} X_i < 1$ and $0 < X_i < 1$ for $i = 1 \ldots d$, where $\alpha_i > 0$, $\beta_i > 0$, $\gamma_i = \beta_i - \alpha_{i+1} - \beta_{i+1}$ for $i = 1 \ldots d-1$ and $\gamma_d = \beta_d - 1$. Note that the generalized Dirichlet distribution is reduced to a Dirichlet distribution when

$\beta_i = \alpha_{i+1} + \beta_{i+1}$:

$$p(X_1, \ldots, X_d)$$

$$= \frac{\Gamma\left(\alpha_1 + \alpha_2 + \ldots + \alpha_d + \alpha_{d+1}\right)}{\Gamma(\alpha_1)\Gamma(\alpha_2)\ldots\Gamma(\alpha_d)\Gamma(\alpha_{d+1})} \left(1 - \sum_{i=1}^{d} X_i\right)^{\alpha_{d+1}-1} \prod_{i=1}^{d} X_i^{\alpha_d - 1} \quad (3.20)$$

where $\alpha_{d+1} = \beta_d$. The mean and variance of the Dirichlet distribution satisfy the following conditions:

$$E(X_i) = \frac{\alpha_i}{\sum_{l=1}^{d+1} \alpha_l} \quad (3.21)$$

$$Var(X_i) = \frac{\alpha_i \left(\sum_{i=1}^{d+1} \alpha_i - \alpha_i\right)}{\left(\sum_{i=1}^{d+1} \alpha_i\right)^2 \left(\sum_{i=1}^{d+1} \alpha_i + 1\right)} \quad (3.22)$$

and the covariance between X_i and X_j is:

$$Cov(X_i, X_j) = -\frac{\alpha_i \alpha_j}{\left(\sum_{i=1}^{d+1} \alpha_i\right)^2 \left(\sum_{i=1}^{d+1} \alpha_i + 1\right)} \quad (3.23)$$

As shown in Fig. 3.2, the Dirichlet distribution offers different symmetric and asymmetric shapes by varying its parameters.

Despite its flexibility, the Dirichlet has a restrictive covariance matrix as shown in Eq. 3.23. According to this equation, any two random variables in $\mathbf{X} = (X_1, \ldots, X_d)$ are negatively correlated which is not always the case in real applications. The general moment function of a generalized Dirichlet

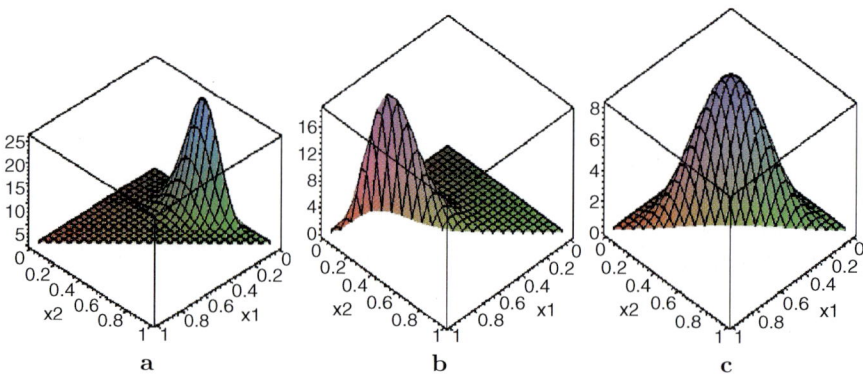

Fig. 3.2a–c. The Dirichlet distribution for different parameters. **a** $\alpha_1 = 8.5$, $\alpha_2 = 7.5$, $\alpha_3 = 1.5$. **b** $\alpha_1 = 10.5$, $\alpha_2 = 3.5$, $\alpha_3 = 3.5$. **c** $\alpha_1 = 3.5$, $\alpha_2 = 3.5$, $\alpha_3 = 3.5$

distribution is [551]:

$$E(X_1^{r_1}, X_2^{r_2}, \ldots, X_d^{r_d}) = \prod_{i=1}^{d} \frac{\Gamma(\alpha_i + \beta_i)\Gamma(\alpha_i + r_i)\Gamma(\beta_i + \delta_i)}{\Gamma(\alpha_i)\Gamma(\beta_i)\Gamma(\alpha_i + \beta_i + r_i + \delta_i)} \qquad (3.24)$$

where $\delta_i = r_{i+1} + r_{i+2} + \ldots + r_d$ for $i = 1, 2, \ldots, d-1$ and $\delta_d = 0$. Then, we can show that the mean and the variance of the generalized Dirichlet distribution satisfy the following conditions [551]:

$$E(X_i) = \frac{\alpha_i}{\alpha_i + \beta_i} \prod_{k=1}^{i-1} \frac{\beta_k + 1}{\alpha_k + \beta_k} \qquad (3.25)$$

$$Var(X_i) = E(X_i)\left(\frac{\alpha_i + 1}{\alpha_i + \beta_i + 1} \prod_{k=1}^{i-1} \frac{\beta_k + 1}{\alpha_k + \beta_k + 1} - E(X_i)\right) \qquad (3.26)$$

and the covariance between X_i and X_j is:

$$Cov(X_i, X_j) = E(X_j)\left(\frac{\alpha_i}{\alpha_i + \beta_i + 1} \prod_{k=1}^{i-1} \frac{\beta_k + 1}{\alpha_k + \beta_k + 1} - E(X_i)\right) \qquad (3.27)$$

Note that the generalized Dirichlet distribution has a more general covariance structure than the Dirichlet distribution [64]. Comparing to the Gaussian distribution, the generalized Dirichlet has smaller number of parameters which makes the estimation and the selection more accurate as we will show in the experimental results. Besides, as a generalization of the Dirichlet, this distribution offers high flexibility and ease of use for the approximation of both symmetric and asymmetric distributions; and can be used in many applications. A generalized Dirichlet mixture with M components is defined as:

$$p(\mathbf{X}|\Theta) = \sum_{j=1}^{M} p(\mathbf{X}|\alpha_j)p_j \qquad (3.28)$$

In this case, the parameters of a mixture for M clusters are denoted by $\Theta = (\alpha, \mathbf{P})$, where $\alpha = (\alpha_1, \cdots, \alpha_M)^T$, $\alpha_j = (\alpha_{j1}, \beta_{j1}, \cdots, \alpha_{jd}, \beta_{jd})$, $j = 1, \cdots, M$ and $\mathbf{P}$ is the mixing parameter vector.

3.4 Learning of Finite Generalized Dirichlet Mixtures

Finite mixture models are usually analyzed by either deterministic or Bayesian methods. Deterministic approaches are in general based on the EM algorithm and its extensions [409]. Bayesian approaches use Gibbs sampler. Interesting discussions about these algorithms can be found in some texts [190] [198]. We now examine the two approaches in the case of generalized Dirichlet mixture.

3.4.1 Deterministic Generalized Dirichlet Mixture Learning

Given a set of N independent vectors $\mathcal{X} = (\mathbf{X}_1, \ldots, \mathbf{X}_N)$, the log-likelihood corresponding to a M-component is:

$$L(\Theta, \mathcal{X}) = \log \prod_{i=1}^{N} p(\mathbf{X}_i|\Theta) = \sum_{i=1}^{N} \log \sum_{j=1}^{M} p(\mathbf{X}_i|\alpha_j)p_j \qquad (3.29)$$

It is well known that the maximum likelihood (ML) estimate:

$$\hat{\Theta}_{ML} = \arg \max_{\Theta} \{L(\Theta, \mathcal{X})\} \qquad (3.30)$$

which cannot be found analytically. The maximization defining the ML estimates is subject to the constraints in Eqs. 3.17 and 3.18. The ML estimates of the mixture parameters can be obtained using expectation maximization (EM) and related techniques [124, 413]. The EM algorithm is a general approach to maximum likelihood in the presence of incomplete data. In EM, the "complete" data are considered to be $Y_i = \{\mathbf{X}_i, \mathbf{Z}_i\}$, where $\mathbf{Z}_i = (Z_{i1}, \ldots, Z_{iM})$, with:

$$Z_{ij} = \begin{cases} 1 & \text{if } \mathbf{X}_i \text{ belongs to class } j \\ 0 & \text{otherwise} \end{cases} \qquad (3.31)$$

constituting the "missing" data. The relevant assumption is that the density of an observation $\mathbf{X}_i$, given $\mathbf{Z}_i$, is given by $\prod_{j=1}^{M} p(\mathbf{X}_i|\alpha_j)^{Z_{ij}}$. The resulting *complete-data log-likelihood* is:

$$L(\Theta, \mathcal{Z}, \mathcal{X}) = \sum_{i=1}^{N} \sum_{j=1}^{M} Z_{ij} \log(p(\mathbf{X}_i|\alpha_j)p_j) \qquad (3.32)$$

Where $\mathcal{Z} = (\mathbf{Z}_1, \ldots, \mathbf{Z}_N)$. The EM algorithm produces a sequence of estimates $\{\Theta^t, t = 0, 1, 2 \ldots\}$ by applying two steps in alternation until some convergence criterion is satisfied:

1. **E-step:** Compute $\hat{Z}_{ij}$ given the parameter estimates from the initialization:

$$\hat{Z}_{ij} = \frac{p(\mathbf{X}_i|\alpha_j)p_j}{\sum_{l=1}^{M} p(\mathbf{X}_i|\alpha_l)p_l}$$

2. **M-step:** Update the parameter estimates according to:

$$\hat{\Theta} = \arg \max_{\Theta} L(\Theta, \mathcal{Z}, \mathcal{X})$$

The quantity $\hat{Z}_{ij}$ is the conditional expectation of Z_{ij} given the observation $\mathbf{X}_i$ and parameter vector Θ. The value Z_{ij}^* of $\hat{Z}_{ij}$ at a maximum of

Eq. 3.32 is the conditional probability that observation i belongs to class j (the a *posteriori* probability); the classification of an observation $\mathbf{X}_i$ is taken to be $\{k/Z_{ik}^* = max_j Z_{ij}^*\}$, which is the Bayes rule. When we maximize the function given by Eq. 3.32, we obtain:

$$p_j^{(t)} = \frac{1}{N} \sum_{i=1}^{N} \hat{Z}_{ij}^{(t-1)} \qquad (3.33)$$

However, we do not obtain a closed-form solution for the α_j parameters. In [58], we proposed an algorithm based on the EM algorithm and the Fisher Scoring method for the estimation of the parameters of the generalized Dirichlet mixture. This method involved the inverse of the $(2 \times d) \times (2 \times d)$ Fisher information matrix, which is not easy to compute, especially for high-dimensional data. In addition, the EM algorithm has some disadvantages. In fact, problems with the EM algorithm can occur in the case of multimodal likelihoods. The increase of the likelihood function at each step of the algorithm ensures its convergence to the maximum likelihood estimator in the case of unimodal likelihoods but implies a dependance on initial conditions for multimodal likelihoods. In this last case, it happens that the EM converges to a saddle point but not to a local maximum. Several extensions to the EM algorithm can be found in the literature to overcome these problems. In [57,60], we proposed a hybrid stochastic expectation maximization algorithm (HSEM) to estimate the parameters of the generalized Dirichlet mixture. The algorithm was called stochastic because it contains a step in which the data elements are assigned randomly to components in order to avoid convergence to a saddle point. The adjective "hybrid" was justified by the introduction of a Newton-Raphson step. The HSEM is based on the SEM algorithm [84] which consists in fact of a modification EM algorithm in which a probabilistic teacher step (Stochastic step or S-Step) has been incorporated. This step consists of a simulation of $\mathcal{Z}$ according to the posterior probability $\pi(\mathcal{Z}|\mathcal{X}, \Theta)$. This posterior probability is chosen to be Multinomial of order one with weights given by the $\hat{Z}_{ij}$ ($\mathcal{M}(1; \hat{Z}_{i1}, \ldots, \hat{Z}_{iM})$). Celeux and Diebolt have proven that the S-Step prevents the estimates from staying near saddle point of the likelihood function [83,84]. The complete algorithm of estimation and selection of a generalized Dirichlet mixture is then given by:

Algorithm. For each candidate value of M:

1. Initialization.
2. E-Step: Compute the *posterior* probabilities:

$$\hat{Z}_{ij} = \frac{p(\mathbf{X}_i|\Theta_j)p_j}{\sum_{l=1}^{M} p(\mathbf{X}_i|\Theta_l)p_l}$$

3. S-Step: For each sample value $\mathbf{X}_i$, draw $\mathbf{Z}_i$ from the multinomial distribution of order one with M categories having probabilities specified by the $\hat{Z}_{ij}$.

4. M-Step:
   a) Update α_j using the Fisher scoring method, $j = 1, \ldots, M$ [57].
   b) Update the $p_j = \frac{1}{N} \sum_{i=1}^{N} \hat{Z}_{ij}$, $j = 1, \ldots, M$.
5. Calculate the associated criterion $MessLen(M)$.
6. Select the optimal model M^* such that:

$$M^* = \arg\min_M MessLen(M)$$

where $MessLen(M)$ is the message length associated to M clusters and given by [654]:

$$MessLen \simeq -\log(h(\Theta)) - \log(p(\mathcal{X}|\Theta)) + \frac{1}{2}\log(|F(\Theta)|) + \frac{N_p}{2}(1 + \log(\kappa_{N_p}))$$

(3.34)

where $h(\Theta)$ is the prior probability, $p(\mathcal{X}|\Theta)$ is the likelihood, $F(\Theta)$ is the expected Fisher information matrix, and $|F(\Theta)|$ is its determinant. N_p is the number of parameters to be estimated and is equal to $(2d+1)M$ in our case. κ_{N_p} is the optimal quantization lattice constant for $\mathbb{R}^{N_p}$ [101] and we have $\kappa_1 = 1/12 \simeq 0.083$, for $N_p = 1$. As N_p grows, κ_{N_p} tends to the asymptotic value given by $\frac{1}{2\pi e} \simeq 0.05855$. We note that κ_{N_p} does not vary much, then we can approximate it by $\frac{1}{12}$.

More details and discussions about the MML principle can be found in [654]. The estimation of the number of clusters is carried out by finding the minimum, with regards to Θ, of the message length $MessLen$. The main problems when using the minimum message length criterion are the computation of the determinant of the Fisher information matrix $|F(\Theta)|$ and the determination of a prior probability density function $h(\Theta)$. It was shown in [59] that the determinant of the Fisher information matrix for a generalized Dirichlet mixture can be approximated as follows:

$$|F(\Theta)| \simeq \frac{N^{M-1}}{\prod_{j=1}^{M} p(j)} \prod_{j=1}^{M} n_j^{2d} \left[\prod_{l=1}^{d} \left(\Psi'(\alpha_{jl})\Psi'(\beta_{jl})\right.\right.$$

$$\left.\left. - \Psi'(\alpha_{jl} + \beta_{jl})\left(\Psi'(\alpha_{jl}) + \Psi'(\beta_{jl})\right)\right)\right] \quad (3.35)$$

Besides, by selecting a symmetric Dirichlet prior with parameters equal to one for the mixing parameters [59],

$$h(\mathbf{P}) = (M-1)! \quad (3.36)$$

and uniform priors for the α_j parameters,

$$h(\alpha) = \prod_{j=1}^{M} h(\alpha_j) = (2de^5)^{-2Md}((2d)!)^M \quad (3.37)$$

a prior probability density function for the Θ is given by [59]:

$$h(\Theta) = (M - 1)! \, (2de^5)^{-2Md} ((2d)!)^M \tag{3.38}$$

using Eqs. 3.38, 3.35 and 3.34, we obtain the following formula for the message length associated to a mixture of generalized Dirichlet distributions [59]:

$$
\begin{aligned}
MessLen(M) = &- \sum_{j=1}^{M-1} \log(j) + 10Md + 2Md\log(2d) - M\sum_{j=1}^{2d} \log(j) \\
&+ \frac{(M-1)\log(N)}{2} - \frac{1}{2}\sum_{j=1}^{M} \log(p_j) + d\sum_{j=1}^{M} \log(n_j) \\
&- \log(p(\mathcal{X}|\Theta)) - \frac{N_p}{2}\log(12) + \frac{N_p}{2} \\
&+ \frac{1}{2}\sum_{j=1}^{M}\sum_{l=1}^{d} \log\left(\left|\Psi'(\alpha_{jl})\Psi'(\beta_{jl})\right.\right. \\
&\qquad\qquad \left.\left. -\Psi'(\alpha_{jl}+\beta_{jl})\left(\Psi'(\alpha_{jl})+\Psi'(\beta_{jl})\right)\right|\right)
\end{aligned}
\tag{3.39}
$$

Note that the well known minimum description length (MDL) criterion proposed by J. Rissanen in [525] and given as follows:

$$MDL(M) = -\log(p(\mathcal{X}|\Theta)) + \frac{N_p}{2}\log(N) \tag{3.40}$$

can be viewed itself as an approximation of the MML criterion. In fact, we have $F(\Theta) = NF^{(1)}(\Theta)$, where $F(\Theta)$ is the Fisher matrix of the entire population and $F^{(1)}(\Theta)$ is the Fisher matrix for a single observation. So, $\log(|F(\Theta)|) = \log(N^{N_p}|F^{(1)}(\Theta)|) = N_p\log(N) + \log(|F^{(1)}(\Theta)|)$, where $|F^{(1)}(\Theta)|$ is the Fisher information for a single observation. For large N, we can remove the terms $\log(|F^{(1)}(\Theta)|)$ and $\frac{N_p}{2}(1 - \log(12))$ from Eq. 3.34. Then, by assuming a flat prior $h(\Theta)$ and drop it from Eq. 3.34, we obtain the well-known MDL selection criterion.

3.4.2 Bayesian Learning of a Generalized Dirichlet Mixture

Given a prior π on the parameters Θ, the S-Step of the SEM algorithm, that we have used in the previous section, can be completed by the simulation of the Θ parameters from $\pi(\Theta|\mathcal{X}, \mathcal{Z})$ [129]. This method is now well-known as Bayesian estimation (See [54] for an introduction to Bayesian statistics). Bayesian learning of finite mixture models estimation is based on the simulation of the vector of parameters Θ from the posterior $\pi(\Theta|\mathcal{X}, \mathcal{Z})$ rather than computing it [62, 526]. The posterior is given by:

$$\pi(\Theta|\mathcal{X}, \mathcal{Z}) \propto p(\mathcal{X}, \mathcal{Z}|\Theta)\pi(\Theta) \tag{3.41}$$

where $\mathcal{Z} = (\mathbf{Z}_1, \ldots, \mathbf{Z}_N)$, $(\mathcal{X}, \mathcal{Z})$ is the complete data, and $\pi(\Theta)$ is the prior distribution. The simulation, in the case of Bayesian mixture estimation, is commonly performed using the Gibbs sampler. Gibbs sampling is based on successive simulation of $\mathcal{Z}$, $\mathbf{P}$ and $\alpha = (\alpha_1, \ldots, \alpha_M)$; and given as follows [392]:

1. Initialization
2. Step t: For $t = 1, \ldots$
   a) Generate $\mathbf{Z}_i^{(t)} \sim \mathcal{M}(1; \hat{Z}_{i1}^{(t-1)}, \ldots, \hat{Z}_{iM}^{(t-1)})$
   b) Generate $\mathbf{P}^{(t)}$ from $\pi(\mathbf{P}|\mathcal{Z}^{(t)})$
   c) Generate $\alpha^{(t)}$ from $\pi(\alpha|\mathcal{Z}^{(t)}, \mathcal{X})$

where $\mathcal{M}(1; \hat{Z}_{i1}, \ldots, \hat{Z}_{iM})$ is a Multinomial of order one with weights given by the $\hat{Z}_{ij}$. In order to perform the sampling, we need to determine $\pi(\mathbf{P}|\mathcal{Z})$ and $\pi(\alpha|\mathcal{Z}, \mathcal{X})$. We start by the distribution $\pi(\mathbf{P}|\mathcal{Z}, \mathcal{X})$ and we have:

$$\pi(\mathbf{P}|\mathcal{Z}) \propto \pi(\mathbf{P})\pi(\mathcal{Z}|\mathbf{P}) \tag{3.42}$$

We know that vector $\mathbf{P}$ is defined on the simplex $\{(p_1, \ldots, p_M) : \sum_{j=1}^{M-1} p_j < 1\}$, then a natural choice, as a prior, for this vector is the Dirichlet distribution [392]:

$$\pi(\mathbf{P}) = \frac{\Gamma(\sum_{j=1}^{M} \eta_j)}{\prod_{j=1}^{M} \Gamma(\eta_j)} \prod_{j=1}^{M} p_j^{\eta_j - 1} \tag{3.43}$$

where $\eta = (\eta_1, \ldots, \eta_M)$ is the parameter vector of the Dirichlet distribution. Moreover, we have:

$$\pi(\mathcal{Z}|\mathbf{P}) = \prod_{i=1}^{N} \pi(\mathbf{Z}_i|\mathbf{P}) = \prod_{i=1}^{N} p_1^{Z_{i1}} \cdots p_M^{Z_{iM}} = \prod_{i=1}^{N} \prod_{j=1}^{M} p_j^{Z_{ij}} = \prod_{j=1}^{M} p_j^{n_j} \tag{3.44}$$

where $n_j = \sum_{i=1}^{N} \mathbb{I}_{Z_{ij}=j}$. Then

$$\pi(\mathbf{P}|\mathcal{Z}) = \frac{\Gamma(\sum_{j=1}^{M} \eta_j)}{\prod_{j=1}^{M} \Gamma(\eta_j)} \prod_{j=1}^{M} P_j^{\eta_j - 1} \prod_{j=1}^{M} P_j^{n_j} = \frac{\Gamma(\sum_{j=1}^{M} \eta_j)}{\prod_{j=1}^{M} \Gamma(\eta_j)} \prod_{j=1}^{M} P_j^{\eta_j + n_j - 1}$$
$$\propto \mathcal{D}(\eta_1 + n_1, \ldots, \eta_M + n_M) \tag{3.45}$$

where $\mathcal{D}$ is a Dirichlet distribution with parameters $(\eta_1 + n_1, \ldots, \eta_M + n_M)$. We note that the prior and the posterior distributions, $\pi(\mathbf{P})$ and $\pi(\mathbf{P}|\mathcal{Z})$, are both Dirichlet. In this case we say that the Dirichlet distribution is a conjugate prior for the mixture proportions.

For a mixture of generalized Dirichlet distributions, it is therefore possible to associate with each α_j a prior $\pi_j(\alpha_j)$. For this we use the fact that the generalized Dirichlet distribution belongs to the exponential family. In fact, if a S-parameter density p belongs to the exponential family, then we can write it as the following [526]:

$$p(\mathbf{X}|\theta) = H(\mathbf{X}) \exp\left(\sum_{l=1}^{S} G_l(\theta)T_l(\mathbf{X}) + \Phi(\theta)\right) \qquad (3.46)$$

In this case a conjugate prior on θ is given by [526]:

$$\pi(\theta) \propto \exp\left(\sum_{l=1}^{S} \rho_l G_l(\theta) + \kappa\Phi(\theta)\right) \qquad (3.47)$$

where $\rho = (\rho_1, \ldots, \rho_S) \in \mathbb{R}^S$ and $\kappa > 0$ are referred as hyperparameters. The generalized Dirichlet distribution can be written as an exponential density and straightforward manipulations give us the following prior:

$$\pi(\alpha_j)$$

$$\propto \exp\left[\sum_{l=1}^{d} \rho_l \alpha_{jl} \kappa \sum_{l=1}^{d} \left(\log\left(\Gamma(\alpha_{jl} + \beta_{jl})\right) - \log\left(\Gamma(\alpha_{jl})\right) - \log\left(\Gamma(\beta_{jl})\right) \right) + \sum_{l=1}^{d} \rho_{l+d}\gamma_{jl}\right]$$

The prior hyperparameters are: $(\rho_1, \ldots, \rho_{2d}, \kappa)$. Having this prior, $\pi(\alpha_j)$, the posterior distribution is then:

$$\pi(\alpha_j|\mathcal{Z}, \mathcal{X}) \propto \pi(\alpha_j) \prod_{Z_{ij}=1} p(\mathbf{X}_i|\xi_j)$$

$$\propto \exp\left[\sum_{l=1}^{d} \alpha_{jl}\left(\rho_l + \sum_{Z_{ij}=1} \log(X_{il})\right)\right.$$

$$+ \sum_{l=1}^{d} \gamma_{jl}\left(\rho_{l+d} + \sum_{Z_{ij}=1} \log\left(1 - \sum_{t=1}^{l} X_{it}\right)\right)$$

$$+ (\kappa + n_j) \sum_{l=1}^{d} \left(\log\left(\Gamma(\alpha_{jl} + \beta_{jl})\right)\right.$$

$$\left.\left. - \log\left(\Gamma(\alpha_{jl})\right) - \log\left(\Gamma(\beta_{jl})\right) \right)\right] \qquad (3.48)$$

We can see clearly that the posterior and the prior distributions have the same form, then $\pi(\alpha_j)$ is really a conjugate prior on α_j. The posterior hyperparameters are:

$$\left(\rho_1 + \sum_{Z_{ij}=1} \log(X_{i1}), \; \ldots, \; \rho_d + \sum_{Z_{ij}=1} \log(X_{id}), \right.$$

$$\rho_{d+1} + \sum_{Z_{ij}=1} \log\left(1 - \sum_{t=1}^{1} X_{it} \right), \; \ldots,$$

$$\left. \rho_{2d} + \sum_{Z_{ij}=1} \log\left(1 - \sum_{t=1}^{d} X_{it} \right), \; \kappa + n_j \right).$$

According the this vector a samples modifies the prior hyperparameters by adding $T_l(\mathbf{X})$ or n_j to the previous values. This information, could be used to get the prior hyperparameters. Indeed, following [327], once the sample $\mathcal{X}$ is known, we can use it to get the prior hyperparameters [39]. Then, we held the hyperparameters fixed at: $\eta_j = 1$, $j = 1, \ldots, M$, $\rho_l = \sum_{i=1}^{N} \log(X_{il})$, $\rho_{l+d} = \sum_{i=1}^{N} \log(1 - \sum_{t=1}^{l} X_{it})$, $l = 1, \ldots, d$, $\kappa = n_j$. Having all the posterior probabilities in hand, the steps of the Gibbs sampler are then:

1. Initialization
2. Step t: For $t = 1, \ldots$
    a) Generate
    $$\mathbf{Z}_i^{(t)} \sim \mathcal{M}(1; \hat{Z}_{i1}^{(t-1)}, \ldots, \hat{Z}_{iM}^{(t-1)})$$
    b) Compute
    $$n_j^{(t)} = \sum_{i=1}^{N} \mathbb{I}_{Z_{ij}^{(t)}=j}$$
    c) Generate $\mathbf{P}^{(t)}$ from Eq. 3.45
    d) Generate $\alpha_j^{(t)}$ $(j = 1, \ldots, M)$ from Eq. 3.48 using the Metropolis–Hastings (M–H) algorithm.

Note that we are using a hybrid Markov chain Monte Carlo (MCMC) algorithm based on both Gibbs and M–H sampling. This approach is better-known as Metropolis-within-Gibbs sampling [526]. This approach is used when one of the parameters is hard to sample and then an M–H step is needed. The M–H algorithm offers a solution to the problem of simulating from the posterior distribution [526]. Starting from point $\alpha_j^{(0)}$, the corresponding Markov chain explores the surface of the posterior distribution. At iteration t, the steps of the M–H algorithm can be described as follows:

1. Generate
$$\tilde{\alpha}_j \sim q(\xi_j | \xi_j^{(t-1)}) \quad \text{and} \quad U \sim \mathcal{U}_{[0,1]}$$

2. Compute

$$r = \frac{\pi\left(\tilde{\alpha}_j | \mathcal{Z}, \mathcal{X}\right) q\left(\alpha_j^{(t-1)} | \tilde{\alpha}_j\right)}{\pi\left(\alpha_j^{(t-1)} | \mathcal{Z}, \mathcal{X}\right) q\left(\tilde{\alpha}_j | \alpha_j^{(t-1)}\right)}$$

3. If $r < u$ then $\alpha_j^{(t)} = \tilde{\alpha}_j$ else $\alpha_j^{(t)} = \alpha_j^{(t-1)}$

Where $\tilde{\alpha}_j = (\tilde{\alpha}_{j1}, \tilde{\beta}_{j1}, \ldots, \tilde{\alpha}_{jd}, \tilde{\beta}_{jd})$. The major problem in this algorithm is the need to choose the proposal distribution q. The most generic proposal is the random walk Metropolis–Hastings algorithm where each unconstrained parameter is the mean of the proposal distribution for the new value. As all the $\tilde{\alpha}_{jl} > 0$, $\tilde{\beta}_{jl} > 0$, $l = 1, \ldots, d$, we have chosen the following proposals:

$$\tilde{\alpha}_{jl} \sim \mathcal{LN}(\log(\alpha_{jl}^{(t-1)}), \sigma_1^2) \tag{3.49}$$

$$\tilde{\beta}_{jl} \sim \mathcal{LN}(\log(\beta_{jl}^{(t-1)}), \sigma_2^2) \tag{3.50}$$

where $\mathcal{LN}(\log(\alpha_{jl}^{(t-1)}), \sigma_1^2)$ and $\mathcal{LN}(\log(\beta_{jl}^{(t-1)}), \sigma_2^2)$ refer to the log-normal distributions with mean $\log(\alpha_{jl}^{(t-1)})$ and variance σ_1^2 and mean $\log(\beta_{jl}^{(t-1)})$ and variance σ_2^2, respectively. Note that (3.49) and (3.50) are equivalent to:

$$\log(\tilde{\alpha}_{jl}) = \log(\alpha_{jl}^{(t-1)}) + \epsilon_1 \tag{3.51}$$

$$\log(\tilde{\beta}_{jl}) = \log(\beta_{jl}^{(t-1)}) + \epsilon_2 \tag{3.52}$$

where $\epsilon_1 \sim \mathcal{N}(0, \sigma_1^2)$ and $\epsilon_2 \sim \mathcal{N}(0, \sigma_2^2)$. With these proposals, the random walk M–H algorithm is composed of the following steps:

1. Generate

$$\tilde{\alpha}_{jl} \sim \mathcal{LN}(\log(\alpha_{jl}^{(t-1)}), \sigma_1^2), \quad \tilde{\beta}_{jl} \sim \mathcal{LN}(\log(\beta_{jl}^{(t-1)}), \sigma_2^2),$$
$$l = 1, \ldots, d \quad \text{and} \quad U \sim \mathcal{U}_{[0,1]}.$$

2. Compute

$$r = \frac{\pi\left(\tilde{\alpha}_j | \mathcal{Z}, \mathcal{X}\right) \prod_{l=1}^d \tilde{\alpha}_{jl} \tilde{\beta}_{jl}}{\pi\left(\alpha_j^{(t-1)} | \mathcal{Z}, \mathcal{X}\right) \prod_{l=1}^d \alpha_{jl}^{(t-1)} \beta_{jl}^{(t-1)}}$$

3. If $r < u$ then $\alpha_j^{(t)} = \tilde{\alpha}_j$ else $\alpha_j^{(t)} = \alpha_j^{(t-1)}$

The Bayesian determination of the number of components M is problematical. The multi-dimensional nature of our data means that reversible jump MCMC techniques [523] would be extremely complicated to operate efficiently, as would the Markov birth-death process approach [586] [81]. In order to determine the number of clusters, we will use Bayes factors [313]. The Bayes factor $B_{M_1 M_2}$ for a finite mixture model with M_1 clusters against another model with M_2 clusters given data $\mathcal{X}$ is given by the following ratio:

$$B_{M_1 M_2} = \frac{p(\mathcal{X}|M_1)}{p(\mathcal{X}|M_2)} \tag{3.53}$$

which is the ratio of the integrated likelihoods. The integrated likelihood is defined by:

$$p(\mathcal{X}|M_k) = \int p(\mathcal{X}|\Theta_{M_k}, M_k)\pi(\Theta_{M_k}|M_k)d\Theta \tag{3.54}$$

Where Θ_{M_k} is the vector of parameters of a finite mixture model with M_k clusters, $\pi(\Theta_{M_k}|M_k)$ is its prior density, and $p(\mathcal{X}|\Theta_{M_k}, M_k)$ is the likelihood function. The main problem now is how to compute the integrated likelihood. In order to resolve this problem, let $\hat{\Theta}_{M_k}$ denotes the posterior mode, satisfying:

$$\frac{\partial \log(\pi(\hat{\Theta}_{M_k}|\mathcal{X}, M_k))}{\partial \Theta_{M_k}} = 0 \tag{3.55}$$

where $\frac{\partial \log(\pi(\hat{\Theta}_{M_k}|\mathcal{X},M_k))}{\partial \Theta_{M_k}}$ denotes the gradient of $\log(\pi(\hat{\Theta}_{M_k}|\mathcal{X}, M_k))$ evaluated at $\Theta_{M_k} = \hat{\Theta}_{M_k}$. The Hessian matrix of minus the $\log(\pi(\hat{\Theta}_{M_k}|\mathcal{X}, M_k))$ evaluated at $\Theta_{M_k} = \hat{\Theta}_{M_k}$ is denoted by $H(\hat{\Theta}_{M_k})$. To approximate the integral given by Eq. 3.54, the integrand is expanded in a second-order Taylor series about the point $\Theta_{M_k} = \hat{\Theta}_{M_k}$, and the Laplace approximation gives:

$$p(\mathcal{X}|M_k) = p(\mathcal{X}|\hat{\Theta}_{M_k}, M_k)\pi(\hat{\Theta}_{M_k}|M_k)(2\pi)^{\frac{N_p}{2}}|H(\hat{\Theta}_{M_k})|^{1/2}$$

where N_p is the number of parameters to be estimated (the dimensionality of Θ_{M_k}) and is equal to $(2d+1)M_k$ in our case, and $|H(\hat{\Theta}_{M_k})|$ is the determinant of the Hessian matrix. For numerical reasons, it is better to work with the Laplace approximation on the logarithm scale. Taking logarithms, we can rewrite Eq. 3.56 as:

$$\log(p(\mathcal{X}|M_k)) = \log(p(\mathcal{X}|\hat{\Theta}_{M_k}, M_k)) + \log(\pi(\hat{\Theta}_{M_k}|M_k)) + \frac{N_p}{2}\log(2\pi)$$
$$+ \frac{1}{2}\log(|H(\hat{\Theta}_{M_k})|) \tag{3.56}$$

If $\hat{\Theta}_{M_k}$ and $H(\hat{\Theta}_{M_k})$ can be found analytically, then we can use Eq. 3.56 to estimate the log-integrated likelihood directly. In many practical situations an analytic solution is not available. Thus, we can use the Laplace-Metropolis estimator which consists of estimating $\hat{\Theta}_{M_k}$ and $H(\hat{\Theta}_{M_k})$ from the Gibbs sampler outputs [368]. Indeed, we estimate $\hat{\Theta}_{M_k}$ as that the Θ_{M_k} in the sample at which $p(\mathcal{X}|\hat{\Theta}_{M_k}, M_k)$ achieves its maximum. The other quantity $H(\hat{\Theta}_{M_k})$ is asymptotically equal to the posterior variance matrix. Then, we could estimate it by the sample covariance matrix of the posterior simulation output [368]. Having Eq. 3.56 in hand the number of components in the mixture model is taken to be $\{M_k/\log(p(\mathcal{X}|M_k)) = \max_{M_l} \log(p(\mathcal{X}|M_l)), M_l = M_{\min}, \ldots, M_{\max}\}$. Note that the MML criterion in Eq. 3.34 is very similar to the criterion given in Eq. 3.56. Indeed, criterion in Eq. 3.56 is reduced to the MML by taking uniform priors over the parameters and by choosing the asymptotic value $\kappa_{N_p} = \frac{1}{2\pi e}$ in Eq. 3.34. Note also that another approximation to the integrated likelihood in Eq. 3.54 is the Bayesian information criterion (BIC) coincides formally (but not conceptually) with the MDL [525]. The BIC can be deduced from Eq. 3.56 by retaining only the terms that increases with N which are $\log(p(\mathcal{X}|\hat{\Theta}_{M_k}, M_k))$ which increases linearly with N and $\log(|H(\hat{\Theta}_{M_k})|)$ which increases as $N_p log(N)$ [95]. The last approximation gives us the BIC criterion:

$$\log(p(\mathcal{X}|M_k)) = \log(p(\mathcal{X}|\hat{\Theta}_{M_k}, M_k)) - \frac{N_p}{2}\log(N) \qquad (3.57)$$

3.5 Experimental Results

This section has three main goals. The first goal is to compare the different approaches that we have presented in Sect. 3.2 when we apply them to differentiate eye patches from non-eye patches. The second goal is to compare the deterministic maximum likelihood approach with the Bayesian approach in the case of the generalized Dirichlet mixture for modeling eye images. The third goal is comparing the performances of the generalized Dirichlet mixture and the Gaussian mixture. For the Bayesian estimation and selection of Gaussian mixtures, we have considered the approach proposed in [39]. For the deterministic estimation of the parameters of the Gaussian mixture, we have added a stochastic step to the well known EM algorithm. We have used the MML criterion developed in [31]. In addition, we have considered a diagonal covariance matrix to avoid numerical problems because of the high-dimensional nature of the vectors that we have used to represent eye image patches as we explain in the next subsection.

3.5.1 Eye Representation

A basic step toward building efficient eye models is choosing an adequate representation of the eye. The representation of the eye region is based on mapping the pixel format to another space through well defined transformations. It is well known, in computer vision, that a good image representation has to be compact, unique and robust to different transformations such as rotation and translation; and also computationally inexpensive. Existing representation methods could be summarized into three groups: brightness-, texture-, shape-, and feature-based methods [215]. Brightness-based methods use the image it-self as a feature descriptor by working with intensity of the pixels. Texture-based methods encode the texture of the eye. With feature-based methods, points and contours are extracted [38] [356]. In our experiments, we have employed both the edge-orientation histograms which provide spatial information [289] and the co-occurrence matrices which capture the local spatial relationships between gray levels [235]. In order to determine the vector of characteristics for each image, we have computed a set a features derived from the co-occurrence matrices. It has been noted that to obtain good results, many co-occurrence matrices should be computed, each one considering a given neighborhood and direction. Considering the following four neighborhoods is sufficient for co-occurrence matrices, in the case of gray level images, to obtain good results in general: $(1;0)$, $(1;\frac{\pi}{4})$, $(1;\frac{\pi}{2})$, and $(1;\frac{3\pi}{4})$ [507]. For each of these neighborhoods, we calculated the corresponding co-occurrence, then derived from it the following features which have been proposed for co-occurrence matrices: Mean, Variance, Energy, Correlation, Entropy, Contrast, Homogeneity, and Cluster Prominence [638]. Besides, a histogram of edge directions is used. The edge information contained in the images is extracted using the Canny edge operator [80]. The corresponding edge directions are quantized into 72 bins of 5° each. We have also introduced the 9-dimensional invariant descriptors used in [215] and widely described in [557] [415]. Using these local descriptors, the co-occurrence matrices and the histogram of edge directions each image was represented by a 119-dimensional vector.

In our experiments, we have used a database containing 9634 image patches of six classes (Closed-Eye, Closed non-Eye, Eye with Glasses, non-Eye with Glasses, Open-Eye, Open non-Eye). Images are gray-scale infrared of 120×64 pixels each and represents the eyes and other parts of the face of different subjects, under different face orientation, and under different illumination conditions. Figure 3.3 shows examples of images from the different classes. In the following, we describe in details two experiments that we have conducted. In the first one, we test our generalized Dirichlet finite mixture model by categorizing the database previously described. In the second experiment, we compare different modeling approaches when applied to distinguish eye from non-eye images. We also proposed the use of finite generalized Dirichlet mixture models as a probabilistic kernel for SVM.

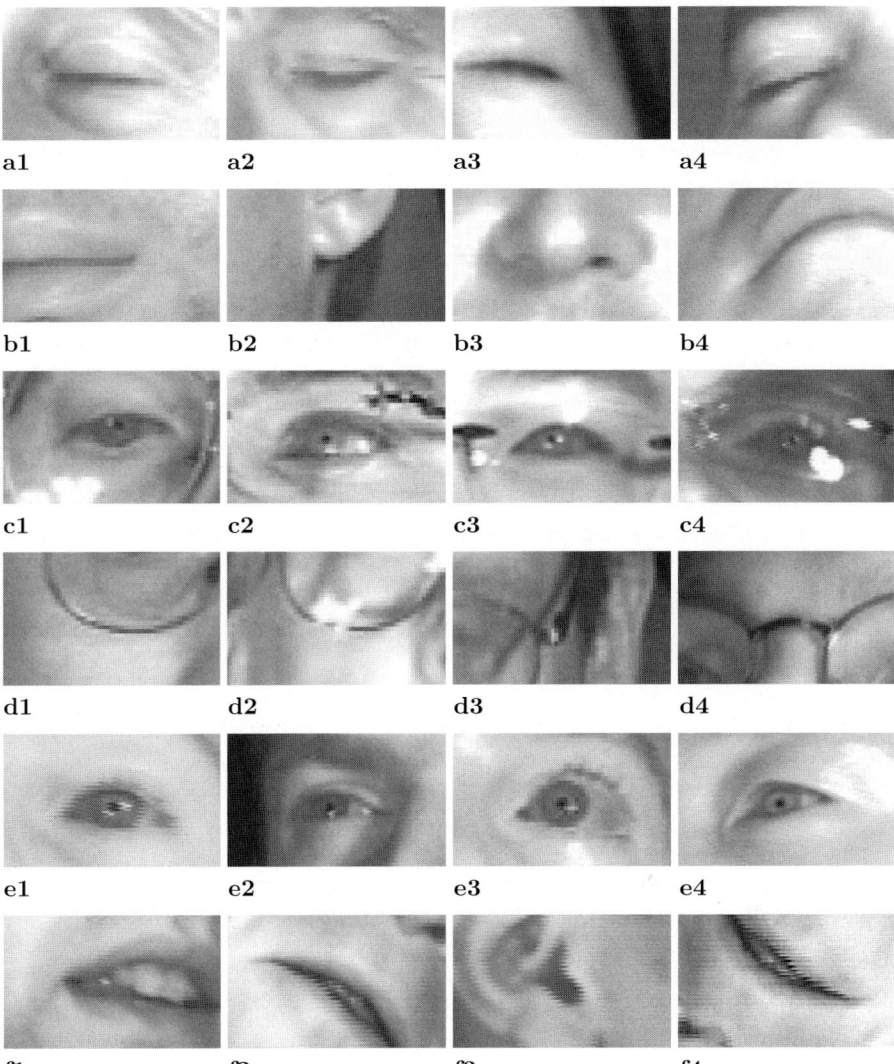

Fig. 3.3a–f. Sample images from each set. **a1–a4** Closed Eye, **b1–b4** Closed non-Eye, **c1–c4** Eye with Glasses, **d1–d4** non-Eye with Glasses, **e1–e4** Open Eye, **f1–f4** Open non-Eye

3.5.2 Image Database Categorization

In this first experiment, the image database was divided into two sets: training set and test set. The repartition of the different classes in the training and test sets is given in Table 3.1. The images in the training set were used to train a classifier. The aim of this experiment is to compare the modeling capabilities of generalized Dirichlet and Gaussian finite mixtures [223, 225] when using both deterministic and Bayesian approaches for the estimation and selection of the parameters. The first step was modeling each class in the training set by a finite mixture. Figure 3.4 shows the number of clusters obtained for the generalized Dirichlet mixture, using both deterministic and Bayesian approaches, representing the different classes in the training set. For the deterministic approach, we have tested both MML and MDL for the selection of the number of clusters. In the case of the Bayesian approach, both Laplace approximation and BIC were tested. Table 3.2 summarizes the number of clusters that we found by using both the generalized Dirichlet and Gaussian mixtures. From this table we can see that deterministic and Bayesian approaches give slightly the same selection and that the Gaussian mixture gives in general more complex models.

In order to perform the assignments of the images in the test set to the different classes, we have used the following rule: $\mathbf{X} \longmapsto \arg\max_k p(\mathbf{X}|\Theta_k)$,

Table 3.1. Repartition of the different classes in the training and test sets

class	Training set	Testing set
Closed-Eye	289	289
Closed non-Eye	680	679
Eye with Glasses	695	695
non-Eye with Glasses	1619	1619
Open-Eye	1082	1083
Open non-Eye	452	452

Table 3.2. Number of clusters, representing the training data set classes, found when using the generalized Dirichlet and the Gaussian mixtures

	C1	C2	C3	C4	C5	C6
Generalized Dirichlet + MML	1	2	2	3	3	1
Generalized Dirichlet + MDL	1	1	1	2	2	1
Generalized Dirichlet + Laplace	1	2	2	3	3	1
Generalized Dirichlet + BIC	1	1	1	2	2	1
Gaussian + MML	2	2	2	3	4	1
Gaussian + MDL	2	2	2	2	4	1
Gaussian + Laplace	2	2	2	2	3	1
Gaussian + BIC	2	2	2	2	3	2

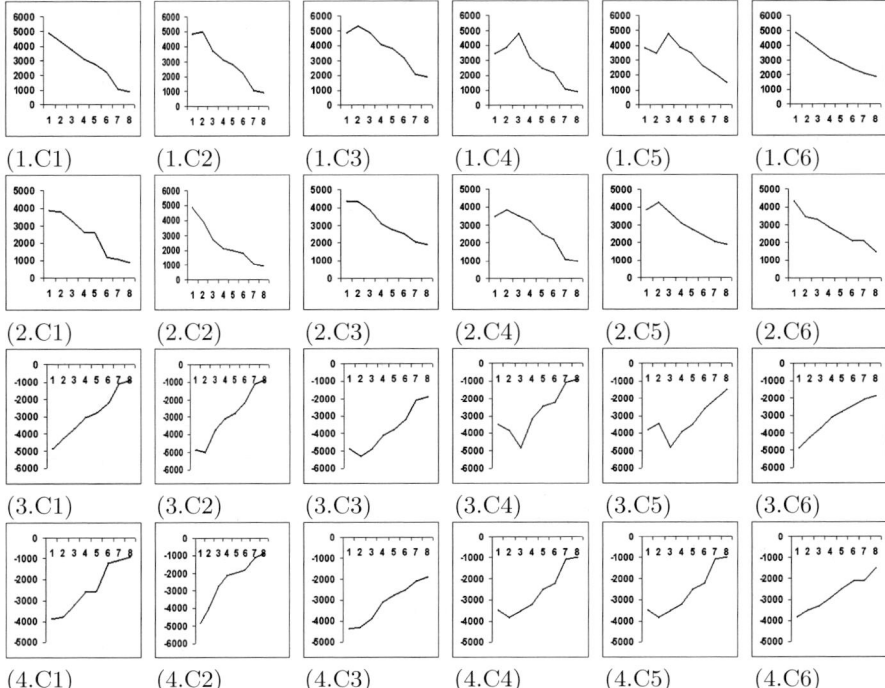

Fig. 3.4. Number of clusters determined to represent each of the six classes in the training set (1.C1–1.C6) Bayesian generalized Dirichlet + Laplace, (2.C1–2.C6) Bayesian generalized Dirichlet + BIC, (3.C1–3.C6) Deterministic generalized Dirichlet + MML, (4.C1–4.C6) Deterministic generalized Dirichlet + MDL

where $\mathbf{X}$ is a 119-dimensional vector of features representing an input test image to be assigned to a class and $p(\mathbf{X}|\Theta_k)$ is a mixture of distributions representing class k, $k = 1, \ldots, 6$. The confusion matrices for our application, when using the generalized Dirichlet mixture, are given in Tables 3.3(a), 3.3(b), 3.3(c), and 3.3(d). In these confusion matrices, the cell ($class\ i$, $class\ j$) represents the number of images from $class\ i$ which are classified as $class\ j$. For instance, the number of images misclassified, when using generalized Dirichlet mixture with Bayesian estimation and selection based on Laplace, was 183 in all, which represents an accuracy of 96.20 percent. Using the same estimation and selection approaches but with Gaussian mixture, the number of misclassified images was 334 (an accuracy of 93.06) as shown in Table 3.4(a). Tables 3.4(b), 3.4(c) and 3.4(d) represent the confusion matrices when using Gaussian mixtures with the three other selection and estimation approaches.

Table 3.3a–d. Confusion matrices using different approaches. **a** Bayesian generalized Dirichlet + Laplace, **b** Bayesian generalized Dirichlet + BIC, **c** Deterministic generalized Dirichlet + MML, **d** Deterministic generalized Dirichlet + MDL

	C1	C2	C3	C4	C5	C6		C1	C2	C3	C4	C5	C6
C1	278	5	4	1	1	0	C1	270	8	3	7	1	0
C2	11	659	5	2	1	1	C2	16	651	6	3	2	1
C3	9	7	672	3	2	2	C3	11	8	665	4	3	4
C4	8	13	25	1567	0	6	C4	21	26	29	1520	10	13
C5	4	7	39	1	1019	13	C5	8	9	40	9	1001	16
C6	1	1	1	2	8	439	C6	2	1	2	2	9	436

| | | | a | | | | | | | | b | | |

	C1	C2	C3	C4	C5	C6		C1	C2	C3	C4	C5	C6
C1	272	6	3	7	1	0	C1	269	8	3	7	1	1
C2	13	654	6	3	2	1	C2	17	648	6	4	2	2
C3	8	7	669	4	3	4	C3	14	8	661	5	3	4
C4	12	18	18	1551	10	10	C4	22	28	29	1516	10	14
C5	8	9	40	3	1015	8	C5	8	9	39	10	1001	16
C6	1	1	1	2	9	438	C6	2	2	2	2	9	435

| | | | c | | | | | | | | d | | |

Table 3.4a–d. Confusion matrices using different approaches. **a** Bayesian Gaussian + Laplace, **b** Bayesian Gaussian + BIC, **c** Deterministic Gaussian + MML, **d** Deterministic Gaussian + MDL

	C1	C2	C3	C4	C5	C6		C1	C2	C3	C4	C5	C6
C1	261	18	7	2	1	0	C1	258	18	3	8	2	0
C2	24	643	7	3	1	1	C2	24	641	8	3	2	1
C3	14	9	662	6	2	2	C3	14	8	660	6	3	4
C4	17	31	59	1497	7	8	C4	17	32	60	1492	9	9
C5	7	9	54	4	991	18	C5	8	9	50	9	989	18
C6	4	3	3	4	9	429	C6	4	2	4	5	9	428

| | | | a | | | | | | | | b | | |

	C1	C2	C3	C4	C5	C6		C1	C2	C3	C4	C5	C6
C1	257	19	3	8	2	0	C1	256	19	3	9	2	0
C2	23	639	10	4	2	1	C2	23	639	10	4	2	1
C3	14	9	658	6	4	4	C3	14	10	655	7	5	4
C4	18	32	60	1489	11	9	C4	18	32	61	1483	14	11
C5	8	9	51	9	988	18	C5	9	9	53	9	984	19
C6	4	3	3	6	8	428	C6	4	2	3	7	8	428

| | | | c | | | | | | | | d | | |

3.5.3 Distinguishing Eye from Non-eye

Our goal in this experiment is to use a set of images labeled as eye or non-eye in order to build a classifier that can determine to which class a novel test image belongs. Figure 3.5 represents 3-dimensional plot of eye patches vs non-eye patches. It is clear that the data is not Gaussian in both cases. One of the most used algorithms for such task is linear discriminant analysis (LDA) [524]. LDA is widely used because of its simplicity. Indeed, with LDA a classifier is built using a closed-form generalized eigenvector solution. In a two-class LDA we try to find the optimal linear subspace, such that when the features are projected onto it the within-class scatter is minimized while the between-class scatter is maximized, in which we will perform the classification. SVM could also be used in this task. An important issue here is the choice of the kernel function. In [577], the author proposed the use of Gaussian mixture kernels. The shapes of the mixture distributions can, however, significantly affect the similarity measurement of the two vectors. In this section, we propose the use of the generalized Dirichlet mixture to generate the kernels, since it offers more shapes than the Gaussian. Generalized Dirichlet mixture kernels are introduced in SVM and applied to develop a classifier system able to automatically differentiate eye images from non-eye images. The SVM is trained for on eye patches and non-eye patches in the feature space. The ultimate goal of this learning process is to find the vectors representing the optimal boundary between the two classes. A generative eye and non-eye models are therefore representing the eye and non-eye distributions, in a compact and efficient way. This application may be of interest in the case of driver behavior analysis. The car may want to know if the driver's eyes are kept closed. This application is also motivated by the need to have a model for closed eyes (in the case of blinking, for example), since the majority of eye tackers only work well for open eyes by tracking the eyes locations [614]. As shown in Table 3.1, our training set has 2066 positive images (eye) and 2751 negative images (non-eye) and the testing set contains 2067 eye images and

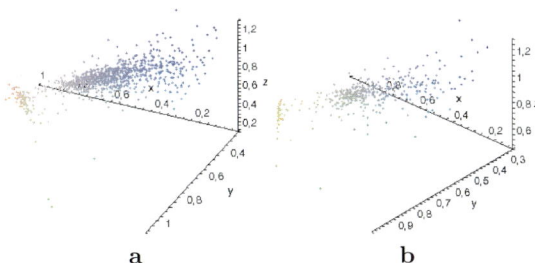

a   **b**

Fig. 3.5a,b. 3-dimensional plot of the training data representing **a** Eye, **b** non-Eye

2750 non-eye images. Table 3.5 shows the testing results with the mixture densities (both Gaussian and Generalized Dirichlet mixtures) using various selection methods to automatically determine the number of mixture components (clusters). Table 3.6 shows the classification results when we used LDA, PCA and SVM with different kernels (linear, polynomial with different degrees, Gaussian with different σ, gaussian mixture, generalized Dirichlet mixture). The Gaussian and generalized Dirichlet mixtures were learned using Bayesian technique. For PCA, we have kept 25 eigenvectors and testing was carried out by using the nearest-neighbor algorithm [150]. We can see clearly that we have achieved the best accuracy result by using Gaussian mixture (93.41%) and generalized Dirichlet mixture kernels (96.35%). This can be explained by the fact that these kernels are learned directly from the data and are not pre-defined like the linear, polynomial or Gaussian kernels.

Table 3.5. Classification accuracies using generalized Dirichlet and Gaussian mixtures

Method	Classification accuracy %
Generalized Dirichlet + MML	95.47
Generalized Dirichlet + MDL	94.04
Generalized Dirichlet + Laplace	96.20
Generalized Dirichlet + BIC	94.31
Gaussian + MML	92.56
Gaussian + MDL	92.27
Gaussian + Laplace	93.06
Gaussian + BIC	92.75

Table 3.6. Classification accuracies using different kernels

Method	Classification accuracy %
PCA	83.21
LDA	68.23
Linear	86.98
Polynomial (degree 2)	86.98
Polynomial (degree 3)	87.19
Polynomial (degree 4)	85.11
Gaussian ($\sigma = 1$)	53.97
Gaussian ($\sigma = 2$)	88.30
Gaussian ($\sigma = 3$)	89.26
Gaussian ($\sigma = 4$)	88.64
Gaussian ($\sigma = 5$)	88.64
Gaussian mixture	93.41
Generalized Dirichlet mixture	96.35

3.6 Conclusion

In this chapter we described some advanced statistical approaches applied to the problem of eye modeling. In particular, we have discussed the use of finite mixture models which are considered as a powerful tool for data modeling. Generalized Dirichlet distribution are introduced to describe the mixture components. For the learning of the parameters, we proposed a deterministic approach based on an extension of the well-known EM algorithm and a pure Bayesian approach based on a Metropolis-within-Gibbs Sampling algorithm. The experimental results indicate that finite generalized Dirichlet mixture models are very efficient. We can conclude also that the introduction of generalized Dirichlet mixture kernels into SVM performs better than classic predefined kernels. This performance can be explained that this kernel is constructed with prior information about the training data. Future works will be devoted to the application of the models developed, in the experimental results, to eyes detection and face recognition.

4

Automatic Eye Position Detection and Tracking Under Natural Facial Movement

Yan Tong and Qiang Ji

4.1 Position of the Problem

Automatic and precise detection and tracking of facial features, and in particular the eyes, are important in many applications of *Passive Eye Monitoring* including Driver Fatigue Detection, Cognitive Driver Distraction and Gaze-Based Interaction. Generally, the facial feature tracking technologies could be classified into two categories: model-free and model-based tracking algorithms. The model-free tracking algorithms [65,217,220,229,384,497,563,624, 627,720] are general purpose point trackers without the prior knowledge of the object. Each facial feature point is usually tracked by performing a local search for the best matching position, around which the appearance is most similar to the one in the initial frame. However, the model-free methods are susceptible to the inevitable tracking errors due to the aperture problems, noise, and occlusion. Model-based methods, on the other hand, focus on explicit modeling the shape of the objects. Recently, extensive work has been focused on the shape representation of deformable objects such as active contour models (Snakes) [314], deformable template method [708], active shape model (ASM) [105], active appearance model (AAM) [102], direct appearance model (DAM) [266], elastic bunch graph matching (EBGM) [682], morphable models [305], and active blobs [558]. Although the model-based methods utilize much knowledge on the face to realize an effective tracking, these models are limited to some common assumptions, e.g. a nearly frontal view face and moderate facial expression changes, and tend to fail under large pose variations or facial deformations in real world applications.

Given these challenges, accurate and efficient tracking of facial feature points under varying facial expression and face pose remains challenging. These challenges arise from the potential variability such as non-rigid face shape deformations caused by facial expression change, the non-linear face transformation resulting from pose variations, and illumination changes in real-world conditions. Tracking mouth and eye motion in image sequences is especially difficult, since these facial components are highly deformable, varying in both shape and color, and subject to occlusion.

In this chapter, a multi-state pose-dependent hierarchical shape model is presented for facial feature tracking under varying face pose and facial expression. The flowchart in Fig. 4.1 summarizes our method. Based on the active shape model, a two-level hierarchical face shape model is proposed to simultaneously characterize the global shape of a human face and the local structural details of each facial component. Multi-state local shape models are further introduced to deal with shape variations of facial components. To compensate face shape deformation due to face pose change, a robust $3D$ pose estimation technique is introduced, and the hierarchical face shape model is corrected based on the estimated face pose to improve the effectiveness of

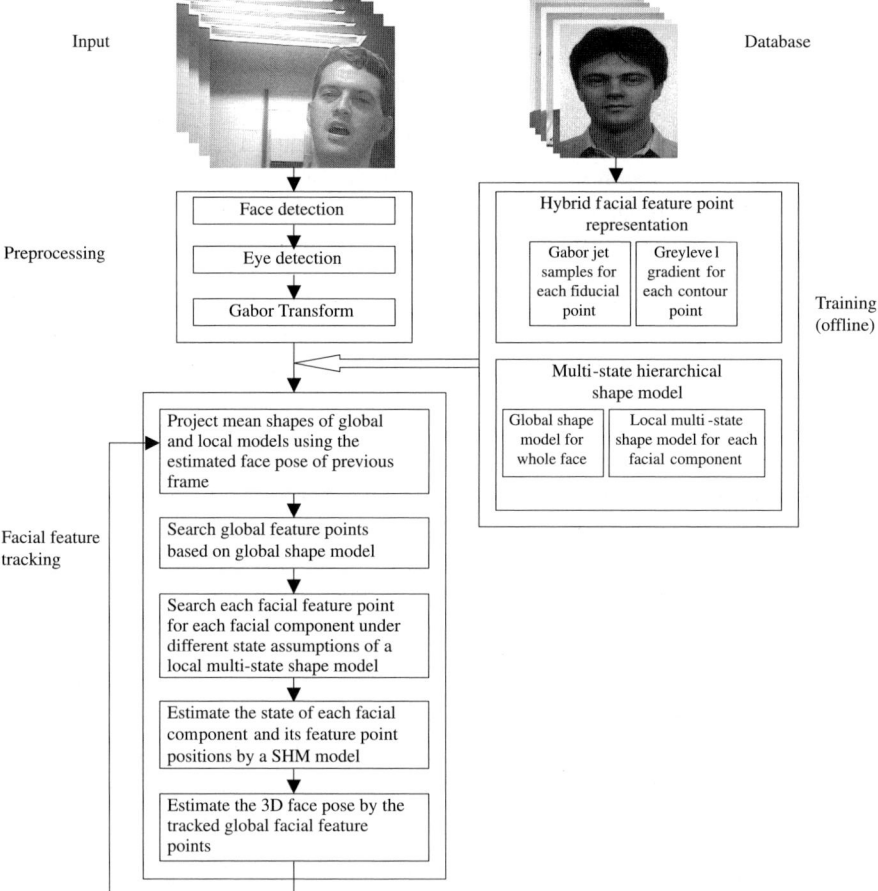

Fig. 4.1. The flowchart of the automatic facial feature tracking system based on the multi-state hierarchical shape model

the shape constraints under different poses. Gabor wavelet jets and gray-level profiles are combined to represent the feature points in an effective and efficient way. Both states of facial components and positions of feature points are dynamically estimated by a multi-modal tracking approach.

The rest of the chapter is arranged as follows. Section 4.2 provides a detailed review on the related work of model-based facial feature tracking approaches. Section 4.3 presents our proposed facial feature tracking algorithm including the hierarchical multi-state pose-dependent face shape model, the hybrid feature representation, and the proposed multi-modal facial feature tracking algorithm. Section 4.4 discusses the experimental results. The chapter concludes in Sect. 4.5, with a summary and discussion for future research.

4.2 Related Work

4.2.1 Facial Feature Tracking in Nearly Frontal View

Extensive recent work in facial component detection and tracking has utilized the shape representation of deformable objects, where the facial component shape is represented by a set of facial feature points.

Wiskott et al. [682] present the elastic bunch graph matching (EBGM) method to locate facial features using object adopted graphs. The local information of feature points is represented by Gabor wavelets, and the geometry of human face is encoded by edges in the graph. The facial features are extracted by maximizing the similarity between the novel image and model graphs.

Recently, statistical models have been widely employed in facial analysis. The active shape model (ASM) [105] proposed by Cootes et al., is a popular statistical approach to represent deformable objects, where shapes are represented by a set of feature points. Feature points are searched by gray-level profiles, and principal component analysis (PCA) is applied to analyze the modes of shape variation so that the object shape can only deform in specific ways that are found in the training data. Robust parameter estimation and Gabor wavelets have also been employed in ASM to improve the robustness and accuracy of feature point search [408, 531]. Instead of using gray-level profiles to represent feature points, multi-scale and multi-orientation Gabor wavelet coefficients are utilized to characterize the local appearance around feature points. The active appearance model (AAM) [102] and direct appearance model (DAM) [266] are subsequently proposed to combine constraints of both shape variation and texture variation.

Unfortunately, the current statistical-model-based facial feature detection and tracking methods are limited to a narrow scope due to the global linear assumptions with PCAs. Research has shown that the facial feature tracking is effective under the assumption of nearly frontal view, while it tends to fail under large pose variations or significant facial expressions in real world applications.

4.2.2 Facial Feature Tracking Under Varying Pose and Expressions

Recent research has been dedicated to model the nonlinearity of facial deformations caused by pose variations or facial expressions. Generally, these approaches could be grouped into three categories: the first group of approaches utilizes a collection of local linear models to deal with the global nonlinearity [98, 103, 106, 242], the second group of technologies employs $3D$ facial models derived from the image sequences [50, 372, 688], and the third group [533, 575] models the nonlinearity explicitly.

1. Multi-model approaches for facial feature tracking
   The multi-modal approach assumes that each shape/appearance corresponding to one specific pose (or expression) could be approximated linearly by a single shape/appearance model, such that a set of $2D$ linear shape/appearance models could be combined together to model the nonlinear variations. Cootes et al. proposed a weighted mixture of Gaussian models [103] to represent the complex shape variation. An EM algorithm is used to estimate the model parameters from a set of training data. After the mixture Gaussian model is obtained, PCA is applied to each Gaussian component for dimensional reduction. Similar to the conventional ASM, feature search is performed once for each mixture component. By projecting back into the original shape space, the probability that the searched shape is generated by the model is computed, and the component that yields the highest probability is selected. Christoudias et al. [98] extend the mixture Gaussian model [223–225] to represent both the object appearance and shape manifold in image space. The mixture model restricts its search to valid shapes and appearance, thus it avoids erroneous matches. However, the main problem for both methods is that performing mixture of Gaussian fitting in original space is time consuming and requires a lot of training data due to high dimensionality with each component, if each component is described by a full covariance matrix. In addition, the best facial features are detected by enumerating each mixture component, which is again time consuming and makes real-time implementation of such methods infeasible.
   Assuming that the model parameters are related to the face pose, Cootes et al. develop a view-based active appearance model [106] to represent the face from a wide range of face poses. The view-based AAM consists of five $2D$ shape models, each of which represents the shape deformation from a specific view point. Each $2D$ model is trained using a different set of feature points from a set of training images taken within a narrow range of head pose for each view. The relationship between face pose angle and

the model parameters can be learned from images taken from different views simultaneously. Initially the best match is achieved by comparing the searching results against the models from all view points. The head pose is then estimated from the model parameters, and the facial features are tracked given the head pose. However, only one pose parameter (pan, tilt or swing) is considered at one time, so that the feature points are assumed to move along circles in $3D$ space and ellipses in 2D. In real world condition, however, the head rotation often involves a combination of the three angles. In addition, enumerating each view to find the best view is also time consuming.

Similarly, Yan et al. [695] extend the direct appearance model (DAM) into multi-view application by combining several DAMs, each of which is trained from a range of face poses. During the facial feature tracking process, the models corresponding to the previous view and the neighboring two views are attempted, and the best matching is chosen as the one with the minimum texture residual error.

Grauman et al. [209] propose a nearest neighbor (NN) method to represent the human body shape across poses. Christoudias et al. [98] extend the NN method to model the manifolds of facial features. Starting from an initial guess, the nearest neighbor search is performed by minimizing the distance with all prototype examples in pixel space and retaining the k-nearest neighbors, then a new example could be generated by using a convex combination of the neighborhood's shape and texture. This method has several advantages: it does not need assumption about the global structure of the manifold, and it could be more naturally extended to shape features having multiple dimensionality. However, it needs many representative prototype examples.

The hierarchical point distribution model (PDM) is proposed by Heap et al. [242] to deal with the highly nonlinear shape deformation, which will result in discontinuous shape parameter space generated by PCA. The hierarchical PDM consists of two levels of PCAs: in the first level, the training data is projected into shape parameter space, and the shape parameter space is divided into several groups, each of which corresponds to a distinct face pose and facial expression combination, by clustering; in the second level, each cluster is projected onto a local PCA space respectively to give a set of overlapped local hyper ellipsoids. A highly nonlinear shape parameters space is generated by the union of the piece-wise local PCAs. Different from the mixture Gaussian approaches, the hierarchical PDM does not have a probabilistic framework. In addition, choosing the optimal number of local clusters is difficult, and substantial data is required for constructing the model. Although hierarchical formulations are employed in [242], all the feature points are at the same level, and their positions are updated simultaneously.

To handle appearance variations caused by the facial expression changes, Tian et al. [617] propose a multi-state facial component model combining the color, shape, and motion information. Different facial component models are used for the lip, eyes, brows respectively. Moreover, for lip and eyes, each component model has multiple states with distinguished shapes. The state of component model is selected by tracking a few control points. The accuracy of their method, however, critically relies on how accurate and reliable the control points are tracked in current frame. Hence, an automatic state switching strategy is desirable. In addition, the facial feature points are manually initialized in the first frame.

The multi-modal approaches have the advantage of handling very large face pose by using a few linear statistical models with different topologies or even different dimensionality for specific view points. These approaches could also be generalized to deal with the non-linear deformation due to facial expressions. The main weakness with current multi-modal approaches is that they could handle facial deformation either from significant facial expression change or from face pose variation, but not both. In contrast, our proposed hierarchical model explicitly accounts for the deformations of the facial components under different states and poses, therefore is able to track the facial features under large variation of facial expression and face pose.

2. Modeling facial features by using a $3D$ face model

The previous approaches on $3D$ deformable models [202, 606, 700] utilize a $3D$ face mesh to model the global rigid motion and the local non-rigid facial motion respectively by a two-step procedure. The $3D$ local models, however, do not explicitly model each facial component. Furthermore, the complexity to obtain a dense point-to-point correspondence between vertices of the face and the face model is considerable. In addition, the local models used in the $3D$ deformable model are not sufficient to handle the high nonlinearity due to significant facial expression changes.

To reduce the computation complexity, Li et al. [372] propose a multi-view dynamic sparse face model to model $3D$ face shape from video sequence, instead of a dense model. The model consists of a sparse $3D$ facial feature point distribution model, a shape-and-pose-free texture model, and an affine geometrical model. The $3D$ face shape model is learned from a set of facial feature points from $2D$ images with labeled face poses. The $3D$ coordinates of each facial feature point are estimated using an orthographic projection model approximation. The texture model is extracted from a set of shape-and-pose-free face images, which is obtained by warping the face images with pose changes onto the mean shape at frontal view. The affine geometrical model is used to control the rotation, scale, and translation of faces. The fitting process is performed by randomly sampling the shape parameters around the initial values. The best matching set of parameters is obtained by evaluating a loss function for all possible combinations. The time complexity of such a method is high, and it could not handle facial deformation due to facial expression change.

Xiao et al. [688] propose a $2D + 3D$ active appearance model, in which they extend the $2D$ AAM to model $3D$ shape variation with additional shape parameters and use a non-rigid structure-from-motion algorithm [689] to construct the corresponding $3D$ shape modes of the $2D$ AAM. Compared with the $3D$ morphable model with a dense $3D$ shape model and multi-view dynamic face model with a sparse $3D$ shape model, the $2D + 3D$ AAM only has $2D$ shape model. The $2D$ shape model has the capability to represent the same $3D$ shape variations as $3D$ shape model, while needing more shape parameters. Since the $2D$ shape model can generate the shape modes that are impossible for $3D$ model, the $3D$ pose obtained by a structure-from-motion method is used to constrain the $2D$ AAM to only generate the valid shape modes corresponding to possible $3D$ shape variations. Compared with the $3D$ shape model approach, the $2D + 3D$ AAM has the advantage of computational efficiency. However, the 2D AAM requires a large number of parameters.

The advantage of employing $3D$ pose information is the ability to render the $3D$ model from new view points. Thus it is useful for $3D$ reconstruction. However, this group of approaches is limited to dealing with the nonlinear variations caused by face pose, and the nonrigid deformation due to facial expression is not considered. In addition, these methods often suffer from significant time complexity, impeding them for real-time applications.

3. Nonlinear Models

Rather than utilizing a set of linear models, Sozou et al. [575] propose a nonlinear polynomial regression point distribution model (PRPDM), which can approximate the non-linear modes of shape variations by using polynomial functions. However, it requires human intervention to determine the degree of polynomial for each mode. Moreover, the PRPDM could only succeed in representing limited non-linear shape variability.

Romdhani et al. [533] introduce nonlinearity into a $2D$ appearance model by using Kernel PCA (KPCA). A view-context-based ASM is proposed to model the shape and the pose of the face under different view points. A set of $2D$ shape vectors indexed with the pose angles and the corresponding gray-level appearance around each feature point are used to model the multi-view face through KPCA. The non-linearity enables the model to handle the large variations caused by face pose, but the computational complexity is prohibitive.

In summary, despite these efforts, previous technologies often focus on only one of the source of nonlinear variations either caused by face pose or by the nonrigid deformation due to facial expression, while ignoring the other. In real applications, the nonlinearity from both of face pose variation and facial expression changes should be taken into account. For example, tracking mouth and eye motion in image sequences is especially difficult, since these facial components are highly deformable, varying in shape, color, and size as a result of simultaneous variation in facial expression and head pose.

4.3 Facial Feature Tracking Algorithm

4.3.1 Hierarchical Multi-state Pose-dependent Facial Shape Model

To model $3D$ facial deformation and $3D$ head movement, we assume that a $3D$ facial model can be represented by a set of dominant facial feature points. The relative movements of these points characterize facial expression and face pose changes. Figure 4.2 shows the layout of feature points including fiducial points and contour points.

Fiducial points are the key points on the human faces. They are located at well-defined positions such as eye corners, top points of eyebrows, and mouth corners. Fiducial points are further divided into global and local fiducial points. The global fiducial points are relatively stable with respect to facial expression change, and their movements are primarily caused by head movements. The local fiducial points are the dominant points located along the boundary of a facial component. Contour points are interpolated between the fiducial points along the boundary of a facial component. They, along with the local fiducial points, are primarily used to characterize facial expression change.

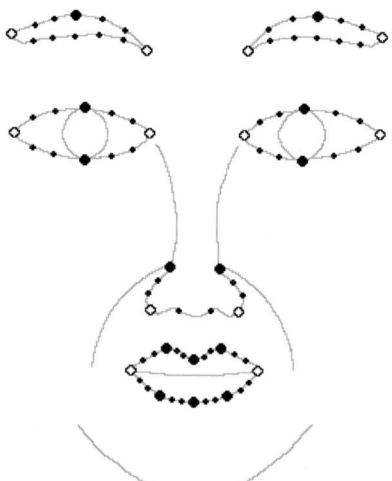

Fig. 4.2. Feature points in the facial model: fiducial points marked by circles (global) and big black dots (local), and contour points marked by small black dots

1. Point Distribution Model

   Given these facial feature points for a particular face view, a point distribution model can be constructed to characterize possible shape variations of human faces. Using the principle of the active shape model (ASM), the point distribution model is constructed from a training set of face images. Facial feature points are marked on each face to outline its structure characteristics, and for each image a shape vector is used to represent the positions of feature points. All face shape vectors are aligned into a common coordinate frame by Procrustes transform [144]. Then the spatial constraints within feature points are captured by principal component analysis (PCA) [105]. A face shape vector $\mathbf{s}$ can be approximated by

   $$\mathbf{s} = \bar{\mathbf{s}} + \mathbf{Pb} \qquad (4.1)$$

   where $\bar{\mathbf{s}}$ is the mean face shape; $\mathbf{P}$ is a set of principal orthogonal modes of shape variation; and $\mathbf{b}$ is a vector of shape parameters.

   The face shape can be deformed by controlling the shape parameters. By applying limits to the elements of the shape parameter vector, it is possible to ensure that the generated shape is an admissible configuration of human face. The ASM approach searches the face shape by an iterative procedure. At each iteration the algorithm seeks to match each feature point locally and then refine all feature point locations by projecting them to the PCA shape space of the entire face. Fitting the entire face shape helps mitigate the errors in individual feature matchings.

2. Hierarchical shape model

   In the conventional active shape model, all the feature point positions are updated (or projected) simultaneously, which indicates that the interactions within feature points are simply parallel. Intuitively, human faces have a sophisticated structure, and a simple parallel mechanism may not be adequate to describe the interactions among facial feature points. For example, given the corner points of an eye, whether the eye is open or closed (or the top and bottom points of the eye) will not affect the localization of mouth or nose. This implies that the eye corners determine the overall location of the eye and provide global shape characteristics, while the top and bottom points of the eye determine the state of the eye (open or closed) and contain local structural details. Generally, facial feature points can be organized into two categories: global feature points and local feature points. The first class characterizes the global shape constraints for the entire face, while the second class captures the local structural details for individual facial components such as eyes and mouth. Based on the two-level hierarchy in facial feature points, a hierarchical formulation of statistical shape models is presented in this section. The face shape vector $\mathbf{s}$ now could be expressed as $(\mathbf{s}_g, \mathbf{s}_l)^T$, where $\mathbf{s}_g$ and $\mathbf{s}_l$ denote global and local feature points respectively. The global and local fiducial points are marked by circles and large black dots respectively as shown in Fig. 4.2. All the contour points (marked as small black dots) are used as local feature points in our facial model. It can be seen from the facial model that the global feature points are less influenced by local structural variations such as eye open or closed, compared to the local feature points. The facial model is partitioned into four components: eyebrows, eyes, nose, and mouth. The two eyes (or eyebrows) are considered as one facial component because of their symmetry.

For the global face shape, a point distribution model can be learned from the training data,

$$\mathbf{s}_g = \bar{\mathbf{s}}_g + \mathbf{P}_g \mathbf{b}_g \qquad (4.2)$$

where $\bar{\mathbf{s}}_g$ is the mean global shape; $\mathbf{P}_g$ is a set of principal orthogonal modes of global shape variation; and $\mathbf{b}_g$ is a vector of global shape parameters.

The local shape model for the ith component is denoted by $\mathbf{s}_{g_i,l_i} = \{\mathbf{s}_{g_i}, \mathbf{s}_{l_i}\}$, where $\mathbf{s}_{g_i}$ and $\mathbf{s}_{l_i}$ represent the global and local feature points belonging to the ith facial component respectively. We have

$$\mathbf{s}_{g_i,l_i} = \bar{\mathbf{s}}_{g_i,l_i} + \mathbf{P}_{g_i,l_i} \mathbf{b}_{g_i,l_i} \qquad (4.3)$$

where $\bar{\mathbf{s}}_{g_i,l_i}$, $\mathbf{P}_{g_i,l_i}$, and $\mathbf{b}_{g_i,l_i}$ are the corresponding mean shape vector, principal orthogonal modes of shape variation, and shape parameters for the ith facial component.

The feature search procedure in the hierarchical shape model is divided into two stages. In the first step, the positions of global feature points are matched, and shape parameters are updated iteratively using the global shape model in (4.2), which is the same as the search process in the active shape model. In the second step, only the local feature points are matched for each facial component, and shape parameters are updated iteratively using the local shape model in (4.3), meanwhile positions of the global feature points remain unchanged. Therefore, in the hierarchical formulation, the global shape model and local shape models form shape constraints for the entire face and individual components respectively. The positions of global feature points will help the localization of local feature points from each facial component. Furthermore, given the global feature points, localization of the local feature points belonging to one facial component will not affect locating the local feature points of other components. For example, given the rough positions of eyes and mouth, whether the eyes are open does not affect locating the feature points on the lips.

3. Multi-state local shape model

Since it is difficult for a single-state statistical shape model to handle the nonlinear shape deformations of certain facial components such as open or closed eyes and mouth, multi-state local shape models are further introduced to address facial expression change. Specifically, for our local facial models, there are three states (open, closed, and tightly closed) for the mouth, two states (open and closed) for the eyes, and one state for the other two components (eyebrows and nose). Given this, for the ith facial component under the jth state, the shape vector becomes $\mathbf{s}_{g_i,l_{i,j}}$, and the local shape model in (4.3) is extended by the multi-state formulation:

$$\mathbf{s}_{g_i,l_{i,j}} = \bar{\mathbf{s}}_{g_i,l_{i,j}} + \mathbf{P}_{g_i,l_{i,j}} \mathbf{b}_{g_i,l_{i,j}} \tag{4.4}$$

where $\bar{\mathbf{s}}_{g_i,l_{i,j}}$, $\mathbf{P}_{g_i,l_{i,j}}$, and $\mathbf{b}_{g_i,l_{i,j}}$ are the corresponding mean shape, principal shape variation, and shape parameters for the ith facial component at the jth state.

Hence a local shape model is built for each state of the facial component. Given the states of facial components, feature points can be searched by the two-stage procedure described in Sect. 4.3.1-2. Therefore, the multi-state hierarchical shape model consists of a global shape model and a set of multi-state local shape models.

4. Pose-Dependent Face Shape Model

The hierarchical face shape model, which we have introduced so far, basically assumes normative frontal face. The shape model (both local and global) will vary significantly, if face pose moves away from the frontal face. To compensate facial shape deformation due to face pose, we propose to estimate the $3D$ face pose and then use the estimated $3D$ pose to correct the hierarchical shape model.

● Robust Face Pose Estimation

Given detected feature points at the previous frame, the three-dimensional face pose can be efficiently estimated. In order to minimize the effect of facial expressions, only a set of rigid feature points that will not move or will move slightly under facial expressions is selected to estimate the face pose. Specifically, six feature points are selected, which include the four eye corners and two fiducial points at the nose's bottom shown in Fig. 4.3a.

In order to estimate the face pose, the $3D$ shape model composed of these six facial features has to be initialized. Currently, the coordinates $\mathbf{X}_i = (x_i, y_i, z_i)^T$ of the six facial feature points in the $3D$ facial shape model are first initialized from a generic $3D$ face model as shown in Fig. 4.3b. Due to the individual difference with the generic face model, the x and y coordinates of each facial feature point in the $3D$ face shape model are adjusted automatically to the specific individual based on the detected facial feature points in the initial frontal face view image. Since the depth values of the facial feature points are not available for the specific individual, the depth pattern of the generic face model is used to approximate the z_i value for each facial feature point. Our experiment results show that this method is effective and feasible in our real-time application.

Based on the personalized $3D$ face shape model and these six detected facial feature points in a given face image, the face pose vector $\alpha = (\sigma_{\text{pan}}, \phi_{\text{tilt}}, \kappa_{\text{swing}}, \lambda)^T$ can be estimated accurately, where

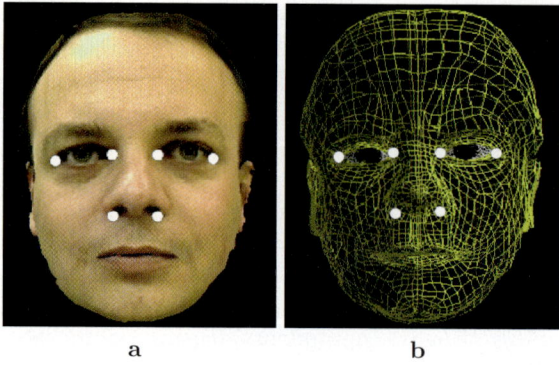

a   b

Fig. 4.3. a A synthesized frontal face image and **b** its $3D$ face geometry with the rigid facial feature points marked by the white dots

$(\sigma_{\text{pan}}, \phi_{\text{tilt}}, \kappa_{\text{swing}})$ are the three face pose angles and λ is the scale factor. Because the traditional least-square method [471] cannot handle the outliers successfully, a robust algorithm based on RANSAC [183] is employed to estimate the face pose accurately.

The pose estimation algorithm is briefly summarized as follows: The procedure starts with randomly selecting three feature points to form a triangle T_i. Under weak perspective projection model [629], each vertex (c_k, r_k) of T_i in the given image and the corresponding point on the face model (x_k, y_k) are related as follows.

$$\begin{pmatrix} c_k - c_0 \\ r_k - r_0 \end{pmatrix} = \mathbf{M}_i \begin{pmatrix} x_k - x_0 \\ y_k - y_0 \end{pmatrix} \qquad (4.5)$$

where $k = 1, 2$, and 3; $\mathbf{M}_i$ is the projection matrix; (c_0, r_0) and (x_0, y_0) are the centers of the triangle in the given image and the reference face model respectively. Given the three detected feature points and the corresponding points on the face model, we can solve the projection matrix $\mathbf{M}_i$ for T_i. Using $\mathbf{M}_i$, the face model is projected onto the given image. A projection error e_i is then computed for all six feature points. e_i is then compared with a threshold e_0, which is determined based on the amount of outliers estimated. $\mathbf{M}_i$ is discarded, if e_i is larger than e_0. Otherwise, a weight ω_i is computed as $(e_i - e_0)^2$ for $\mathbf{M}_i$. After repeating the above for each triangle formed by the six feature points, we will get a list of matrices $\mathbf{M}_i$ and their corresponding weights ω_i. From each projection matrix $\mathbf{M}_i$, a face pose vector α_i is computed uniquely after imposing some consistency constraints. Then the final face pose vector can be obtained as

$$\alpha = \frac{\sum_{i=1}^{K} \alpha_i * \omega_i}{\sum_{i=1}^{K} \omega_i} \qquad (4.6)$$

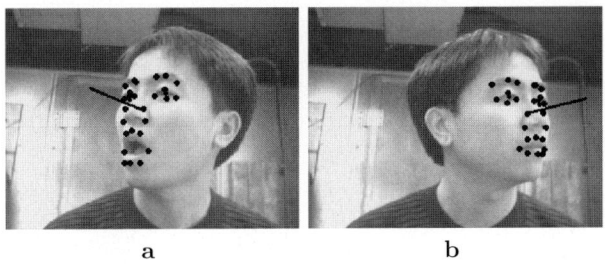

Fig. 4.4a,b. The face pose estimation results under different facial expressions: face normal is represented by the dark line, and the detected facial feature points are marked by the dark dots

Figure 4.4 shows some face pose estimation results, where the face normal is perpendicular to the face plane and represented by the three estimated Euler face pose angles.

- Face Shape Compensation

  Given the estimated $3D$ face pose, the hierarchical model is modified accordingly. Specifically, for each frame, the mean global and local shapes are modified by projecting them to the image plane using the estimated face pose through (4.5). The modified mean shapes are more suitable for the current pose and provide better shape constraints for the feature search. Moreover, the projected mean shapes offer good initialization to avoid being trapped into local minima during the feature search process.

4.3.2 Multi-modal Facial Feature Tracking

A multi-modal tracking approach is required to enable the state switching of facial components during the feature tracking process. Since the global feature points are relatively less influenced by local structural variations, it is assumed that the state switching of facial components only involves local shape models, and the global shape model in (4.2) remains unchanged. In this work, the switching hypothesized measurements (SHM) model [662] is applied to dynamically estimate both the component state s_t and feature point positions of each facial component at time instant t.

For feature points of a facial component, the hidden state $\mathbf{z}_t$ represents their positions at time instant t. The hidden state transition can be modeled as

$$\mathbf{z}_{t+1} = \mathbf{F}\mathbf{z}_t + \mathbf{n} \qquad (4.7)$$

where $\mathbf{F}$ is the state transition matrix, and $\mathbf{n}$ represents the system perturbation with the covariance matrix $\mathbf{Q}$.

The switching state transition is modeled by a first order Markov process to encourage the temporal continuity of the component state. For the local

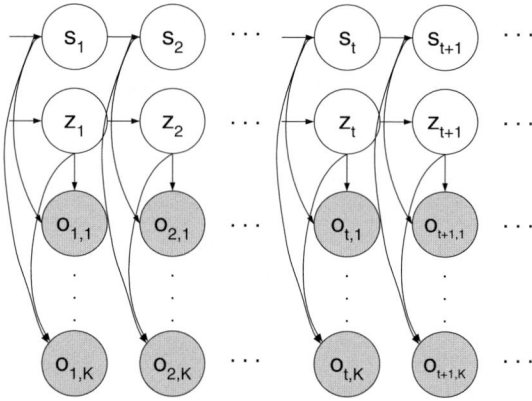

Fig. 4.5. Directed acyclic graph specifying conditional independence relations for switching hypothesized measurements model, where s_t is the component state, $\mathbf{z}_t$ is the facial feature point positions, and $\mathbf{o}_{t,j}$ are the measurements at time t

shape model of ith component with K possible states, Fig. 4.5 illustrates the SHM model by a representation of dynamic Bayesian network, where the nodes s_t and $\mathbf{z}_t$ are hidden nodes, and the shaded nodes $\mathbf{o}_{t,j}, j \in 1, \ldots, K$ are the observation nodes representing the measurements at time t. Since the component state s_t is unknown, feature point positions are searched once under each hypothesized component state. Under the assumption of the jth state of the component, a hypothesized measurement $\mathbf{o}_{t,j}$ represents the feature point positions of the facial component obtained from the feature search procedure at time t. Given the component state s_t, the corresponding hypothesized measurement $\mathbf{o}_{t,s_t}$ could be considered as a proper measurement centering on the true feature positions, while every other $\mathbf{o}_{t,j}$ for $j \neq s_t$ is an improper measurement generated under a wrong assumption. The improper measurement should be weakly influenced by true feature positions and have a large variance. To simplify the computation, the measurement model is formulated as

$$\mathbf{o}_{t,j} = \begin{cases} \mathbf{Hz}_t + \mathbf{v}_{t,j} & \text{if} \quad j = s_t \\ \mathbf{w} & \text{otherwise} \end{cases} \qquad (4.8)$$

where $\mathbf{H}$ is the measurement matrix; $\mathbf{v}_{t,j}$ represents the measurement uncertainty assuming as a zero mean Gaussian with the covariance matrix $\mathbf{R}_j$; and $\mathbf{w}$ is a uniformly distributed noise. The state transition matrix $\mathbf{F}$, system noise $\mathbf{n}$, and measurement matrix $\mathbf{H}$ are defined in the same way as in a Kalman filter [210].

At time t, given the state transition model, measurement model, and all hypothesized measurements, the facial feature tracking is performed by maximizing the posterior probability:

$$p(s_t = i, \mathbf{z}_t | \mathbf{o}_{1:t}) = p(s_t = i | \mathbf{o}_{1:t}) p(\mathbf{z}_t | s_t = i, \mathbf{o}_{1:t}) \tag{4.9}$$

where $\mathbf{o}_m = \{\mathbf{o}_{m,1}, \ldots, \mathbf{o}_{m,K}\}$, $1 \le m \le t$.

Let $\beta_{t,i} = p(s_t = i | \mathbf{o}_{1:t})$ with $\sum_i \beta_{t,i} = 1$, and assume that $p(\mathbf{z}_t | s_t = i, \mathbf{o}_{1:t})$ is modeled by a Gaussian distribution $N(\mathbf{z}_t; \mu_{t,i}, \mathbf{P}_{t,i})$, with the mean $\mu_{t,i}$ and covariance matrix $\mathbf{P}_{t,i}$. Then at time $t + 1$, given the hypothesized measurements $\mathbf{o}_{t+1}$, the parameters $\{\beta_{t+1,i}, \mu_{t+1,i}, \mathbf{P}_{t+1,i}\}$ are updated in the SHM filtering algorithm [657] as below:

$$\beta_{t+1,i} = p(s_{t+1} = i | \mathbf{o}_{1:t+1}) = \frac{\sum_j^K \gamma_{i,j} \beta_{t,j} N(\mathbf{o}_{t+1,i}; \mathbf{H}\mu_{t+1|t,j}, \mathbf{S}_{t+1,i|j})}{\sum_i^K \sum_j^K \gamma_{i,j} \beta_{t,j} N(\mathbf{o}_{t+1,i}; \mathbf{H}\mu_{t+1|t,j}, \mathbf{S}_{t+1,i|j})} \tag{4.10}$$

where $\gamma_{i,j}$ is the state transition probability:

$$\gamma_{i,j} = p(s_{t+1} = i | s_t = j) \tag{4.11}$$

with $\sum_i^K \gamma_{i,j} = 1$.

$$\mu_{t+1|t,j} = \mathbf{F}\mu_{t,j} \tag{4.12}$$

$$\mathbf{P}_{t+1|t,j} = \mathbf{F}\mathbf{P}_{t,j}\mathbf{F}^T + \mathbf{Q} \tag{4.13}$$

$$\mathbf{S}_{t+1,i|j} = \mathbf{H}\mathbf{P}_{t+1|t,j}\mathbf{H}^T + \mathbf{R}_i \tag{4.14}$$

$$\mathbf{G}_{t+1,i|j} = \mathbf{P}_{t+1|t,j}\mathbf{H}^T\mathbf{S}_{t+1,i|j}^{-1} \tag{4.15}$$

where $\mathbf{G}$ is the Kalman gain matrix.

$$\mu_{t+1,i|j} = \mu_{t+1|t,j} + \mathbf{G}_{t+1,i|j}(\mathbf{o}_{t+1,i} - \mathbf{H}\mu_{t+1|t,j}) \tag{4.16}$$

$$\mathbf{P}_{t+1,i|j} = \mathbf{P}_{t+1|t,j} - \mathbf{G}_{t+1,i|j}\mathbf{H}\mathbf{P}_{t+1|t,j} \tag{4.17}$$

$$\beta_{t+1,i|j} = \frac{\gamma_{i,j}\beta_{t,j}N(\mathbf{o}_{t+1,i}; \mathbf{H}\mu_{t+1|t,j}, \mathbf{S}_{t+1,i|j})}{\sum_j^K \gamma_{i,j}\beta_{t,j}N(\mathbf{o}_{t+1,i}; \mathbf{H}\mu_{t+1|t,j}, \mathbf{S}_{t+1,i|j})} \tag{4.18}$$

$$\mu_{t+1,i} = \sum_j^K \beta_{t+1,i|j}\mu_{t+1,i|j} \tag{4.19}$$

$$\mathbf{P}_{t+1,i} = \sum_j^K \beta_{t+1,i|j} \left[\mathbf{P}_{t+1,i|j} \right.$$
$$\left. + (\mu_{t+1,i|j} - \mu_{t+1,i})(\mu_{t+1,i|j} - \mu_{t+1,i})^T \right] \tag{4.20}$$

Therefore, the hidden state $\mathbf{z}_{t+1}$ and the switching state s_{t+1} could be estimated as

$$\hat{s}_{t+1} = \underset{i}{\text{argmax}}\, \beta_{t+1,i} \tag{4.21}$$

$$\hat{\mathbf{z}}_{t+1} = \mu_{t+1,\hat{s}_{t+1}|\hat{s}_t} \tag{4.22}$$

Since the measurement under the true hypothesis of the switching state usually shows more regularity and has smaller variance compared with the other hypothesized measurements, the true information (the facial component state and feature point positions) could be enhanced through the propagation in the SHM filter. Moreover, for a facial component with only one state, the multi-modal SHM filter degenerates into a unimodal Kalman filter.

Given the multi-state hierarchical shape model, the facial feature detection and tracking algorithm performs an iterative process at time t:

1. Project the mean shapes of global model $\bar{\mathbf{s}}_g$ and local shape models $\bar{\mathbf{s}}_{g_i,l_i}$ using the estimated face pose α_{t-1} from the previous frame.
2. Localize the global facial feature points $\mathbf{s}_g$ individually.
3. Update the global shape parameters to match $\mathbf{s}_g$, and apply the constraints on $\mathbf{b}_g$.
4. Generate the global shape vector $\mathbf{s}_g$ as (4.2), and then return to the Step 2 until convergence.

   Enumerate all the possible states of the ith facial components. Under the assumption of the jth state:
5. Localize the local feature points individually.
6. Update the local shape parameters to match $\mathbf{s}_{g_i,l_{i,j}}$, and apply the constraints on $\mathbf{b}_{g_i,l_{i,j}}$.
7. Generate the shape vector $\mathbf{s}_{g_i,l_{i,j}}$ as (4.4), and then return to Step 5 until convergence.
8. Take the feature search results $(\mathbf{s}_{g_i}, \mathbf{s}_{l_{i,j}})^T$ for the ith facial component under different state assumptions, as the set of hypothesized measurements $\mathbf{o}_{t,j}$. Estimate the state of the ith facial component and the positions of its feature points at time t through the SHM filter as (4.21) and (4.22).
9. Estimate the $3D$ face pose α_t by the tracked six rigid facial feature points.

4.3.3 Hybrid Facial Feature Representation

For feature detection, a hybrid feature representation, based on Gabor wavelet jets [115] and gray-level profiles [105], is utilized in this work to model the local information of fiducial points and contour points respectively.

1. Wavelet-based representation
   Multi-scale and multi-orientation Gabor wavelets [115] are employed to model local appearances around fiducial points. Gabor-wavelet-based feature representation has the psychophysical basis of human vision and achieves robust performance for expression recognition [615, 715], face recognition [682], and facial feature representation [297, 408] under illumination and appearance variations.
   For a given pixel $\mathbf{x} = (x, y)^T$ in a gray scale image I, a set of Gabor coefficients $J_j(\mathbf{x})$ is used to model the local appearance around the point.

The coefficients $J_j(\mathbf{x})$ are resulted from convolutions of image $I(\mathbf{x})$ with the two-dimensional Gabor wavelet kernels ψ_j, i.e.,

$$J_j(\mathbf{x}) = \sum \sum I(\mathbf{x}')\psi_j(\mathbf{x} - \mathbf{x}') \qquad (4.23)$$

Here kernel ψ_j is a plane wave restricted by a Gaussian envelope function:

$$\psi_j(\mathbf{x}) = \frac{\mathbf{k}_j^2}{\sigma^2} \exp\left(-\frac{\mathbf{k}_j^2 \mathbf{x}^2}{2\sigma^2}\right) \left[\exp\left(i\mathbf{k}_j \cdot \mathbf{x}\right) - \exp\left(-\frac{\sigma^2}{2}\right)\right] \qquad (4.24)$$

with the wave vector

$$\mathbf{k}_j = \begin{pmatrix} k_{jx} \\ k_{jy} \end{pmatrix} = \begin{pmatrix} k_v \cos \varphi_u \\ k_v \sin \varphi_u \end{pmatrix} \qquad (4.25)$$

where $k_v = 2^{-(v+1)}$ is the radial frequency in radians per unit length; $\varphi_u = \frac{\pi}{6}u$ is the wavelet orientation in radians, rotated counter-clockwise around the origin; $v = 0, 1, 2$; $u = 0, 1, \ldots, 5$; $j = u + 6v$; and $i = \sqrt{-1}$ in this section. In this work, $\sigma = \pi$ is set for a frequency bandwidth of one octave.

Thus the set of Gabor kernels consists of three spatial frequencies and six different orientations, and eighteen Gabor coefficients in the complex form are used to represent the pixel and its vicinity. Specifically, a jet vector $\mathbf{J}$ is used to denote $(J_0, J_1, \ldots, J_{17})$, where $J_j = a_j \exp(i\phi_j)$, a_j and ϕ_j are the magnitude and phase of the jth Gabor coefficient. The Gabor wavelet jet vector is calculated for each marked fiducial point in training images. Given a new image, the fiducial points are searched by the sample jets from the training data. The similarity between two jet vectors is measured with the following phase sensitive distance function:

$$D_\phi(\mathbf{J}, \mathbf{J}') = 1 - \frac{\sum_j a_j a_j' \cos(\phi_j - \phi_j' - \mathbf{d} \cdot \mathbf{k}_j)}{\sqrt{\sum_j a_j^2 * \sum_j a_j'^2}} \qquad (4.26)$$

where jet vectors $\mathbf{J}$ and $\mathbf{J}'$ refer to two locations with relative small displacement $\mathbf{d}$. The displacement between the two locations can be approximately estimated as [184, 611]:

$$\mathbf{d}(\mathbf{J}, \mathbf{J}') = \begin{pmatrix} d_x \\ d_y \end{pmatrix} \approx \frac{1}{\Gamma_{xx}\Gamma_{yy} - \Gamma_{xy}\Gamma_{yx}} \times \begin{pmatrix} \Gamma_{yy} & -\Gamma_{yx} \\ -\Gamma_{xy} & \Gamma_{xx} \end{pmatrix} \begin{pmatrix} \Phi_x \\ \Phi_y \end{pmatrix} \qquad (4.27)$$

if $\Gamma_{xx}\Gamma_{yy} - \Gamma_{xy}\Gamma_{yx} \neq 0$, where

$$\Phi_x = \sum_j a_j a_j' k_{jx}(\phi_j - \phi_j') \,,$$

$$\Gamma_{xy} = \sum_j a_j a_j' k_{jx} k_{jy} \,,$$

Φ_y, Γ_{xx}, Γ_{yx}, and Γ_{yy} are defined accordingly.

The phase sensitive distance defined in (4.26) changes rapidly with location, which helps accurately localize fiducial points in the image. Compensated by the displacement in (4.27), the search of fiducial points can achieve subpixel sensitivity.

2. Profile-based representation

   Gray-level profiles (gradients of pixel intensity) along the normal direction of the object boundary are used to represent contour points [105]. The Mahalanobis distance function is used to search these feature points. For the lth contour point,

   $$D_M(\mathbf{g}, \bar{\mathbf{g}}_l) = (\mathbf{g} - \bar{\mathbf{g}}_l)^T \mathbf{C}_l^{-1} (\mathbf{g} - \bar{\mathbf{g}}_l) \qquad (4.28)$$

   where $\mathbf{g}$ is a gray-level profile in the given image; $\bar{\mathbf{g}}_l$ is the mean profile of the lth contour point computed from the training data; and $\mathbf{C}_l^{-1}$ is the corresponding covariance matrix obtained by training.

   The profile-based representation is computationally efficient, and thus leads to fast convergence in the feature search process. However, the gray-level gradients are not sufficient to identify all the facial feature points. For example, the profile of the mouth bottom point may not be distinctive due to the shadow or beard below the lower lip. On the other hand, the magnitudes and phases in wavelet-based representation provide rich information of local appearances, and therefore lead to accurate feature point localization, but with relatively high computation complexity. To balance the searching effectiveness and computational efficiency, in this work the fiducial points are modeled by Gabor wavelet jets, and the contour points, which are relatively less important to the face shape deformation, are represented by gray-level profiles. Compared to wavelet-based representation for all the feature points, the hybrid representation achieves similar feature search accuracy and enhances the computation speed by 60% in our experiments.

3. Feature Detection

   Given the hybrid representation for each feature point, we can perform feature detection. Feature detection starts with eye detection. An accurate and robust eye detector is desirable to help estimate the position, size, and orientation of the face region, and thus improve the shape constraints for the feature search. In this work, a boosted eye detection algorithm is employed based on recursive non-parametric discriminant analysis (RNDA) features proposed in [660,661]. For eye detection, the RNDA features provide better accuracy than Harr features [652], since they are not constrained with rectangle-like shape. The features are sequentially learned and combined with Adaboost to form an eye detector. To improve speed, a cascade structure is applied. The eye detector is trained on thousands of eye images and more non-eye images. The resulting eye detector classifier uses less than 100 features.

   The eye localization follows a hierarchical principle: first a face is detected, then the eyes are located inside the detected face. An overall

94.5% eye detection rate is achieved with 2.67% average normalized error (the pixel error normalized by the distance between two eyes) on FRGC 1.0 database [494]. Given the knowledge of eye centers, the face region is normalized and scaled into a 64×64 image, such that the eyes are nearly fixed in same positions in each frame.

Given the normalized face region, the other facial feature points as illustrated in Fig. 4.2 are detected based on the hybrid feature representations: the contour points are searched by minimizing the Mahalanobis distance defined in (4.28); and the fiducial points are detected by minimizing the phase sensitive distance function in (4.26).

4.4 Experimental Results

Twenty-six fiducial points and fifty-six contour points are used in our facial model (see Fig. 4.2). The global shape model, multi-state local shape models, Gabor wavelet jets of fiducial points, and gray-level profiles of contour points are trained using 500 images containing two hundred persons from different races, ages, face poses, and facial expressions. Both feature point positions and facial component states are manually labeled in each training image. For ASM analysis, the principal orthogonal modes in the shape models stand for 95% of the shape variation. The test sequences consist of ten sequences, each of which consists 100 frames. The test sequences contain six subjects including five Asians and one European, who are not included in training data. The test sequences are 24-bit color images collected by a USB web camera under real world conditions with 320×240 image resolution. The system can reliably detect and track the face in a range of 0.25–1.0 meters from the camera. Since the face region is normalized and scaled based on the detected eye positions, the system is invariant to the scale change. Our C++ program can process about seven frames per second on a Pentium 4 2.8GHz PC.

4.4.1 Experiments on Facial Feature Tracking

Figures 4.6–4.8 exhibit the results of facial feature tracking under pose variations and face deformations due to varying facial expressions. To make the results clear, only the fiducial points are shown, since they are more representative than the contour points. Figure 4.6 shows the results of the proposed method and the results without using the pose modified mean shapes. Compared to the results in Fig. 4.6b, the feature points are more robustly tracked in Fig. 4.6a under large pose variations, which demonstrates that the projection of mean shapes through face pose estimation helps improve shape constraints in the feature search process. Figure 4.7 shows the results of the proposed method and the results without using the multi-state local shape

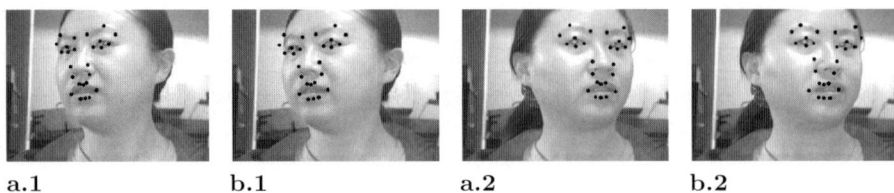

a.1   b.1   a.2   b.2

Fig. 4.6a,b. Feature tracking results: **a** by the proposed method and **b** by the proposed method without using modified mean shapes

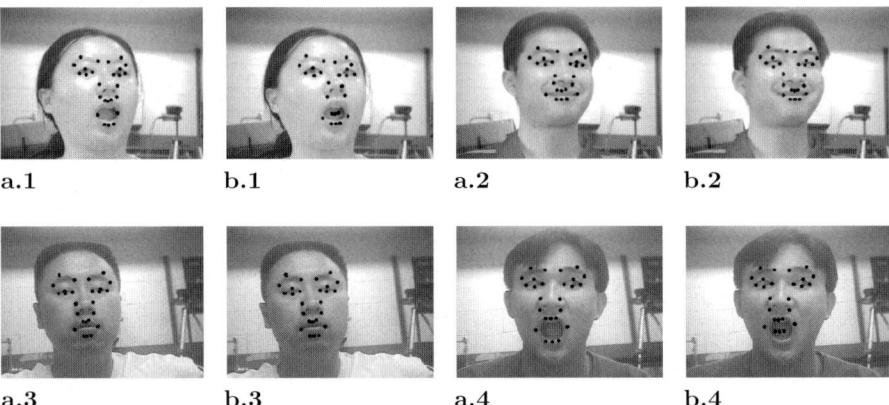

a.1   b.1   a.2   b.2

a.3   b.3   a.4   b.4

Fig. 4.7a,b. Feature tracking results: **a** by the proposed method and **b** by the proposed method without using the multi-state local shape models

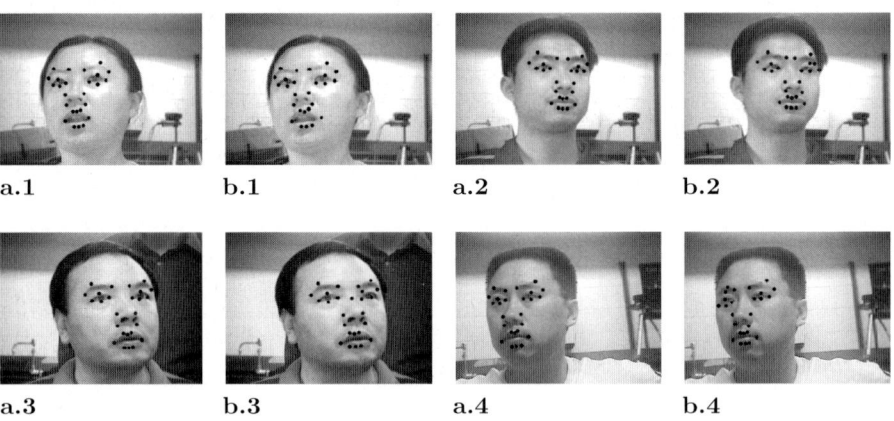

a.1   b.1   a.2   b.2

a.3   b.3   a.4   b.4

Fig. 4.8a,b. Feature tracking results: **a** by the proposed method and **b** by the proposed method without using the hierarchical shape model

models (i.e. a single-state local shape model is used for each facial component). It can be seen from Fig. 4.7 that the multi-state models substantially improve the robustness of facial feature tracking, especially when eyes and mouth are open or closed (e.g. mouth is the first and fourth images, and eyes in the second and third images). This demonstrates that the state switching in local shape models helps in dealing with nonlinear shape deformations of facial components.

Figure 4.8 shows the results of the proposed method and the results without using the hierarchical shape model (i.e. all the feature points are simultaneously updated in the feature search process). Compared to the results in Fig. 4.8b, the feature points are more accurately tracked in Fig. 4.8a (e.g. mouth in the first image, right eye in the second and third images, and left eyebrow in the fourth image) using the hierarchical shape model, since the two-level hierarchical facial shape model provides a relatively sophisticated structure to describe the interactions among feature points.

4.4.2 Quantitative Evaluation

The results of facial feature tracking are evaluated quantitatively besides visual comparison. Fiducial and contour feature points are manually marked in 1000 images from the ten test sequences for comparison under different face pose and facial expressions. Figure 4.9 show a set of testing images from an image sequence, where the face undergoes the face pose change and facial expression change simultaneously. For each feature point, the displacement (in pixels) between the estimated position and the corresponding labeled position is computed as the feature tracking error. Figure 4.10 shows error distribution of the feature tracking results of different methods. Compared to the feature tracking result using single state local model and using multi-state local models without the hierarchical models, the use of the multi-state hierarchical shape model averagely reduces the feature tracking error by 24% and 13% respectively. Besides the comparison of the average pixel displacement, Fig. 4.11 illustrates the evolution of the error over time for one image sequence, where the face undergoes both significant facial expression and large face pose change simultaneously from frame 35 to frame 56. By employing the multi-state pose dependent hierarchical shape model, the proposed method substantially improves the robustness of facial feature tracking under simultaneous facial expression and face pose variation.

4.4.3 Comparison with Mixture of Gaussian Model

Figure 4.12 compares the feature tracking results of the proposed method with the results of the mixture Gaussian method by Cootes et al. [103]. It can be seen that the proposed multi-state pose-dependent hierarchical method outperforms mixture Gaussian approach under pose variations (e.g.

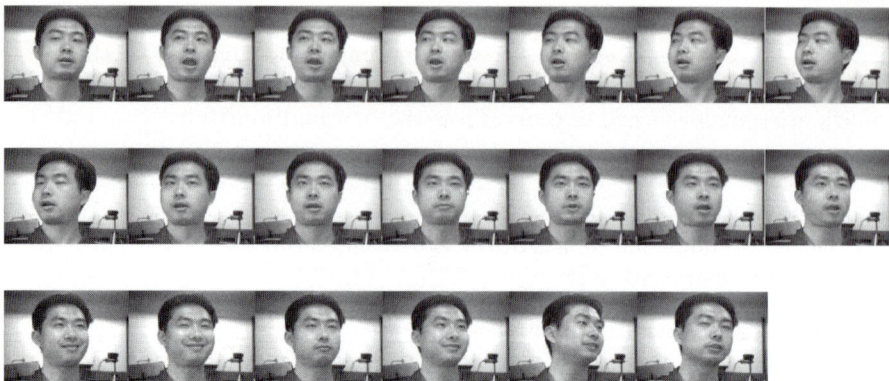

Fig. 4.9. A set of testing images from an image sequence, where the face undergoes the face pose change and facial expression change simultaneously

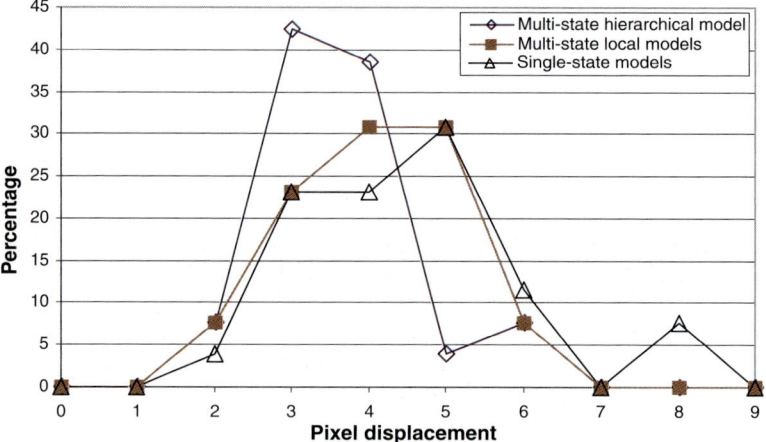

Fig. 4.10. Error distribution of feature tracking. Diamonds: results of the proposed method. Triangles: results of the proposed method without the multi-state local shape models. Squares: results of the proposed method without the hierarchical shape model

the left eye and eyebrow in the first and third images, the mouth in the second image, and the right eye, right eyebrow and nose in the fourth image). Figure 4.13 shows the error distribution of the feature tracking results using the proposed method and the mixture Gaussian method respectively under face pose change only. Furthermore, Fig. 4.14 compares the facial feature tracking results using the proposed method and the mixture Gaussian method under simultaneous facial expression and face pose change. Compared to the mixture Gaussian method, the proposed method reduces the feature tracking error by 10% and 23% respectively under only face pose change and the

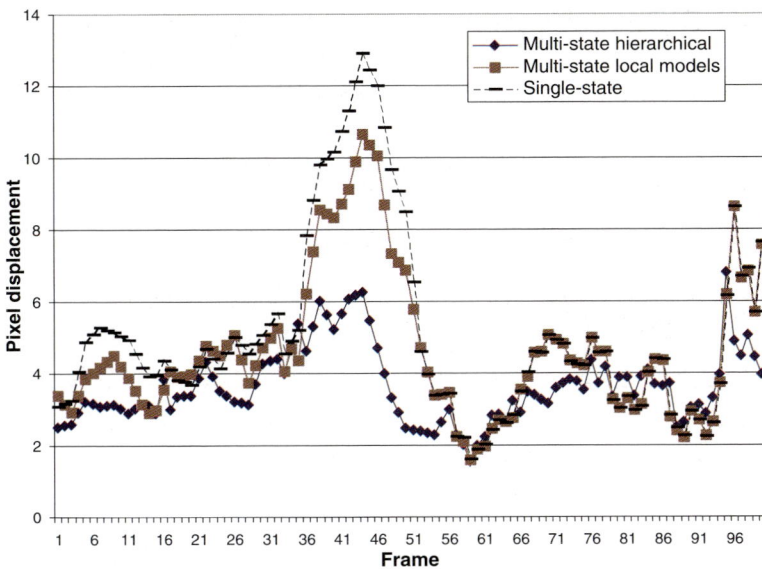

Fig. 4.11. Error evolution of feature tracking for an image sequence. Diamonds: results of the proposed method. Bars: results of the proposed method without the multi-state local shape models. Squares: results of the proposed method without the hierarchical shape model

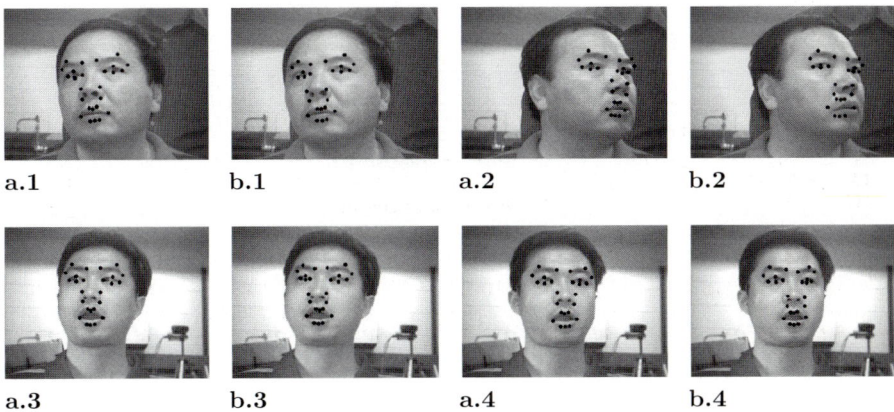

Fig. 4.12. Feature tracking results: **a** proposed method and **b** mixture Gaussian model

combination of facial expression and face pose change. Therefore, the proposed method is more appropriate to handle the real world conditions, where the facial shape change due to both of facial expression and face pose. In addition, the proposed method is more efficient than the mixture Gaussian method, which runs at about 5.8 frame/second under same condition.

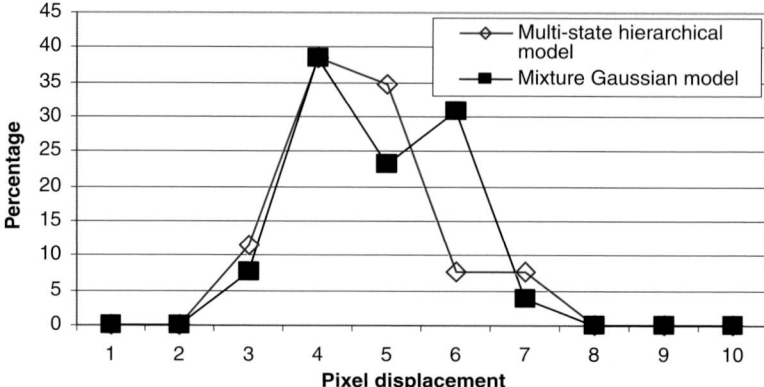

Fig. 4.13. Error distribution of feature tracking under face pose change. Diamonds: results of the proposed method. Squares: results of the mixture Gaussian method

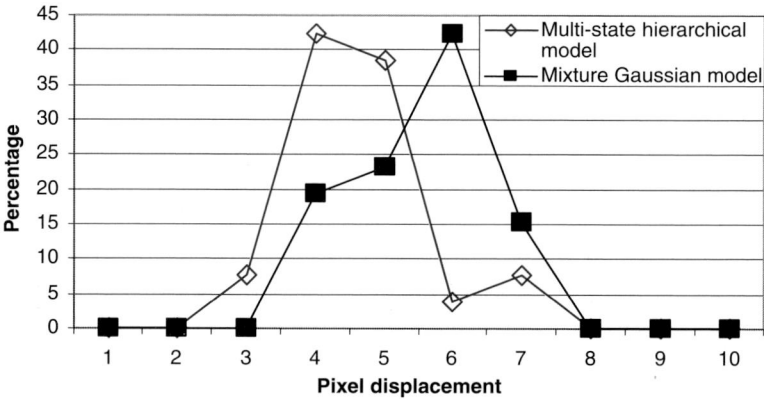

Fig. 4.14. Error distribution of feature tracking under both of face pose and facial expression change. Diamonds: results of the proposed method. Squares: results of the mixture Gaussian method

4.5 Conclusion

In this chapter, a multi-state pose-dependent hierarchical face shape model is successfully developed to improve the accuracy and robustness of facial feature detection and tracking under simultaneous pose variations and face deformations. The model allows to simultaneously characterize the global shape constraints and the local structural details of human faces. Shape constraints for the feature search are significantly improved by modifying mean shapes through robust face pose estimation. In addition, Gabor wavelet jets and gray-level profiles are integrated for effective and efficient feature representation. Feature point positions are dynamically estimated with multi-state local shape models using a multi-modal tracking approach. Experimental results demonstrate that the proposed method significantly reduces the feature tracking error, compared to the classical feature tracking methods.

In the current work, we ignore the relationships between different facial components by decoupling the face into different local models, because those relationships are complex, dynamic, and very uncertain. Moreover, incorrect modeling of such relationships will lead to the failure in detection of facial feature. In the future, we would like to combine the relationships between different facial components into the facial feature tracking by exploiting spatial-temporal relationships among different action units.

Part II

Mid-level Building-Blocks – Gaze Calibration,
Gaze & Eye Pose Tracking,
and Eye Images Interpretation

5

Simple-to-Calibrate Gaze Tracking Method

Takehiko Ohno, Kenji Hara, and Hirohito Inagaki

5.1 Position of the Problem

Personal calibration is a procedure which offsets the impact of personal differences on gaze tracking performance. During the calibration, the user has to look precisely at five to twenty markers that are shown the computer screen in succession. While the calibration marker is displayed on the screen, the user should gaze at the marker without eye movement until it disappears. This is a troublesome and wearying task. If she doesn't look at the exact position of the marker, or if there is any involuntary eye movement or eye blink, the result may contain gaze tracking error. The difficulty of calibration explodes when the number of calibration markers increases. Estimating human eye gaze without personal calibration is one of the most important goals in the field of gaze tracking technology. For researchers who want to capture the user's eye movement by a gaze tracking system, it is well known that the high cost of traditional personal calibration procedures makes eye tracking studies difficult. For example, it is extremely difficult to observe a child's eye movement because of the personal calibration barrier. Since infants cannot understand the observer's instructions, personal calibration often fails. A lot of effort is needed to attract the children's attention to the calibration markers to accomplish personal calibration. The problem of personal calibration is not limited to children. For example, it is difficult to measure the eye movements of a large number of subjects because calibration is so time consuming.

Difficulties in setting up a gaze tracking system also limit the application area of gaze tracking. For example, eye gaze could be used very effectively to support Human–Computer Interaction (see Chap. 8 for detail). However, interacting with the computer by gaze, which is called *gaze-based interaction*, requires a really practical gaze tracking system that can be used as easily as the keyboard or mouse. The public will not welcome the use of gaze-based interaction in everyday activities if the gaze tracking system requires long and laborious setup procedures. To encourage the use of gaze tracking systems in daily life, we need a gaze tracking system that satisfies the following three requirements.

- Minimum setup difficulty. The ideal gaze tracking system has no personal calibration procedure. If personal calibration is required, it should be as simple as possible.
- No eye-detection camera or other attachments to the user's head. Even if these attachments are light, they can prevent the continuous use of gaze tracking.
- Free movement of the user's head. Many gaze tracking systems prevent the user from making free head movements because a fixed camera is used to detect the user's eye. This limitation prevents users from becoming comfortable with the gaze tracking system. The gaze tracking system should accept ordinary head movements.

To satisfy these three requirements, we focus on the vision-based gaze tracking approach. Different gaze tracking methods like Electro Oculography (EOG) and the magnetic coil method (the user wears a wire wound contact lens) have been proposed. They do not use vision techniques. However, because these methods do not satisfy the second condition, they are not adequate for everyday use. We use the corneal reflection method, which is known to be an accurate gaze detection method, for gaze tracking. It is one of the most popular gaze tracking methods and is used widely in many commercial products. When we project an near-infrared point light onto the eye, a reflection image (glint) appears on the corneal surface. An example of the eye image taken by a near-infrared sensitive camera is shown in Fig. 5.1. The glint on the cornea is named the first Purkinje image or in short, the Purkinje image. The combination of the Purkinje image and the pupil makes it possible to determine gaze direction.

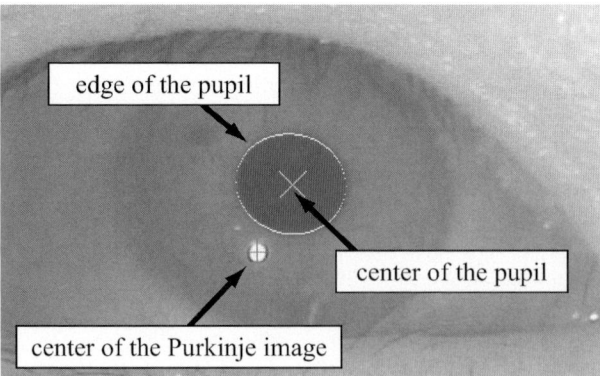

Fig. 5.1. An example of eye image: This image was taken by an infrared-light sensitive camera. Purkinje image (glint) appears on the corneal surface

In this chapter, we describe two gaze tracking methods and prototype gaze tracking systems developed to satisfy the three requirements. Both methods assume that there is a computer screen in front of the user; the gaze point on the screen is calculated by the gaze tracking system. The first method, which is called *the Two-Point Calibration (TPC)* gaze tracking method, places two calibration markers on the screen [466, 468]. It accepts the free movement of the user's head. The other method, which is called *the One-Point Calibration (OPC)* gaze tracking method, is an extension of the TPC method. It places only one calibration marker on the screen [464]. Even though they still require personal calibration, they drastically reduce the burden of the calibration procedure.

5.1.1 Why is Personal Calibration Required?

To understand the complexity of personal calibration, we first describe the purpose of personal calibration in gaze tracking. In the majority of cases, no gaze tracking system can estimate the gaze direction accurately without personal calibration. There are many factors that trigger gaze measurement error. The main causes of gaze measurement error are:

1. Personal difference in eyeball size and shape. Figure 5.2 shows the structure of the human eyeball. Among all adults, there is a 10% difference in eyeball radius. That is, measurement error is likely if we assume a constant eyeball size in setting some of the system parameters. There are also individual differences in the shape of the corneal surface.
2. Refraction at the corneal surface. When using the pupil for gaze detection, we should consider that there is refraction at the corneal surface because the region between the cornea and the crystalline lens is filled with a fluid (aqueous humor). The observed pupil differs from the real position, so that the calculated gaze contains some measurement error.

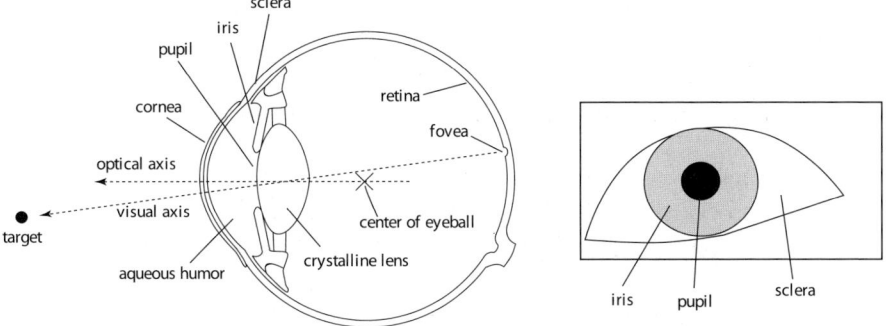

Fig. 5.2. Cross section and front view of human eyeball: For a user looking at a target, the visual axis differs from the optical axis

3. Difference between optical axis and visual axis. The optical axis is defined by the center of the pupil and the center of the corneal curvature. This, however, differs from eye gaze, which is called the visual axis. Visual axis is defined as the vector from the fovea (the highest resolution area on the retina) to the center of the crystalline lens. Since it is difficult to observe the exact position of the fovea by any of the cameras likely to be used in a gaze tracking system, the visual axis can not be determined. To calculate the visual axis from the optical axis, it is necessary to compensate this difference.

4. Screen position difference. If the size, the position, and the resolution of the computer screen is not fixed, it is not possible to determine the gaze position on the computer screen. In this case, it is necessary to display several calibration markers on the screen for estimating the parameters related to the computer screen. The presence of unknown parameters generally increase the number of calibration markers needed.

5. Eye positioning error. When the gaze tracking system allows free head movement, the user's eye position is generally different from the initial position used for personal calibration. This often causes gaze measurement error because the gaze tracking parameters differ from the initial calibration condition.

6. Refraction at the surface of eyeglasses. When the user wears eyeglasses, refraction appears at the surface of the glasses. It is necessary to consider that observed pupil position and its size are different from the real ones if the gaze tracking system user wears eyeglasses.

The basic idea of the proposed gaze tracking methods is to reconstruct the user's eyeball position and the gaze direction accurately in a 3-D coordinate system from the observed eye image. Screen size, resolution and position are given initially, therefore, screen position estimation is not necessary for personal calibration. The effect of refraction at the corneal surface is compensated when calculating the gaze direction. The residual gaze measurement error is compensated by the personal parameters derived by personal calibration.

5.2 Existing Personal Calibration Methods

A lot of research has been done on gaze tracking methods [145,423]. However, most of them focus on reducing the restrictions placed on the user's head movement (e.g. [45,484,602,702]). Conventional gaze tracking methods based on corneal reflection use second or higher-order approximation for personal calibration. For example, Morimoto et al. use a second order polynomial calibration method which requires nine calibration points [420].

Shih et al. proposed a calibration-free gaze tracking method [565] in which multiple cameras and multiple point light sources are used to estimate gaze direction. Correction of the refraction at the corneal surface is also done for accurate eyeball modeling. They performed computer simulations to confirm their method. Morimoto et al. proposed another calibration-free gaze tracking method that uses at least one camera and two or more infrared light sources [419]. The accuracy as determined using simulated synthetic eye images was about 3 degrees (view angle). In both Shih and Morimoto's models, the difference between the visual axis and the estimated gaze direction is not considered; therefore, some measurement error may still remain.

Vision-based gaze tracking methods based on approaches other than corneal reflection have also proposed. For example, Matsumoto et al. used a real-time stereo vision technique to measure head position and gaze direction where the limbus, the boundary of the sclera and the iris are used for estimating gaze direction [399]. It does not require personal calibration, but gaze measurement error becomes significant when the user wears eyeglasses because iris position can not be accurately estimated.

Holman et al. proposed a gaze tracking method that places several Eye Contact Sensors on the target plane [258]. Each sensor could detect if the user was looking at the sensor or not. The system arranges light-emitting diode (LED) arrays at two different places, around the camera lens and far from the lens, to detect the bright and the dark pupil images. From the difference between two images, it is possible to detect the pupil robustly with simple image processing [154, 420] By integrating the result of each sensor, it is possible to roughly estimate gaze direction. Because it does not require personal calibration, it is reasonable for some applications like puplic displays. Different from other gaze tracking methods, low accuracy of gaze tracking limits the application area.

As described in Sect. 5.1.1, there are many factors that may cause gaze detection error, and a lot more research effort is needed to achieve a calibration-free gaze tracking method that matches the accuracy of existing gaze tracking methods that use extensive personal calibration.

5.3 Two-Point Calibration Gaze Tracking Method

As described in Sect. 5.1.1, the basic idea of the proposed gaze tracking method is to reconstruct the user's eyeball position; the gaze direction is then accurately estimated in a 3-D coordinate system. The residual error is compensated by personal calibration. This procedure consists of four stages.

1. Eye positioning. To accept free head movement, we use an eye positioning unit and a gaze tracking unit. The former detects the user's eye position for controlling the direction of the gaze tracking unit, while the latter uses a near-infrared light sensitive camera to determine the user's eye gaze. In eye positioning, the user's rough eye position is estimated by the eye positioning unit.

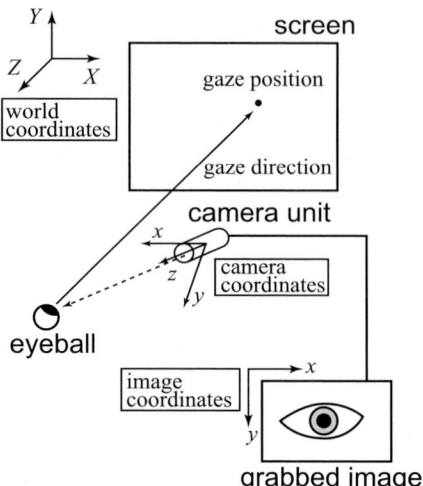

Fig. 5.3. Three coordinate systems of gaze detection: Image coordinates are the 2-D coordinates yielded by the captured image. World coordinates and camera coordinates are right-handed and 3-D

2. Gaze direction calculation. With the geometric eyeball model, the gaze direction is derived from the eye image taken by the gaze tracking unit.
3. Gaze compensation with personal calibration. To suppress gaze measurement error, gaze direction is compensated by the personal parameters determined in personal calibration. At the beginning of gaze tracking, personal calibration is required to calculate the personal parameters for each user.
4. Gaze position estimation. Finally, the gaze position on the computer screen is calculated from the compensated gaze direction and the eye position.

In this section, we detail each stage. First of all, we define the three coordinate systems illustrated in Fig. 5.3. World coordinates and camera coordinates are right-handed and 3-D. The origin of the world coordinates is defined at the corner of the gaze tracking system, and the origin of the camera coordinates is the center of the camera. The image coordinates lie in the 2-D coordinate system defined by the eye image captured by the camera.

5.3.1 Eye Positioning with Stereo Camera

The first step in gaze tracking is to determine the eye position by the eye positioning unit. We use a stereo vision technique for eye positioning. The eye positioning unit uses a stereo camera, which is calibrated by Tsai's camera calibration algorithm [631]. With Tsai's algorithm, the camera's intrinsic parameters including focal length, radial lens distortion and the principal point i.e. the intersection of the camera's Z axis and the camera's charge coupled device (CCD) image plane, and extrinsic parameters including translation and rotation parameters to permit transformation between the world and the camera coordinates are determined. With these parameters, an object in the image coordinates can be transformed into a point in world coordinates when the distance between the camera and the object is given.

To calculate the user's 3-D eye position, the eye positioning unit first detects the user's eye in each camera image by image processing. The distance between the camera and the eye is then derived from the stereo image. Next, the eye position **c** in world coordinates is calculated from the distance between the camera and the eye, and also the eye position in the bitmap image taken by the camera. We define u_z as the Z coordinate of the eyeball surface (actually, we use the Purkinje image position as the eyeball surface) in world coordinates.

For accurate gaze tracking, precise eye positioning is essential. To improve the accuracy of eye position in the current implementation described in Sect. 5.3.4, we recalculate the 3-D eye position from the rotation parameters of pan and tilt mirrors, the Z coordinate of the eyeball surface derived from the stereo camera, and the Purkinje image position captured by the gaze tracking unit.

5.3.2 Gaze Direction Calculation with the Geometric Eyeball Model

Next, gaze direction is derived from the eye image taken by the near-infrared camera. Figure 5.4 illustrates the geometric eyeball model used in calculating gaze direction. In the model, user's eye gaze is defined as the gaze direction vector **v** and the gaze position **c**. We define the gaze position **c** as the center of the corneal curvature. The gaze direction vector **v** is defined as the vector that passes through the center of the corneal curvature and the center of the pupil, which is described as

$$\mathbf{v} = \mathbf{s} - \mathbf{c} . \tag{5.1}$$

where **s** is the position of the center of the pupil.

To determine the user's gaze, we need to calculate the center of the pupil **s** and the center of the corneal curvature **c** from the 2-D eye image taken by the camera. The basic strategy of the proposed method is to estimate the position of **s** and **c** with high accuracy. Because the eyeball is small (its diameter is about 25 mm), even slight measurement error yields large error in predicted gaze position on the computer display. For example, the distance between **s** and **c** is about 4.5 mm, so 0.1 mm of pupil position error yields a view angle error of 1.2 degrees, which becomes 14.7 mm in a 700 mm distant display.

To calculate **v**, the eyeball model estimates the real center position of the pupil from the observed pupil image. Because there is refraction at the

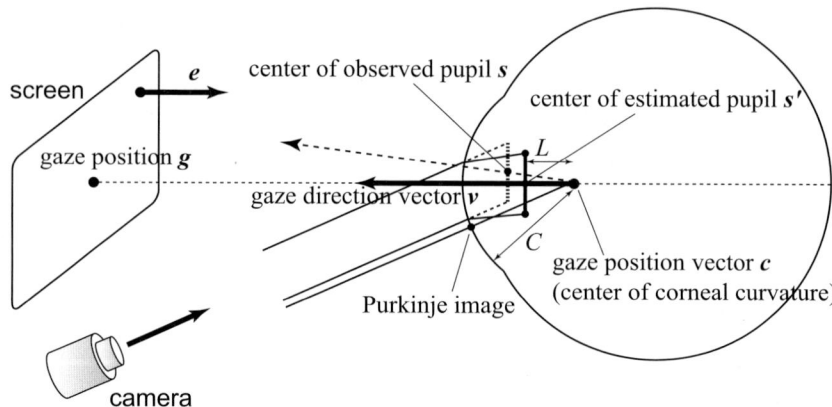

Fig. 5.4. Geometric eyeball model used in calculating gaze direction: The gaze direction is defined from the center of the pupil and the center of the corneal curvature

corneal surface, the observed pupil position differs from that of the real pupil. Therefore, using the observed pupil position for gaze direction calculation yields some measurement error. Actually, the measurement error changes when the eyeball rotates. This means that the relationship between gaze direction and eyeball rotation is non-linear if the gaze direction is derived from the observed pupil image. To compensate the measurement error, we need non-linear fitting of gaze position which requires more than two points for personal calibration. In the geometric eyeball model, the real position of the pupil is used to avoid this non-linear calibration problem.

First, an eye image taken by the camera is captured by a frame grabber, and the pupil and the Purkinje image are detected by image processing. Due to need for accurate modeling of the eyeball, subpixel image detection is used to identify the center of the pupil and the Purkinje image. To estimate the center of the pupil, we first detect the edge points of the pupil. They are searched for from the center of the pupil to the circumference along radial lines, and the highest gradient position of the brightness is defined as the edge. Next, the set of observed edge points is subjected to least square fitting to determine the center of the pupil. The estimated ellipse is described as

$$\begin{bmatrix} x^2 & xy & y^2 & x & y & 1 \end{bmatrix} \begin{bmatrix} 1 & a_1 & a_2 & a_3 & a_4 & a_5 \end{bmatrix}^T = 0 \,, \tag{5.2}$$

where $a_1, \ldots, a_5$ are ellipse parameters determined by the fitting.

The center of the ellipse s_0 in the image, which is equal to the center of the pupil, can be derived from (5.2) as

$$s_0 = \left(\frac{2a_3a_4 - a_2a_5}{a_2{}^2 - 4a_3}, \frac{2a_5 - a_2a_4}{a_2{}^2 - 4a_3} \right). \tag{5.3}$$

After detecting the center of the pupil, the Purkinje image, which is a bright point near the pupil, is also detected. The center of the Purkinje image $\mathbf{u}_0$ in the image is defined as the center of mass of the bright region.

Next, the gaze direction from the pupil and the Purkinje image is calculated by the geometric eyeball model. The eyeball model has two parameters that are user specific:

C the radius of corneal curvature, and   (5.4)

L the distance between the pupil and the center of corneal curvature.   (5.5)

In general, C and L depend on the user; however, they are defined as constant values in the model. The gaze detection error caused by personal differences is compensated by personal calibration. This is necessary because it is not feasible to estimate them.

In the geometric eyeball model, the center of the corneal curvature in the camera coordinates is derived from the Purkinje image in the captured image. Given the Z coordinate of the Purkinje image system as u_z, the position of the Purkinje image in the world coordinates $\mathbf{u}$ is derived from the center of the Purkinje image $\mathbf{u}_0$ in image coordinates with Tsai's algorithm (see [468] for detail).

Because the Purkinje image is a reflection on a spherical surface, the observed Purkinje image is a virtual image appearing smallest at the focal point of the surface, where the focal length is one half the radius of the corneal curvature. That is, the real Purkinje image $\mathbf{u}'$, which exists on the corneal surface, is derived from the virtual Purkinje image $\mathbf{u}$ as

$$\mathbf{u}' = \frac{u_z - C/2}{u_z}\mathbf{u} .$$ (5.6)

From now, we use $\mathbf{u}$ instead of $\mathbf{u}'$ to describe the position of the real Purkinje image. Using the estimated Purkinje image, the center of the corneal curvature $\mathbf{c}$ is derived as

$$\mathbf{c} = \mathbf{u} + C\,\frac{\mathbf{u}}{\|\mathbf{u}\|} ,$$ (5.7)

where $\|\mathbf{u}\|$ is the norm of $\mathbf{u}$.

Here, we describe a method to estimate the real pupil position from the observed pupil image. Because there is refraction on the corneal surface, observed pupil position is different from the real pupil position. To compensate this refraction, we use a set of edge points of the observed pupil. Consider that n edge points $(p_1, \ldots, p_n)$ are sampled from the circumference of the ellipse given by (5.2). For each point, p_i, we estimate its real position, p_i', by compensating the refraction on the corneal surface. We estimate an ellipse by fitting the set of real positions, and take it to represent the real pupil.

We estimate edge point $\mathbf{p}'$ of the real pupil from the observed edge point $\mathbf{p}$ by the following procedure. Let $\mathbf{p}_0$ be the projection of $\mathbf{p}$ to a plane that contacts the corneal surface at the Purkinje image $\mathbf{u}$; we assume paraperspective projection. From u_z (Z coordinate of Purkinje image in world coordinates) and a edge point p_i in image coordinates, $\mathbf{p}_0$ in world coordinates is calculated by Tsai's algorithm.

Let M be the distance between $\mathbf{u}$ and $\mathbf{p}_0$, which is derived as

$$M = \|\mathbf{p}_0 - \mathbf{u}\|. \tag{5.8}$$

With M and C, the position of the pupil edge point $\mathbf{p}$ on the corneal surface is given by

$$\mathbf{p} = \mathbf{p}_0 + \left(C - \sqrt{C^2 - M^2}\right) \frac{\mathbf{p}_0}{\|\mathbf{p}_0\|}. \tag{5.9}$$

As illustrated in Fig. 5.5, the corneal surface refracts the ray from the camera at $\mathbf{p}$. Therefore, when we estimate the refracted ray at $\mathbf{p}$, it is possible to compensate the refraction. Let $\mathbf{t}$ be the unit vector of the refracted ray at $\mathbf{p}$, $\mathbf{n}$ the unit normal vector of the curvature at $\mathbf{p}$, n_1 the refractive index of air, and n_2 that of aqueous humor.

Here, $\mathbf{n}$ is described as

$$\mathbf{n} = \frac{\mathbf{p} - \mathbf{c}}{\|\mathbf{p} - \mathbf{c}\|}. \tag{5.10}$$

Given the unit vector which passes through the center of the camera and $\mathbf{p}$ as $\mathbf{r}$, $\mathbf{t}$ is derived using Snail's law as

$$\mathbf{t} = \frac{n_1}{n_2} \left\{ \mathbf{r} - \left((\mathbf{r}, \mathbf{n}) + \sqrt{\left(\frac{n_2}{n_1}\right)^2 - 1 + (\mathbf{r}, \mathbf{n})^2} \right) \mathbf{n} \right\}, \tag{5.11}$$

where $(\mathbf{r}, \mathbf{n})$ is the inner product of $\mathbf{r}$ and $\mathbf{n}$.

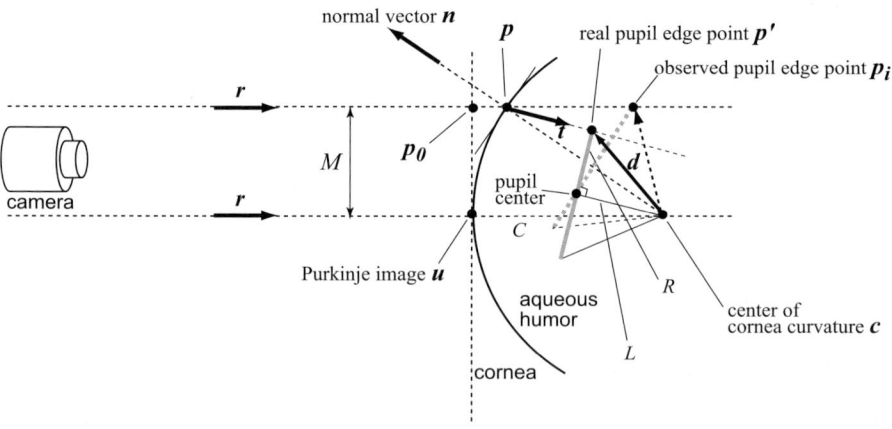

Fig. 5.5. Detection of the real pupil center from the virtual pupil image: Real pupil center in the 3-D coordinate system is calculated from a set of edge points of the observed pupil

The real edge point of the pupil exists on vector $\mathbf{t}$. Therefore, $\mathbf{t}$ can be determined given the distance between the center of corneal curvature and the estimated edge point, D. We define D as

$$D = \sqrt{L^2 + R^2} \, , \tag{5.12}$$

where L is the distance between the center of the corneal curvature and the center of the pupil, and R is the radius of the observed pupil.

Given the vector from the center of the corneal curvature to the estimated pupil edge as $\mathbf{d}$, $\mathbf{p}$, $\mathbf{t}$, $\mathbf{c}$ and $\mathbf{d}$ satisfy the following equation.

$$\mathbf{p} + m\,\mathbf{t} = \mathbf{c} + \mathbf{d}, \tag{5.13}$$

where m is the distance between $\mathbf{p}$ and estimated pupil edge. Here,

$$\|\mathbf{d}\| = D, \tag{5.14}$$

therefore, m is derived as

$$m = (\mathbf{c} - \mathbf{p}, \mathbf{t}) - \sqrt{(\mathbf{c} - \mathbf{p}, \mathbf{t})^2 - (\|\mathbf{c} - \mathbf{p}\|^2 - \|\mathbf{d}\|^2)} \, . \tag{5.15}$$

Finally, estimated pupil edge point $\mathbf{p}'$ is derived as

$$\mathbf{p}' = \mathbf{p} + m\mathbf{t} \tag{5.16}$$

$$= \mathbf{p} + \left\{ (\mathbf{c} - \mathbf{p}, \mathbf{t}) - \sqrt{(\mathbf{c} - \mathbf{p}, \mathbf{t})^2 - (\|\mathbf{c} - \mathbf{p}\|^2 - \|\mathbf{d}\|^2)} \right\} \mathbf{t} \, . \tag{5.17}$$

From the set of estimated pupil edge points, the center of the real pupil $\mathbf{s}'$ is estimated by ellipse fitting.

To calculate the gaze point on the computer screen, we project the gaze vector $\mathbf{v} = \mathbf{s}' - \mathbf{c}$ from the center of corneal curvature $\mathbf{c}$ to the plane of the screen. The plane of the computer screen is specified by normal vector $\mathbf{e}$ and scalar parameter d, as

$$(\mathbf{e}, \mathbf{x}) + d = 0 \, , \tag{5.18}$$

where $\mathbf{x}$ is any point on the plane. The gaze position $\mathbf{g}$ on the plane is derived as

$$\mathbf{g} = -\frac{(\mathbf{e}, \mathbf{c}) + d}{(\mathbf{e}, \mathbf{v})} \mathbf{v} + \mathbf{c} \, . \tag{5.19}$$

5.3.3 Gaze Compensation with Personal Calibration

To suppress the gaze measurement error, only gaze direction $\mathbf{v}$ is compensated with the personal parameters determined in two-point personal calibration. This is because the major component of measurement error lies in gaze direction. If the gaze position contains some error, it is directly reproduced on

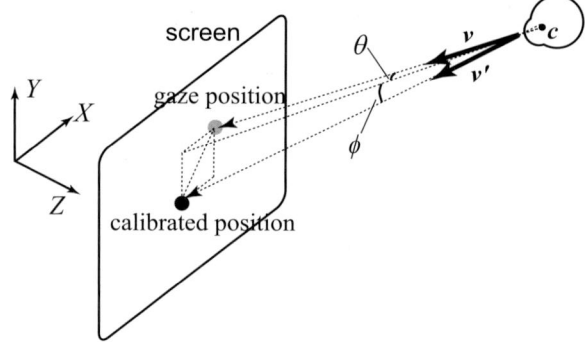

Fig. 5.6. Personal calibration of the user's gaze direction vector. Pan and tilt rotation of gaze direction are compensated by the calibration

the screen. On the other hand, if the gaze direction vector has some error, it is amplified on the screen. When the distance between the pupil and the screen is 600 mm, it is about 130 times the length from the center of the corneal curvature to the center of the pupil. Therefore, a gaze direction error of 0.1 mm becomes a 13 mm error on the screen.

To compensate the gaze direction vector (Fig. 5.6), we need to calibrate its angle and scale. This requires at least two calibration points. In the calibration procedure, the gaze direction vector is described in the polar coordinate system as

$$\mathbf{v} = \begin{bmatrix} x \\ y \\ z \end{bmatrix} \longrightarrow \mathbf{v}_\theta = \begin{bmatrix} l \\ \theta \\ \phi \end{bmatrix} = \begin{bmatrix} \sqrt{x^2 + y^2 + z^2} \\ \tan^{-1}(x/z) \\ \sin^{-1}\left(y/\sqrt{x^2 + y^2 + z^2}\right) \end{bmatrix}. \tag{5.20}$$

Next, calibration matrix W is defined as a 3×4 matrix in the homogeneous space,

$$W = \begin{bmatrix} 1 & 0 & 0 & 0 \\ 0 & w_1 & 0 & w_2 \\ 0 & 0 & w_3 & w_4 \end{bmatrix}. \tag{5.21}$$

In personal calibration, calibration matrix W is determined from the user's gaze data. The calibration matrix W has four unknown parameters $w_1, w_2, \ldots, w_4$, so at least two calibration points on the screen are required to determine those parameters. To determine W, the user looks at two calibration markers on the screen. The personal calibration procedure is described later in this section.

Once W is calculated, the calibrated gaze direction vector $\mathbf{v}'_\theta$ is given by

$$\mathbf{v}'_\theta = W \cdot \mathbf{v}_\theta \tag{5.22}$$

$$= \begin{bmatrix} l \\ w_1\theta + w_2 \\ w_3\phi + w_4 \end{bmatrix}. \tag{5.23}$$

Calibrated gaze vector $\mathbf{v}'$ is derived, in rectangular coordinate system, from $\mathbf{v}'_\theta$ as

$$\mathbf{v}'_\theta \longrightarrow \mathbf{v}' . \tag{5.24}$$

Finally, calibrated gaze position on the screen can be computed by using $\mathbf{v}'$ instead of $\mathbf{v}$ in (5.19). However, if we use $\mathbf{v}'$, the effect of compensation becomes large when θ and ϕ are large. This means that in the surrounding area of the screen, gaze position is over compensated. To avoid this problem, we use the following method to derive the gaze position on the screen. First, the compensation angle $\Delta\mathbf{v}_\theta = [0 \ \Delta\theta \ \Delta\phi]^T$ is calculated as

$$\Delta\mathbf{v}_\theta = W \cdot \mathbf{v}_\theta - \mathbf{v}_\theta \tag{5.25}$$

$$= \begin{bmatrix} 0 \\ w_1\theta + w_2 - \theta \\ w_3\phi + w_4 - \phi \end{bmatrix} . \tag{5.26}$$

Next, compensated gaze position $\mathbf{p}'$ on the screen is derived as

$$\mathbf{p}' = \mathbf{p} + \begin{bmatrix} m \cdot \sin\Delta\theta \cdot \cos\Delta\phi \\ m \cdot \sin\Delta\phi \end{bmatrix} , \tag{5.27}$$

where m is the distance between the gaze position and the non-calibrated gaze point on the screen, which is described as

$$m = \|\mathbf{p} - \mathbf{c}\| . \tag{5.28}$$

The calibration procedure is briefly described below. This method requires at least two calibration points to derive calibration matrix W. If there are additional calibration points, the stability of calibration is improved because the variance of measurement error is decreased. As illustrated in Fig. 5.7, the two calibration markers appear in different corners of the screen. While gazing at the marker, the user's eye position and gaze direction are calculated. We use the average of the eye position and the gaze direction for a short period (i.e. 800 ms in the current implementation), and if the variance of the eye position or the gaze direction is too large, it is judged that there is some measurement error, and the system retries calibration.

Let us consider that the user is looking at calibration marker p_1 on the screen. Calibration marker position $\mathbf{p}_1$ in world coordinates is given. From the measured eye position $\mathbf{c}_1$ and $\mathbf{p}_1$, the actual gaze direction $\mathbf{v}'_1$ is estimated as

$$\mathbf{v}'_1 = \mathbf{p}_1 - \mathbf{c}_1 , \tag{5.29}$$

which is different from the observed gaze direction $\mathbf{v}_1$. There is another calibration marker p_2 on the screen, and its position vector $\mathbf{p}_2$ is given. Therefore, the actual eye gaze direction $\mathbf{v}'_2$ is estimated while the user is looking at p_2 as

$$\mathbf{v}'_2 = \mathbf{p}_2 - \mathbf{c}_2 , \tag{5.30}$$

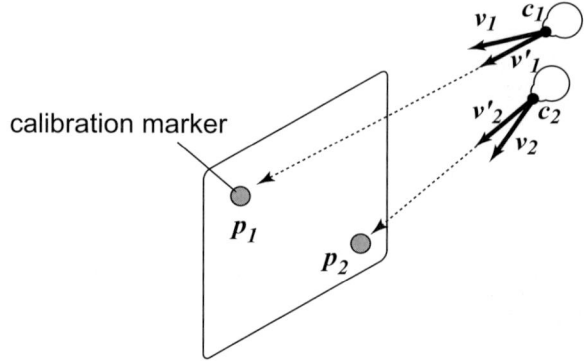

Fig. 5.7. Personal parameter calculation for two-point personal calibration: Two markers are presented on different corners of the screen successively, and the eye position and the gaze direction are measured while the user looks at each calibration marker in turn

from the measured eye position c_2 and p_2.

From the set of v_1, v_1', v_2 and v_2', the calibration matrix W is calculated in the polar coordinate system.

5.3.4 Implementation

We implemented a prototype system based on the TPC gaze tracking method. Its external appearance is shown in Fig. 5.8. It consists of the eye positioning unit and the gaze tracking unit. The eye positioning unit, which detects the user's eye position in 3-D coordinate system, has two NTSC cameras with a focal length of 6 mm. It is a stereo-camera, and eye position in the 3-D coordinate system is estimated by the unit. The gaze tracking unit, which detects the user's gaze direction, has a near-infrared-sensitive NTSC camera with 75 mm focal length lens. Near the gaze tracking unit, there are pan and tilt mirrors controlled by two ultrasonic motors to catch the user's one eye (right or left, which is selected initially) by the gaze tracking camera. There is another ultrasonic motor which controls the focus of the gaze tracking camera. Underneath the gaze tracking unit, there is an infrared LED array, whose wavelength is 850 nm. Its luminosity can be manually changed by the user.

A Windows XP personal computer with Intel Pentium 4 3.2 GHz CPU and 1 GB memory, controls the prototype system. It also controls the three ultrasonic motors through RS-232C serial interfaces. The two face images taken by the eye positioning unit, and the one eye image taken by the gaze tracking unit are captured by a frame grabber (Euresys PICOLO Tetra) at a resolution of 640 × 480 pixels in 8 bit gray scale. The sampling rate is 30 frames per second.

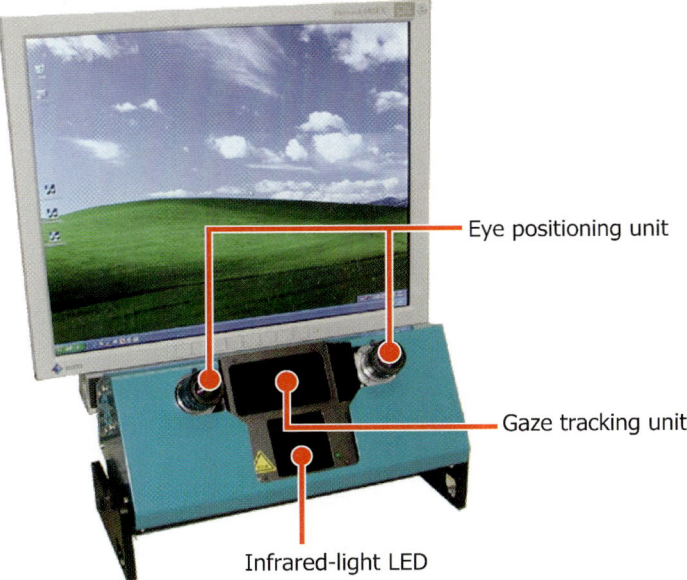

Fig. 5.8. External appearance of the prototype TPC gaze tracking system: It consists of the eye positioning unit, infrared-light LED array, and the gaze tracking unit

Because the prototype gaze tracking system is designed for PC operation, the acceptable distance from the user to the display is between 60 cm and 80 cm. When the distance is 60 cm, the acceptable eye area is about 15 cm × 15 cm. When the distance is 80 cm, the acceptable eye area is about 30 cm × 30 cm. We believe that this limitation reasonable for everyday computer operation. The measurable area can be expanded by changing the focal lengths and the convergence angle of the stereo camera.

A 18.1 inch LCD monitor EIZO L675 with resolution of 1280 × 1024 is used with the gaze tracking system. Its screen is 303 mm high and 376 mm wide. Screen parameters including screen size, resolution, and relative position between the screen and the gaze tracking camera are given initially.

Figure 5.9 shows an example of non-calibrated and calibrated gaze data. The user first calibrated the system with the TPC method, and then looked at the twelve markers arranged in a grid on the computer screen. The non-calibrated gaze positions also form a regular grid, so adding linear transformation to the TPC method would compensate the gaze detection error.

In the user study, we confirmed that the accuracy of the prototype system is about 0.8 deg in view angle, which is equal to 32 pixels on the screen. Six volunteers participated the experiment. Two subjects had normal vision, and the rest of them wore eyeglasses. In the user study, they calibrated the system with the two calibration markers, and then twelve confirmation markers appeared at different positions on the screen one after the other. For

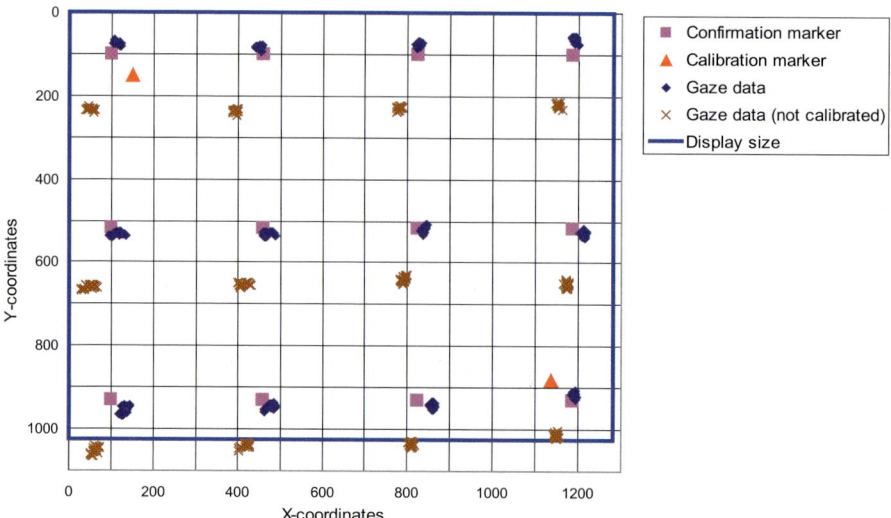

Fig. 5.9. An example of calibrated and non-calibrated gaze data; the user looked at the twelve confirmation markers on the screen: The gaze data was calibrated by two calibration markers prior to the trial. Display size was 376 mm (width) × 303 mm (height), its resolution was 1280 × 1024 pixels. The distance between the user and the screen was 760 mm. The average accuracy was 0.59 deg (view angle)

each marker, the subject was asked to press the space bar while looking at the marker. For the subjects examined, the accuracy ranged from 0.6 deg to 1.0 deg in view angle. There was no significant difference with the use of eyeglasses.

5.4 One-Point Calibration Gaze Tracking Method

In extensive testing we found that even two-point calibration was perceived by users to be troublesome. Some felt that it was hard to gaze at the exact positions on the screen. Others could not use the gaze tracking system because of eye detection error during calibration. For users wearing eyeglasses, reflections on the surface of the eyeglasses sometimes prevented precise eye detection while looking at the lower corner of the screen. To overcome these problems, we developed the One-Point Calibration (OPC) gaze tracking method, which can roughly estimate eye gaze direction. It requires the user to look at just one calibration point at the top of the screen at the beginning of the gaze tracking session.

The most important difference between the OPC method and the TPC method is that pupil position is modified by the personal parameter s, which is derived from personal calibration. The basic idea of the OPC method is to estimate the eyeball shape using the reference points created by the two infrared lights near the camera, and to modify the eyeball model to compensate the personal difference. Fig. 5.10 illustrates the two LED arrays used in the OPC method. Here, we define D as the distance between the centers of the two LED arrays.

Because it can be assumed that the cornea is a convex mirror, the two Purkinje images captured are virtual images that appear on the mirror. In general, when the size of the object is p, the focal length of the convex mirror is f and the distance between the object and the convex mirror is c, the size

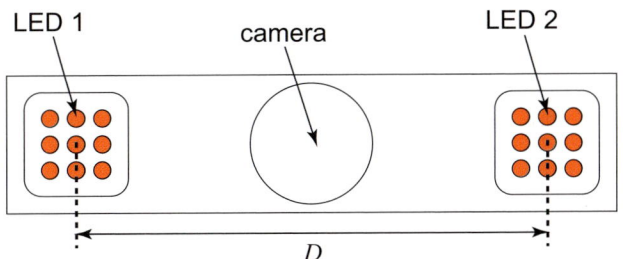

Fig. 5.10. Two LED arrays used in the OPC gaze tracking method: The two LED arrays are arranged horizontally near the camera

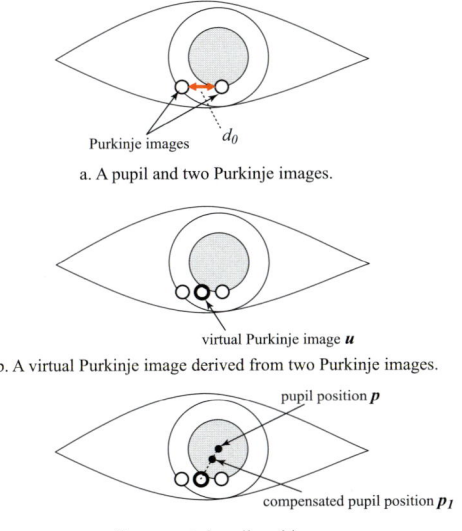

a. A pupil and two Purkinje images.

b. A virtual Purkinje image derived from two Purkinje images.

c. Compensated pupil position.

Fig. 5.11a–c. Two Purkinje images appear on the corneal surface: Virtual Purkinje image is derived from two Purkinje images, and pupil position is compensated

of the virtual image on the convex mirror is defined as

$$\frac{f}{f-c} \cdot p ,\qquad(5.31)$$

where the focal length f of the convex mirror is defined as $f = -r/2$ when the radius of the convex mirror is r.

The distance between the two Purkinje images observed by the camera is defined as d. Given that the radius of the corneal curvature is C in the eyeball model, the focal length of the corneal curvature is described as $-C/2$. Therefore, given the distance between the camera and the Purkinje image is e, the distance between two Purkinje images in model d_0, which is described in Fig. 5.11a, is derived as

$$d_0 = \frac{f}{f-e} \cdot D = \frac{C\,D}{2e+C} .\qquad(5.32)$$

If the user's radius of corneal curvature equals C, d is equal to d_0. However, because of personal differences in eyeball shape and the refraction created by eyeglasses if worn, d generally does not equal d_0. This result implies that the gaze direction is not correct if we use the observed distance between the pupil and the Purkinje image in the geometric eyeball model which assumes that the radius of the corneal curvature is C.

To correct for these differences, we introduce a coefficient of correction s which is defined as

$$s = \frac{d_0}{d} .\qquad(5.33)$$

In the second step, the gaze direction is calculated using s. To calculate the gaze direction, only one Purkinje image is required. Therefore, we assume a virtual Purkinje image that exists at the center of the two Purkinje images (Fig. 5.11b). On the captured eye image, the position of the virtual Purkinje image is defined as $\mathbf{u}$ and the center of the pupil is defined as $\mathbf{p}$. Here, the compensated pupil position $\mathbf{p}_1$, which is illustrated in Fig. 5.11c, is defined as

$$\mathbf{p}_1 = \mathbf{u} + s(\mathbf{p} - \mathbf{u}) .\qquad(5.34)$$

Using $\mathbf{p}_1$ instead of $\mathbf{p}$ for calculating gaze direction compensates the effect of eyeball difference (and the effect of eyeglasses if worn).

The remaining major factor behind gaze detection error is the difference between the optical axis and the visual axis of the eyeball. If the user wears eyeglasses, it also affects the gaze direction. This is compensated from the gaze direction observed while the user looks at the calibration marker. Fig. 5.12 illustrates how the calibration marker appears at the top of the display. We use the same calibration procedure described in Sect. 5.3.3.

In Fig. 5.12, the position of the calibration marker is defined as $\mathbf{m}$ and the position of the camera is defined as $\mathbf{a}$. When the user looks at the calibration

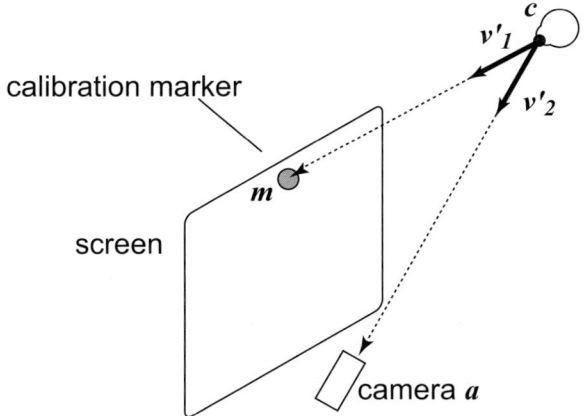

Fig. 5.12. Personal calibration of the user's gaze direction: While the user looks at the calibration marker, two gaze direction vectors are calculated to derive the personal parameters

marker, the calculated gaze direction vector is defined as $\mathbf{v}_1$, and the correct gaze direction vector $\mathbf{v}_1'$ is defined as

$$\mathbf{v}_1' = \mathbf{m} - \mathbf{c} \,. \tag{5.35}$$

To use the calibration procedure, we need another gaze direction vector. While the user looks at the calibration marker, it is possible to calculate another gaze direction vector $\mathbf{v}_2$, which assumes that the user is looking at the center of the camera. When the user looks at the camera, the Purkinje image exists at the center of the pupil if the optical axis is equal to the visual axis. Therefore, gaze direction vector $\mathbf{v}_2$ is estimated from the pupil position on the captured eye image, which equals the Purkinje image position. On the other hand, the correct gaze direction vector $\mathbf{v}_2'$ is defined as

$$\mathbf{v}_2' = \mathbf{a} - \mathbf{c} \,. \tag{5.36}$$

From two gaze direction vectors $\mathbf{v}_1$ and $\mathbf{v}_2$, the calibration matrix W described in (5.21) is calculated.

5.4.1 Implementation

We implemented a prototype system that detects the user's eye gaze with the OPC method. Figure 5.13 overviews the prototype system. We used an NTSC video camera (SONY EVI-D70) that was sensitive to near-infrared light to detect the user's eye. An IR-pass filter was attached to the camera. The eye images taken by the camera were captured by a frame grabber (Matrox Meteor-II) as 8 bit gray-scale images, with resolution of 640×480. Two near-infrared (750 nm) LED arrays were attached to the camera; the distance between the arrays was 230 mm.

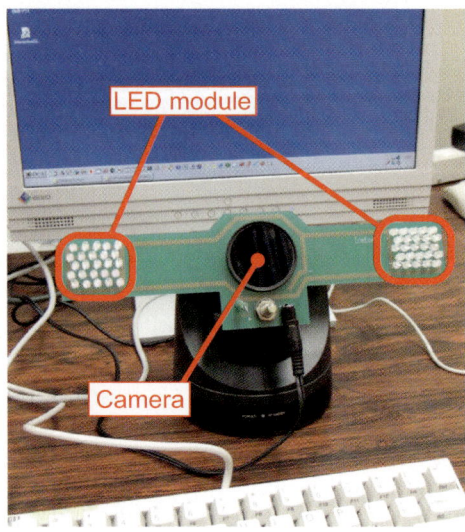

Fig. 5.13. External appearance of the OPC gaze tracking system: Two LED arrays are arranged near the near-infrared light sensitive camera

A Windows XP personal computer with Pentium 4 2 GHz processor was used to control the gaze tracking system. It was connected to the camera via a RS-232C cable to monitor the focus parameter of the camera. A 18.9 inch LCD display (EIZO L767) whose resolution was set to 1280×1024 (376 mm wide and 303 mm high) was placed on the desk, and the camera with LED module was located in front of the display. During gaze detection, the user was required to move his head within a 4 cm square (horizontal and vertical directions) because the camera direction was fixed.

The accepted depth between the user and the display ranged from 60 cm to 90 cm. When the depth changed, the auto-focus function of the camera refocused the lens on the eye so that the gaze could be detected. The camera had a pan/tilt control function, but it was not used in the current implementation.

The accuracy depends on the user. We conducted a preliminary user study with five subjects to confirm the accuracy of the prototype system. Two participants had normal vision, and three wore eyeglasses. For each participant, personal calibration with the single calibration marker was performed. Next, twelve confirmation markers were presented on the screen, and the user was asked to press the space key while looking at the marker. The average accuracy was 2.0 deg (view angle). However, for one user, we found that the accuracy was not sufficient. We confirmed that for this subject the TPC method yielded sufficient gaze tracking accuracy. We are trying to identify the reason for this personal difference so as to improve the accuracy of gaze tracking.

One of the remarkable features of the prototype system is its hardware simplicity. We used a commercial auto-focus camera to catch the user's eye image. The LED module is not a commercial product, but it can be assembled easily. People believe that gaze tracking requires expensive hardware. However, the prototype system works well even if no special hardware is used. If more accurate gaze tracking is needed, the TPC gaze tracking method should be used. Thanks to the auto-focus camera, the user does not have to consider the distance to the screen. The weak point of the system is that eye position is limited to within a 4 cm square. If free head movement of the user is essential, the system described in Sect. 5.3.4 is suggested. We have not yet implemented the OPC method in a head free system. However, integrating the OPC method with the head free system will make one-point-calibration, head-free gaze tracking possible.

5.5 Conclusion

We described two gaze tracking methods that provide extremely simple personal calibration. The two-point calibration gaze tracking method is an accurate gaze tracking method that requires only two points for personal calibration while it accepts free head movement. The one-point calibration gaze tracking method estimates the rough gaze position with just one calibration point. We need only five seconds for initial setup. With this method, people have no time to feel that the calibration procedure is bothersome because they have to look at just one point. These methods will change our approach to the Human–Computer Interaction; gaze tracking is not peculiar in terms of human behavior. In the current implementation, the OPC method fails to well support for HCI becase of its low accuracy. But further development of this method may prove successful. On the other hand, applications that do not require accurate gaze tracking have also been proposed (e.g. attentive user interfaces [644]). They are strong candidates for using the OPC method. It is a very natural and easy-to-use way for not only understanding human behavior but also interacting with computers.

6

Free Head Motion Eye Gaze Tracking Techniques

Carlos H. Morimoto and Flávio L. Coutinho

6.1 Position of the Problem

In this chapter we look at a passive eye monitoring system, or simply eye tracker (ET), as a mono-camera system that measures the position and orientation of the subject's eye. The most common image based ET technique is known as the pupil-corneal reflection (PCR) technique, because it uses a reflection on the surface of the cornea (the first Purkinje image) generated by an external light source that is used as a "reference" point. The two dimensional image vector defined by the center of the pupil (or the iris) and the corneal reflection is used to estimate the gaze direction, after a simple calibration procedure that defines the mapping from image coordinates to screen coordinates[1]. To facilitate pupil detection and tracking in real-time, an active differential lighting scheme is commonly used [421]. By placing a light source near the camera optical center, a bright pupil image is generated, while a second light source, distant from the camera center, generates a dark pupil image. The two light sources can be synchronized with the video frame rate, so that at every other frame the camera captures a bright or a dark pupil image, and the pupil can be easily segmented as a high contrast region from the subtraction of the bright and dark pupil images.

Although it is claimed that the corneal reflection method tolerates small head motion, the accuracy of the ET decays considerably as the head moves away from the calibrated position [423], and that is why most data is acquired using byte bars or chin rests to immobilize the head. The second big problem is the constant need for recalibration every time the geometry of the camera, monitor, and user changes.

In this chapter we describe several techniques developed to overcome these problems. The next section describes several multiple camera techniques, followed by Sect. 6.3 with the description of two single camera methods that

---

[1] We assume that the objects being visualized by the user are on a computer screen.

we believe will be more appropriate in practice, once high resolution cameras become more easily available to overcome their single current problem, i.e., their narrow field of view. In Sect. 6.4 we conduct several experiments that demonstrate the robustness of the cross-ratio single camera ET technique, and in Sect. 6.5 we introduce a few improvements that allows the technique to work with a more accurate eye model. In Sect. 6.6 we discuss implementation issues to develop a real-time (30 frames per second) prototype. Section 6.7 presents experimental results of the prototype using real users, and Sect. 6.8 concludes this chapter.

6.2 Multiple Cameras Techniques

The most common approach to allow free head motion in image based ETs is to compute the position and orientation of the eye in 3D using multiple cameras. Some approaches combine wide and narrow field of view (FoV) cameras. The wide FoV camera is used to detect the face and position the narrow FoV camera to track the eye with high resolution. Wang and Sung [656] and Newman *et al.* [439] give examples of systems that first compute the face pose in 3D, and then compute the eye gaze. Newman *et al.* [439] locate the 3D position of the corners of the eye from stereo, and computes the 3D gaze direction from the orientation of the eyeball. Some of the eye parameters have to be trained, per person. Although the system runs in real time, its accuracy is low, about 5°(degree). Wang and Sung [656] also combine a face pose estimation system with a narrow FoV camera to compute the gaze direction. They assume that the iris contour is a 2D circle and estimate its normal direction in 3D. To compute the point of regard using real images, a second image of the eye from a different position is used. Their tests using synthetic images and real images from 3 subjects show that the accuracy of the system is bellow 1°.

Eye trackers that use stereo cameras are able to track the position and orientation of the eye in 3D, and have proved to be very accurate. Beymer and Flickner [45] use a separate wide FoV stereo system to detect the face in 3D. Once a face is detected, this information is used to steer a second narrow FoV stereo pair of cameras. The large pupil images are then used to iteratively fit a 3D eye model to compute its orientation using features extracted from the image and back-projecting the model onto the image.

This model fitting allows the computation of the orientation of the optical axis of the eye, or the line of gaze (*LoG*), which is the line that contains the center of the pupil and the center of the eye (or cornea). But to compute the direction that a person is actually looking at, one has to consider the line of sight (*LoS*), which is the line that contains the center of the pupil and the fovea, the region in the retina that concentrates all of the color sensitive photoreceptors (the cones), and is responsible for our most accurate vision. The fovea is a very small region, about 1° in diameter, and is about 5° off the *LoG*. To compensate for this difference, Beymer and Flickner use a one-time calibration step per user to estimate some intrinsic parameters of the eye, such as the radius of the cornea and the angular offset of the *LoS*. The authors report a 0.6° accuracy for one person at a distance of 622 mm from the monitor.

Shih and Liu [564] do not have a system to position the narrow field of view stereo cameras, but similar to Beymer and Flickner, their method is based on a simplified eye model. They use multiple cameras and multiple point light sources to estimate the optical axis of the eye. Using the simplified Le Grand eye model, they show that using 2 calibrated cameras and at least 2 point light sources with known positions, it is possible to compute the *LoG*. The offset of the *LoS* to the *LoG* is also obtained from a one-time calibration procedure per user, and usually takes 2–3 seconds. In their implementation they use 3 IR LEDs, and process 30 frames a second with an accuracy of under 1°.

6.3 Single Camera Techniques

In this section, we describe two simpler techniques that require a single camera to compute the point of regard on a computer screen. The first method is based on the theory of convex mirrors, and the second uses invariance properties of projective geometry. Although these simpler methods use narrow FoV cameras, once high resolution cameras become more common and affordable, we believe that single fixed camera solutions will become the standard remote eye tracker for most applications since they are simpler to implement and setup than multiple camera systems.

6.3.1 Convex Mirror Technique

This method was developed by Morimoto *et al.* [419]. From the theory of convex spherical mirrors it is known that a point light source L creates a virtual upright image I behind the convex mirror with center C, vertex V and radius r, as can be seen in Fig. 6.1. The focal point F of the mirror is located midway on the segment VC ($|VF| = r/2$).

For an observer (camera) positioned at O, the position of the image I is seen by the observer as reflected from the point G on the surface of the mirror. The position of I can be geometrically determined from the intersection of any two of the following light rays leaving the object (shown with arrows in Fig. 6.1):

1. The light ray from L to C is reflected straight back to L.
2. The light ray from L towards F is reflected parallel to the principal axis.
3. Light rays parallel to the principal axis are reflected through F.

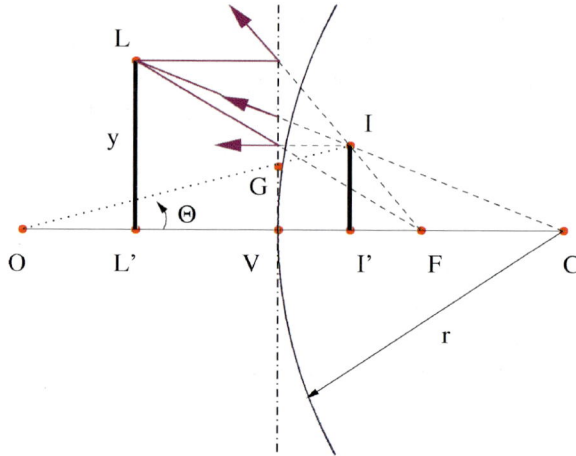

Fig. 6.1. Convex mirror image formation process: The position of the image can be defined by the intersection of light rays

Let y be the height of L, defined as the distance $|LL'|$, and y' the height of its image I (distance $|II'|$). Let d be the distance $|L'V|$ and d' the distance $|I'V|$. The convex mirror equation for paraxial rays is

$$\frac{1}{f} = \frac{1}{d} + \frac{1}{d'} \tag{6.1}$$

where the focal distance $f = |VF| = r/2$.

It is important to notice that even if the mirror rotates around an arbitrary axis through C, the image still remains constant.

6.3.1.1 Eye Gaze Estimation

Assuming the surface of the cornea as a spherical mirror, it is possible to compute the position of its center by first recovering its principal axis, defined as the line that connects the centers of the camera and the cornea. In order to do these computations, consider that the camera is fully calibrated, and that the positions of multiple point light sources L_i are known.

Figure 6.1 shows that the camera sees the reflection of an arbitrary light source L_i (or simply L in the figure) at G_i. The vector $\overrightarrow{OL_i}$ is known (from system calibration) and the direction of the vector $\overrightarrow{OG_i}$ can be measured using the camera image, but not its magnitude (though the focus of the camera might give an estimate of this value). These two vectors define a plane Π_i that contains the center C of the mirror.

Shih *et al.* [564] have demonstrated that multiple light sources can be used to recover the principal axis of the mirror. Consider a second light source L_j non-coplanar with Π_i. L_j defines a plane Π_j which intersects with Π_i along the principal axis of the mirror. If other non-coplanar lights L_k are added, they will define other planes Π_k, all of which will intersect along the principal axis. Therefore, using a single camera and any arbitrary number of lights, it is not possible to recover the position of the cornea without further knowledge.

One special configuration is created when a light source and O are at the same position, i.e., when a light source L_O is placed **on** the optical center of the camera. In this configuration, $G_O \equiv V$, and therefore the principal axis can be recovered using a single light source.

Hence, using a single camera and one on-axis light source, or at least 2 off-axis light sources non-collinear with O, it is possible to recover the principal axis of a cornea, but not the exact 3D position of its center.

To recover the center of the cornea, at least two cameras and two light sources are required. The method presented by Shih *et al.* [564] considers that each camera contributes with its own principal axis, and the center of the cornea is computed from the intersection of these axis, on a common coordinate frame (i.e., the cameras must be fully calibrated).

If the radius of the cornea is known (according to [423], it is about 7.4 mm), then the center of the cornea can be computed using a single camera and two light sources, one on-axis (O) and a second off-axis (L). From the mirror Eq. 6.1, it is easy to derive that when the object is placed far from the mirror $(d \gg f)$ then the image I is formed at F, i.e., $d' \sim f$.

Let the known coordinates of O and L be $O = (0,0)$, $L = (L_x, L_y)$. The direction of the vectors $\overrightarrow{GO}$ and $\overrightarrow{VO}$ can be estimated from the camera image. To estimate $C = (C_x, 0)$ note that $C \in \overline{VO}$, and $I = \overline{CL} \cap \overline{GO}$. The equation of the line $\overline{CL}$ can be defined as

$$y - L_y = (x - L_x)\frac{-L_y}{C_x - L_x} \ . \tag{6.2}$$

Since I is formed at $r/2$ between V and C, the coordinate of $I = (C_x - r/2, \tan\theta(C_x - r/2))$, and because $I \in \overline{CL}$

$$\left[\tan\theta\left(C_x - \frac{r}{2}\right) - L_y\right](C_x - L_x) = -L_y\left(C_x - \frac{r}{2} - L_x\right) \ . \tag{6.3}$$

The coordinate C_x can then be computed by solving the following second order polynomial:

$$C_x^2 + a\ C_x + b = 0 \ ; \tag{6.4}$$

where

$$a = -\left(L_x + \frac{r}{2}\right) \quad \text{and} \quad b = \frac{r}{2}\left(L_x - \frac{L_y}{\tan\theta}\right) \ .$$

The minimum positive solution of C_x is used later to compute the position of the actual reflection at G.

The direction of gaze is defined as the vector that connects the center of the cornea to the pupil center. Because the pupil lies behind the cornea surface its image is refracted, mainly at the outer cornea boundary with air. The position of the pupil can therefore be approximated by considering the cornea as a concave spherical surface and, for an average eye, the pupil lies 3.5 mm inside the cornea. In practice, Morimoto *et al.* [419] suggest a calibration procedure to estimate the *LoG*, because this measure is user dependent. The authors presented results from simulations using ray traced images of the Gullstrand eye model. Their accuracy was about 3° of visual angle, for a user distant 60 cm from the computer screen.

6.3.2 Cross Ratio

The second method that is described in this section was first introduced by Yoo *et al.* in [703]. Their method uses 4 near IR LEDs around the monitor screen to project these corners on the surface of the cornea. A 5th LED is placed near the CCD camera lens to create a bright pupil image, and help segmenting the pupil. They assume the cornea is flat, so it is easy to use the cross ratio invariance property of perspective transformations to compute the gazed point on the screen with an accuracy of about 2°. The greatest advantage of this method is that it does not require camera calibration.

More recently, Yoo and Chung [702] have refined the geometric model to compensate the displacement of the corneal reflections due to the curvature of the cornea surface, where the previous model simply assumed it was planar. Their new system actually makes use of two cameras instead of only one. Both cameras are mounted on a pan-tilt unit. One of the cameras has a wide field of view and is responsible for locating the user's face and keep the other camera (a narrow field one) always pointing to one of the user's eyes.

To better understand the results and some of the implementation issues described in Sect. 6.7, we give a brief description of Yoo and Chung's method next (only the narrow field-of-view camera will be considered).

As we have seen in Fig. 6.1, the corneal reflection G is the image of a light source L on the surface of the cornea. This model assumes paraxial rays, where the observer and light sources are placed very far from the mirror. Under these conditions, a reasonable approximation is to consider that all the corneal reflections are coplanar, as shown in Fig. 6.2. In this figure, 4 light sources (near IR LEDs) are placed at the corners of the computer screen, and a 5th LED is placed near the optical center of the camera.

Let L_i be one of the four LEDs attached to the monitor ($1 \leq i \leq 4$), E the center of the cornea and O the center of projection of the camera. G is the reflection of the LED fixed at the camera's optical axis and V_i are the virtual projections of the corneal reflections of the light sources. V_i are coplanar to G. The real corneal reflections, lying on the cornea surface, are not displayed in Fig. 6.2, but we will represent them as R_i.

When an image of the eye is captured, G is projected to g, and the virtual points V_i **would** be projected onto v_i. The real corneal reflections R_i

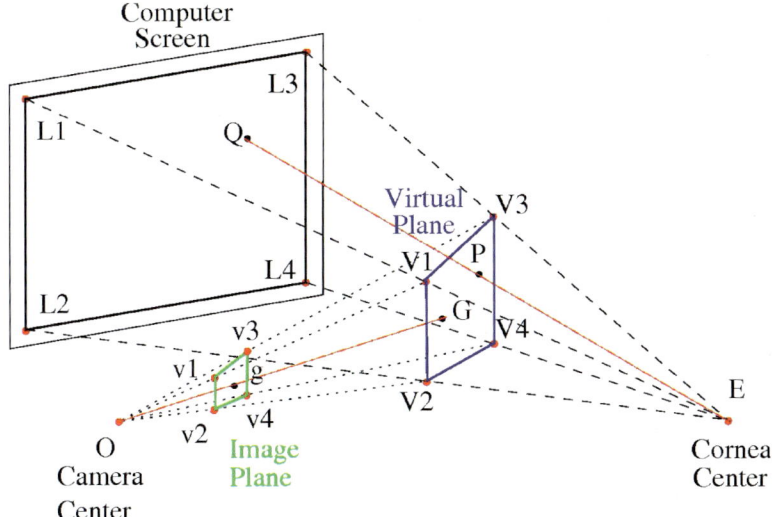

Fig. 6.2. Geometry of the system: Four LEDs are placed on the corners of the computer screen. The reflections of such LEDs on the cornea are assumed to be coplanar, and this virtual plane, as seen by the camera, is used to compute the line of gaze, that contains the points E, P, and Q

are projected to r_i (also not shown in the figure). Assuming a typical user position and motion in front of the computer screen, Yoo and Chung [702] have demonstrated that the image position of the virtual reflections v_i can be approximated from the image positions of the real projected reflections r_i according to the following equation:

$$v_i = g + \alpha(r_i - g) \tag{6.5}$$

where α is a constant close to 2, and can be accurately determined through a calibration process. A detail derivation of Eq. 6.5 is given in [702].

6.3.2.1 Estimating the Gaze Direction

Given a captured image of the eye, showing the pupil (p) and the five corneal reflections (r_1 to r_4 and g) it is possible to calculate the image of the virtual points v_1, v_2, v_3, v_4 using (6.5). The projection of the virtual points v_i are the result of two projective transformations of the point light sources L_1, L_2, L_3 and L_4. The pupil center (P) located on the cornea surface can also be considered as the projection of the point of regard (Q) on the virtual plane and the point p is the result of the application of two projective transformations over Q. This way, as mentioned earlier, it is possible to use the cross-ratio to estimate Q.

The cross-ratio of 4 points (P_1, P_2, P_3, P_4) is defined as follows:

$$\text{cr}(P1, P2, P3, P4) = \frac{|P_1 P_2||P_3 P_4|}{|P_1 P_3||P_2 P_4|} \tag{6.6}$$

where $|P_i P_j|$ is the distance between points P_i and P_j. The important property of the cross-ratio is that if the line containing P_1, P_2, P_3 and P_4 are subject to any projective transformation, the ratio remains constant [237].

Figure 6.3 shows the points used to estimate the gaze by application of the cross-ratio. After calculating the virtual projections v_1, v_2, v_3 and v_4 it is possible to calculate the point m by the intersection of the diagonals of the quad $v_1 v_2 v_3 v_4$, which corresponds to the projection of the central point on the screen M, and the ideal points, i_1 and i_2, from the two pairs of lines that are defined from the virtual points. The ideal points are then used to compute the coordinates (p_x, p_y), and (m_x, m_y), from the lines defined by p, m and the ideal points. Finally, from the groups of 4 collinear points, such as $(v_1, p_x, m_x, \text{and } v_2)$ and $(v_1, p_y, m_y, \text{and } v_4)$, it is possible to compute the point of regard Q on the computer screen using the cross-ratio property.

For the derivation of (6.5), Yoo and Chung assumed the pupil to be located on the cornea surface. But because it is really located behind the cornea relative to the camera, the α value of 2.0 is not be suitable for the estimation of the virtual projections in practice. To deal with this issue, the authors use different α values to estimate each of the four LEDs virtual projections and each one is calculated by the following equation:

$$\alpha_i = \frac{|p - g|}{|r_i - g|} . \tag{6.7}$$

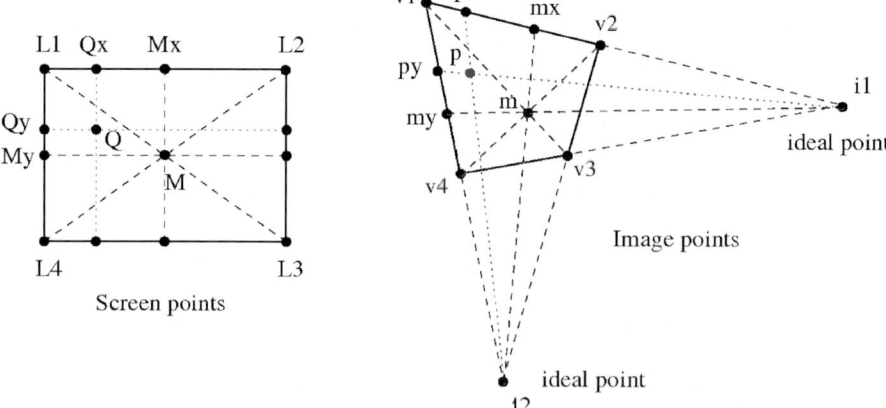

Fig. 6.3. Gaze estimation using cross-ratio: First, i_1 and i_2 are computed using the virtual points v_i. Then the coordinates (p_x, p_y) and (m_x, m_y) are computed and used to determine the screen gaze coordinates (Q_x, Q_y)

Where p is the pupil position in the image when the user is gazing directly at L_i. Therefore, to compute α_i, the user must look at the computer screen corner that contains L_i, and repeat the process for each corner to calibrate the system. The idea behind this procedure is that when a user looks at one of these lights, the pupil center should match the virtual projection of the corresponding LED. We have seen though, from previous works, that the point Q corresponds to the *LoG* and not the *LoS*. This issue is discussed further in the next section.

As we have seen, the single camera convex mirror technique requires a full calibrated camera to work and, even using synthetic data, the accuracy of the method is about 3°. The cross ratio technique presented in [703] does not require a calibrated camera or any kind of system calibration, and its accuracy is also better than the mirror technique, about 2°. In [702], the accuracy is improved using a calibration procedure that computes the correction function for the virtual corneal reflections.

All these single camera techniques are based on very simple models of the eye that, for example, do not consider the difference between the *LoS* and *LoG* axes. In the next section, we investigate the performance of this most promising technique, based on the cross-ratio invariance property, using simulation data.

6.4 Simulation of the Single Camera Techniques

The objective of the simulation experiments presented in this section is to verify the accuracy of the single camera methods as the head moves freely in front of the computer screen. In [423], a similar simulation was used to estimate the accuracy of calibration models for the more traditional pupil-corneal reflection (PCR) ET. The idea is that, by using a similar procedure, we will also be able to compare these new techniques with the PCR-ET.

The mirror technique and three cross-ratio techniques were tested. The first cross-ratio technique consists of using the corneal reflections directly, similar to the method described in [703]. In the second method, each virtual projection is computed using a constant α value of 2.0, which is the approximate expected value for the correction factor as seen in the previous section. Finally, the third method is the most recent proposed by Yoo and Chung [702] in which a calibration procedure is performed to determine an α_i value for each of the four virtual projections. For more realistic results, we also considered the angle offset between the *LoS* and the *LoG*.

The performance simulation of each method consists of translating the eye 10 cm in every direction ($\pm x$, $\pm y$, and $\pm z$ in a orthogonal coordinate system centered at the camera), generating a total of 7 possible positions (including the initial one). For each position the eye is rotated so that it observes 48 test points placed at the center of each element of a 8×6 grid that covers the entire screen. For each eye position and observed point, the locations of the corneal reflections are computed and projected along with the pupil position to the image plane, in order to calculate the virtual projections. The gaze point is calculated using the cross-ratio property, and this result is compared to the real observation point (the target point on the screen).

The cornea is modeled as a sphere of radius $0.77\,$cm, and its center E is positioned at $(0, 27, 60)$. The camera center O is placed at the origin $(0, 0, 0)$, and the monitor's LEDs were positioned at $(-18.3, 27.4, 0)$, $(18.3, 27.4, 0)$, $(18.3, 0, 0)$, and $(-18.3, 0, 0)$, where all coordinates are given in centimeters. The pupil center P has been considered to be located on the cornea surface and the vector connecting E to P is defined as the optical axis, with initial value of $(0, 0, -1)$. This setup approximates a common viewing situation, where the user is sitting about $60\,$cm in front of a $19\,$inch computer monitor screen, and was also used in [423], to evaluate the performance of the calibration of a PCR ET.

To model the eye axes difference, the optical axis of the eye (or LoG) is rotated to define the LoS. We use $4.5°$ pan and $3.0°$ tilt, that result in an angular difference of approximately $5.5°$ between the axes (which is in agreement with the real angular difference of $4°$ to $8°$ observed in the human eye [687]). In the following simulations, the LoG is used to evaluate the accuracy of the model (LoG) and the LoS to predict the accuracy of the technique in practice (LoS). In Sect. 6.7 we show experimental results using a real-time prototype that confirm the results of these simulations.

6.4.1 Mirror Technique Simulation Results

Table 6.1 shows the simulation results for the mirror technique described in Sect. 6.3.1. Observe that the position of the cornea center is estimated very accurately, resulting in accurate gaze estimations using the LoG. This technique is sensitive to depth changes of the user's head because the image

Table 6.1. Mirror technique simulation results: Results using an off axis light source placed at $(10, 0, 0)$ for estimation of the cornea center, considering the LoS and LoG axes

Position	Estimated Pos.	LoG		LoS	
		Avg. Error	Std. Dev.	Avg. Error	Std. Dev.
$(0, 27, 60)$	$(0, 27.04, 60.09)$	0.201	0.1313	6.195	0.2475
$(10, 27, 60)$	$(10.01, 27.03, 60.08)$	0.1932	0.1266	6.33	0.2868
$(-10, 27, 60)$	$(-10.01, 27.03, 60.07)$	0.1876	0.1245	6.308	0.4587
$(0, 37, 60)$	$(0, 37.04, 60.06)$	0.167	0.1062	6.476	0.3072
$(, 17, 60)$	$(0, 17.03, 60.1)$	0.2312	0.153	6.107	0.3304
$(0, 27, 70)$	$(0, 27.02, 70.05)$	0.1525	0.09825	7.051	0.2096
$(0, 27, 50)$	$(0, 27.06, 49.88)$	0.2712	0.1822	5.375	0.3039

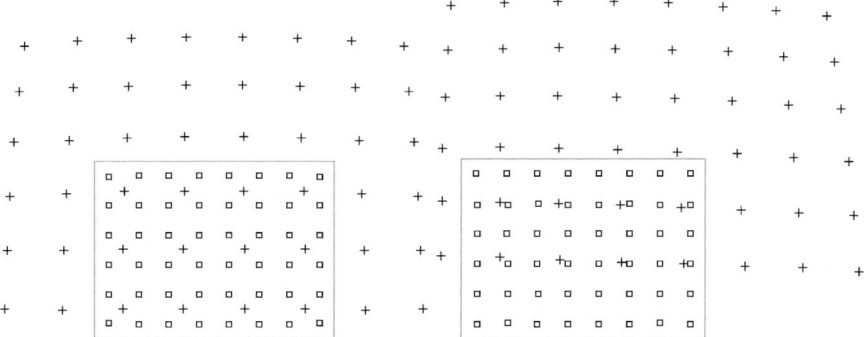

Fig. 6.4. Corneal reflection simulation results: Estimated gaze points (*crosses*) and real screen targets (*squares*) for the initial position of the eye. The corneal reflections are assumed to be coplanar, i.e., no correction factor is computed. The left image shows the results for the *LoG* and the right image the results for the *LoS*

of the light source may move away from the focus point, in particular when the user's moves closer to the camera, which is the case for eye positions $(0, 17, 60)$ and $(0, 27, 50)$. Because we do not compensate for the difference of axes, the accuracy for the *LoS* is considerably reduced. Because it is an angular offset, as the head moves away from the camera, the results get worse, as is the case for the eye position at $(0, 27, 70)$.

6.4.2 Corneal Reflection Simulation Results

Results obtained by direct use of the corneal reflections to estimate the gaze point can be seen in Fig. 6.4, and column "Reflection" of Table 6.2. The left image shows the results considering the *LoG* and the right image shows the results considering the *LoS*. Note that both results were very poor, and we guess that the results could be improved by an affine transformation, to achieve the accuracy reported in [703].

6.4.3 Using a Constant $\alpha = 2.0$

Figure 6.5a shows the results of the cross-ratio ET method using a constant α value of 2.0 to correct the position of the virtual corneal reflections. The left image shows the results considering the *LoG* and the right image shows the results considering the actual *LoS*. Observe that the results for the *LoG* are very good, and that the results of the *LoS* could be improved using a constant offset. We also guess that something similar was done in the real system implementation described in [702]. The column "$\alpha = 2$" in Table 6.2 shows the average error of the method as the head moves in all 6 directions.

Table 6.2. Cross-ratio techniques simulation results: The "Reflection" column shows the results for the simplest cross ratio technique, that assumes a flat cornea surface. The "$\alpha = 2.0$" column shows the results using a constant correction factor of 2.0, the "Calibrated α_i" uses calibration to correct for each corneal reflection. None of these methods compensate for the difference of axes of the eye. Column "β Calibration" shows the results of our new method, that uses a single calibration β parameter to correct for the cornea surface and the difference of axes. All values are in cm

pos	Reflection		$\alpha = 2.0$		Calibrated α_i		β Calibration	
	LoG	LoS	LoG	LoS	LoG	LoS	4 points	9 points
initial	16.1 ± 5.3	22.5 ± 5.7	0.3 ± 0.2	6.0 ± 0.4	0.7 ± 0.3	4.2 ± 1.0	1.0 ± 0.3	0.5 ± 0.3
x (−)	16.1 ± 5.4	23.2 ± 6.8	0.3 ± 0.2	6.1 ± 0.3	0.7 ± 0.3	4.1 ± 0.9	1.1 ± 0.3	0.6 ± 0.3
x (+)	16.1 ± 5.3	22.1 ± 4.9	0.3 ± 0.2	6.2 ± 0.6	0.7 ± 0.3	4.5 ± 1.2	1.1 ± 0.4	0.6 ± 0.3
y (−)	17.3 ± 6.1	23.8 ± 6.7	0.3 ± 0.2	5.9 ± 0.3	0.8 ± 0.3	4.1 ± 1.0	1.0 ± 0.3	0.4 ± 0.2
y (+)	15.2 ± 4.8	21.9 ± 4.9	0.3 ± 0.2	6.3 ± 0.5	0.7 ± 0.3	4.5 ± 1.1	1.2 ± 0.3	0.7 ± 0.3
z (−)	15.7 ± 4.9	20.7 ± 5.0	0.4 ± 0.3	5.2 ± 0.4	0.8 ± 0.4	3.5 ± 1.0	1.0 ± 0.5	1.0 ± 0.4
z (+)	16.5 ± 5.7	24.2 ± 6.2	0.2 ± 0.1	6.9 ± 0.3	0.7 ± 0.3	5.0 ± 1.1	1.6 ± 0.2	1.0 ± 0.2

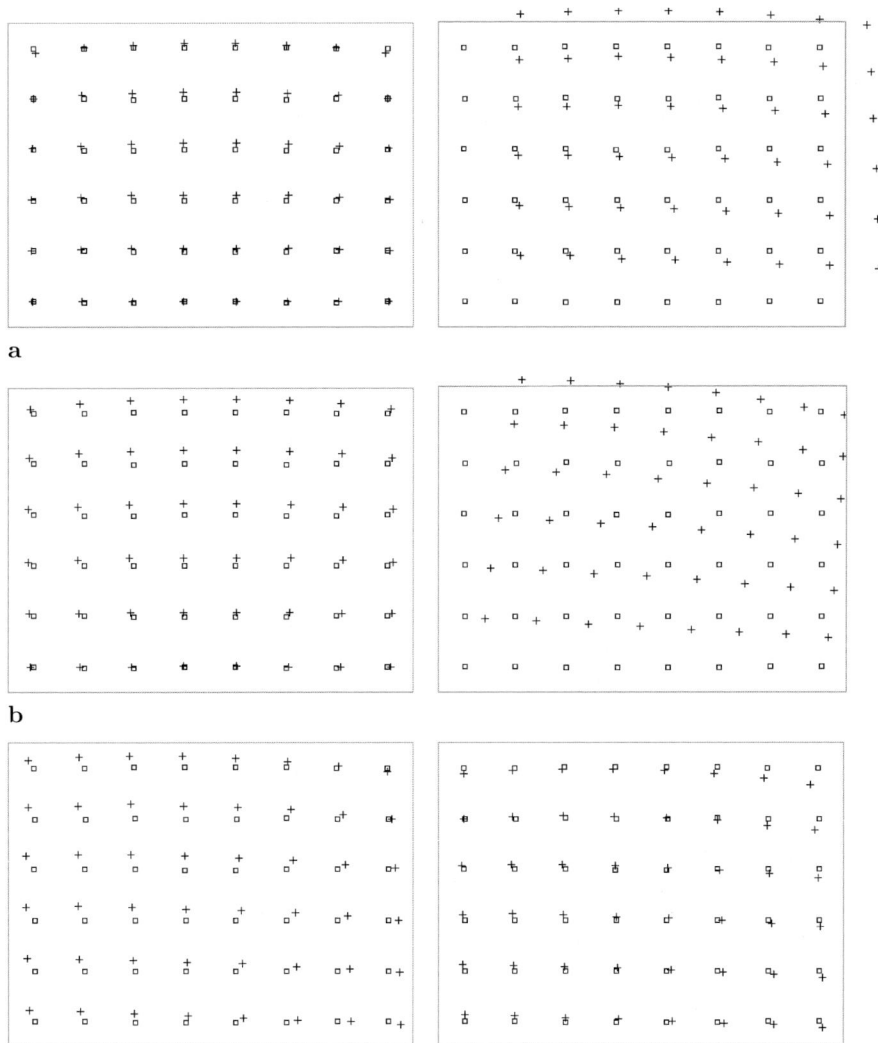

Fig. 6.5a–c. Simulation results using corrected virtual points: Estimated gaze points (*crosses*) and real target points (*squares*) for the initial position of the eye. **a** Results using a constant $\alpha = 2.0$. *Left* and *right* images show the *LoG* and *LoS* respectively. **b** Results using calibrated α_i parameters. *Left* and *right* images show the *LoG* and *LoS* respectively. **c** Results using β calibration. *Left* and *right* images use 4 and 9 calibration points respectively

6.4.4 Calibrated Cross-Ratio Method

Figure 6.5b shows the results of the calibrated cross-ratio ET method proposed in [702], that computes an α value to correct each virtual corneal reflection independently. The left image shows the results considering the *LoG* and the right image shows the results considering the actual *LoS*. The column "Calibrated α_i" in Table 6.2 shows the average error of the calibrated cross-ratio method as the head moves in all 6 directions. Once again, it is clear that the method needs to compensate for the angle offset in the case of the *LoS*, i.e., the calibration procedure does not compensate for the difference between the *LoG* and *LoS* axis.

6.5 β Calibration: A New Cross-Ratio Technique

From the result of the simulations presented in the previous section we can conclude that:

- it is necessary to compute the virtual projections of the LEDs when estimating the point of regard to compensate for the curvature of the cornea;
- the approximation of the virtual projections using a single constant α value produces reasonably good results;
- the angle offset between the *LoG* and *LoS* of the eye must be considered and can be compensated by adding an offset vector to the estimated points. However this solution might be sensitive depth changes of the eye position.

Based on these observations we have developed a variation of the method proposed in [702], where a different constant is used, and its value is computed from a simple calibration procedure. The novelty of this procedure is that it also takes into account the true *LoS* for gaze estimation.

The idea is to calculate one constant β value (very similar to the α value we have discussed so far) that generates a set of estimated points that can also describe a translation of the real observed points (in this case, the calibration points). Using n calibration points, consider

$$C = \{C_i | i \in [1..n]\} \tag{6.8}$$

as the set of such points, and

$$E_\beta = \{E_{\beta i} | E_{\beta i} = T(C_i, \beta), i \in [1..n]\} \tag{6.9}$$

the set of the estimated points (where $T(C_i, \beta)$ corresponds to the application of the cross-ratio technique for calibration data relative to point C_i and the given β value). We also define

$$D_\beta = \{\overrightarrow{d_{\beta i}} | \overrightarrow{d_{\beta i}} = (C_i - E_{\beta i}), i \in [1..n]\} \tag{6.10}$$

as the set of offset vectors and $\overrightarrow{m_\beta}$ as the average offset vector. In the ideal case the value of β should transform each element of D_β to be equal to $\overrightarrow{m_\beta}$. Since there is some variation of the offset vectors, the best value of β is the one that generates a D_β set with elements as uniform as possible. This value of β can be obtained by minimizing the following sum:

$$sum(\beta) = \sum_{i=1}^{n} |\overrightarrow{d_{\beta i}} - \overrightarrow{m_\beta}| \tag{6.11}$$

The plot of $sum(\beta)$ versus β shows a curve that starts with a negative slope, reaches a minimum value for the sum, and then starts to grow again. Placing the eye at different positions, the general form of the plot remains the same, so we can use a bisection method for the computation of the optimum β value that generates the minimum sum. Once the value of β is calculated, the vector $\overrightarrow{m_\beta}$ is taken as the average offset to be used to correct estimation errors due to the axes difference of the eye.

To verify the behavior of the β calibration cross-ratio method, we simulated it under the same conditions used for the other methods presented in the previous section. Two tests were made, one using 4 calibration points that correspond to the positions of the LEDs on the screen (thus similar to the α_i calibration proposed in [702]), and the other using 9 calibration points corresponding to the center of each element obtained by dividing the screen as a regular 3×3 grid. Results of the first test can be seen in left image of Fig. 6.5c and column "β Calibration: 4 points" of Table 6.2. Results of the second test are shown in the right image of Fig. 6.5c and column "β Calibration: 9 points" of Table 6.2. Notice that the β calibration method works well and achieves better performance using 9 calibration points because they offer a more detailed sample of the screen.

Notice in Table 6.2 that the precision of the cross-ratio techniques that corrects for the curvature of the cornea (using the parameters α or β) are more sensitive to depth changes, even though the results show great improvement over the traditional pupil-corneal reflection eye tracker presented in [423]. Table 6.3 shows the comparison of head movement tolerance using Morimoto and Mimica's results from [423] and the new β calibration cross ratio method with 9 calibration points. Observe that the cross-ratio method is more robust to head translations.

In Fig. 6.6 it is possible to observe how the β calibration method using 9 calibration points is affected by large translations (30 cm in each direction). Observe that errors for translations in the z axis grow almost twice as fast than for translations in the x and y axes. For these two axes, even for translations up to 20 cm, we have an average error close to or below 1 cm (that corresponds to approximately $1°$ of visual angle).

Table 6.3. Comparison with a traditional eye tracker: The simulation results of a traditional pupil-corneal-reflection eye tracker, as reported in [423], is compared to the results of the β calibration cross-ratio eye tracker. Observe that the accuracy of the cross-ratio technique holds very well for translations in all directions. The (+) and (−) indicate, respectively, positive and negative translations of 10 cm along the specified axis

Position	Average error		
	traditional	cr (4 points)	cr (9 points)
initial	0.80 cm	1.0 cm	0.48 cm
$x\ (-)$	0.99 cm	1.13 cm	0.55 cm
$y\ (-)$	2.17 cm	0.95 cm	0.39 cm
$z\ (+)$	4.05 cm	1.63 cm	0.94 cm

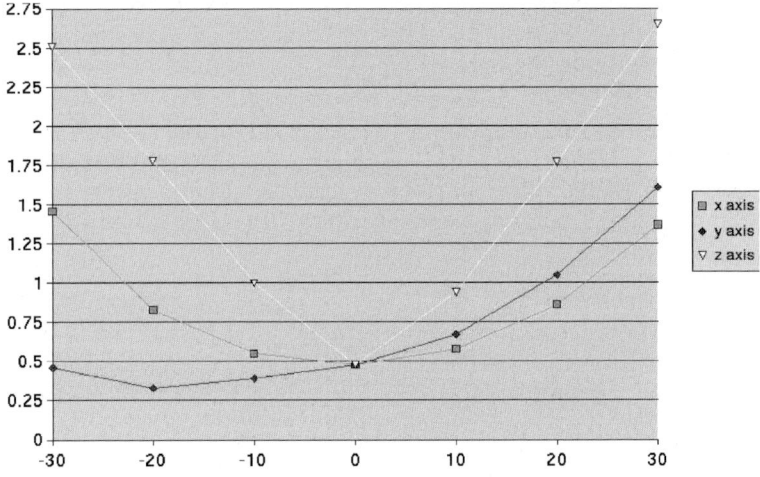

Fig. 6.6. Error as a function of eye translations: Average simulation errors (in cm) using the β calibration cross-ratio technique, with nine calibration points

6.6 Implementation Issues

The prototype was developed in a Linux platform, a micro-computer with an Athlon 1.4 GHz CPU with 512 Mb of RAM, an Osprey 100 video capture card, a 17 inches LCD monitor and a camera with two sets of IR LEDs (one on the optical axis of the camera, and the other on the monitor's corners as can be seen in Fig. 6.7). A custom external circuit synchronizes the activation of each set of LEDs with the even/odd frames of the interlaced NTSC video signal. The software basically executes the following steps:

1. image acquisition;
2. image processing;
3. gaze estimation.

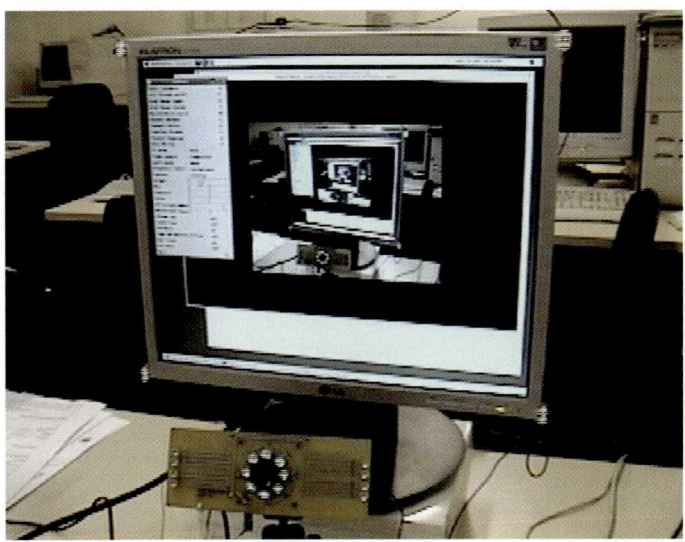

Fig. 6.7. Eye gaze tracker system: The picture shows the monitor with the IR LEDs attached to its corners and the camera with a set of LEDs around its optical axis

The image acquisition module was developed using the video4linux API (V4L) which offers a standard interface for a large number of capture devices. A lot of the image processing was facilitated by the use of the OpenCV library. In the image processing step, the captured image is deinterlaced in order to obtain two images of the eye: one with a bright pupil and the corneal reflection generated by the set of LEDs of the camera; the other with a dark pupil and the four reflections of the LEDs placed on the corners of the screen.

Next, a rough estimation of the pupil location is computed using a reduced resolution image of the bright and dark pupil images. By subtracting the dark pupil image from the bright one and thresholding the difference image, a connected component labeling algorithm selects the most appropriate pupil candidate based on the blob's geometric properties. The location of the pupil is defined by its center of mass, and it is used to define a region of interest (ROI) in the full size deinterlaced images for further processing.

The corneal reflections (CRs) are detected next. Since the CRs appear as small bright regions in the image we apply a threshold over the deinterlaced images. Next we apply morphological operations and label each resulting connected component. Big components and those located too far from the pupil center are discarded, and the resulting ones become our CR candidates. Since the implemented technique for gaze tracking requires a fixed number of CRs to be tracked (4 reflections in the dark pupil image and 1 CR in the bright pupil image), the excess or lack of CRs are conditions that we need to solve. Unfortunately, when CRs are missing we are currently not able to recover. When too many CRs are detected in the bright pupil image we simply consider the largest one. For the dark pupil image we need to choose the four correct CRs. To do this we take all possible combinations of four CRs and choose the combination which forms a quad that is closest to a rectangle.

Once we have detected all expected CRs, we do precise estimation of the pupil in the full sized images. It is similar to the initial estimation, but instead of calculating the center of mass of the blob that corresponds to the pupil, we take its contour and remove its portions which overlap with the detected CRs. Finally we use the resulting contour to fit an ellipse that best matches the pupil.

Having all the feature points that are necessary for the computation of the point of regard, we estimate the observed point as already discussed: first computing the virtual projections and then applying the algorithm based on the cross-ratio. For comparison purpose we have implemented both Yoo and Chung's calibrated α_i method and our β calibration method.

6.7 Experimental Results

In this section we discuss the results of our real time implementation of the β calibration cross-ratio technique. Five users participated in our experiments. The test procedure consisted in each user looking at 48 points (the centers of each element of a 8×6 grid covering the screen) and comparing the estimated gaze point with the real point. To observe the tracker's tolerance to user head motion each user executed the test procedure, after initial calibration, 3 times and after each execution they were asked to move a little in order to change their positions in space, and the camera was also moved to keep the eye within its narrow field of view. The results are shown in Table 6.4 (first test trial) and Table 6.5 (other trials) where we can observe a small deterioration of the precision after the execution of the first trial. The average error for all users in all executions of the test was 2.48 cm (about 2.4°) with standard deviation of 1.12 cm when using Yoo and Chung's calibrated α_i method for

Table 6.4. Experimental results using a fixed camera: Gaze estimation results for each user with the camera at the same position used for calibration. Users marked with (*) e (**) were wearing, respectively, contact lenses and glasses

1st test User	Calibrated α_i Error	Std dev	β Calibration Error	Std dev
user 1*	1.51 cm	0.69 cm	1.23 cm	1.39 cm
user 2	2.51 cm	0.82 cm	0.73 cm	0.29 cm
user 3**	1.75 cm	0.48 cm	0.73 cm	0.39 cm
user 4	3.15 cm	1.13 cm	0.88 cm	0.48 cm
user 5	2.98 cm	0.86 cm	0.78 cm	0.43 cm

Table 6.5. Experimental results after moving the camera: Gaze estimation results for each user after the camera is moved away from the position used for calibration (the user must reposition his face). Users marked with (*) e (**) were wearing, respectively, contact lenses and glasses

| other tests | Calibrated α_i | | β Calibration | |
User	Error	Std dev	Error	Std dev
user 1*	1.98 cm	0.96 cm	1.15 cm	0.96 cm
user 2	2.53 cm	1.03 cm	0.98 cm	0.62 cm
user 3**	1.8 cm	0.59 cm	0.9 cm	0.55 cm
user 4	2.95 cm	1.15 cm	0.98 cm	0.62 cm
user 5	3.36 cm	1.11 cm	0.93 cm	0.51 cm

virtual projection estimation. For our β calibration method, the average gaze estimation error was 0.95 cm (about 0.91°) with standard deviation of 0.7 cm. In Fig. 6.8 we can see the estimated points for all trials of all users according to the virtual projection estimation method used. We can observe that in our method, the estimated gaze points are more concentrated around the real observed points. One of the users who participated in the tests wear glasses and another contact-lenses. Despite the fact that CR detection was more difficult for these users, we did not observe any influence of the corrective lenses in the estimation results.

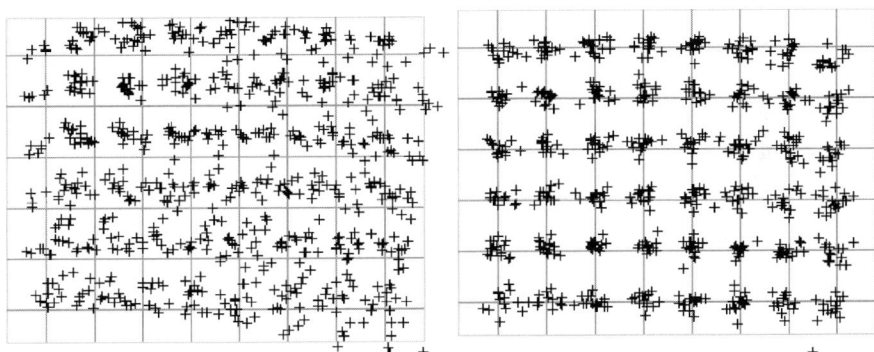

Fig. 6.8. Global comparison: Estimation results for all users on the 3 test trials. *Crosses* represents the estimated points and line intersections the real observed points. The *left* image shows the result using the calibrated α_i parameters, and the *right* image shows the results using the β calibration

6.8 Conclusion

In this chapter we have presented an overview of recent methods developed to overcome the two main limitations of current image based eye gaze trackers: the very limited head motion allowed by the system and need for constant recalibration. While most solutions to these problems suggest the use of multiple cameras, we have focused on single camera solutions. We argue that the cross-ratio technique offers the best practical solution since it requires a single non-calibrated camera to work. Once high resolution cameras become more common and affordable, they will offer a compact, easy to implement, and easy to use solution to eye gaze tracking applications. We have conducted extensive experiments using synthetic data that demonstrate the robustness of the cross-ratio techniques to head motion and the need to correct the eye line of gaze (optical axis) to the true line of sight. Based on these simulations, we have proposed an extension to the cross-ratio technique that uses a simple calibration procedure called β calibration. This procedure combines two correction effects: the eye angle angle offset that corrects the true line of sight and the curvature of the cornea surface. We have developed a real-time prototype that process 30 fps. Our experimental results using this prototype shows an accuracy of about 1° of visual angle and great tolerance for head motion, as predicted by our simulation results.

Extraction of Visual Information from Images of Eyes

Ko Nishino and Shree K. Nayar

7.1 What Do Eyes Reveal?

Our eyes are crucial to providing us with an enormous amount of information about our physical world. What is less often explored is the fact that the eye also conveys equally rich information to an external observer. Figure 7.1 shows two images of eyes. One can see a building in the example on the left and a person's face in the example on the right. At the same time, we can see that the mapping of the environment within the appearance of the eye is complex and depends on how the eye is imaged. In [633], highlights in an eye were used to locate three known point sources. The sources were then used to apply photometric stereo and relight the face. Extensive previous work has also been done on using eyes to estimate gaze [154, 468, 588, 603, 693, 706] and to identify people from their iris textures [116, 185][1]. However, virtually no work has been done on using eyes to interpret the world surrounding the person to whom the eye belongs.

In this chapter, we provide a comprehensive analysis of exactly what information is embedded within a single image of an eye. We believe this work is timely. Until recently, still and video cameras were relatively low in resolution and hence an eye in an image simply did not have a sufficient number of pixels to represent useful information. Recently, however, CCD and CMOS

---

[1] The texture of the iris is known to be a powerful biometric for human identification [116,185,315]. It is important to note that sensors used for scanning the iris use special lighting to ensure that the reflections from the cornea (appearance of the external world) are minimized. In unstructured settings, however, the corneal reflections tend to dominate the appearance of the eye and it is exactly this effect we seek to exploit. Clearly, even in an unstructured setting, the texture of the iris will contribute to the appearance of the eye. In our work, we do not attempt to eliminate the contribution of the iris; this problem is significant by itself and will be addressed separately in the future.

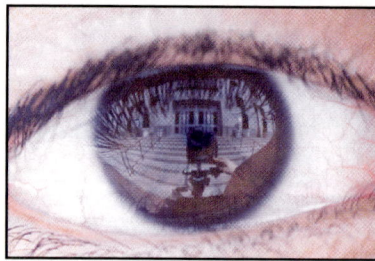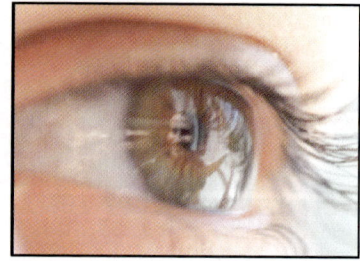

Fig. 7.1. What does the appearance of an eye tell us about the world surrounding the person and what the person is looking at?

image detectors have made quantum leaps in terms of resolution. At this stage, one can legitimately ask the following question: What does an image of an eye reveal about the world and the person and how can this information be extracted?

Our key observation is that the combination of the cornea of an eye and the camera capturing the appearance of the eye can be viewed as a catadioptric (mirror + lens) imaging system. We refer to this as the *corneal imaging system*. Since the reflecting element (the cornea) is not rigidly attached to the camera, the corneal imaging system is inherently an uncalibrated one. We use a geometric model of the cornea based on anatomical studies to estimate its 3D location and orientation[2]. This is equivalent to calibrating the corneal imaging system. Once this is done, we show that we can compute a precise wide-angle view of the world surrounding the person. More importantly, we can obtain an estimate of the projection of the environment onto the retina of the person's eye. We refer to this as the retinal image. From this retinal image of the surrounding world, we are able to determine what the person is looking at (focus of attention) without implanting an image detector in his/her eye.

We present a detailed analysis of the characteristics of the corneal imaging system. We show that, irrespective of the pose of the cornea, the field of view of the corneal system is greater than the field of view observed by the person to whom the eye belongs. We find that the spatial resolution of the corneal system is similar in its variation to that of the retina of the eye; it is highest close to the direction of gaze. It also turns out that this imaging system is a non-central one; it does not have a single viewpoint but rather a locus of viewpoints. We derive the viewpoint locus of the corneal system and show how it varies with the pose of the cornea. When both eyes of a person are captured in the same image, we have a catadioptric stereo system. We derive

---

[2] The computed corneal orientation may be used as an estimate of the gaze direction. However, it is important to note that gaze detection is not the focus of our work – it is just a side product.

the epipolar geometry for such a system and show how it can be used to recover the 3D structure of objects in front of the person.

Our framework for extracting visual information from eyes has direct implications for various fields. In visual recognition, the recovered wide-angle view of the environment can be used to determine the location and circumstance of the person when the image was captured. Such information can be very useful in security applications. The computed environment map also represents the illumination distribution surrounding the person. This illumination information can be used in various computer graphics applications including relighting faces and objects as we have shown in [450]. The computed retinal image can reveal the intent of the person. This information can be used to effectively communicate with a computer or a robot [55], leading to more advanced human-machine interfaces. The image of a person's face can tell us their reaction, while the appearance of their eyes in the same image can reveal what they are reacting to. This information is of great value in human affect studies [162, 625].

7.2 Physical Model of the Eye

As can be seen in Fig. 7.2a, the most distinct visual features of the eye are the cornea and the sclera. Figure 7.2b shows a schematic of a horizontal cross-section of the right eyeball. The cornea consists of a lamellar structure of submicroscopic collagen fibrils arranged in a manner that makes it transparent [119, 315, 675]. The external surface of the cornea is very smooth. In addition, it has a thin film of tear fluid on it. As a result, the surface of the cornea behaves like a mirror. This is why the combination of the cornea and

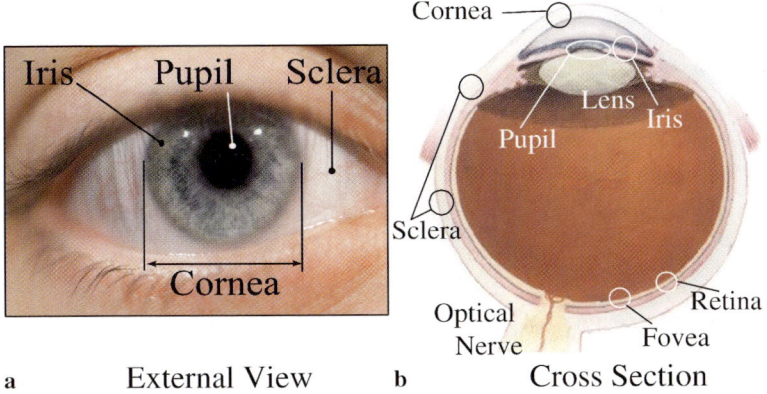

a   External View   b   Cross Section

Fig. 7.2. a An external view of the human eye. The sclera and the cornea (behind which the pupil and the iris reside) are the most visually distinct components of the eye. **b** A horizontal cross-section of the right human eyeball. Despite its fairly complex anatomy, the physical dimensions of the eye do not vary much across people

a camera observing it form a catadioptric system, which we refer to as the *corneal imaging system*.

To interpret an image captured by this system, we need a geometric model of its reflecting element, i.e. the cornea. In the fields of physiology and anatomy, extensive measurements of the shape and dimensions of the cornea have been conducted [243,315]. It has been found that a normal adult cornea is very close to an ellipsoid, as shown in Fig. 7.3. In Cartesian coordinates (x, y, z), an ellipsoid can be written as [17]

$$pz^2 - 2Rz + r^2 = 0, \tag{7.1}$$

where $r = \sqrt{x^2 + y^2}$, $p = 1 - e^2$ where e is the eccentricity and R is the radius of curvature at the apex of the ellipsoid. Now, a point $\mathbf{S}$ on the corneal surface can be expressed as

$$\mathbf{S}(t, \theta) = (S_x, S_y, S_z) = \left(\sqrt{-pt^2 + 2Rt} \cos \theta, \sqrt{-pt^2 + 2Rt} \sin \theta, t \right) \tag{7.2}$$

where $0 \leq \theta < 2\pi$ (see Fig. 7.3). It turns out that the parameters of the ellipsoid do not vary significantly from one person to the next. On average, the eccentricity e is 0.5 and the radius of curvature R at the apex is 7.8 mm [315].

The boundary between the cornea and the sclera is called the *limbus*. The sclera is not as highly reflective as the cornea. As a result, the limbus defines the outer limit of the reflector of our imaging system. From a physiological perspective, the cornea "dissolves" into the sclera. However, in the case of an adult eye, the limbus has been found to be close to circular with radius r_L of approximately 5.5 mm. Therefore, in Eq. 7.2, the parameter t ranges from 0 to t_b where t_b is determined using $-pt_b^2 + 2Rt_b + r_L^2 = 0$ and found to be 2.18 mm.

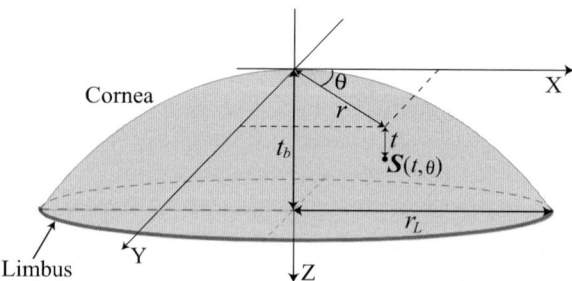

Fig. 7.3. The cornea is modeled as an ellipsoid whose outer limit corresponds to the limbus. For a normal adult cornea, the approximate eccentricity and the radius of curvature at the apex are known. The Cartesian coordinates of corneal surface points can be parameterized with the azimuth angle θ and the height t along the Z axis

In the remainder of this chapter, we will use the above geometric model of the cornea, which well approximates a normal adult cornea. It is important to note that slight changes in the parameters of the model will not have a significant impact on our results. However, if the cornea significantly deviates from a normal adult cornea, for instance due to diseases such as keratoconus, it will be necessary to measure its shape. This can be done by using structured light [214] and the measured shape can be directly used in our method.

7.3 Finding the Pose of an Eye

To explore the visual information captured by the corneal imaging system, we first need to calibrate the system. This corresponds to estimating the 3D location and orientation of the cornea from the image of an eye. We accomplish this by first locating the limbus in the image.

7.3.1 Limbus Localization

The limbus in an image is the projection of a circle in 3D space. Even in the extreme case when the gaze direction is perpendicular to the optical axis of the camera, the depth of this circle is only 11 mm (the diameter of the limbus). Hence, we can safely assume the camera projection model to be weak-perspective; orthographic projection followed by scaling. Under this assumption, the limbus is imaged as an ellipse.

Let (u, v) denote the horizontal and vertical image coordinates, respectively. An ellipse in an image can be described using five parameters which we denote by a vector $\mathbf{e}$. These parameters are the center (c_u, c_v), the major and minor radii $r_{\max}$ and $r_{\min}$ and the rotation angle ϕ of the ellipse in the image plane. We localize the limbus in an image $I(u, v)$ by searching for the ellipse parameters $\mathbf{e}$ that maximize the response to an integro-differential operator applied to the image after it is smoothed by a Gaussian g_σ:

$$\max_{\mathbf{e}=(c_x, c_y, r_{\min}, r_{\max}, \phi)} \left| g_\sigma(r_{\max}) * \frac{\partial}{\partial r_{\max}} \oint_{\mathbf{e}} I(u, v) \, ds \right.$$
$$\left. + g_\sigma(r_{\min}) * \frac{\partial}{\partial r_{\min}} \oint_{\mathbf{e}} I(u, v) \, ds \right| . \tag{7.3}$$

Note that the filter responses are integrated along the arc ds of the ellipse. Daugman [116] proposed the use of a similar integro-differential operator to detect the limbus and the pupil as circles in an image of a forward looking eye. In contrast, our algorithm detects the limbus as an ellipse, which is necessary when the gaze direction of the eye is unknown and arbitrary.

We provide an initial estimate of $\mathbf{e}$ to the algorithm by drawing an ellipse in the image. This is accomplished with a simple user interface for positioning, resizing and rotating a predefined ellipse by dragging the center or the arc of it. We also let the user specify the arc range of this initial ellipse to be used for evaluating Eq. 7.3. In many cases, due to the large intensity gradient change between the cornea and the sclera, this arc range setting can be very crude. However, for cases when the eyelids significantly occlude the limbus or when the cornea itself occludes one side of the limbus, it becomes important to discard such occlusions from being evaluated in Eq. 7.3.

Same as [116], starting with the initial ellipse parameters, we adopt a discrete iterative implementation for obtaining the optimal ellipse parameters that maximize Eq. 7.3. We use simplex method for the iterative search. Alternatively, one can use other gradient based optimization algorithms. Furthermore, instead of manually specifying the arc range, robust estimators can be used to discard non-limbus edges.

Figure 7.4 shows results of the limbus localization algorithm applied to different images of eyes. Note that the limbus is accurately located for arbitrary gaze directions, despite the complex texture of the iris and partial occlusions by the eyelids.

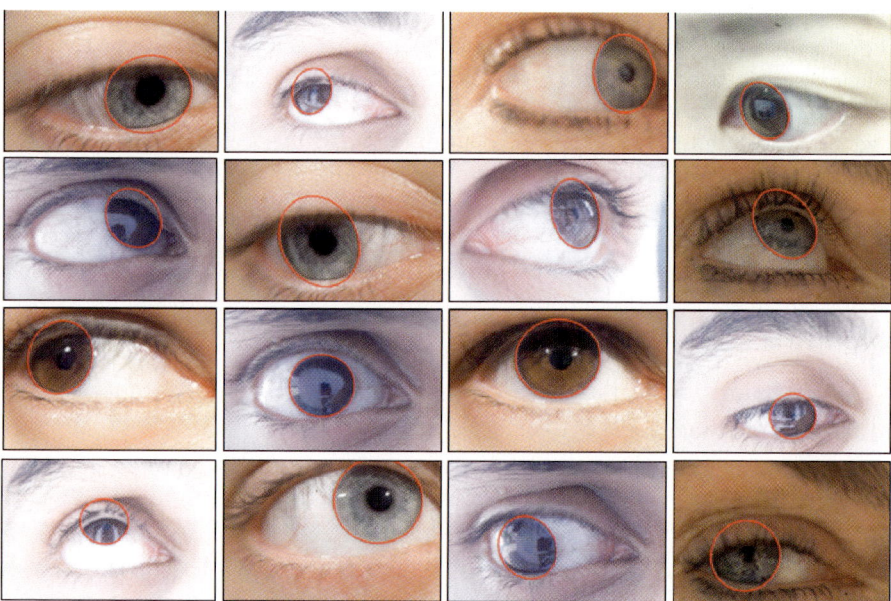

Fig. 7.4. Detected limbuses (*red ellipses*) in images of eyes. The ellipse detector successfully locates the limbus for different unknown gaze directions, despite the complex texture of the iris and partial occlusions by the eyelids

7.3.2 3D Location of the Cornea

Once the limbus has been detected, we can find the 3D location of the cornea in the camera's coordinate frame. We assume the internal parameters of the camera are calibrated *a priori*[3].

As shown in Fig. 7.5, under weak-perspective projection, the limbus in 3D space is first orthographically projected onto a plane parallel to the image plane. This plane lies at the average depth of the limbus from the camera. Since the limbus is a circle in 3D space, this average depth plane always passes through the center of the limbus. As a result, the major axis of the estimated ellipse in the image corresponds to the intersection of the limbus and this average depth plane. The length of this line segment is the diameter of the limbus. Therefore, the distance to the average depth plane, d, can be computed as

$$d = r_L \frac{f}{r_{\max}}, \tag{7.4}$$

where $r_L = 5.5\,\mathrm{mm}$, f is the focal length in pixels and $r_{\max}$ is known from the limbus detection in Sect. 7.3.1. Note that the depth d, the ellipse center (c_u, c_v) in the image and the focal length f determine the 3D coordinates of the center of the limbus.

Figure 7.6 shows results of an experiment we conducted to verify the accuracy of depth (eye distance) estimation. We captured 10 images of a person's

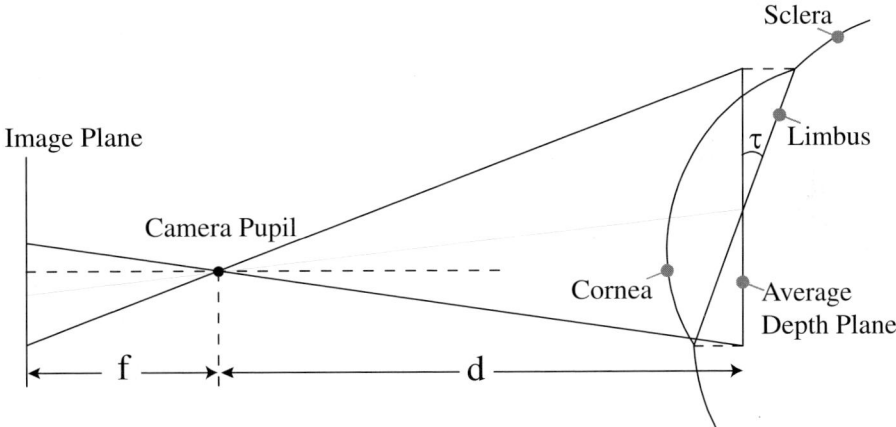

Fig. 7.5. Weak-perspective projection of the eye onto the image plane. The 3D location and orientation of the cornea in camera space can be computed from the imaged limbus

---

[3] All eye images in this chapter were captured with a Kodak DCS 760 camera with 6 M pixels. Close-up views are obtained using a Micro Nikkor 105 mm lens. We implemented the rotation based calibration method described in [584] to find the internal parameters of the camera.

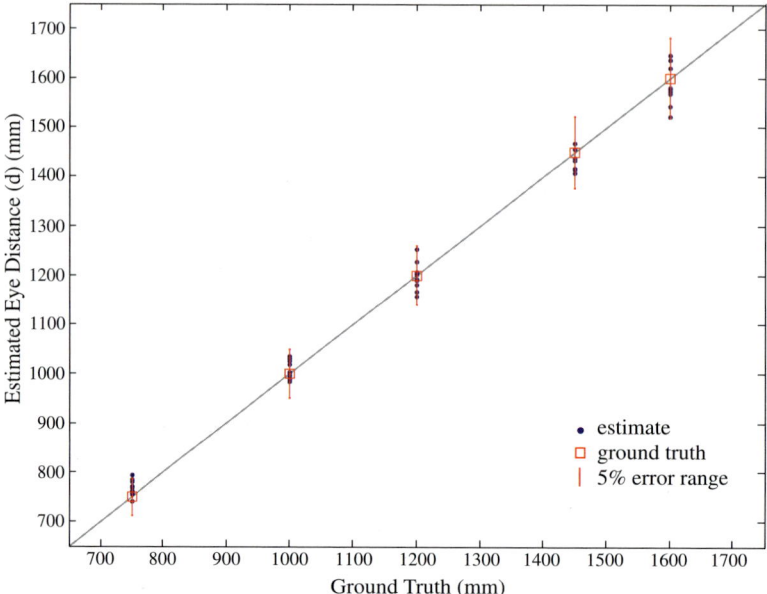

Fig. 7.6. Estimation results of the distance to the eye (d of Eq. 7.4). Images of eyes taken at 5 different distances and 10 different gaze directions for each distance were used. At each of 5 distances, 10 images were captured for different gaze directions. Regardless of the gaze direction, the depth d is estimated with good accuracy (always less than 5% error and RMS error 1.9%)

right eye at 5 different distances ranging from 75 cm to 160 cm. For each of the 10 images, the person was asked to change his gaze direction. Regardless of the gaze direction, the depth d was estimated with good accuracy in all cases with an RMS error of 1.9%.

7.3.3 3D Orientation of the Cornea

From the image parameters of the limbus we can also compute the 3D orientation of the cornea. The 3D orientation is represented using two angles (ϕ, τ). ϕ is the rotation of the limbus in the image plane which we have already estimated. Consider the plane in 3D on which the limbus lies. τ is the angle by which this plane is tilted with respect to the image plane (see Fig. 7.5) and can be determined from the major and minor radii of the detected ellipse:

$$\tau = \arccos \frac{r_{\min}}{r_{\max}} \, . \tag{7.5}$$

Note, however, that there is an inherent two-way ambiguity in the estimate of τ; for instance, an eye looking downward and upward by the same amount will produce the same ellipse in the image. Recently, Wang et al. [657] showed

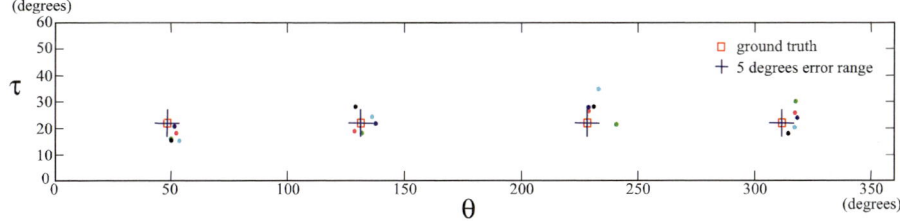

Fig. 7.7. Estimation results of the orientation of the cornea (ϕ, τ). Images of 5 people looking at markers placed at 4 different known locations on a plane were used. In most cases, both angles are estimated within 5 degrees error and the RMS errors was $3.9°$ for ϕ and $4.5°$ for τ

that anthropometric properties of the eye ball can be used to automatically break this ambiguity. In our experiments, we manually break the ambiguity.

The computed orientation of the cornea (the angles ϕ and τ) represents the direction of the optical axis of the eyeball. Despite the fact that the actual gaze direction is slightly offset[4], the optical axis is commonly used as an approximation of the gaze direction. We will therefore refer to the computed 3D corneal orientation as the gaze direction.

To verify the accuracy of orientation estimation, we used images of 5 people looking at markers placed at 4 different known locations on a plane. Figure 7.7 shows the estimation results of the corneal orientation (ϕ, τ). Estimates of both ϕ and τ are mostly within the $5°$ error range and the RMS errors were $3.9°$ and $4.5°$, respectively. These errors are small given that the limbus itself is not a sharp discontinuity.

7.4 The Corneal Imaging System

In the previous section, we showed how to calibrate the corneal catadioptric imaging system. We are now in a position to investigate in detail the imaging characteristics of this system. These include the viewpoint locus, field of view, resolution and epipolar geometry (in the case of two eyes) of the system.

7.4.1 Viewpoint Locus

Previous work on catadioptric systems [18] has shown that only the class of conic mirrors can be used with a perspective camera to configure an imaging system with a single viewpoint. In each of these cases, the entrance pupil of the camera must be located at a specific position with respect to the

---

[4] The actual gaze direction is slightly nasal and superior to the optical axis of the eyeball [315]. This means the optical axis does not intersect the center of the fovea.

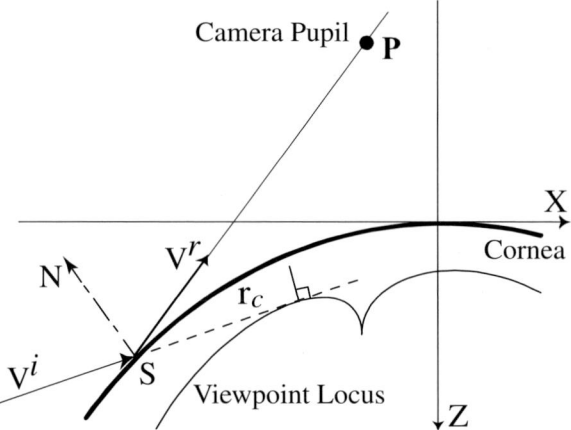

Fig. 7.8. The locus of viewpoints of the corneal imaging system is the envelope of tangents produced by the incident rays that are reflected by the cornea into the pupil of the camera. One can see that the shape of this locus will depend on the relative orientation between the cornea and the camera

mirror. Although the mirror (cornea) in our case is a conic (ellipsoid), the camera observing it is not rigidly attached to it. Therefore, in general, the corneal imaging system does not have a single viewpoint but rather a *locus of viewpoints*. Such non-single viewpoint systems have been explored in other contexts [107, 594].

Consider the imaging geometry shown in Fig. 7.8. The reference frame is located at the apex of the cornea. Let the pupil of the camera be at **P**. An incident ray $\mathbf{V}^i(t,\theta)$ can be related to the surface normal $\mathbf{N}(t,\theta)$ at the reflecting point $\mathbf{S}(t,\theta)$ and the reflected ray $\mathbf{V}^r(t,\theta)$ via the law of specular reflection

$$\mathbf{V}^i = \mathbf{V}^r - 2\mathbf{N}(\mathbf{N}\cdot\mathbf{V}^r)\,,\tag{7.6}$$

$$\mathbf{V}^r = \frac{\mathbf{P}-\mathbf{S}}{|\mathbf{P}-\mathbf{S}|}\,.\tag{7.7}$$

As seen in Fig. 7.8, the viewpoint corresponding to each corneal surface point $\mathbf{S}(t,\theta)$ must lie along the incident ray $\mathbf{V}^i(t,\theta)$. Therefore, we can parameterize points on the viewpoint locus as

$$\mathbf{V}(t,\theta,r) = \mathbf{S}(t,\theta) + r\mathbf{V}^i(t,\theta)\,,\tag{7.8}$$

where r is the distance of each viewpoint from the corneal surface. In principle, the viewpoints can be assumed to lie anywhere along the incident rays (r can be arbitrary). However, a natural representation of the viewpoint locus is the *caustic* of the imaging system, to which all the incident rays are tangent [107]. In this case, the parameter r is constrained by the caustic and can be determined by solving [75]

$$\det J(\mathbf{V}(t, \theta, r)) = 0 , \tag{7.9}$$

where J is the Jacobian. Let the solution to the above equation be $r_c(t, \theta)$[5]. Then, the viewpoint locus is $\{\mathbf{V}(t, \theta, r_c(t, \theta)) : 0 \leq t \leq t_b, 0 \leq \theta < 2\pi\}$. Note that this locus completely describes the projection geometry of the corneal system; each point on the locus is a viewpoint and the corresponding incident ray is its viewing direction.

Figure 7.9 shows the viewpoint loci of the corneal imaging system for four different eye-camera configurations. As can be seen, each computed locus is a smooth surface with a cusp. Note that the position of the cusp depends on the eye-camera configuration. The cusp is an important attribute of the viewpoint locus as it can be used as the (approximate) effective viewpoint of the system [594]. Also note that the viewpoint locus is always inside the cornea; its size is always smaller than the cornea itself.

7.4.2 Field of View

Since the cornea is an ellipsoid (convex), the field of view of the corneal imaging system is bounded by the incident light rays that are reflected by the limbus (outer limit of the cornea).

A point on the limbus and its corresponding incident ray direction can be written as $\mathbf{S}(t_b, \theta)$ and $\mathbf{V}^i(t_b, \theta)$, respectively, where t_b was defined in Sect. 7.2. Here, $\mathbf{V}^i(t_b, \theta)$ is a unit vector computed using Eq. 7.6. Let us define the field of view of the corneal system on a unit sphere. As we traverse the circular limbus, $\mathbf{V}^i(t_b, \theta)$ forms a closed loop on the unit sphere. Hence, the solid angle subtended by the corneal FOV can be computed as the area on the sphere bounded by this closed loop:

$$\text{FOV}(\mathbf{P}) = \int_0^{2\pi} \int_0^{\arccos V_z^i(t_b, \theta)} \sin\varphi \, d\varphi \, d\theta$$

$$= \int_0^{2\pi} (-V_z^i(t_b, \theta) + 1) \, d\theta . \tag{7.10}$$

Here, V_z^i is the Z coordinate of $\mathbf{V}^i$ and θ and φ are azimuth and polar angles defined in the coordinate frame of the unit sphere. From Eq. 7.6 we know that the above FOV depends on the camera position $\mathbf{P}$. For gaze directions that are extreme with respect to the camera, the camera may not see the entire extent of the cornea. We are neglecting such self-occlusions here as we have found that they occur only when the angle between the gaze direction

---

[5] The solution is very lengthy and there is no simple expression for it. We used Mathematica to solve Eq. 7.9.

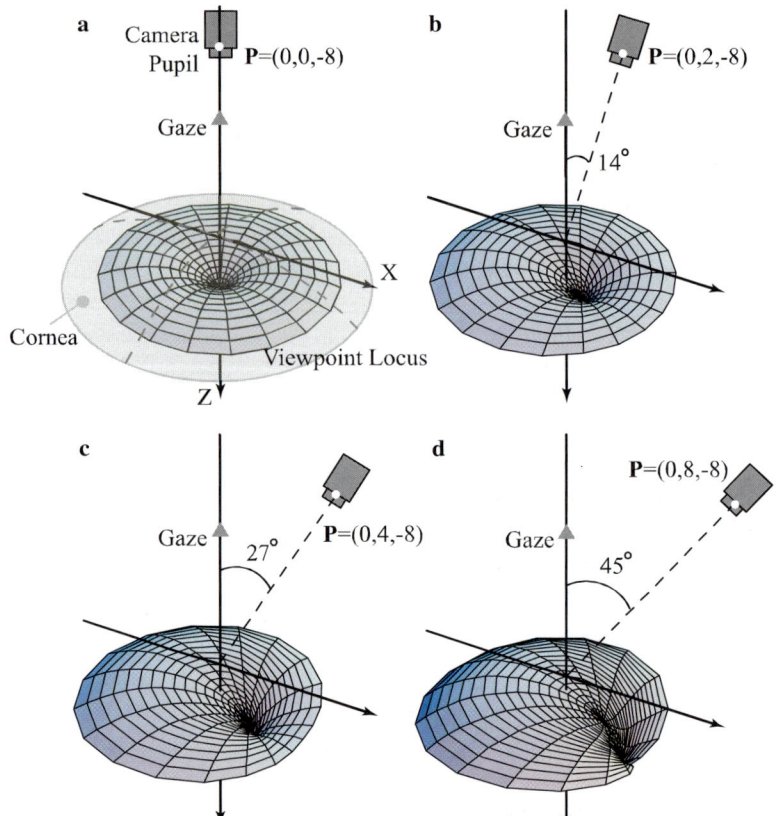

Fig. 7.9a–d. The viewpoint loci of the corneal imaging system for different relative orientations between the eye and the camera. The camera observes the eye from **a** $0°$ (camera pupil $\mathbf{P} = (0, 0, -8)$), **b** $14°$ ($\mathbf{P} = (0, 2, -8)$), **c** $27°$ ($\mathbf{P} = (0, 4, -8)$) and **d** $45°$ ($\mathbf{P} = (0, 8, -8)$). The viewpoint locus is always smaller than the cornea and has a cusp whose location depends on the relative orientation between the eye and the camera

and the camera exceeds $40°$ for a reasonable eye-camera distance. Also note that we are not taking into account other occlusions such as those due to other facial features including the eyelids and the nose.

Figure 7.10 depicts how the field of view varies with different eye-camera configurations. Notice how the field of view gradually decreases as the viewing angle of the camera increases (P_x increases). However, the field of view remains large for all viewing angles and for this particular setting ($P_z = 8$) it is always larger than a hemisphere ($2\pi \approx 6.28$). We found that for all the cases we encountered in the later examples, the corneal FOV was always larger or almost equal to a hemispherical field of view.

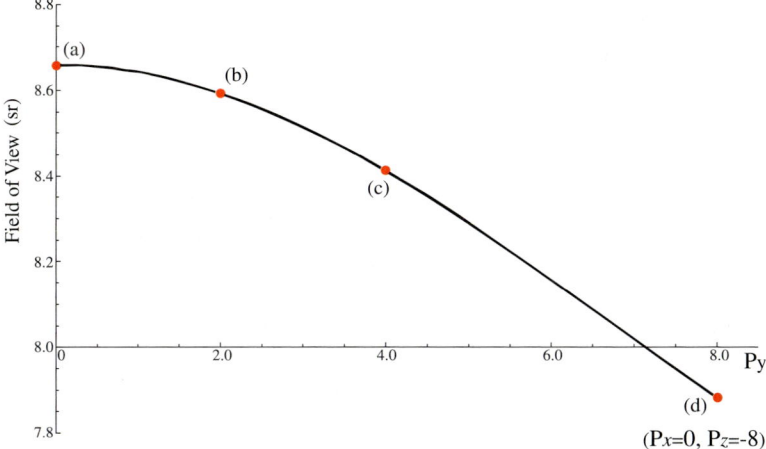

Fig. 7.10a–d. The field of view of the corneal imaging system as a function of the location of the camera pupil ($\mathbf{P} = (0, P_y, -8)$). The *red dots* **a**–**d** correspond to the four configurations used in Fig. 7.9. Note how the field of view gradually decreases as the viewing angle of the camera increases. The field of view of the corneal imaging system is generally large. In this example setting, it always exceeds a hemispherical field of view

Figure 7.11 depicts the corneal FOVs for the four different eye-camera configurations as shaded regions of unit spheres. The monocular field of view of human perception is roughly $120°$ [675]. In Fig. 7.11, the red contour on each sphere represents the boundary of this human FOV. It is interesting to note that, for all the eye-camera configurations shown, the corneal FOV is always greater than, and includes, the human FOV. That is, the corneal system generally produces an image that includes the visual information seen by the person to whom the eye belongs.

7.4.3 Resolution

We now study the spatial resolution of the corneal imaging system. As shown in Fig. 7.12, we use the same corneal coordinate system as before with the camera pupil located at $\mathbf{P} = (P_x, 0, P_z)$. We can assume that the camera pupil lies on the XZ plane without loss of generality. However, the image plane has another degree of freedom; tilting around the Z axis as depicted in Fig. 7.12. The angle between the pixel where the light ray hits and the center of projection $\alpha(t, \theta)$ as well as the angle between the pixel and the Z axis $\beta(t, \theta)$ can be computed for each corneal surface point.

Now, let us consider an infinitesimal area δA on the image plane. This small area is a projection of a scene area by the reflection on the cornea. The resolution of the imaging system at this particular pixel (the center of area

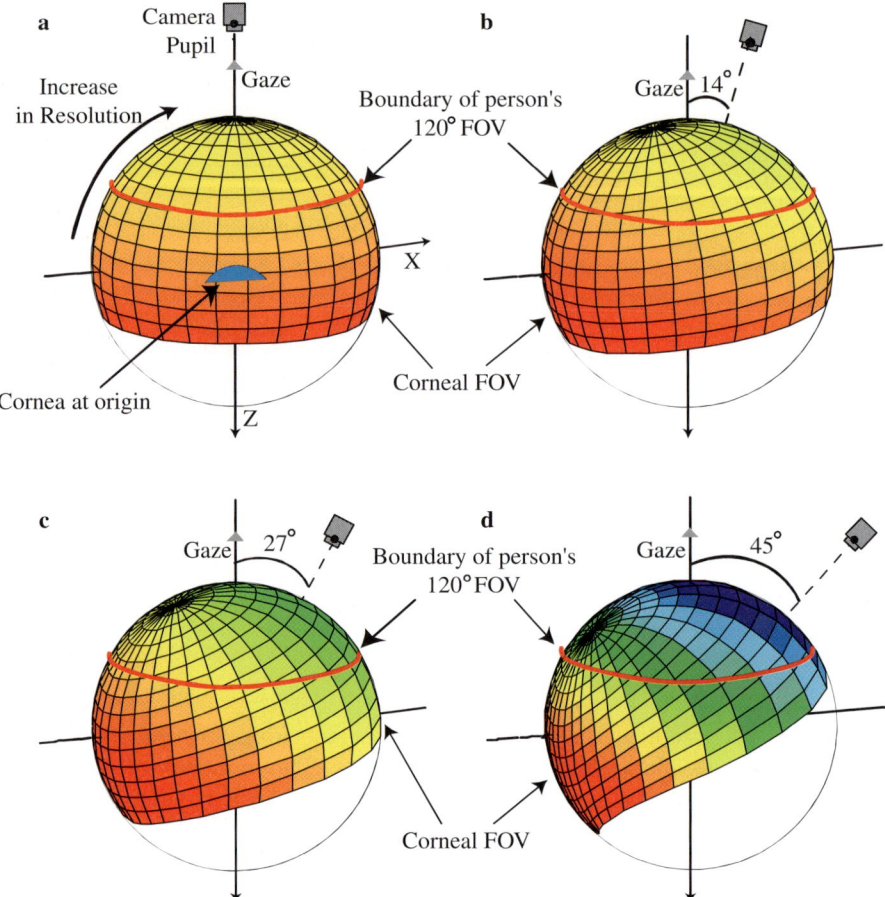

Fig. 7.11. The field of view and the spatial resolution of the corneal imaging system for the four different eye-camera orientations ((a)–(d)) used in Fig. 7.9. The shaded regions on the spheres represent the corneal FOV. The colors within the shaded regions represent the spatial resolution (resolution increases from red to blue). The red contour shows the boundary of the FOV of the human eye itself (120°). Note that the corneal FOV is always greater than this human FOV and the resolution is highest around the gaze direction and decreases towards the periphery

δA) can be defined as the ratio of the infinitesimal area δA on the image plane to the solid angle $\delta\Omega$ subtended by the corresponding scene area δB from the viewpoint locus.

First, the area perpendicular to the chief ray that projects to surface area δA on the image plane is

$$\delta\tilde{A} = \delta A \cos\alpha(t,\theta) \ . \tag{7.11}$$

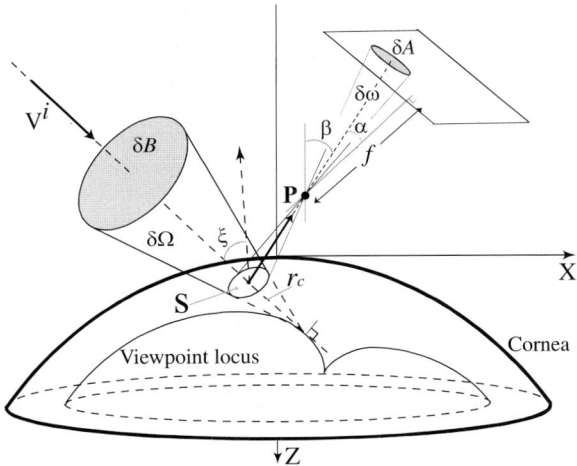

Fig. 7.12. The resolution of the corneal imaging system can be defined as the ratio of the solid angles δw and $\delta \Omega$. This resolution varies over the field of view of the corneal system

The solid angle δw subtended by this area from the camera pupil is

$$\delta w = \left(\frac{f}{\cos \alpha} \right)^2 \delta \tilde{A} \tag{7.12}$$

Then, the surface area of the cornea that corresponds to this area can be computed as

$$\delta \tilde{B} = \delta w \left(\frac{S_z(t, \theta) - P_z}{\cos \beta} \right)^2 \cos \xi(t, \theta) \, . \tag{7.13}$$

where S_z and P_z are the Z coordinates of the corneal surface point and camera pupil. Finally, the solid angle $\delta \Omega$ from the viewpoint locus can be computed as

$$\delta \Omega = \frac{1}{r_c^2(t, \theta)} \frac{\delta \tilde{B}}{\cos \xi(t, \theta)} \, . \tag{7.14}$$

Therefore, the resolution we gain at a pixel whose corresponding incident light ray hits corneal surface point $\mathbf{S}(t, \theta)$ can be computed as

$$\frac{\delta A}{\delta \Omega} = \frac{\left(f \frac{\cos \beta(t, \theta)}{\cos \alpha(t, \theta)} \right)^2 r_c^2(t, \theta)}{\cos \alpha(t, \theta)(S_z(t, \theta) - P_z)^2} \, . \tag{7.15}$$

Note Eq. 7.15 coincides the resolution equation derived in [594] when $\alpha(t, \theta) = \beta(t, \theta)$, i.e. when the optical axis of the camera and the cornea are aligned.

Figure 7.13 shows the spatial resolution characteristic of the corneal imaging system along the XZ cross section of the cornea. The resolution variations

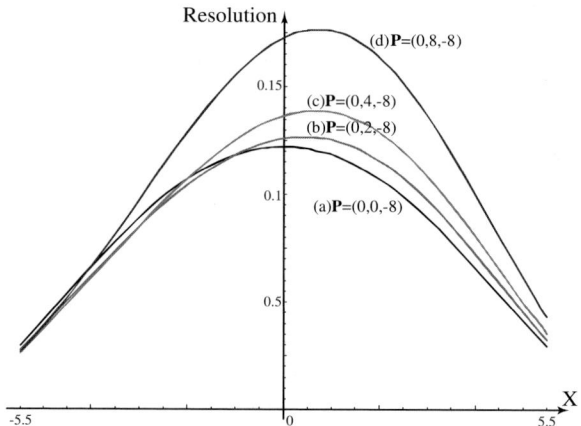

Fig. 7.13. Spatial resolution characteristic of the corneal imaging system on the corneal surface. The resolution variation along the cross section of the cornea sliced by the XZ plane is depicted. The resolution is highest around the gaze direction and the maximum resolution increases as the viewing angle of the camera increases ((a) to (d))

are depicted for the four different eye-camera configurations used in Fig. 7.9. Note that the resolution is highest around the gaze direction (optical axis of the cornea which is the Z axis) and the maximum resolution increases as the viewing angle of the camera increases.

In Fig. 7.11, the colors shown within each FOV represent the spatial resolution of the system. Here, resolution increases from red to blue. Notice how the resolution changes inside the red contour which corresponds to the human FOV. The highest resolution is always close to the gaze direction and decreases towards the periphery of the FOV. It is well known that the resolution of the human eye also falls quickly[6] towards the periphery [315]. The above result on resolution shows that, irrespective of the eye-camera configuration, the corneal system always produces an image that has roughly the same resolution variation as the image formed on the retina of the eye.

7.4.4 Epipolar Geometry of Eyes

When both eyes of a person are captured in an image, the combination of the two corneas and the camera can be viewed as a catadioptric stereo system [433, 435]. The methods we developed to calibrate a corneal imaging system can be used to determine the relative orientations between the two

---

[6] The spatial acuity of the retina decreases radially. This reduction is rapid up to 3° from the center and is then more gradual up to 30°. Then, it once again decreases quickly to the periphery [315].

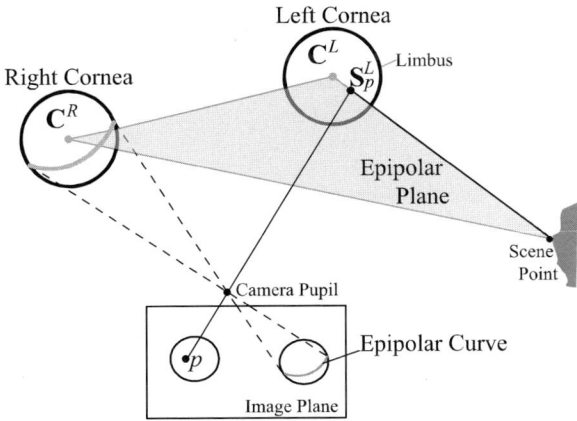

Fig. 7.14. The epipolar geometry of the corneal stereo system. The epipolar plane intersects the eye at a curve which projects onto the image plane as a curve (the epipolar curve)

corneas and the camera. Here, we derive the epipolar geometry of a corneal stereo system [479].

Consider the corneal stereo system shown in Fig. 7.14. Once we locate the two limbuses in the image (Sect. 7.3), we can compute the 3D coordinates of the cornea (center of the limbus) $\mathbf{C}^R$ and $\mathbf{C}^L$ and the orientation (θ^R, τ^R) and (θ^L, τ^L) for the right and left cornea, respectively. Now, let us denote the orientations of the major axes of the right and left limbuses in the image with unit vectors $\mathbf{A}^R = [\cos\phi^R \;\; \sin\phi^R \;\; 0]^T$ and $\mathbf{A}^L = [\cos\phi^L \;\; \sin\phi^L \;\; 0]^T$. Then the corneal surface points of the right cornea can be expressed in the camera's coordinate frame (origin at the camera pupil and Z axis perpendicular to the image plane) as

$$\mathbf{S}^R(t, \theta) = \mathbf{T}(\mathbf{A}^R, \tau^R)(\mathbf{S}(t, \theta) - [0\; 0\; t_b]^T) + \mathbf{C}^R \;, \tag{7.16}$$

where $\mathbf{T}(\mathbf{A}, \tau)$ is the 3×3 matrix for rotation by τ around the axis $\mathbf{A}$, $t = t_b$ is the plane that the limbus lies on and the superscript T stands for transpose. $\mathbf{S}$ and $\mathbf{C}$ are column vectors. $\mathbf{S}^L(t, \theta)$ can also be derived with the same equation by substituting $\mathbf{A}^R, \tau^R$ and $\mathbf{C}^R$ accordingly. Then, the surface normals for each point on each cornea can also be computed as,

$$\mathbf{N}^R(t, \theta) = \mathbf{T}(\mathbf{A}^R, \tau^R)\mathbf{N}(t, \theta) \;. \tag{7.17}$$

where, $\mathbf{N}(t, \theta)$ is the surface normal of $\mathbf{S}(t, \theta)$ computed in the corneal coordinate frame (Fig. 7.3).

Now let us consider a 2D image point $\mathbf{p}^L$ which lies in the imaged left cornea:

$$\mathbf{p}^L = [p_x^L \; p_y^L \; -f]^T \;, \tag{7.18}$$

where f is the focal length. Here, the image coordinates (p_x, p_y) are normalized with respect to the width, height and the center of projection of the image. The surface point on the left cornea corresponding to this image point $\mathbf{S_p^L} = \mathbf{S}^L(t_{\mathbf{p}^L}, \theta_{\mathbf{p}^L})$ can be computed by solving

$$\gamma \frac{-\mathbf{p}^L}{|\mathbf{p}^L|} = \mathbf{S}^L(t_{\mathbf{p}^L}, \theta_{\mathbf{p}^L}) \tag{7.19}$$

for $\gamma, t_{\mathbf{p}^L}$ and $\theta_{\mathbf{p}^L}$ which satisfies $0 \le t_{\mathbf{p}^L} < t_b$ and the corresponding γ is the smallest positive of the solutions.

As we have seen in Sect. 7.4.1, the viewpoint locus is small compared to the distance between the two eyes. Therefore, as shown in Fig. 7.14, for scenes that are not extremely close to the eyes, we can safely assume the viewpoint locus of each cornea to be a point located at the center of the limbus, i.e. $\mathbf{C}^R$ and $\mathbf{C}^L$. As a result, as shown in Fig. 7.14, the epipolar plane corresponding to point $\mathbf{p}^L$ in the left cornea intersects the right cornea at a curve. The points on this 3D curve must satisfy the constraint

$$\left(\frac{\mathbf{C}^R - \mathbf{C}^L}{|\mathbf{C}^R - \mathbf{C}^L|} \times \frac{\mathbf{S_p^L} - \mathbf{C}^L}{|\mathbf{S_p^L} - \mathbf{C}^L|} \right) \cdot (\mathbf{S}^R(t, \theta) - \mathbf{C}^R) = 0 . \tag{7.20}$$

The above equation can be solved to obtain the azimuth angles $\tilde{\theta}(t)$ that correspond to the 3D curve. We now have the 3D curve on the right cornea: $S^R(t, \tilde{\theta}(t))$. The projection of this 3D curve onto the image plane is a 2D curve

$$\left(\frac{f S_x^R(t, \tilde{\theta}(t))}{S_z^R(t, \tilde{\theta}(t))}, \frac{f S_y^R(t, \tilde{\theta}(t))}{S_z^R(t, \tilde{\theta}(t))} \right) , \tag{7.21}$$

where the subscripts denote the Cartesian coordinates of $\mathbf{S}$. For each point in the image of the left cornea, we have the corresponding epipolar curve within the image of the right cornea. These epipolar curves can be used to guide the process of finding correspondences between the corneal images and recover the structure of the environment.

7.5 The World from Eyes

We are now in a position to fully exploit the rich visual information captured by the corneal imaging system. We first show that a single image of an eye can be used to determine the environment of the person as well as what the person is looking at.

7.5.1 Where Are You?

Once we have calibrated the corneal imaging system as described in Sect. 7.3, we can trace each light ray that enters the camera pupil back to the scene

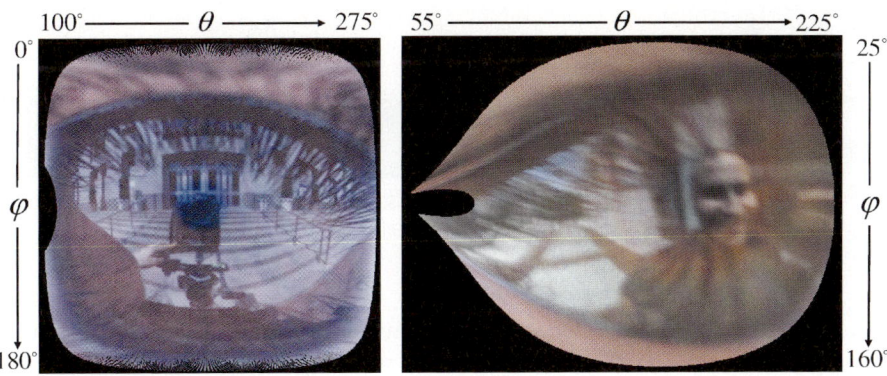

Fig. 7.15. Spherical panoramas computed from the eye images shown in Fig. 7.1. Each spherical panorama captures a wide-angle view of the world surrounding the person to whom the eye belongs. It reveals the location and circumstance of the person when the image was captured

via the corneal surface using Eq. 7.6. As a result, we are able to recover the world surrounding the person as an environment map. We will represent this map as a spherical panorama.

Figure 7.15 shows spherical panoramas computed from the images of eyes shown in Fig. 7.1. These panoramas are cropped to include only the corneal FOV. As predicted in Sect. 7.4.2, the corneal FOV is large and provides us a wide-angle view of the surrounding world. This allows us to easily determine the location and circumstance of the person at the time at which their image was captured. For instance, we can clearly see that the person in the left example is in front of a building with wide stairs. We see that the person in the right example is facing another person. Since these panoramas have large fields of view, one can navigate through them using a viewer such as Apple's QuickTime VR^{TM}.

Note that the computed environment map includes the reflections from the iris as well. It is preferable to subtract this from the environment map for better visualization of the scene surrounding the person. However, the problem of separating reflections from the iris and the cornea is significant by itself and will be pursued in future work. Also, notice that the dynamic range and absolute resolution of the recovered environment maps are limited by those of the camera we use.

If we have multiple images of the eye taken from the same location[7] as shown in Fig. 7.16a, we can merge multiple spherical panoramas to obtain a even wider-angle view of the surrounding environment. However, since the

---

[7] Note that, since 3D coordinates and orientation of the cornea are computed in the camera's coordinate frame, if the camera is moved, its location has to be known in order to register the computed spherical panoramas.

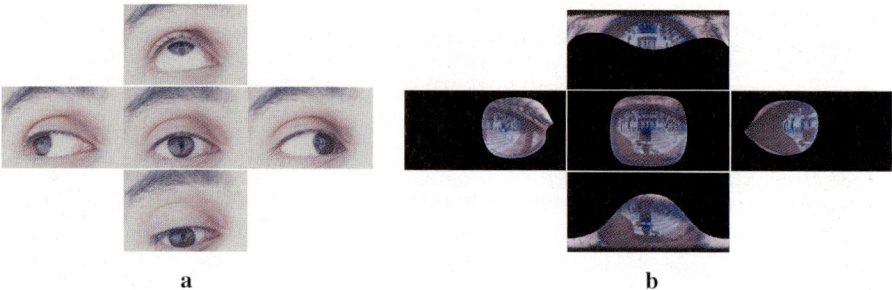

Fig. 7.16. a 5 images of an eye with different extreme gaze directions taken from the same viewpoint. **b** Spherical panoramas computed from each eye image in **a**

estimates of corneal orientation include a few degrees of errors (see Sect. 7.3.3) simply overlaying spherical panoramas computed from different eye images can introduce misalignment artifacts. In order to precisely register these multiple spherical panoramas, we developed a simple registration algorithm.

The registration of multiple spherical panoramas can be accomplished by minimizing the distortion in the overlaid spherical panoramas. Since the pixel values of each spherical panorama are computed from the 3D coordinates and orientation of the cornea, this minimization should be accomplished by varying the three parameters: d and (ϕ, τ). The canonical way to measure the distortion would be to use colors at each point on the environment map. However, since the iris texture underlies each pixel value which we do not attempt to separate out in this work, direct use of pixel values would not be a robust metric of distortion. Instead, we manually specify corresponding points inside the corneal region in each eye image pair and minimize the angular distance between those points on the environment map. Although three corresponding points will provide a closed form solution, for robustness, we specify n (≥ 3) points and accomplish iterative least square minimization.

Figure 7.17 shows a spherical panorama computed by merging spherical panoramas computed from the 5 images in Fig. 7.16a. Figure 7.16b shows individual spherical panoramas computed from each eye image. Eyelids and eyelashes in the middle eye image were masked out before registration. Figure 7.17 clearly covers a larger field of view compared to the left spherical panorama in Fig. 7.15, where only the middle image in Fig. 7.16a is used. Notice that we can see windows located above the entrance of the building, the blue sky above the building, a black pole which was right next to the person's right eye, the green loan field on the extreme right side, etc. These provide more information to understand the surrounding world of the eye.

Fig. 7.17. Registered and merged spherical panoramas computed from the eye images in Fig. 7.16a. A wider-angle view of the surrounding environment can be obtained compared to the spherical panorama computed from a single eye image (left of Fig. 7.15)

7.5.2 What Are You Looking at?

Since we know the gaze direction, each eye image can be used to determine what the person is looking at from the computed environment map. The gaze direction tells us exactly which point in the environment map lies along the optical axis of the eye. A perspective projection of the region of the environment map around this point gives us an estimate of the image falling on the retina centered at the fovea. We will call this the foveated retinal image.

Figure 7.18 shows several spherical panoramas and foveated retinal images computed from images of eyes captured in various settings[8]. We show the foveated retinal images with a narrow field of view (45°) to better convey what the person is looking at. In scenario (I), we see that the person is in front of a computer monitor and is looking at the CNN logo on a webpage. The person in scenario (II) is meeting two people in front of a building and is looking at one of them who is smiling at him. The person in scenario (III) is playing pool and is aiming at the yellow ball. Note that although the eye-camera configurations in these three cases are very different, all the computed foveated retinal images have higher resolution around the center ($\approx$ gaze direction) as predicted by our analysis in Sect. 7.4.3. The above examples show that the computed foveated retinal images clearly convey where and what the person is looking at in their environment.

---

[8] Note that these results are obtained from single images.

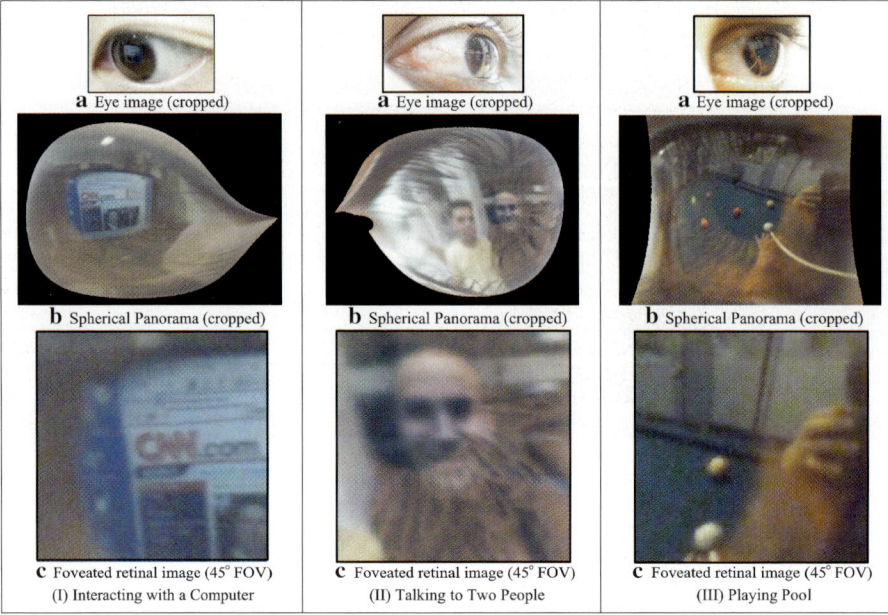

Fig. 7.18a–c. Examples of spherical panoramas and foveated retinal images computed from images of eyes. Three different examples are shown. Each example includes **a** a cropped image of the eye, **b** a cropped image of the computed spherical panorama and **c** a foveated retinal image with a 45° field of view. The spherical panoramas and foveated retinal images clearly convey the world surrounding the person and what and where the person is looking. This information can be used to infer the person's circumstance and intent

7.5.3 What is its Shape?

As described in Sect. 7.4.4, when both eyes of a person are captured in an image, we can recover the structure of the environment using the epipolar geometry of the corneal stereo system. Since our computed environment maps are inherently limited in resolution, one cannot expect to obtain detailed scene structure. However, one can indeed recover the structures of close objects that the person may be interacting with.

Figure 7.19a shows an image of the eyes of a person looking at a colored box. In Fig. 7.19b, the computed epipolar curves corresponding to four corners of the box in the right (with respect to the person in the image) cornea are shown on the left cornea. Note that the epipolar curves pass through the corresponding corner points. A fully automatic stereo correspondence algorithm must contend with the fact that the corneal image includes the texture of the iris. Removal of this iris texture from the captured image is a significant problem by itself and is beyond the scope of this chapter. Here, we circumvented the problem by manually specifying the correspond-

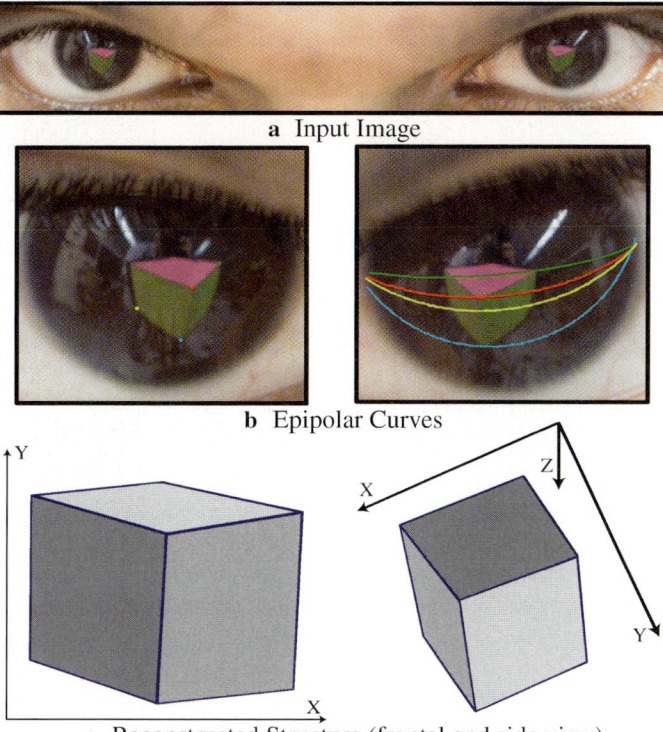

a Input Image

b Epipolar Curves

c Reconstructed Structure (frontal and side view)

Fig. 7.19. a An image of two eyes looking at a box. **b** The epipolar curves on the left cornea corresponding to four corners of the box in the right cornea. **c** A reconstructed wire-frame of the box

ing points on the computed epipolar curves. From these correspondences, we reconstructed a wire-frame of the box, as shown in Fig. 7.19c. This ability to recover the structure of what the person is looking at can be very useful in communicating with a machine such as a robot. For instance, the robot can be programmed by capturing a video of a person's eyes while he/she manipulates an object.

7.6 Implications

We presented a comprehensive framework for extracting visual information from images of eyes. Using this framework, we can compute: (a) a wide-angle view of the environment of the person; (b) a foveated retinal image which reveals what the person is looking at; and (c) the structure of what the person is looking at. Our approach is a purely image-based one that is passive and non-invasive. We believe our framework has direct implications for the following fields.

Visual Recognition

From the computed environment map, one can easily determine the location of the person as well as infer the circumstance he/she was in when the image was captured. Such information can be of great value in surveillance and identification applications.

Human-Machine Interfaces

Our results can extend the capability of the eye as a tool for human-machine interactions [55, 272, 285]. The eye can serve as more than just a pointer; it is an imaging system that conveys details about the person's intent. In the context of teaching robots [277, 311], the ability to reconstruct the structure of what the person is looking at is powerful.

Computer Graphics

The environment map recovered from an eye not only visualizes the scene surrounding the person but it also tells us the distribution of the lights illuminating the person [120]. Such illumination information can be very useful, for instance, to relight objects in the scene or the face of the person [50, 121, 393, 633]. Interested readers are referred to [450]. The extracted illumination information can also be used for illumination-invariant face recognition [448].

Human Affect Studies

By capturing the eyes as well as the facial/body expressions in an image, it is possible to record what the person is looking at and his/her reaction to it with perfect synchronization. Such data is valuable in human affect studies. Such studies give us important insights into human emotions [159, 162] and the way social networks work [625, 666].

7.7 Conclusion

We have shown that eyes can serve as much more than just visual sensors for the person to whom the eye belongs. We showed that the combination of the cornea of an eye and the camera viewing the eye form a catadioptric imaging system. We can self-calibrate this corneal imaging system by locating the limbus in the image and computing the 3D coordinates and orientation of the cornea in the camera's coordinate frame. Once this self-calibration is done, we can compute a wide-angle view of the surrounding environment and an approximation of the projection on the retina of the person. Furthermore, we showed that from two eyes in an image, we can compute the structure of the object in the scene. In other words, we can exploit the visual information embedded in the appearance of eyes which convey rich information about the person's intent and circumstances. We believe these results have strong implications in various fields.

Part III

Gaze-based Interactions

8

Gaze-based Interaction

Takehiko Ohno and Riad I. Hammoud

8.1 From Eye Movement to Interaction

Nowadays, there are variety of computers available in our everyday life. Personal computers (PC) exist in the office and the home to support people's activities. Cellular phones and Personal Digital Assistants (PDA) now carry huge amounts of information, and also work as a powerful data terminal connected to a wireless network for supporting their user's ubiquitous life. TV sets in the living room are now ready to work with a set top box connected to a broadband network to access huge amounts of online content. In the car, audio systems and satellite navigation systems require complicated user operation. However, operation style with those variety of computers is still limited. Traditionally, users are required to use their hands to control an input device like a keyboard, mouse, pen, touch screen or remote controller for computer operation. Even though the computer style is diversified, we have few choices to control them.

Interacting with the computer by human eye gaze, or shortly, gaze-based interaction, provides a new interaction of style with the computer. The fundamental idea of gaze-based interaction is to measure the user's gaze position on the computer screen with the eye tracking system, and applications react to the user's gaze at some level. Eye tracking was initially a tool for psychologists to conduct experiments; during the last decade, a tremendous effort has been made on developing robust gaze tracking tools for various applications. Current gaze tracking systems are mainly used in two different fields: in medical diagnostics and for interaction [146].

Because the eye moves quickly to acquire information on the computer screen, the gaze position provides the key information to estimate user's intent, and it reduces the operation time and operation workload. Moreover, a user is able to interact with the computer even if both hands are busy with different tasks. A variety of applications have been proposed for supporting our everyday activities.

Nielsen described traditional user interfaces as belonging to "Command-Based" interfaces where users have been required to pay close attention to the control of their computer [444]. That is, users are required tp provide the appropriate commands in the appropriate syntax. Many gaze-based applications including eye keyboard, menu selection and other gaze-controllable Graphical User Interface (GUI) belong to the command-based interface, where users are required to point the target with their eyes, and then select it explicitly. On the other hand, there are different types of user interfaces, which can be called "non-command" interfaces [282, 444]. They allows users to focus on the task rather than on operating the computer [444].

In other words, the applications can be selective; the gaze is used as a control medium, or gaze-contingent; that is, the system is aware of the user's gaze and may adapt its behavior based on the visual attention of the user [146].

In this section, we focus on interactive applications that utilize gaze tracking in real time.

8.1.1 Nature of Eye Movements

Before starting to describe the detail of gaze-based applications, we first introduce the features of eye gaze from the point of view of gaze-based interactions.

1. Fast. When we look at a target, the eye moves quickly to catch the target with the the high-resolution area of the retina (called *fovea*). This is called a saccadic eye movement (or shortly, saccade), which is known as the very fast movement. The peak angular speed of the eye during a saccade achieves about 700 degrees per second. During a saccade, people are not able to control the eyes direction or target position, and if the final position is different from the target, an additional saccade occurs.
2. Unstable. Precise control of gaze position is difficult because the size of the fovea restricts the accuracy of eye targeting. In addition, the eye moves slightly even if it stays on the target. This is called a *micro saccade*. Once the eye stops on the target, it stays for a short period, generally from 100 msec to 300 msec (eye fixation). It sometime lasts more than 1000 msec. However, longer dwells on the same position are difficult and causes fatigue.
3. Uncontrollable. In general, the eye jumps by saccade to achieve the target; the eye does not move continuously except when following a slowly moving target. Intentional and exact eye movement is difficult if there is no target at the destination. For example, when users control the mouse cursor by gaze, they feel labored in comparison with hand operation.

When we build gaze applications, it is necessary to consider that the nature of eye movement is completely different from hand motor control.

Therefore, different design criteria and different approaches are required for designing applications.

8.2 Eye Typing and Various Selection Tasks

Early work on interactive eye gaze tracking applications [272,364,609] focused primarily on users with disabilities for whom the control of the eyes may be the only means of communication. The first applications were "eye typing" systems, where the user could produce text through gaze inputs. Over the years, a number of eye typing systems have been developed (for a review, see [389]). In those systems, the user selects a character or a word from the on-screen keyboard by gaze. Recent methods enable the user to type up to 25 words per minute through continuous gestures [664] rather than the more common on-screen keyboards [234]. In addition to text entry, there are gaze-controlled applications for drawing [263], music [264], games, Internet browsing, email, environmental control etc. Such applications are typically included in commercial eye communication systems targeted at people with disabilities [234, 358]. They belong to the command-based interface, where the user is required to select the command explicitly by gaze.

8.2.1 Midas Touch

When designing gaze-controlled applications, we must deal with a significant problem in both design and implementation, the difficulty of interpreting user eye movement [548]. Also, gaze data coming from the gaze tracking system will contain some measurement error. Therefore, applications that apply gaze for traditional WIMP (Window, Icon, Menu, and Pointer) interface design cause serious uncontrollability.

The most naive approach is to control applications to use eye position as a direct substitute for a mouse. Using the eye as an input device is essential for disabled people who have no other means of controlling their environment. However, the eye is a perceptual organ and has not evolved to be a control organ. It is therefore important that gaze applications do not force the users to use it as such. Another important issue with gaze-based applications is the so called *Midas touch* [284] problem: everywhere you look something gets activated. The main reason for this is that gaze only provides information about coordinates on the screen, but not the users intentions with the gaze. At least one additional bit (i.e. mouse click) of information is needed to avoid the Midas touch.

A variety of solutions have been developed [232, 643, 694]), however, the Midas touch remains a common problem in gaze-controlled applications.

One common solution is to use dwell-time activation – that is, an area such as a button gets selected after the user has dwelled her gaze on the area for a certain time limit [232, 284], i.e. 1000 msec [284]. It should be longer than the normal fixation, which is less than 300 msec in general. However, overly long dwell times make selection speed too slow.

The problem with dwell-time approach is that it requires the user to wait for a period for each selection. To overcome this problem, Ohno [463, 467] proposed a very fast menu selection method, which is called "Quick Glance Selection Method" (QGMS). In QGMS, there is a menu which consists of the command name area and the selection area When the user looks at the command name area, a marker appears in the selection area, which helps the user's next eye movement, proceed to the target selection area. Once the user looks at the selection area, the menu is selected. Direct gaze at the selection area also causes selection. Frequent use of the menu enables the user to remember the exact position of the each menu label. Therefore, the skilled user can select the menu by just looking at the selection area, which enables very fast selection.

Salvucci et al. proposed a probabilistic method for object selection [547, 548]. In their method, each object is assigned a score, and the probability of object selection is calculated from the eye movement pattern in real-time. Once the probability exceeds a threshold, the object is selected. They designed a GUI system which is controlled with the proposed method. They called it"gaze-added interface" which means that gaze controllability is added to the normal user interface design.

Zhai et al. [711] developed a technique that avoids the two most common problems of gaze input – the inaccuracy of the measured point of gaze and the Midas touch problem – but still benefits from the speed of eye pointing. They used a combination of eye and hand pointing. Typically, the user looks at an item to interact with. Eyes are fast, and they are on the target long before the user moves the mouse. MAGIC (manual and gaze input cascaded) pointing warps the mouse cursor automatically to the vicinity of the target, thus saving the user from large mouse movements. The hand is used to accurately select the target with the mouse, thus avoiding both the Midas touch and the inaccuracy problems. To the user MAGIC pointing appears as a manual task, with reduced physical effort. However, if the cursor moves along with the gaze all the time, it may disturb the user. Zhai et al. [711] conducted an experiment that compared the conventional mouse-only setup with two different methods of gaze-enhanced MAGIC pointing: (1) a liberal approach where the cursor follows the eye constantly, and (2) a more conservative approach where the cursor is not warped until the user touches the mouse in order to move it. Surprisingly, the moving cursor did not disturb the users. They preferred the liberal technique because the conservative technique required more practice and caused uncertainty of not knowing where the cursor would appear. The users appreciated the reduced physical effort of MAGIC pointing so much that some even felt disappointed after the experiment when they realized the automatic cursor warping no longer worked.

8.3 Interaction Without Selection

Instead of using the eyes as an explicit control medium, an application can make use of the information of the user's eye movements subtly in the background without disturbing the user's normal viewing. such an interface belongs to the non-command interfaces as they do not require an explicit command selection by the user.

8.3.1 Interactive Video Applications

Gaze direction is used in several interactive video systems [216, 218, 224]. Bolt [56] introduced a system called the Gaze-Orchestrated Dynamic Windows. It consisted of multiple video images, a "World-of-Windows" as Bolt called it. Having several video streams playing simultaneously is like trying to watch several TV channels at once, more than a person can absorb. The video stream the user was looking at was zoomed in; the size of the video increased, and the level of sound of that video was amplified. Accordingly, the other windows faded out, decreasing in video size and sound level. The work on active window control by gaze was continued by Jacob [284] with the Listener Window system that used gaze direction to control the selection of the active task window, instead of e.g. a mouse click. To avoid erroneous window activations, the system used a delay that allowed a brief glance to other windows without activating them. Fono and Vertegaal [186] experimented with EyeWindows media browser and desktop EyeWindows. They found that using eyes to indicate the focus but letting the user perform the actual selection with a key press works considerably better than using the gaze alone. Eye input (alone) was the fastest selection method, when compared to manual input or combined eye and manual input. However, eye input alone is too prone to false activations. For example, if automatic activation of a window is used, the user may not be able to glance at other windows while giving commands to the current window, as the newly activated window would steal the input. Fono and Vertegaal also found that the benefit of using eye input for focus activation increases as the number of windows increases.

8.3.2 Attentive Displays and Appliances

Gaze-contingent displays also utilize the user's eye movement without command selection. According to [286], Tong and Fisher [626] were among the first to exploit the information of the user's focus of attention in flight simulators with large screens. The system provided a high resolution, max wherever the user was fixating and low resolution for the periphery, thus creating an illusion of a large high-resolution display. By reducing the level of detail in the periphery, it was possible to save transmission bandwidth and processing power. Displaying lower resolution in the periphery does not impair the quality of viewing because the human peripheral vision has limited resolution as well; only items under focus are seen in detail. Various similar gaze-contingent multiresolutional displays have been developed since, for a review, see [29, 148, 520].

Attentive user interfaces [644] benefit from the information of the user's area of interest (AOI), and change the way information is displayed or the way the application behaves depending on the assumed attentive state of the user.

For some applications, simply detecting the presence of eyes or recognizing eye contact can enhance the interaction substantially. EyePliances [562] are appliances and devices that respond to human eye contact. The Attentive Television, with an embedded low cost eye contact sensor, is an example of such EyePliance. The Attentive TV will automatically pause a movie when the eye contact is lost, deducing nobody is watching the video. An advanced version of the Attentive TV, the EyeDisplay [646], is able to detect gaze direction at low resolution; it is able to detect up to 6 tiles of media (regions on the display) at a time (see Fig. 8.1). The viewer selects the media stream by looking at it and pressing the play button.

EyeBlog [128] is an eye-contact aware video recording and publishing system, which is able to automatically record face-to-face conversations. EyeBlog uses Eye-Contact Sensing Glasses (ECSGlasses [561]) that have a wireless eye contact sensing camera integrated into the frame. The ECSGlasses do not detect the gaze direction but are able to detect when a person is looking at the wearer of the glasses. EyeBlog triggers recording automatically as soon as it detects the eyes of the other person, and automatically saves video recordings

Fig. 8.1. EyeDisplay is able to detect which of the six video tiles the viewer is looking at. (Courtesy of Roel Vertegaal, Queen's University)

of the conversations. Shell et al. [561] have implemented a number of variations of EyePliances. For example, Look-To-Talk EyePliances with embedded
speech recognition know when they are spoken to. Such device reacts to e.g.
a "turn on/off" command only when a person is looking at it. If EyePliances
become common, they could potentially be of great benefit, especially for
people with motor disabilities. Another major benefit of attentive devices is
that they hold the potential of reducing inappropriate intrusions. They not
only know when they are the target of a remote command, but they may also
be able to make assumptions of the user's attentive state. For example, an
attentive cell phone does not disturb a user (wearing ECSGlasses) engaged
in a conversation with another person [561]. The phone can automatically
quiet the ringing tone, or even communicate to the caller that the person is
busy. There are a number of variations of EyePliances and other eye-aware
applications and devices not reported here, see [275] for a review.

8.3.3 Intention Estimation

By monitoring the user's eye movements the application "knows" more about
the user's state and intentions and is able to react in a more natural way,
thus helping the user to work on the task instead of interacting with the computer. The Little Prince Storyteller [583] is a good example of an attentive
application. In it, an embodied agent read aloud the story of Little Prince by
Antoine de Saint-Exupéry. The world of the story was illustrated on a computer screen. As the user watched the images on the screen, the application
made assumptions of the locus of the user's interest based on the user's gaze
patterns and fixations. Objects that got more visual attention were zoomed
in for a closer look and the agent told more about them. Therefore, instead of
just reading aloud the story linearly, the agent was able to change the order
in narration according to the user's interest – without any special effort from
the user.

Several applications have exploited the information of the user's natural
eye movements during reading. The Reading Assistant [567] aids people with
reading disabilities by recognizing words the user has trouble with by analyzing the gaze behavior. The system highlights the difficult word and speaks it
aloud with synthesized speech. The Translation Support System [600] assists
in the demanding task of translating text from Japanese to English. It detects
patterns in the user's eye movements, such as reading, searching (scanning),
and hesitating (pausing). When the system detects the reader has scanned
through the current corpus, it automatically retrieves new information. If the
reader hesitates, the system provides new keywords and corpus that may help
in translating the word. Similarly, iDict [274] aids the user in reading foreign
text. To consult printed or electronic dictionaries forces the user to interrupt
the reading, which also interrupts the line of thought and may thus even
affect the comprehension of the text. iDict aims at reducing the user's cognitive load by proactively providing information that may facilitate reading and

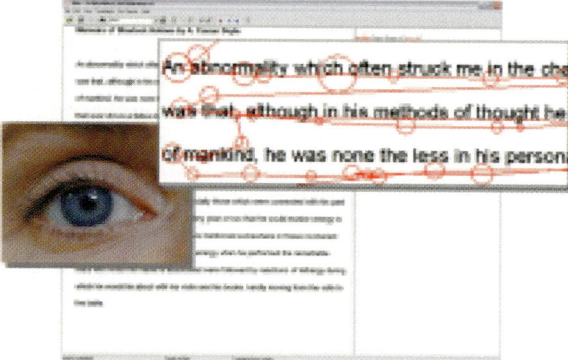

Fig. 8.2. iDict detects deviations in the normal reading based on the gaze behavior. (Courtesy of Aulikki Hyrskykari, University of Tampere)

understanding of the text [276]. iDict detects deviations in the normal reading process based on the gaze behavior (see Fig. 8.2). For example, the user may pause on a difficult word or phrase, or may make regressive fixations to already read portions of text. The system automatically consults embedded electronic dictionaries and shows the best guess for an instant translation. The instant translation is shown right above the problematic word to allow a convenient quick glance. A syntactic and lexical analysis of the text is conducted to improve the linguistic accuracy of the translation. For example, iDict is able to detect the word class, thus translating verbs as verbs and nouns as nouns. If the user is not satisfied with the instant translation, a glance to the Dictionary Frame (located on right side of the text) shows the full dictionary entry for the word.

The Simple User Interest Tracker, SUITOR [388], tracks the user's attention through multiple channels: keyboard input, mouse movements, web browsing, and gaze behavior. SUITOR automatically finds and shows potentially relevant information in a timely but unobtrusive manner in a scrolling display at the bottom of the screen. Gaze direction reveals information not available from other channels. For example, if the system monitors only the mouse activity as the user scrolls a web page, the system cannot accurately asses if the user is interested in the whole page or only a part of the page. Gaze direction reveals which parts of the page the user has read and how much time he or she has spent studying each part. This improves the accuracy of the provided extra information as it helps in determining the relevant keywords that are used in fetching potentially interesting new information.

EyePrint is a technique that uses the user's eye movement to create passive traces of document comprehension for digital documents [461, 462]. The eye gaze traces generated from the user's eye movements become metadata of the document; they can be used in subsequent searches, document brows-

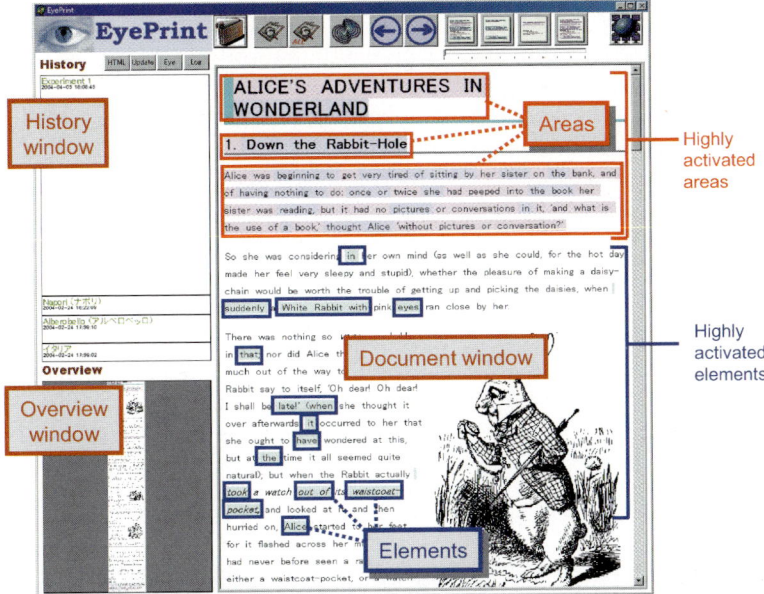

Fig. 8.3. Screen shot of the EyePrint: The digital document is displayed in the document window. Highly activated areas are highlighted

ing, and also zooming. Figure 8.3 illustrates an example of the EyePrint browser. The concept of EyePrint is to recreate the trace of the user's document comprehension process from the eye movement data, and to keep it in the computer. Instead of trying to retrieve the trace from human memory during information retrieval, it is possible to recreate the trace from both human memory and data in the computer.

8.4 Communication with Gaze

The area of gaze applications is not limited to human–computer interaction; it also covers the computer-mediated communications (CMC). Gaze direction plays an important role in face-to-face situations and conversations between humans [317]; it shows if a person's attention is directed at another person or an object. In remote video conferencing, the information on gaze direction is disturbed, and the direct eye contact is often lost, because the camera and the screen are not aligned. The person on the screen seems to look towards the viewer only if she looks directly at the camera. GAZE [645] is an attentive video conferencing system that conveys information about the participants' visual attention. Each participant's gaze direction is tracked via an eye tracking device. The participants meet in a virtual 3D meeting room, which shows an animated snapshot of each participant in a separate video panel. The panel of each participant is rotated towards the person the participant is looking at. This helps the participants to follow who is talking to whom. A lightspot, with different color for each participant, is used to indicate which object the participant is looking at on the shared virtual table. The lightspot helps in resolving ambiguous references (e.g. "look at this"). A newer version of the system, GAZE-2 [647], shows live video of the participant. GAZE-2 uses the information of the participants' attention to also optimize the bandwidth of the streaming media. The active person, who is being looked at, is broadcast in higher resolution than the others, as illustrated in Fig. 8.4.

Weak Gaze Awareness is a video mediated communication system that conveys gaze information to a shared space in the remote location [465]. Unlike existing video mediated communication systems, the system conveys limited and primitive visual cues, gaze position and face direction. The appearance of those cues on the display depends on the distance between the screen and the user's eyes, which allows the user to control the appear-

Fig. 8.4. GAZE-2 indicates who is looking at whom by rotating video panels. (Courtesy of Roel Vertegaal, Queen's University)

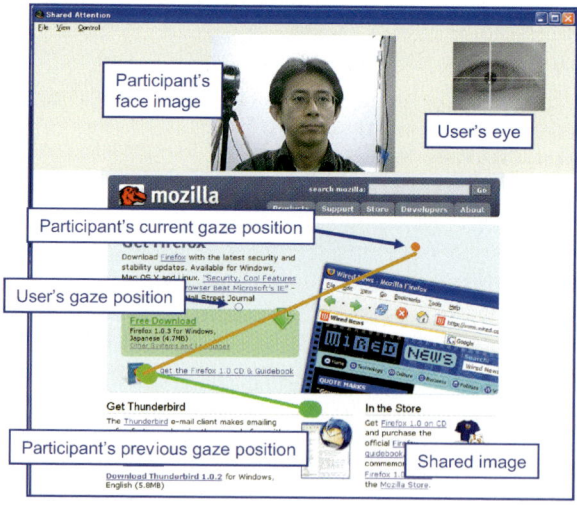

Fig. 8.5. In Weak Gaze Awareness, participant's face image is displayed in the upper area of the screen and his gaze position is displayed in the shared space

ance of her gaze. The last one second of the user's gaze position is represented as continuous lines drawn in the shared space, as illustrated in Fig. 8.5.

Gemmel et al. [195] as well as Jerald and Daily [294] applied a different kind of method in their attempt to solve the problem of the lost eye contact. They manipulated the real-time video by modifying the image of the eyes in the original image so that the person's eyes seemed to look in the correct direction, thus creating the illusion of direct eye contact. They preferred the real video stream over animated avatars because the real video shows the facial expressions as they appear.

Qvarfordt et al. [504, 505] studied eye-gaze patterns in a remote dialogue situation where the participants cannot see each others eyes. An experimental system, RealTourist, allows a tourist and a tourist consultant to interact remotely. They both see the same map displayed on their screens. In addition, the tourist consultant sees the tourist's eye movements superimposed onto the map. The experiment showed that the spatial information of the user's eye movements greatly facilitated the remote conversation. The user's gaze not only revealed items the user was interested in but also helped in resolving references to objects and in interpreting unclear statements. It helped the tourist consultant to ensure they both talked about the same thing and to asses if the tourist had understood instructions. The information from the user's point of gaze facilitated establishing common ground for the conversation. Motivated by the results from the RealTourist experiment, Qvarfordt and Zhai [506] developed a virtual tourist consultant application, iTourist. As a tourist looks at a map and photos from different places, iTourist makes assumptions on what the tourist might be interested in. It then adapts its output accordingly, providing pre-recorded verbal information of the places of interest. The user study showed that, despite occasional mistakes, most of the time iTourist was successful in describing places the users really were interested about. As demonstrated by iTourist, information kiosks also could benefit from gaze tracking.

EyeGuide [153] took the tourist consulting into the real world environment. They experimented with using a lightweight wearable eye tracker for monitoring the user's gaze direction when viewing large public displays, such as a subway map. EyeGuide monitors a traveler's gaze behavior and detects when the traveler appears to be lost. When the user looks at locations on the map, the system provides contextual spoken hints to help the user navigate the map (e.g. "Exit here for the JFK airport"). Instructions are "whispered" to the user's ear via an earphone, to preserve privacy. Using eye gaze tracking in the real world introduces new challenges: the user's eye direction has to be detected in 3D, taking into account the head and body movements. Special calibration techniques and head position and orientation tracking are needed. The experimental system did not allow the user to move around during viewing, to preserve sufficient accuracy of the measured point of gaze.

The applications presented above save the user from an explicit command to select objects, and may therefore help the user to concentrate on absorbing information instead of controlling the application.

8.5 Discussion

As described above, variety of gaze-based applications have been proposed. However, there are still several factors that obstruct the use of these applications in our everyday activities.

- Discomfort of eye tracking. Early eye tracking was known as a painful methodology for psychologists because participants had to wear a heavy head-gear type gaze tracking tool, or their head position was fixed with a chin stand. In addition, a complicated setup procedure including personal calibration limits the application area (see Sect. 5). Recent improvement of gaze tracking technology reduces the discomfort and the complexity of gaze tracking, they are not yet affordable to everyone.
- User limitation. It is well known that gaze tracking fails for a part of users. Especially, for people who wears eyeglasses or contact lenses, accuracy of gaze tracking often drops in, or in worst case, no gaze tracking data is acquired. Covering pupil with the lower eyelid also makes the gaze tracking difficult. Robust gaze tracking methods are require to settle the user limitation.
- Inappropriate interface design. As described in Sect. 8.1.1, eye movement is a complicated procedure that requires the researchers careful interface design. We can look at an object with no effort, but it requires a lot of effort to "select" the object explicitly. It will make the interaction slow and fatigued. This problem sometimes occurs in command-based interfaces. Especially, it is obvious when the application is designed for the frequent and long computer operation.

Those problems become serious when the application area of gaze-based interaction expands. On the other hand, improvement of gaze tracking technology makes the application area wide, and the expansion of application area drives the invention of novel and usable gaze tracking technologies. The high cost of gaze tracking had known as another serious problem, but current powerful personal computers and low-cost video cameras make the gaze tracking much cheaper and easier. Expansion of application area also makes the price of commercial gaze tracking lower. To become affordable, mainstreaming gaze interaction for a mass market is needed [233]. We believe there is a great potential for attentive desktop and home applications, gaze-enhanced games, etc.

8.6 Conclusion

In this chapter, we have briefly reviewed a number of applications that utilize the user's gaze for human–computer interaction and computer-mediated communication . Different from other traditional interaction methods, eye movement provides the unique modality; it moves very fast, it reflects the user's intention, and it also works independently of other modalities. More than forty five years have passed since the mouse was invented in 1961, and it is now used widely. Its style is changed variously and it supports our ubiquitous activities. We believe eye gaze has a great potential to support our everyday life in the different operation style. In the next decade, gaze-based interaction will be used in diverse environments including entertainment, wearable and mobile systems, home control, and intelligent transport systems (ITS).

9

Gaze-Contingent Volume Rendering

Mike G. Jones and Stavri G. Nikolov

9.1 Introduction

Gaze-contingent display (GCD) [148, 486, 520] has been used in a number of guises for many different applications. One of the original drivers of its development was the need, in flight simulators, to reduce the computational cost of generating images, in order that sufficiently high display rates could be achieved. By tracking the pilot's fixation point, higher resolution rendering could be employed only around a small fraction of the display area; outside of this region, blurred images could be shown and, provided that the tracking system responded sufficiently quickly to the pilot's eye movements, no visible deterioration would be experienced.

Volumetric rendering [139, 316, 365] is computationally expensive. Initial approaches to displaying volume data involved massive simplification of the data. In multiplanar slicing, one, or a small set of, discrete planes cutting through the volume are rendered. Surface fitting (SF) involves pre-computing a polygonal representation of surfaces within the volume. Whilst these polygons may be rendered at great speed on all contemporary graphics hardware, they represent only an approximation of the data structures and cannot preserve all the detail [378]. Direct volume rendering (DVR) uses the raw voxel samples within the rendering algorithm. Interactively rendering a modestly sized data set of 256^3 voxels will require upwards of a billion samples per second. If computational resources can be targeted, via GCD, to a region within the volume, rather than spread over the whole volume, there may be considerable overall savings.

There is, however, another, and perhaps more important benefit to employing GCD in the exploration of volumetric data: features of interest may be buried deep within a volume. If a rendering method is employed that does not take account of fixation point, then each spatial location within the volume is generally treated equally. Feature overlaps feature and the resultant image may well resemble an amorphous blob. Although this is somewhat harsh a generalisation – there are many powerful and readily implemented techniques to enhance the appearance of features such as surfaces within the volume – the basic concept applies. There is an opportunity for more effective display if an observer's fixation point can be measured in real time and added to the mix. Additionally, a volumetric GCD could be used to guide viewers to selected regions within the volume, by enhancing these to attract attention.

A GCD requires two main components: a gaze-tracking system and a rendering system that can modify its output in response to a stream of gaze position estimates. The next section reviews past research of these components in isolation and when combined to form a GCD. Then follow sections describing the approach that we have taken to develop gaze-contingent volumetric rendering, the results obtained to-date and an outline of possible future research directions.

This work has been in development for a number of years, with the authors moving between different institutions. The CGD system has had several different implementations over this period. In many ways, the challenge in achieving a working volumetric gaze-contingent display has been one of achieving sufficient performance and fidelity for each of the component subsystems, in order that the integrated GCD function satisfactorily. There are many components that must work, work well, and most importantly work within tight temporal constraints. Gaze-contingent displays are closed feedback loops and small delays in any sub-system or rendering artifacts can have significant effects on the overall system performance.

9.2 Review of Past Work

9.2.1 Review: Eye and Gaze Tracking

Technologies for tracking an observer's eyes find application in a number of areas. Such research may be aimed at better understanding and characterising the human visual process itself, at analysing the contribution vision makes to performance of a complex task, estimating the features of interest within a scene, or to control another device or display – the so called, gaze-contingent paradigm [25]. Approaches to gaze-tracking are well reviewed in [146].

The phrase *eye-tracking* is occasionally used, within the literature, to refer to different things. In this chapter, we use the term to refer to identifying and tracking the position of an observer's eyes, and the term *gaze-tracking* to refer to the process of measuring where an observer's gaze is directed, that is, what the observer fixates upon (see the book introduction for terminologies and definitions).

Gaze-tracking is most commonly applied in a 2-D fashion, measuring, for example, where on a screen that an observer looks. A term commonly used to denote this is the point-of-regard (POR). In different environments a researcher might wish to measure where on a shelf that shoppers look as they are moving along a supermarket aisle; or where car drivers look as they navigate a road. Whilst such studies may estimate fixation points in 3-D space, it is important to note that such positions are inferred directly from the direction of gaze as the first surface encountered proceeding from the eye(s) along this direction vector. This approach does not allow for estimating true 3-D gaze position within semi-transparent objects.

In many fields of science the display of partially transparent (or opaque) objects has many advantages. Medical and geophysical imaging commonly use volumetric data which may be rendered semi-transparently so that, for example, extended structure may be seen within an organ rather than being obscured by the organ boundary. Information display may be enhanced by showing graphs or other graphical representations with transparency, again so that features may be overlapped and yet still remain perceptible. Applying gaze-tracking to such displays may have a number of purposes: for medical image rendering a gaze-contingent display can be constructed in which features are shown most clearly as each is fixated in turn; control interfaces can be envisaged in which a user fixates a node in order, perhaps, to select display of detailed information pertinent to that node. For these applications, true 3-D gaze-tracking is required. This has occasionally been tackled and examples are found in [146, 172].

9.2.2 Review: Gaze Contingent Displays

A number of publications have reviewed GCDs in significant detail and the reader is referred to examples such as [146, 148, 446, 486, 520]. Whilst we do not here attempt to match these works, a brief summary of past work is appropriate.

Adding gaze-contingence to a display involves considerable effort and expense. Questioning why this should be done should often prefix any discussion of how it should be achieved. One of the main motivations for GCD – somewhat ironically in light of the computational expense involved in our present iteration of volumetric gaze-contingent rendering – has been in reducing the computational burden of display. The human eye is inherently multiresolutional, with much greater acuity within the fovea than in the surrounding region [488], and so when a resolution suitable for the fovea is deployed throughout the display, much visual information is wasted.

A particular application of GCD that continues to receive research interest is to mitigate the effects of lower bandwidth communications, such as, for example, found over the Internet. Images, or video, can be transmitted, prioritising a fixated location within the image. Both still image [333] and video image [192] systems have been reported. The image compression standard JPEG2000 allows region information to be included within the compression process and so readily supports this technique [607].

Significant understanding of human perception and congnition has been acquired through use of GCDs [510]; one specific application field is that of reading research. Although the eyes exhibit high acuity only within the foveal region, a person is not generally disturbed by the greatly reduced visual acuity and lack of colour sensitivity in the surrounding region, nor by the continual random eye movements that supplement the conscious larger scale movements. GCDs allow oculomotor and pereptual processes to be probed, testing for example, the bounds of the temporal window, from a short time before saccade begins to a time after it has completed, during which visual information is suppressed and the degree to which this occurs [380, 381, 540]. A further variant on the GCD concept is aimed at directing a user's gaze appropriately. This technique has been applied by adding red dots for a period sufficiently short to be scarcely perceived consciously [26].

A range of display technologies have been used to implement GCDs. Several provide wide fields-of-view, such as those finding application in flight simulators. Most commonly, displays are capable of reproducing full resolution images across their extent, but alternatives exist, where a high resolution display covering a restricted spatial extent is superimposed over a larger coarse resolution display [30].

9.2.2.1 Nomenclature in GCDs

Within the literature a variety of terms have been adopted to characterise GCDs. Table 9.1 lists some common terms and their meanings.

Multi-modality is a term used with two distinct meanings. In some contexts it refers to a user exploring data via a multiplicity of human sensors (often such displays/interfaces are termed *multi-sensorial*), such as aural and visual presentation. Within the context of this chapter, however, we use the term to denote data that have been obtained from more than one imaging mechanism or imaging sensor [446]. A good example of such multi-modality data are the CT, MRI and histological data sets provided under the US National Library of Medicine's Visible Human Project [610].

A significant body of work has addressed evaluating the effects of, and reducing, artifacts in GCDs. Investigations often aim better to understand

Table 9.1. Common terms relevant to gaze-contingent display

Foveated [display]	An alternative name for gaze-contingent [display] that suggests targeting information display to the eye's foveal region, where acuity is highest
GCMRD	Gaze-contingent multi-resolution display
GCMMD	Gaze-contingent multi-modality display
POR	Point of regard
AOI	Area of interest (either the area displayed in higher detail; or, in visual psychology, the area attended to)
AUI	Attentive user interface

human perception and cognition; this cannot accurately be evaluated if instrument effects mask the real information of interest. For a summary, see [381, 519, 520].

The accuracy that a gaze-tracker can provide has a direct influence on the degree and extent of any region enhancement applied within a GCD. In 2-D, inaccurate tracking will require that higher resolution regions must extend further, in order that lower resolution parts do no inadvertently overlap with the foveal region. In 3-D, inaccurate tracking may preclude operation of a volumetric GCD entirely: enlarging the region to compensate is not an option, since this inappropriately extends the obscuring depth through which the user is trying to see.

9.2.2.2 Boundary Form

The form of the boundary between regions displayed in GCD has been investigated by many groups. A good review of this work, within the scope of GCMRDs, can be found in [520]: the approaches divide into those using discrete levels of image resolution, e.g. [381], and those using a continuum of resolutions, e.g. [194, 487]. The Reingold paper notes that a number of authors have reported that sharp-edged boundaries do not negatively impact on a range of performance criteria. With the increases seen in computational power over the past decades, arbitrary variable resolution displays can be realised at high frame rates [193].

Texture mapping and its multi-resolution extension, Mipmapping, are the most common method by which gaze-contingent rendering is now implemented, with all of the calculation performed on the graphics adapter's graphics processor unit (GPU). Duchowski, Cournia and Murphy provide a good overview of approaches in their review paper [148]; also, see [447]. Given the extensive programmability of current graphics adapters, using languages such as Cg or GLSL [541], it is unlikely that a non-GPU approach would offer significant benefits.

9.2.2.3 Associated Gaze-Contingent DVR Work

Gaze tracking and volume rendering have been linked in an non-GCD manner. Lu, Maciejewski and Ebert automated rendering parameter selection based on attention given by a user to different displayed regions [383]. Kim and Varshney validated their saliency-guided volume visualisation using a gaze-tracker to ensure that their enhancements elicited viewer attention [321].

Gaze-contingent volume rendering *per se* has received attention. Levoy and Whitaker's work did not, however, measure where *within* a volume that an observer's gaze was directed – and thence apply spatially-variant rendering in relation to that point – but rather considered the 2-D screen POR. Higher resolution rendering was applied to the part of the volume that projected onto the region of the screen and this was blended with a lower resolution global rendering [366]. The authors have also previously published outlines of gaze-contingent volume visualisation, initially whilst lacking the necessary graphic and computational resources for real-time rendering [304], and later when such resources started to become available at affordable costs [447].

9.2.3 Review: Rendering of Volume Data

Extracting isosurfaces from volume data and rendering the resultant polygonal meshes forms the basis for surface fitting (SF) approaches to rendering of volume data. The seminal work in this field was the *Marching Cubes* algorithm developed by Lorensen and Cline [378]. This has seen numerous implementations and developments over the years. Two examples are: the use of tetrahedral, rather than cubic, sampling grids (*Marching Tetrahedra* [628]); and variations that provide a natural front-to-back ordering of the polygons that supports transparent rendering.

Whilst not targeted at making visible features buried within a volume, level-of-detail (LOD) rendering is an important area that provides multi-resolution volume rendering [691]. In particular, the requirement that images must be rendered within tight time constraints is shared with GCD. A good reference to this field is [385]. This model-based resolution-reduction process can be seen as an alternative to screen-based approaches (e.g. local pixel averaging) [146].

Volume rendering development has occurred in parallel with the development of hardware able to support it. Throughout this process, pre-computed implementations have always been available: when true on-the-fly volume rendering could render a given size of data in real-time, pre-computed approaches could extend this a little further. As lighting models [400] and artifact reduction methods have developed over time, this has remained true, and for challenging volume rendering tasks, pre-computation of a varying proportion of the expressions that compute the final image is beneficial – though often places constraints, such as restricting the allowable viewpoints, on the process.

In direct volume rendering (DVR), transfer functions (TFs) map raw voxels onto the colours and degrees of opacity used to composite the final image [683]: if the original data themselves contain colour information then that may be used directly, obviating the need for a TF that includes colour mapping. Usually, these TFs are implemented using a lookup table: for 8-bit intensity samples only 256 entries are needed. Using a simple 1-D mapping from intensity to opacity can, for example, allow the bright features in the raw data to be represented with significant opacity, whilst making transparent the darker features. Extended low opacity regions give rise to homogenous images, however, in which individual features can not be clearly seen. Much more effective display may be achieved using 2-D TFs, in which the degree of opacity is a function not just of value, but also of the gradient of value. This emphasises the surfaces, boundaries and detail within the volume. 2-D TFs are now the most commonly employed form (higher dimensionality TFs may be developed [281]), in which the additional dimension is usually provided by the magnitude of the gradient of the data value. Gradients peak at boundaries within the volume and so 2-D TFs allow emphasis of surface information [365].

Constructing appropriate TFs to illustrate the features found within a volume is a challenging task and many approaches have been developed, including manual, semi-automatic and automatic methods. A good outline of these, in the form of teams competing to render common data sets, can be found in [493].

Intensity and the (magnitude-)gradient of intensity are not independent variables and this observeration has formed the basis for research to develop straightforward and effective methods for TF specification [281, 330].

Volume rendering, in common with most forms of graphical display, is generally now implemented using the programmable hardware present on current graphics cards. The texture mapping hardware within such cards efficiently performs the resampling necessary to project slices through the volume onto the display device [411]. A distinction can be made based on whether these slices occur parallel to a face of the data volume, or parallel to the screen. The first can be achieved using bilinear interpolation; the second requires trilinear interpolation. It is worth mentioning that real-time ray-casting-based approaches also exist [674].

Adding semantic information can significantly aid the visualisation process. Segmenting and classifying features within a data volume allows emphasis or suppression of particular components within the rendering image. Two-level volume rendering uses explicit segmentation to allow specific objects to be selected. Appropriate and independent rendering methods can then be applied and combined to produce a final image [213, 239]. This can be seen as display of local information within the context of another, but here, *local* and *context* are used in a semantic, rather than a spatial, sense.

Over recent years, a number of researchers have explored rendering methods that take account of locality within the volume relative to a selected point. *Focus + Context* visualisation is a term that has come to embody many of these approaches [238]. It is often used to refer to magnification-style rendering, in which a region of the volume is enlarged within the display – either warping or over-painting the surrounding standard magnification region [100, 659]. Whilst integration of such a display mode within a GCD would constitute an interesting experiment, having the displayed spatial structure distort in response to fixations in 3-D is likely to present very significant challenges and may cause user disorientation. Alternative uses of the term

Focus + Context include those that provide sampling rate [659], and opacity [238, 425, 445], increases within the focal region.

Finally, several non-gaze-tracked techniques have been developed that aimed at more effectively conveying volume structural information to the viewer. *Non-photorealistic rendering* is a general term for methods that incorporate ideas of illustration and presentation. Application of these approaches to volume rendering [521] is now an active research field. Methods have been developed to ensure visibility of features within a volume through adjustment of the appearance of overlying structure [72, 650] and via automatic selection of an appropriate viewpoint [651].

9.3 Stereoscopic Display and Gaze-tracking

Figure 9.1 shows the display hardware, which is comprised of a mirror stereoscope and stereoscopic eye-tracking system . Display is provided by a ViewSonic VX2025wm monitor, which provides 1680 × 1050 pixels over a 434 mm × 272 mm area, 176° viewing angle and 8 ms grey-to-grey response time. The left image shows the two eye pieces through which the user views the display. Each eye piece is, optically, about 300 mm from the screen and observes a screen area of 200 mm × 200 mm.

The first mirrors encountered are *cold mirrors*, which reflect visible light and transmit infrared wavelengths. The visible light path proceeds from the left and right halves of the LCD display, via conventional mirrors, then via the cold mirrors, to the eye pieces. Around the eye pieces (see middle image of Fig. 9.1) are the ends of the fibre optics, which channel light from near-IR LEDs (see Sect. 9.3.1). Via the cold mirrors, video cameras image the eyes illuminated using these fibre optics. The right hand image of Fig. 9.1 shows one of the authors using the system (since this picture was taken the CRT monitor shown has been replaced with the LCD monitor described in the text).

Fig. 9.1. Display hardware showing mirror stereoscope and twin video cameras used for 3-D gaze-tracking

In any form of video-based gaze-tracking a major challenge is in separating the effects of eye translation from eye rotation. For illustration, consider an image of the pupil shown within successive frames of a video sequence (Fig. 9.7 includes such an image). The position may change either because the observer's head moved (and hopefully the eyes moved with it); or because the observer's gaze direction changed. Purkinje eye-trackers, by exploiting the optical path that light takes into the eye and following single or multiple reflections within the eye, gain their great accuracy because they are naturally able to distinguish displacement from rotation. The two video-based approaches we evaluate here differ in the method by which they provide compensation for head (and thus eye) displacement.

9.3.1 Eye-proximal Illumination

For the eye-proximal illumination method we use a fixed pattern of illumination sources around each eye. Using infrared illumination and video imaging allows good control of the illumination environment in a non-intrusive fashion that does not distract from the stereoscopic display presented to the observer. Each eye is imaged by a single video camera; the fields of the two cameras do not overlap (a single video camera whose field covered both eyes would be an alternative, but the geometry of our display/tracker made difficult such an implementation). The simple model upon which this approach is based considers the eye as a sphere on which the pupil appears as a dark circle. The specular reflections of the illumination sources from the surface do not change with rotation of the sphere around its centre but the image of the pupil does change position. With displacement of the sphere, however, the specular reflection pattern does change since the curvature of the sphere is significant with respect to the distance between the positions of the illuminating sources. Figure 9.2 shows simulated specular reflections within this model as the sphere is displaced (9.2a reference position; 9.2b displacement along positive x-axis; 9.2c displacement along positive y-axis; 9.2d displacement along positive z-axis).

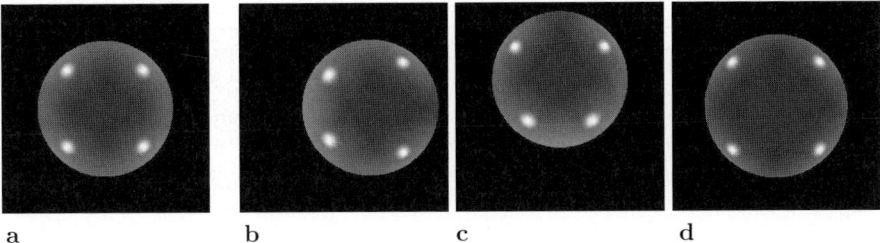

a   b   c   d

Fig. 9.2a–d. Simulated specular reflections from a sphere as it is displaced relative to a fixed pattern of four illumination points

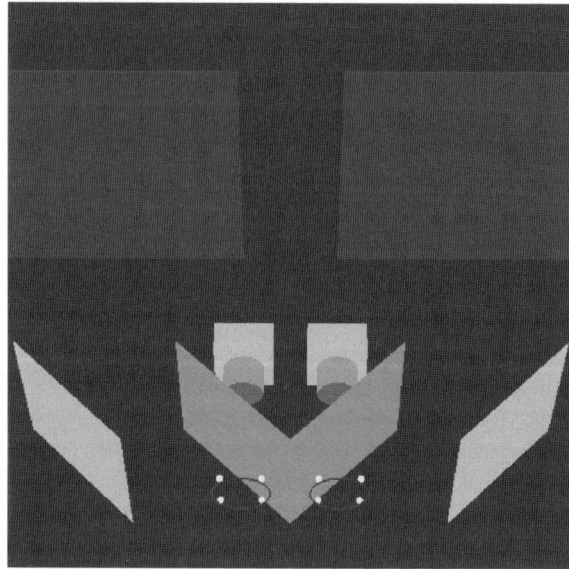

Fig. 9.3. Eye-proximal illumination and dual monoscopic imaging

Figure 9.3 shows schematically the arrangement used for stereoscopic display and eye imaging. In the figure the following shading is used: screen(s) – dark grey; cold mirrors – mid grey; standard mirrors – light grey. Video cameras are shown in line drawing. Cold mirrors are mirrors that reflect visible light and transmit [near-]infrared light. The large cold mirrors used for the design shown in Fig. 9.3 had to be custom manufactured (Edmund Optics, Barrington, NJ, USA). The eye-proximal illumination is represented by two dark rings (eye-pieces), just inside the perimeter of which are the ends of four 0.5 mm diameter plastic optical fibres that channel illumination from GaAlAs infrared LEDs (880 nm peak illumination).

Illumination for the stereoscopic system used diffused high power near-IR LEDs that globally light the face and eyes of the observer but do not produce obvious specular reflections. These LEDs are positioned along the outside edges of the large cold mirrors.

Additional infrared transmissive filters have been used to block visible light from the video cameras.

Figure 9.4 represents the display geometry. The left image shows the overlapping left and right eye frustums for the stereoscopic display region within which the data volume is rendered. The middle image emphasises the video camera views of the eyes that provide gaze tracking; the right image, the fixation point triangulated from the two eye gaze vectors.

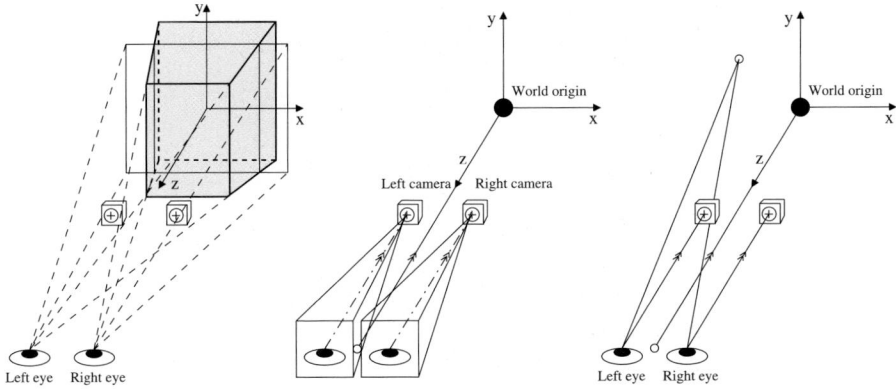

Fig. 9.4. 3-D geometry for fixation estimation

9.3.2 3-D Gaze-tracking and Taking Advantage of Motion

The key challenge to tracking gaze in three dimensions is in achieving good axial accuracy, that is along the direction of gaze. Most simply, 3-D fixation position is calculated using triangulation of gaze vectors from the left and right eyes. However, the relatively short baseline for triangulation in comparison with the distance of features being scrutinised for common display geometries makes basic axial fixation accuracy very poor. A typical interocular distance is 65 mm; the viewing distance may be upwards of 300 mm. There are a number of physiological issues that compound the basic geometrical considerations, such as: do the eyes fixate precisely upon a single point in 3-D space; do the stereoscopically displayed features contain sharp detail that can be fixated by the two eyes; what is the temporal reponse of the eyes as the fixation point is changed, both during smooth pursuit eye motion and following saccadic jumps? These considerations are further discussed in later sections of this paper.

One of the most promising routes to improving axial accuracy is through exploiting motion. Figure 9.5 shows how this may be achieved, considered both from the standpoint of gaze vectors sampled at discrete instants in time and of measured gaze vector velocity (whose measurement via video imaging will likely also involve discrete time sampling and thus these arguments are very closely related).

For the upper two images in Fig. 9.5, an observer watching a rotating cube can be considered equivalent to an observer being rotated around a fixed cube (in reality, the two situations are not identical since the oculomotor system provides some compensation for head movement). Thus, if the same point is fixated at two instants, then triangulation may be performed using gaze vectors obtained at different times. Appropriate selection of gaze vectors can be used to lengthen the triangulation baseline.

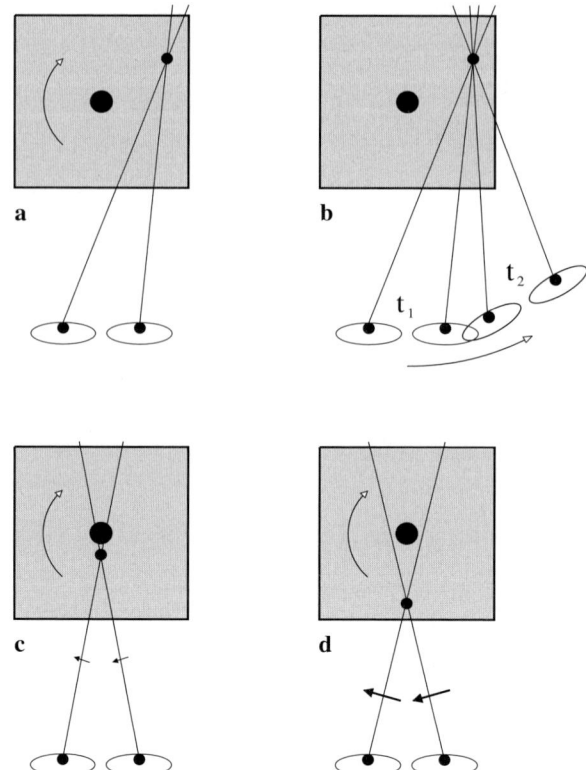

Fig. 9.5a–d. Exploiting eye motion to improve axial fixation accuracy

The lower two images in Fig. 9.5 represent the angular velocity of changing gaze vectors as the observer fixates a point close to (left), and distant from (right), the axis of rotation. This angular velocity can, given estimation of the approximate position of the fixated point, be used to improve the axial accuracy of the fixation position estimate.

In order fully to exploit these possibilities we thus consider gaze-tracking as a true dynamic tracking problem and use the same approaches as are used for related tracking applications. The fixated point is considered a target with process noise and measurement noise, and Kalman filtering is applied to obtain estimates of its changing position.

The above arguments are based on an assumption that the observer fixates the same point over time. This assumption is reasonable given the very tight coupling that exists between the retina and primary stages of human visual processing and the oculomotor system: smooth eye motion occurs only when following a smoothly moving point and thus in the absence of detected saccadic motions, it can be inferred that a single point was being fixated. This argument is likely to be weakened somewhat, however, if the fixated feature changes its form significantly as the effective viewpoint moves. Finally, we note that there will be significant motion tracking issues associated with the oculomotor control itself. To a first approximation this might be considered as a simple feedback loop with associated phase lag, so that the eyes will fixate just behind slowly moving features, but significantly behind more rapidly moving features.

9.3.3 Calibration

Calibration was performed during an explicit procedure that preceded measurement of gaze-tracking accuracy. Each subject was required to fixate upon a target positioned within the three-dimensional extent of the stereoscopic display. During the first part of the calibration process the target was moved randomly to each of the positions in a regular $3 \times 3 \times 3$ grid as shown in Fig. 9.6, jumping abruptly between positions then remaining stationary at each point for 2 s. During the second part of the calibration process the target moved smoothly between positions, with a range of velocities, constant for each segment of movement.

M in Fig. 9.6 is the mid-ocular point; the eyes are separated by a distance d_e; M is a distance of d_v from the centre of the target grid; and the inter-target distance (equal along each of the x, y and z axes) is d_t. The values used

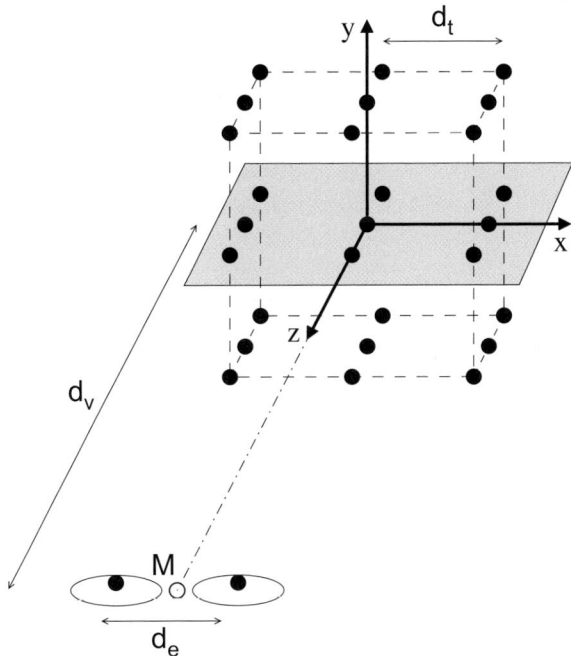

Fig. 9.6. Calibration target geometry and coordinate system

for our experiments were: $d_v = 0.3\,\mathrm{m}$; $d_t = 0.1\,\mathrm{m}$. d_e is subject dependent and found as part of the calibration process. These values provide for system calibration using an angular target span range of $2 \arctan\left(d_t/(d_v - d_t)\right)$.

Data acquired from the calibration phase was then used to set the parameters of the gaze geometry model used for fixation estimation. These parameters were found using the multidimensional downhill simplex method from Numerical Recipes [503], minimising an error term calculated as the mean square distance of the reconstructed fixation positions relative to the known target positions across the calibration sequence. Segments of the acquisition video sequence immediately following abrupt changes in target position or velocity were removed from the data used for minimisation, to allow for the settling time as the observer fixated upon each new position or adapted to the changed target velocity.

The error term over which minimisation occurs is:

$$\sum_{n=0}^{N-1} \left(\sum_{t \in T_n} \left| \overrightarrow{F_n(t)} - \overrightarrow{targ_n} \right| \right) ,$$

where, $\overrightarrow{F_n(t)}$ is reconstructed fixation point, n; and $\overrightarrow{targ_n}$ is the corresponding known target position.

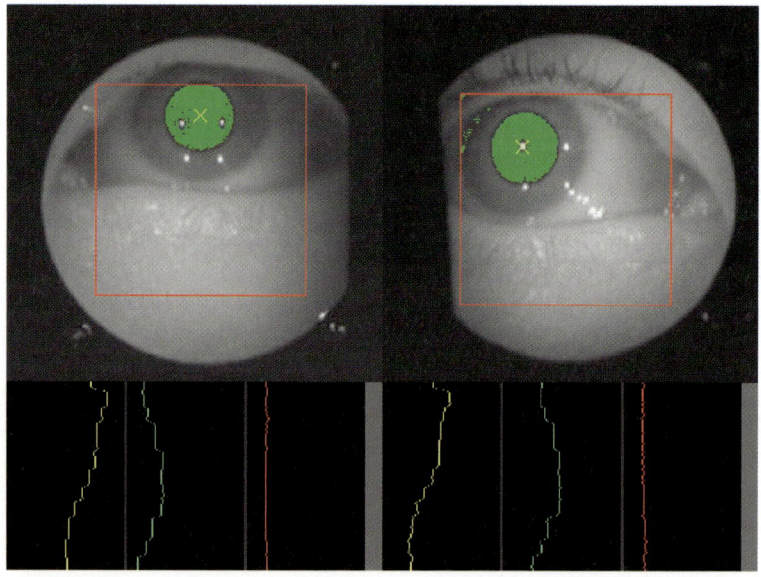

Fig. 9.7. Screen shot of gaze-tracker during real-time operation

9.3.4 Real-time Gaze-tracker Operation

Figure 9.7 shows a screen-shot taken from the gaze-tracker whilst processing video images. A small subset of diagnostic features have been included within this screen-shot. The green shaded regions on the main video images show the front-end thresholding applied during image processing to initialise the estimate of pupil position. Image processing is restricted to within the red rectangle. The yellow crosses indicate the final fixation position. There is scope further to optimise gaze-tracking in future by incorporating more advanced pupil detection algorithms. An example of one that is already publicly available is the Starburst algorithm [370] developed as part of the openEyes project.

In the traces below the video images *now* is indicated by the bottom of the plot: the traces scroll up as time passes. Three example time plots are included: the yellow and green plots show, respectively, horizontal and vertical pupil centre displacements; the red plot shows the cumulative area of the segmented pupil region. Such time plots, together with the ability to record video sequences and play them back with precise time information one field at a time, proved very beneficial in optimising the tracking system and incorporating filters for detecting, for example, eye saccades. Such saccadic movements can be clearly identified via the sudden displacements seen within the yellow and green plots.

Figure 9.8 shows reconstructed fixation positions following calibration. For clarity reconstructed positions for only six target locations (above, below, left, right, front and back) are shown and each fixation is colour-coded according to the target that was highlighted at the instant that it was captured.

9.4 Spatially-variant Volume Renderer

An appropriate starting point for presenting volume rendering is to consider the raw image data fed into the rendering process. Figure 9.9 shows orthogonal slices taken from two of the imaging modalities that constitute the Visible Human Project data. The top set of images shows CT data and the bottom set, colour histological data. The aspect ratios of these images are distorted, but the colour and grey-level characteristics are correct. The transverse images (left column) represent the plane in which the body was physically sectioned. The images have been resampled to give an in-plane resolution of 0.66 × 0.66 mm; slice separation is 1.0 mm. A mutual information (MI) registration algorithm [252] was used to displace and scale the CT image slices to improve their alignment with the histology images.

Some basic examples of volume rendering are shown in Fig. 9.10. The left image uses a transfer function that makes all voxels fully opaque: only the

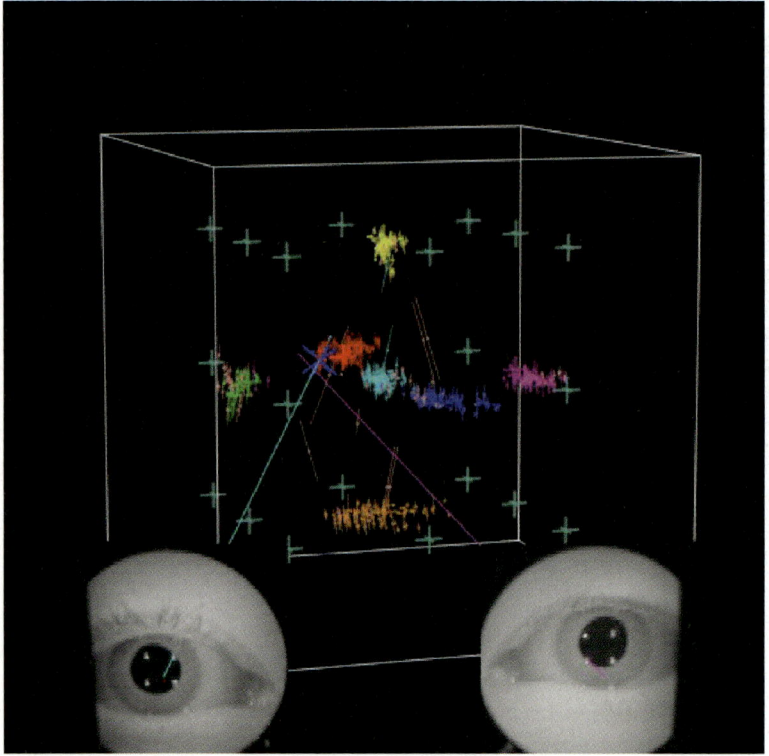

Fig. 9.8. Reconstructed fixation positions following calibration

volume surface is seen. The middle image makes transparent the darker vox-els. These happen to be mostly comprised of the coloured ice that surrounds the visible human head and so this region disappears. The right image applies a highly transparent transfer function and so features are shown throughout the volume, albeit each with reduced contrast.

The work reported here has been developed and tested on an NVIDIA Quadro FX 3450 card, a mid-range professional graphics adapter. The card provides 256 MB of graphics memory, a 256-bit 27.2 GB/s memory interface, x16 PCIe bus, 4.2 Gpixel/s fill rate, and 32-bit-per-component floating point colour composition.

The bulk of the graphical calculations were performed in OpenGL [684] and its shader language, GLSL [541]. The operating system was 64-bit Linux using a 2.6.20 kernel, 2 GB RAM, dual-core processor, 64-bit NVIDIA graphics driver version 1.0-9746 and X Window version 6.9.0.

DVR was implemented using 3-D texture mapping – employing proxy geometry slices parallel to the screen. Internally, tri-linear interpolation re-samples the original voxel sample locations to texel locations. Texture proxy slices were spaced regularly at a distance corresponding to half of the min-

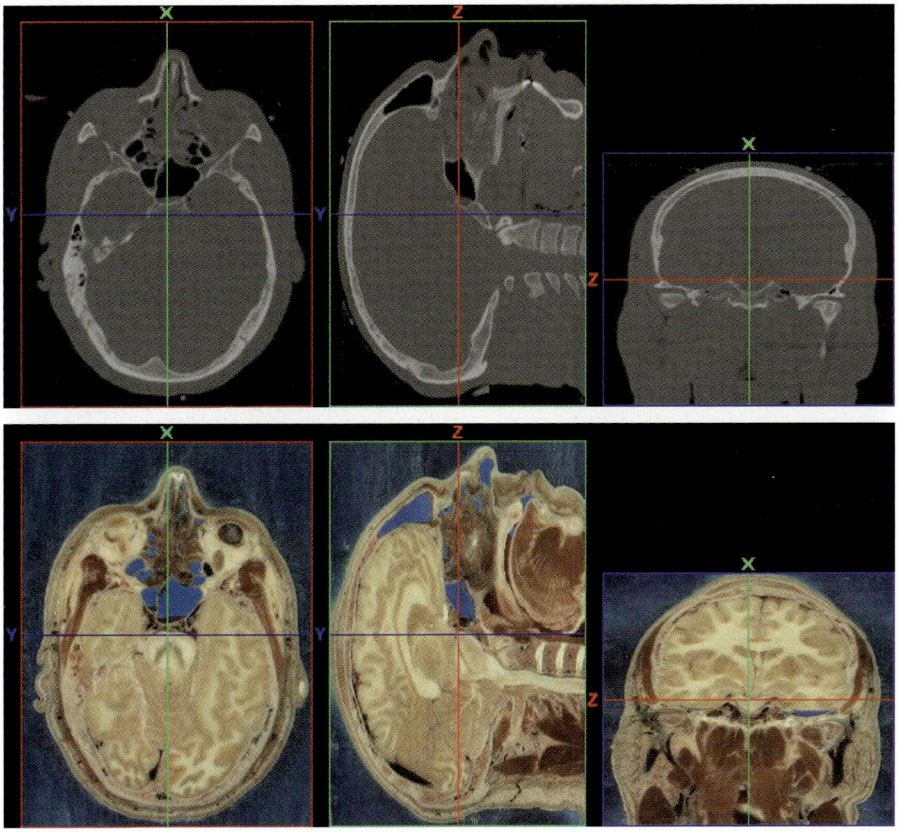

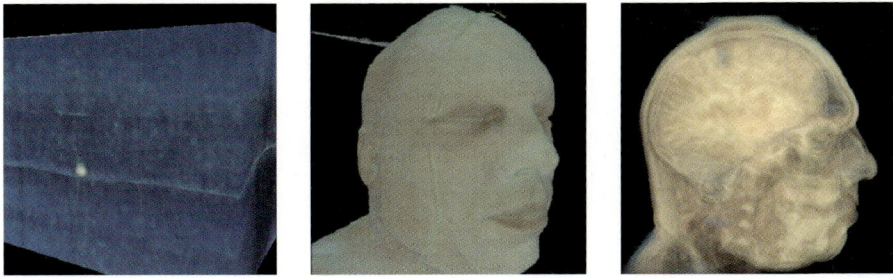

Fig. 9.9. Raw data displayed as orthogonal planes for CT (*top*) and colour histological (*bottom*) modalities

Fig. 9.10. Conventional volume rendering of colour histology data using a range of transfer functions from fully opaque (*left*) through to highly transparent (*right*)

imum voxel sample spacing along the x, y and z axes; this produces visible, but not overly disturbing, artifacts and allows much more rapid rendering than the finer slice spacing needed to remove the artifacts' visibility entirely.

9.4.1 Composition

Composition forms a final image by overlaying multiple semi-transparent source images. The standard equations, applied iteratively to calculate an updated image as each semi-transparent layer is added on top, are:

$$\alpha := \alpha_i + (1 - \alpha_i)\,\alpha \tag{9.1}$$

$$c := \alpha_i c_i + (1 - \alpha_i)\;. \tag{9.2}$$

Here, α represents the partial opacity value held in the frame-buffer; and α_i, the partial opacity value of the overlaid semi-transparent layer. For c_i, substitute each of three colour components, $\{r_i, g_i, b_i\}$. A variant, sometimes encountered, pre-multiplies each colour component by its corresponding alpha value. Given current graphics adapter architectures there is no speed benefit to doing this and, indeed, some quality detraction if the pre-multiplied colour values are more severely quantised as they are passed to the graphics adapter than the values that would be stored within the graphics adapter during on-the-fly calculation.

9.4.2 Transfer Functions for GCD and Region-enhancement

To develop the TFs that we apply to GCD, we begin with a conventional TF. This may be represented as an assignment:

$$\{r_i, g_i, b_i, \alpha_i\} = \mathcal{B}\,(y_i, y_i') \tag{9.3}$$

We adopt the script letter $\mathcal{B}$ to denote this TF in anticipation of our subsequent treatment of this TF as comprising the *base* volume rendered component; y_i is the value of sample i; y_i' is the magnitude of the gradient of the sample.

A variety of methods could be adopted to enhance the appearance of the location around which a user fixates. The degree of (partial-)opacity of each voxel could be scaled; alternatively, relative enhancement could be achieved by using thresholding to make fully transparent voxels outside the enhancement region whose opacity mapping falls below a set value. A further method is to apply a threshold α_t to each α_i value from (9.3) and discard those fragments that fall below this threshold, rather than composite them using (9.1) and (9.2): α_t is made spatially variant so that only the most significant features are displayed from the non-enhanced regions – providing the spatial context – but all features are shown within the enhanced region [302].

The approach that has been used here is to use two transfer functions. Simply put, one is applied throughout the volume outside of the enhancment region; one is applied within the enhancement region. More correctly, a weighting value ω, which lies in the range $[0.0, 1.0]$, is applied to blend between them.

Some discussion of the use of values for ω such that $0 < \omega < 1$ is warranted. As noted in the introduction, one common, but not entirely consistent, finding of GCD research is that the abruptness of the boundary between modified and surrounding regions is not significant: smoothly-blended boundaries are perceived no differently from hard-edged boundaries. A good overview of past research into this area is provided in [520]. However, a critical difference for this work is that the boundary exists in three dimensions and the user is therefore looking through its front surface. Under this condition, the abruptess of the boundary certainly does have an effect upon the perception of the gaze-contingent display.

Using shorthand to represent the complement of ω: $\bar{\omega} = 1 - \omega$, blended transfer function mapping can be represented as:

$$\{\alpha_i, r_i, g_i, b_i\} = \omega \mathcal{S}\left(y_i, v_i'\right) + \bar{\omega} \mathcal{B}\left(y_i, y_i'\right) \tag{9.4}$$

Here, $\mathcal{S}$ is used to represent the TF comprising the *supplemental* component of volume rendering.

Figure 9.11 represents a typical TF pair used for GCD, overlaying coloured trapezoidal patches on a 2-D log-intensity histogram that shows the distribution of (y_i, y_i') pairs within the raw volume data (see, for example, [329]). The patch colours match the rendered colours, but given the highly transparent TFs suitable for the base component $(\mathcal{B}(y_i, y_i'))$, the patches are shown opaquely on Fig. 9.11 or they would be scarcely visible. The actual α values for each patch follows a triangular profile with increasing y_i, from 0.0 at the

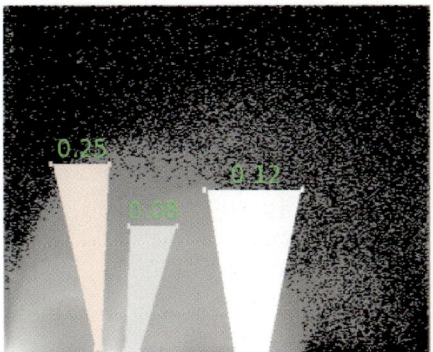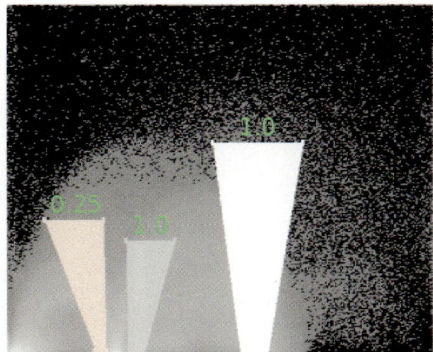

Fig. 9.11. TFs used for GCD. Each image represents a 2-D TF (see text for details). The *left hand image* shows the TF applied in the surrounding volume and the *right image*, the TF applied in the enhancement region

left and right boundaries, peaking in the middle. The peak value is the figure appearing above each patch.

With regards to the spatial form of ω, the authors have previously presented some rudimentary investigations into different weighting functions [303]. A number of approaches can be taken. For example, a simple spherical form could be used with enhancement peaking in the centre; alternately, emphasis could peak around the horoptor, the curved surface that passes through the fixation point and for which stereoscopic disparity is zero. Within this work, we have applied a simple clamped ellipsoidal function, defined by:

$$d = \sqrt{\left(\frac{P_x - F_x}{e_x}\right)^2 + \left(\frac{P_y - F_y}{e_y}\right)^2 + \left(\frac{P_z - F_z}{e_z}\right)^2} \qquad (9.5)$$

$$\omega = \begin{cases} 1.0, & d \le d_{\min} \\ \frac{d_{\max} - d}{d_{\max} - d_{\min}}, & d_{\min} < d < d_{\max} \\ 0.0, & d_{\max} < d. \end{cases}$$

Here, $\mathbf{P}(= (P_x, P_y, P_z))$ is the position at which ω is calculated, $\mathbf{F}$ is the fixation point, and the components of $\mathbf{e}$ specify the extent of the enhanced region along the three axes (oriented towards the mid-ocular point as shown in Fig. 9.14). $d_{\min}$ and $d_{\max}$ thus define the extent of the fully-enhanced region ($\omega = 1$) and the abruptness of transition to non-enhanced region.

The effects of spatially-variant blending of transfer functions are shown in Fig. 9.12. All these images have been rendered using 2-D TFs, in comparison with the 1-D TFs used to render the images in Fig. 9.10. The left image shows conventional DVR. The middle image shows blended-TF enhancement applied as represented by (9.4). The right hand image shows a multi-modality extension in which the $\mathcal{S}(y_i, y_i')$ term of (9.4) has been

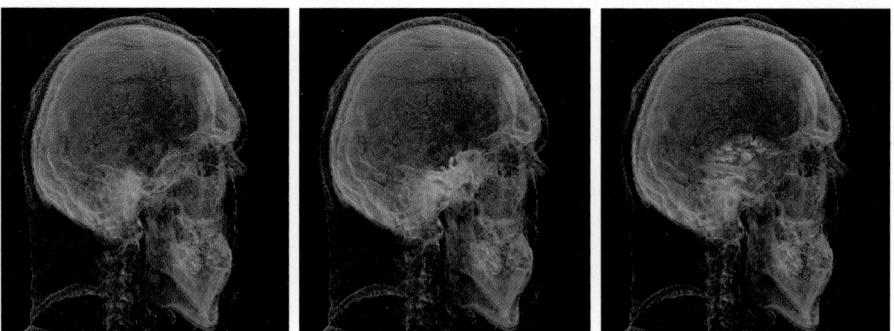

Fig. 9.12. Direct volume rendering variations: conventional; blended-TF single-modality enhancement; blended-TF multi-modality enhancement

```
uniform sampler3D volTexture;
uniform sampler2D tfTextureA;
uniform sampler2D tfTextureB;
uniform sampler3D wTexture;
uniform vec3 lightDir;
void main(void)
{
  float weight = texture3D(wTexture,gl_TexCoord[3].stp).r;
  vec4 vCol = vec4(texture3D(volTexture,gl_TexCoord[0].stp));
  vec3 vDir = vec3(vCol.g-0.5, vCol.b-0.5, vCol.a-0.5);
  float mag = 1.17 * length(vDir);
  vec3 norm = normalize(vDir);
  float diffuseTerm = clamp(dot(lightDir, norm), 0.0, 1.0);
  float l = 0.5 + 0.5*diffuseTerm;
  vec4 ctexA = vec4(texture2D(tfTextureA,vec2(vCol.r, mag)));
  vec4 ctexB = vec4(texture2D(tfTextureB,vec2(vCol.r, mag)));
  vec4 ctex = ctexB * weight + ctexA * (1.0 - weight);
  gl_FragColor = vec4(ctex.r*l, ctex.g*l, ctex.b*l, ctex.a);
}
```

Fig. 9.13. Blended TF GLSL fragment shader

replaced with components bringing in the colour of a second modality (directly) and an opacity value produced by applying a TF mapping to this second modality.

9.4.3 GLSL Shader Program

Figure 9.13 provides the source code for a fragment shader typical of those employed during this work. It implements (9.4), performing blending between two 2-D TFs, with support for ambient and diffuse lighting.

weight calculates ω using tri-linear interpolation into the weighting texture, wTexture. Simple lighting is used with a light position represented by an appropriately transformed direction vector, lightDir: diffuseTerm is calculated using a standard diffuse lighting model, with the surface normal interpolated from pre-calculated X, Y and Z gradient components stored, respectively, in the g, b and a components of volTexture. The two transfer functions, $\mathcal{B}$ and $\mathcal{S}$ are sampled as ctexA and ctexB.

9.4.4 Spatially-Variant Weighting

The weighting applied to the TF blending program varies with position within the data volume relative to the fixation point. The enhancement function is centred at the fixation point, oriented towards the mid-ocular point. Two approaches were tried for calculating ω: using CPU (C++) calculation of ω values at slab vertex locations; and using GPU texture mapping, within the fragment shader, into a 3-D texture that represents ω's spatial distribution. The first method links more straightforwardly with the C++ code that feeds vertex arrays to the graphics adapter (see subsection 9.4.5) and requires one fewer 3-D texture unit; the second requires a `glReadPixel` read back of ω values re-sampled into the eye (vertex) coordinate system, but leverages the GPU's performance. We note that graphics adapters can efficiently interpolate per-vertex specified values, so the first method is still achieves a continuum of ω values at each fragment location. A third option could be to write a GPU vertex program to calculate ω according to a chosen function: a requirement similar to that described above to retrieve calculated values for generating the vertex arrays applies.

For per-vertex CPU calculation of ω, a simple function was used to calculate a 3-D ellipsoidal weighting function that decreased linearly with distance, from $\omega = 1.0$ at its centre (as per [303]). It would be possible to set up a subsequent vertex program to pass through a user-defined variable to the fragment shader. A ready-made alternative, however was explicitly to set the fog coordinate, supported in OpenGL since version 1.4 [684]. This is a single variate value that can be specified on a per-vertex basis using a vertex array, for efficient computation.

For GPU calculation, a 3-D weighting texture was generated in advance of visualisation. The centre of the texture represented the fixation point and, typically, the position of maximum enhancement. Suitably oriented and scaled, via a texture matrix, this allowed an arbitrary enhancement pattern to be positioned around the enhancement position. A $64 \times 64 \times 64$ texture provided more than adequate resolution for any plausible enhancement function (the graphics adapter uses tri-linear interpolation to calculate actual weighting values at each pixel location). Clamping of the texture values, together with a boundary of $\omega = 0.0$ weighting around the outer layer of voxels that make up the texture, ensured that no enhancement was applied to any voxel outside the texture.

9.4.5 Pre-Calculation

Running a blended two-dimensional transfer function shader program over volumetric textures presents a challenge even for a top-end graphic card. Rather than limit the investigations to the size of datasets that could be rendered in real-time, the alternative approach of pre-calculating significant components of the final displayed image was taken.

Pre-calculation does allow for a great increase in the size of data sets that can be examined using GCD and in the quality of the displayed images that may be obtained. However, it does have a serious drawback in that, for the methods of pre-calculation employed here, the viewpoint for

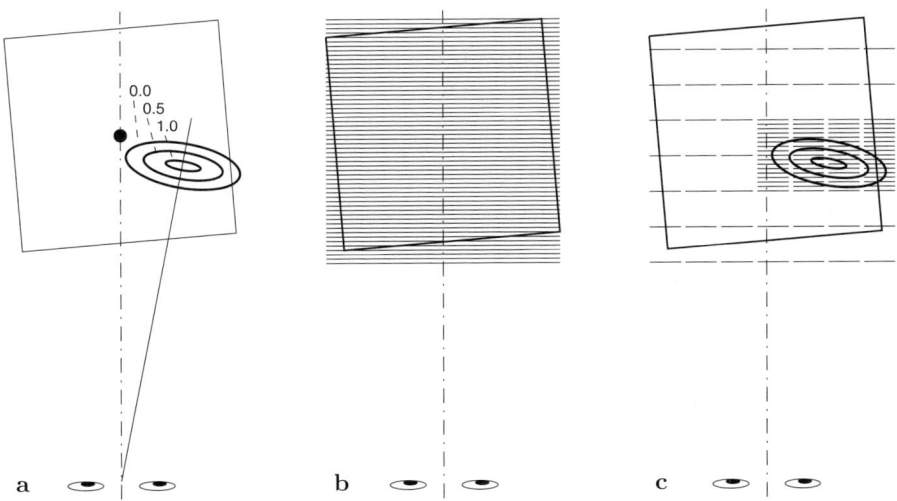

Fig. 9.14a–c. Region-enhancement slice and tile composition geometry

each frame of a rendered sequence is fixed. Although the enhancement region may be moved freely within the volume, this can only be done while, for example, watching a data volume rotate back and forth around a fixed path.

Two forms of pre-calculation were employed. In the first, the data volume is decomposed into two regions: a base and a supplemental region. The base region comprises those parts of the data set where $\omega = 0.0$; the supplemental region, those parts where $\omega > 0.0$. The second form of decomposition extends this approach, to include an additional region where $\omega = 1.0$.

It is only in the region where $0.0 < \omega < 1.0$, that real-time volume rendering must be done.

Figure 9.14 represents the approach. Figure 9.14a shows a plan view of the enhancement region positioned within the volumetric data set. Contour lines illustrate the degree of enhancement increasing from the baseline value of 0.0 at the enhancement boundary to the maximum of 1.0 deep inside the region where the user fixates. Figure 9.14b shows the conventional texture proxy slices utilised in 3-D texture mapping. The method adopted for this work is to composite a set of these proxy slices during pre-rendering computation. In the figure, eight proxy slices are composited into a single precomputed slab. These slabs may be broken up into squares for volumetric rendering: in plan view this is represented by the horizontal breaks seen in Fig. 9.14c. Wherever enhancement is used, however, the pre-calculated segments must be replaced with the full set of composited slices as shown in Fig. 9.14c.

For real-time image generation, pre-calculation generates a series of images, each of which corresponds to a set of slices that make up a slab through the volume.

9.4.5.1 Data, Composition and Display Resolutions

A variety of different resolutions must be considered in optimising rendering speed. The display should preserve the detail present in the original data, but without excessively oversampling the data, adding computational expense for no visual benefit. Using separate resolutions for the composition and display phases of the process aids this. Composition occurs at a resolution selected to achieve a desired display rate. The resultant image is then used as a texture and expanded to occupy the desired region of the display. Since composition processes a multiplicity of slabs and slices, this stage consumes much larger computational resources than the final 2-D texture mapping. This approach also avoids the need to set the graphics adapter resolution to suit the composition resolution.

9.4.5.2 Tradeoffs in Slab Thickness

There is a tradeoff to be made in selecting an appropriate slab thickness for pre-computed rendering. In locations outside the enhancement region, slabs must be composited in real-time during display. Using thinner slabs means that more must be composited and the bandwidth required for this process (as well as the memory required to store this higher number of slab images) can become the limiting factor. Conversely however, each slab tile boundary within which the enhancement region extends must be rendered using the full fragment program of Fig. 9.13; as the slabs become thicker, the efficiency of the rendering decreases since for a greater proportion of the process $\omega = 0.0$ (and so texture mapping replicates what was already done to pre-compute the slab images).

Selecting a suitable slab thickness may be done by balancing these considerations with support of real numbers. The PCIe bus (version 1.0) used for this work supports a theoretical bandwidth of 4 GB/s over the 16 channels used for the main graphics adapater connection. Measurements of real texture download times suggest that a rate somewhat nearer 3 GB/s was being achieved in practice.

An alternative approach was actually explored of performing slab composition in software using SIMD instructions. Since the slab images have already had all the 'heavy lifting' work incorporated, in modeling transfer function, trilinear interpolation and lighting effects, the remaining composition is very simple: multiply the frame buffer pixel components by the complement of the incoming pixel alpha; add the incoming pixel components. However, some 15–20 instructions still proved necessary to implement this strategy, and on the

AMD processor used for this work the equivalent composition bandwidth was measured at around $1.5\,GB/s$, a figure competitive with, but not ahead of, the GPU approach (GPU composition also includes the overhead of the actual texture composition, as well as the transfer time to get the texture from system memory onto the graphics adapter).

9.5 Volumetric Gaze-Contingent Display

Gaze-contingent display is achieved when the components of Sections 9.3.4 and 9.4 are linked. Getting GCD to work effectively, however, does need some tuning. With poor set-up it is very feasible for the act of looking at a feature of interest within the volume immediately to result in the feature of interest disappearing, perhaps because inaccurate gaze-tracking has indicated that another locality should be enhanced, or perhaps because, though the enhancement is correctly centred on the feature, enhancement occurs throughout a region and the front of this region gains sufficient opacity to obsure the actual fixation point.

It is known that time lags have a detrimental effect upon user experience with GCDs. Loschky and McConkie experimentally validated that 5 ms of latency served as a no-delay baseline condition [380]. With the gaze-tracking video cameras and graphics hardware available to us for this study, we have not been able to approach this degree of responsiveness and would estimate that our system currently exhibits a latency of between 80 ms and 150 ms (dependent upon the occurrence of an eye movement relative to a video frame capture time, and upon the complexity and extent of the spatially variant enhancement applied). Given that 80 ms represents the upper limit of the time lags measured by Loschky and McConkie, we have not been in a position to perform 3-D (volumetric) GCD investigations analogous to their 2-D investigations. Such a study remains future work.

9.5.1 Gaze-Contingent Single-Modality Exploration

At this time we are able to report only preliminary experiences of using volumetric GCD. We were able to fixate upon a feature within the display and display enhancement occured within a region around this point. There was, however noticeable noise in the fixation point measurments and a degree of inaccuracy evident in the absolute positions. It took a moderate degree of concentration to compensate for these tracking problems – although fixating slightly askew of where one is interested in scrutinising somewhat defeats the purpose of GCD – and then it was possible to explore features within a volume with some success. One problem that we suspect underlay this difficulty was in the difference between the tracking accuracy achieved for fixating calibration point targets and that achieved when peering within more diffuse and complex structures. In future work we recognise that we shall have to develop tracking calibration procedures that utilise more realistic displays for measurement of tracking position and accuracy. The problem really is not one of providing more geometrically accurate tracking, but rather of characterising the visual features within the display that draw each eye to orientate in a particular direction. If diffuse features are scrutinised then the concept of an identifiable fixation point in 3-D is somewhat strained.

Figure 9.15 shows the gaze-contingent single-modality mode for different fixation positions. We do point out that the images shown within Fig. 9.15, whilst each rendered within a time interval sufficiently short for real-time GCD, have been generated at synthesised fixation positions. The positions have been chosen to illustrate the appearance of GCD (the same positions are used in Fig. 9.15 in order that comparison between corresponding images may be made).

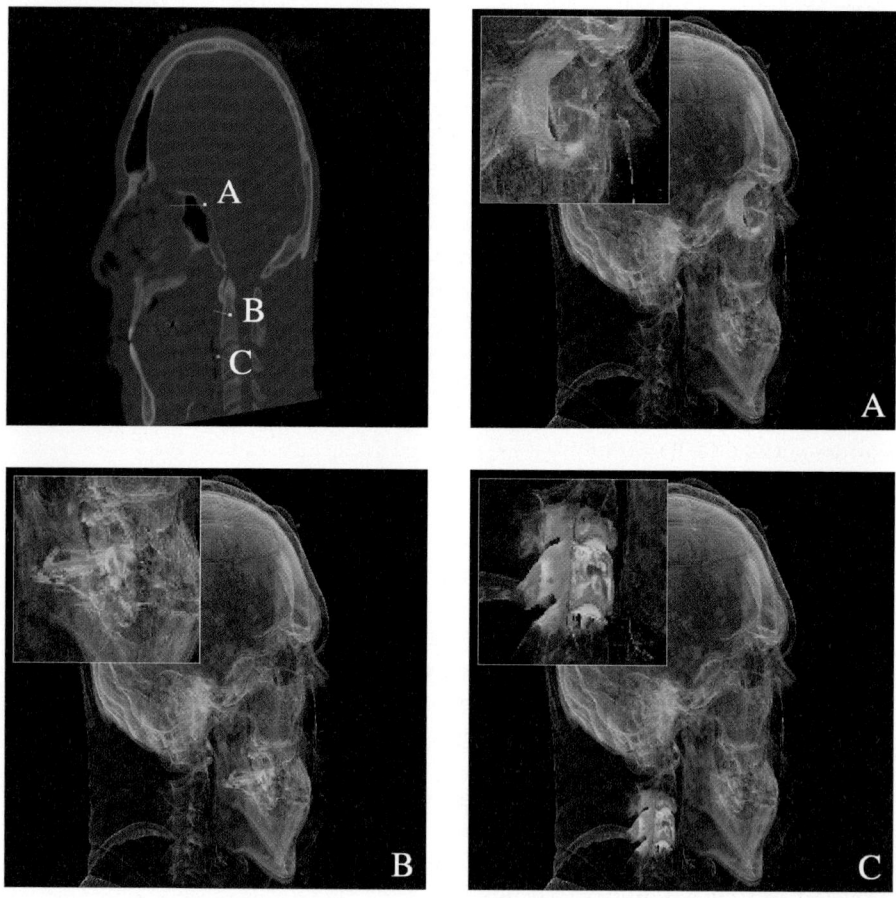

Fig. 9.15. Gaze-contingent single-modality exploration

The volume data set rendered in Fig. 9.15 was $272 \times 360 \times 236$ voxels in size. The images were rendered to a resolution of 512×512 pixels using the pre-calculation method of Sect. 9.4.5. Composition of the slabs that produced the final image, using no enhancement, took 20 ms. Enabling enhancement resulted in a 50 ms rendering time for a spherical enhancement region with a radius equal to 8% of the maximum dimension (from x, y and z) of the volume. During display 97.4% of the area of the pre-calculated slabs was used; the remaining areas required spatially-variant volume rendering.

The approach adopted for developing suitable TFs for GCD was to consider an "opacity budget" to be spent as one progresses from the front to the back of the volume. The majority of the expenditure should occur within the enhancement region, but with noticeable amounts of expenditure made in the surrounding volume in order that a suitably informative spatial context be provided. In developing this concept a distinction was identified between considering mean or peak opacity – to a location in the volume – and an occlusion budget. The occlusion budget could be expressed in terms of the fraction of a region that could be allowed to contain displayed image pixels with opacity greater than a given threshold: it may be very acceptable to have a fully opaque pixel directly in front of the centre of the enhancement region, but if most of the pixels that overlie the region are similarly opaque, that would not be acceptable. The research of several other groups has influenced our approach and we gratefully acknowledge their work [72,321,330,521,651].

Another aspect of our approach was to consider this to be a preliminary study and, as such, not to set the goal of immediately identifying the most computationally efficient and effective method of automatic (dual) TF generation. That will be a target of future work, but at this stage, significant, and time-consuming, pre-processing of the volume, in order better to understand the statistical distribution of information throughout it is accepted, such that the benefits of GCD may be demonstrated.

9.5.2 Gaze-contingent Multi-modality Exploration

Perhaps one of the most useful modes of gaze-contingent volumetric display is that using multi-modality rendering, in which a base modality provides a spatial context, with a supplementary modality added only within the enhancement region. For an alternative, VR-based approach to a similar problem, see [704].

The images in Fig. 9.16 show this mode for different fixation points within the volume, using CT volume data to provide the base modality and colour histology images to provide the supplementary modality. This combination supports the gaze-contingent paradigm well, since the CT data are naturally sparse relative to the more varied histology data and so provide an informative, but not over-dominant, spatial context.

Implementing spatially variant rendering for two volumetric data sets with support for lighting and independent two-dimensional transfer func-

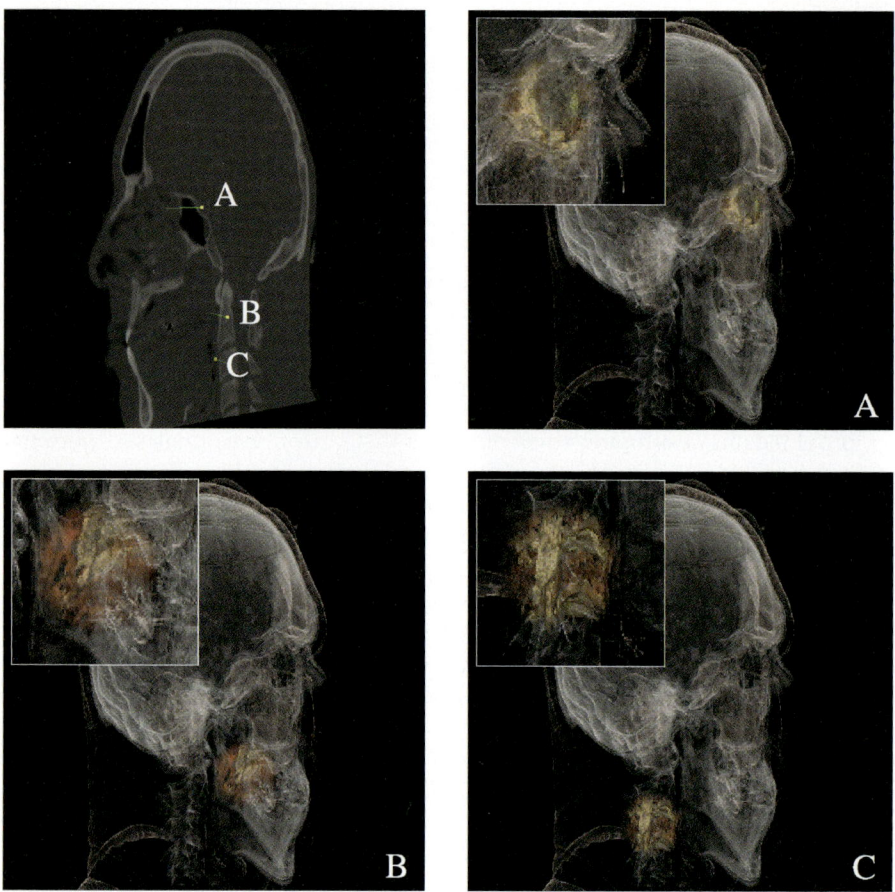

Fig. 9.16. Gaze-contingent multi-modality exploration

tions presents some challenges. The FX 3450 graphics card, whilst targeted at the professional market, still only supported simultanous use of four textures (as returned by `glGetIntegerv(GL_MAX_TEXTURE_UNITS, &number_texture_units);`). A basic implementation would require two 2-D textures for the TFs, one 3-D texture for the weighting function, and, for each of the volume data sets, either one or two 3-D textures depending on which components were stored: luminance, red, green, blue, magnitude-gradient, x-gradient, y-gradient, z-gradient.

Accordingly, it was necessary to place multiple texture regions into a single texture unit. The method adopted was to place the two TFs adjacently in one unit and to utilise a hybrid pre-calculated direction gradient. With this approach, hybrid x-gradient, y-gradient and z-gradient components are calculated for both data sets by taking the maximum of each component from central difference gradients calculated for the individual data sets. Essentially

Table 9.2. Mapping volume data, transfer functions and weighting function to texture units

Texture unit	Dimensionality	Role
0	3-D	CT luminance (Y); gradient $(\nabla_x, \nabla_y, \nabla_z)$
1	3-D	Histology (R, G, B)
2	2-D	TF1 $(\mathcal{B})$, TF2 $(\mathcal{S})$ (α only for TF2)
3	3-D	Weighting function (ω)

this means that if either data set exhibits a high gradient in a given location, then this high value will be present in the hybrid gradient data. Given accurate registration of the data sets, this should be a reasonable approach.

Table 9.2 shows the texture units that were applied to implement this.

This technique provided one of the most natural forms of display. During visualisation it was possible to fixate upon the upper cervical vertabrae and subsequently obtain perception of the blood vessels, airways and nerve bundles that pass around these structures.

Unfortunately, this method did prove to be too computationally expensive realistically to claim that interactive GCD had been achieved. Using the same enhancement region as produced a 50 ms rendering time for single-modality GCD resulted in a 130 ms time for multi-modality GCD. We speculate that adding a second large 3-D texture into the rendering process causes excessive cache-thrashing. Whilst alternative mappings of data components to texture units may exhibit better performance, perhaps a more promising way forward might be to move away from standard 3-D texture mapping approaches entirely. Volumetric GCD appears to work best when the context is provided by sparse features distributed throughout the volume. Utilising a volume texture to hold such data, most of which will contribute little opacity to the final display, is highly inefficient.

9.6 Conclusion and Future Work

We believe that the system presented in this chapter is the first volumetric GCD to be described in the literature. However, whilst the display that has been prototyped shows some benefits of GCD in exploration of volumetric data, it by no means represents a full exploration of the technique. The compromises adopted in order to achieve real-time performance significantly limited the scope of the display which could be experienced. A much more powerful GCD might offer a blend of visual and haptic control, for example, in which points or regions of interest could be fixated, zoomed, volumes spun and translated.

There is considerable scope for optimising the rendering process. Tailoring the 3-D spatial resolution around the fixation point and allowing colour resolution reduction away from the foveal region, as encouraged in [520], are both methods that could supplement the existing texture mapped approach. Temporal blur [135] is an artifact that needs to be addresed.

Improvements in passive eye monitoring accuracy will also assist overall instrument performance. Incorporating information about the distribution of salient features throughout the volume is an idea already applied [519], which could benefit volumetric gaze-tracking.

Finally, we again note the potential to adapt our volumetric GCD for dynamically guiding visual exploration of 3-D data, as has already been done for 2-D data [26], and which has been done statically for 3-D data [321].

A Temporal Multiresolution Pyramid
for Gaze-Contingent:
Manipulation of Natural Video

Martin Böhme, Michael Dorr, Thomas Martinetz, and Erhardt Barth

10.1 Introduction

A gaze-contingent display manipulates some property of the (static or moving) image as a function of gaze direction (see [148] for a review). Among the image properties that can be manipulated are contrast, colour, and spatial frequency content. Alternatively, parts of the display may be masked depending on gaze position. Gaze-contingent displays were first used in reading research [406, 510, 512] and have since been used in many psychophysical and perceptual studies (e.g. [108, 381, 382]).

One of the most popular applications of gaze-contingent displays is to "foveate" an image or video, i.e. to simulate the effect of the variable resolution of the human retina, which is highest at the fovea and falls off towards the periphery. If the foveation is adjusted to match the resolution distribution of the retina, the effect is not noticeable for the observer, but the resulting images can be compressed more efficiently because they contain less high-frequency content [192, 333, 560]. Foveation requires an algorithm that can adjust the spatial frequency content of different parts of the image depending on their position relative to the point of regard. This type of algorithm may also be used to visualize the effect of diseases of the eye, e.g. glaucoma [492]; these visualizations can help educate students or family members of patients about the effects of such diseases.

Foveation is also a popular technique in active vision [24], where it is used to reduce the amount of data that needs to be processed, in analogy to the human visual system. In computer graphics, a similar technique, known as "Eccentricity Level of Detail (LOD)", renders objects that lie in the visual periphery with less detail than objects in the centre of the visual field [515].

In this chapter, we present a gaze-contingent display that manipulates not the spatial, but the temporal resolution of a video. The basic effect of temporal filtering is to blur the moving parts of an image while leaving the static parts unchanged (see Fig. 10.1 for an example).

Fig. 10.1. *Top*: Image from one of the video sequences. *Bottom*: Same image with gaze-contingent temporal filtering applied. The *white square* at the centre *left* (below the *white sail*) indicates the point of regard. Increasing amounts of temporal filtering are visible towards the periphery, i.e. with increasing distance from the square. For example, the two men walking at the left of the image are slightly blurred, and the person who is about to leave the image at the right edge has almost vanished completely

Gaze-contingent temporal filtering can be used for similar applications as spatial filtering, e.g. to reduce the bandwidth requirements for video storage and processing. However, the primary application for temporal filtering, in our view, is as a tool for psychophysical research on the spatio-temporal characteristics of the human visual system. Gaze-contingent temporal filtering allows the temporal frequency content of visual stimuli to be adjusted individually at each point in the retinal coordinate system. This makes it possible to conduct very specific experiments on the processing of temporal information and the way in which this processing varies across the visual field. Moreover, the gaze-contingent effect may be applied to both synthetic and natural stimuli, and subjects can view the scene freely, without having to maintain a steady fixation as in other experimental setups. We discuss these issues in more detail in Sect. 10.5.

Our particular research interest lies in the effect that gaze-contingent temporal filtering has on eye movements. Movement or change in the periphery of the visual field is a strong cue for eye movements, and indeed the results of our experiments indicate that gaze-contingent temporal filtering reduces the number of saccades with large amplitude [136] but can remain unnoticed by the observer [135]. We will present a summary of these results, but our main focus in this chapter is on the temporal filtering algorithm. Our long-term goal is to find ways of guiding an observer's eye movements [279]. This requires two components: We must prevent saccades to undesired locations (using techniques such as temporal filtering) and encourage saccades to desired locations (using suitable stimuli).

Foveation can be implemented in hardware, using either a "retina-like" sensor [53, 552] or a conventional sensor combined with a special lens [352]. This approach has been used in robotics; gaze-contingent displays, however, typically use a software implementation to perform foveation (or, more generally, gaze-contingent spatial filtering). The naive approach of filtering with a low-pass filter kernel of appropriate size at each pixel location is too slow to run in real time. There exist several different approaches to reducing the computational burden: Foveation can be expressed as a wavelet operator and combined with the fast wavelet transform to yield a fast implementation [88]. Another approach is to reduce the complexity of convolving with the filter kernel using a recursive filtering algorithm [605]. A third possibility is to decompose the input image into a multiresolution pyramid and to interpolate between adjacent levels of the pyramid to achieve the desired amount of filtering [492].

Several authors have presented schemes for performing gaze-contingent spatial filtering on the graphics processing unit (GPU) [147, 447]; see also Chap. 9. This offloads the CPU to perform other tasks and may allow higher-resolution images to be processed in real time.

Our algorithm for gaze-contingent temporal filtering is similar to that of Perry and Geisler's algorithm for spatial filtering [492] in that it is based on interpolating between the levels of a multiresolution pyramid. However, in our case, the pyramid is built along the temporal dimension. Because it is not feasible to hold the entire multiresolution representation of the video in memory, we shift a temporal window along the pyramid as the video plays and compute the contents of this window in real time. On a standard personal computer, our algorithm achieves real-time performance (30 frames per second) on high-resolution videos (1024 by 576 pixels). Note that the computational requirements for temporal filtering are greater than for spatial filtering because the pyramid reduces resolution (and thus bandwidth) only along one dimension and because the interpolation between levels of the pyramid cannot be interleaved with the upsampling steps.

A limitation of temporal filtering compared to spatial filtering is that the temporal pyramid requires a certain amount of "lookahead" in the video stream because the filters used in the pyramid are non-causal. This means that the algorithm is only useful for pre-recorded video material; on a live video stream from a camera, the algorithm would introduce a latency of several seconds, making it impractical for applications such as head-mounted displays with video-see-through. Note, however, that this latency applies only to the video stream; the latency of the gaze-contingent effect, i.e. the time elapsed between an eye movement and the display update, depends only on the time required to process and display one frame.

10.2 Multiresolution Pyramids

For readers who are not familiar with multiresolution pyramids, we present a summary of their properties in this section.

A multiresolution pyramid contains multiple versions of an image or signal at different resolutions or sampling rates [76, 288]. Usually, the resolution of the original signal is successively reduced by a factor of two. For example, an image with an original resolution of 1024×768 pixels would be successively reduced to 512×384 pixels, 256×192 pixels, and so on.

Such a multiresolution representation can be used for various purposes, such as data compression, scale-independent object detection or texture mapping (where the multiresolution pyramid is typically called a *mipmap*). In this chapter, we will interpolate between the levels of the pyramid to change the temporal resolution gradually across the visual field.

A multiresolution pyramid is built by low-pass filtering and subsampling the original signal repeatedly. (In practice, these two operations of filtering and subsampling are typically combined by computing the filter result only at those locations that remain after subsampling.) Figure 10.2 shows this process for a one-dimensional signal; the resulting tapered data structure resembles a pyramid, hence the name.

Subsampling by a factor of two halves the Nyquist frequency, i.e. the maximum frequency that can be represented at a given sampling rate. For our application, the signal must therefore be low-pass filtered before subsampling to ensure that it does not contain any frequency components above the Nyquist frequency for the new sampling rate. If this is not done, aliasing may occur, i.e. frequency components above the new Nyquist frequency will appear in the subsampled signal as components of a different, lower frequency. (Note that, in some applications, aliasing is not of concern because the aliased frequencies cancel out when the pyramid levels are reassembled into a single signal.)

Figure 10.3a illustrates the effect of the filtering and subsampling steps on the spectrum of the signal. The original signal may contain frequency components up to its Nyquist frequency, which, by convention, is normalized to a frequency value of 1; beyond this point, the spectrum repeats periodically. The Nyquist frequency after subsampling corresponds to a frequency

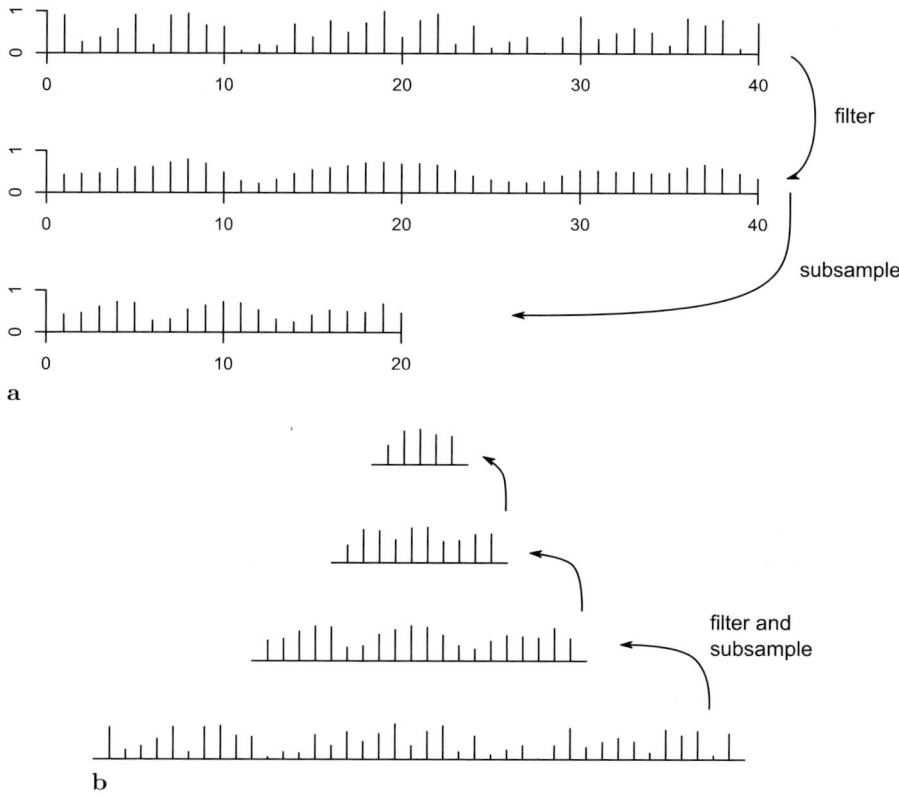

Fig. 10.2a,b. Computation of a multiresolution pyramid. **a** A level in the pyramid is computed from a higher-resolution level by applying a low-pass filter, then subsampling. **b** Repeated filtering and subsampling yields the complete pyramid

of 0.5 in the plot. To avoid aliasing, the low-pass filter should remove all frequency components above this point while leaving the remaining frequency components untouched. Figure 10.3b shows the result after applying the filter. The signal can then be subsampled, which causes a periodic repetition of the spectrum beyond the new Nyquist frequency, see Fig. 10.3c.

To be able to compute the pyramid efficiently, an approximation to the ideal low-pass filter is usually used in practice. The 5-tap binomial filter $\frac{1}{16}\,[1\ 4\ 6\ 4\ 1]$ is a popular choice. Its transfer function is plotted in Fig. 10.4. It is evident that the filter attenuates the frequency components above 0.5 but does not eliminate them entirely. When the filtered signal is subsampled, some aliasing will therefore occur. Also, the filter begins attenuating below the desired cutoff frequency of 0.5. However, these deviations from the ideal transfer function are usually not a real problem in practical applications.

For our purpose of producing a gaze-contingent display, the lower-resolution levels of the pyramid have to be interpolated back to the resolution of

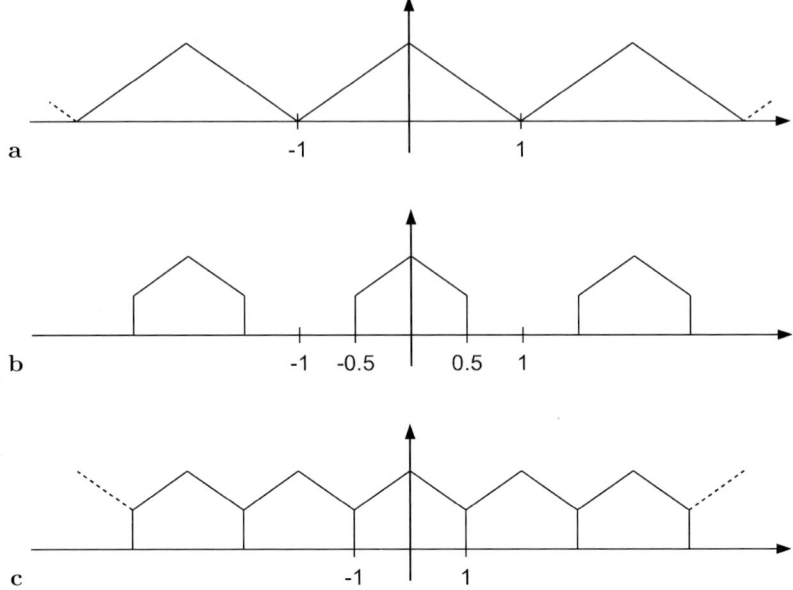

Fig. 10.3a–c. The effect of the low-pass filtering and subsampling operations on the spectrum of the signal. **a** Spectrum of the original signal. The frequency axis is normalized so that a frequency of 1 corresponds to the Nyquist frequency. **b** Spectrum after low-pass filtering with a cutoff frequency of 0.5. **c** Spectrum after subsampling. The new Nyquist frequency has again been normalized to 1

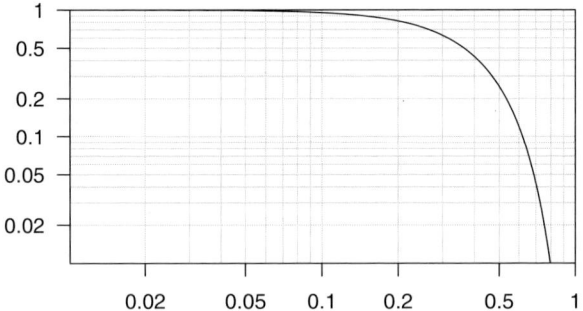

Fig. 10.4. Transfer function of the 5-tap binomial filter $\frac{1}{16}$ [1 4 6 4 1]

the original signal. This is typically done by inserting zeros to upsample the signal to the resolution of the next level, then low-pass filtering to interpolate at the locations where the zeros were inserted. This process is iterated until the resolution of the original signal is reached. Figure 10.5 illustrates the process.

   The multiresolution pyramid described here is also referred to as a Gaussian pyramid in the literature [76] because its levels are effectively ob-

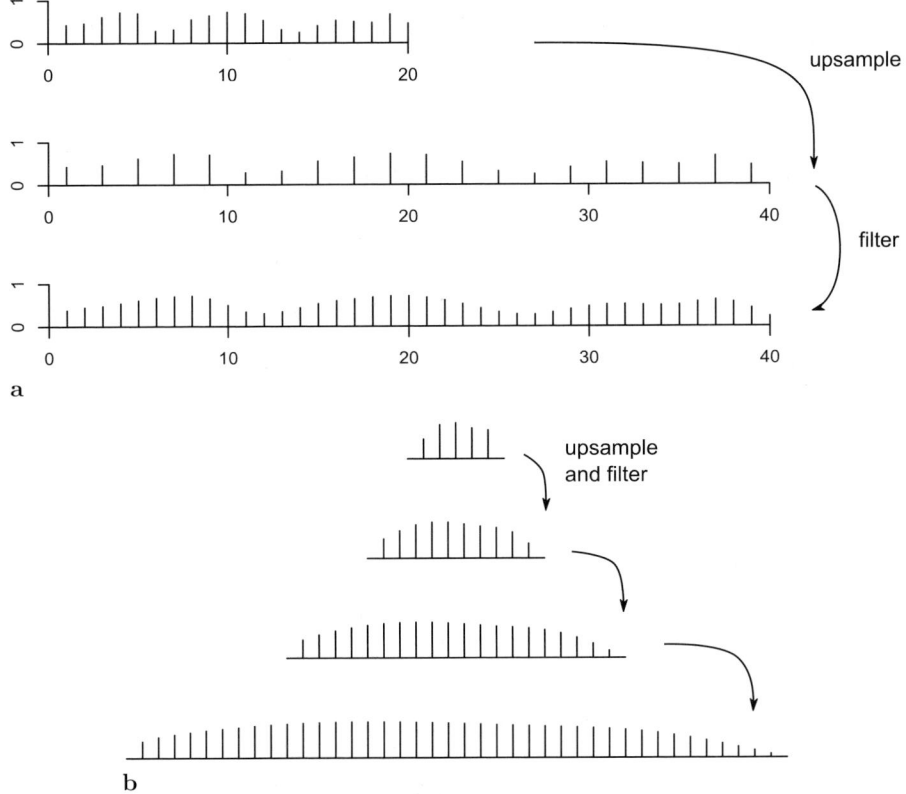

Fig. 10.5a,b. Interpolating a pyramid level to the resolution of the original signal. **a** A single step in the interpolation process: The signal is upsampled by inserting zeros and then low-pass filtered to produce interpolated values. **b** This process is repeated until the resolution of the original signal has been reached

tained by filtering with Gaussian(-like) low-pass filters of increasing kernel width. The technique is related to the Discrete Wavelet Transform [390], which splits a signal's spectrum up into several non-overlapping frequency bands.

10.3 Algorithm

The gaze-contingent temporal filtering algorithm takes a resolution map $R(x, y)$ that specifies the desired temporal resolution at each point in the visual field, relative to the point of regard. For each frame in the video, the resolution map is centred on the current gaze position (obtained from the eye tracker), and the amount of filtering specified by the map is applied at each pixel. This is achieved by interpolating between the levels of a temporal multiresolution pyramid.

For a spatial multiresolution pyramid, the complete data for all pyramid levels can be kept in memory at the same time. This means that the whole of pyramid level l can be computed before pyramid level $l + 1$ (which depends on level l). If we wanted to take the same approach for the temporal multiresolution pyramid, then, because one level of the pyramid spans the entire image sequence, we would have to store this image sequence along with all of its multiresolution versions in main memory – which is clearly not feasible for all but the shortest of videos. Alternatively, we could precompute all of the pyramid levels and read them from disk, one video stream per pyramid level. However, we found that decoding these video streams actually takes more CPU time than computing the multiresolution pyramid on the fly. (Storing the video in uncompressed form is not an option because this would require excessive disk bandwidth.) For this reason, we compute the multiresolution pyramid as the video is being displayed, keeping only those frames of each pyramid level in memory that are needed at any given time.

10.3.1 Notational Conventions

The input video is given as a sequence of images $I(0), I(1), \ldots$ (the sequence is assumed to be infinite). The images contain a single channel (colour videos can be processed by filtering the colour channels individually); they have a width of w pixels and a height of h pixels. The pixel at position (x, y) of image $I(t)$ is referred to as $I(t)(x, y)$ $(x, y \in \mathbf{N}, 0 \leq x < w, 0 \leq y < h)$. (The convention that is chosen for the direction of the image axes is irrelevant for our purposes.) Operations that refer to entire images, such as addition of images or multiplication by a constant factor, are to be applied pixelwise to all pixels in the image.

The individual levels of the multiresolution pyramid are referred to as P^0 to P^L (i.e. the pyramid contains $L + 1$ levels). $P^l(n)$ refers to the n-th image at level l. P^0 is identical to the input image sequence, and P^{l+1} is obtained by low-pass filtering P^l temporally and then subsampling temporally by a factor of 2, i.e. dropping every other frame, as described in Sect. 10.2. Hence, $P^{l+1}(n)$ corresponds to the same point in time as $P^l(2n)$, and for an input image $I(t)$ at time t, the corresponding image in pyramid level P^l is $P^l(\frac{t}{2^l})$ (see Fig. 10.6). Of course, such an image only exists if t is a multiple of 2^l. By Λ_t, we designate the index of the highest pyramid level that contains an image corresponding to $I(t)$; Λ_t is thus the maximum integer value that fulfils $0 \leq \Lambda_t \leq L$ and $t \bmod 2^{\Lambda_t} = 0$.

Because each frame of the output video is generated by interpolating between all of the levels of the multiresolution pyramid, we need to upsample each level to the full video frame rate. We refer to these upsampled pyra-

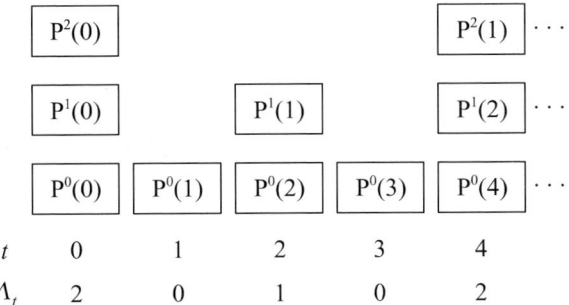

$P^2(0)$				$P^2(1)$	$\cdots$
$P^1(0)$		$P^1(1)$		$P^1(2)$	$\cdots$
$P^0(0)$	$P^0(1)$	$P^0(2)$	$P^0(3)$	$P^0(4)$	$\cdots$
t 0	1	2	3	4	
Λ_t 2	0	1	0	2	

Fig. 10.6. A temporal multiresolution pyramid with three levels. The temporal resolution (frames per second) is halved from level to level. Below the pyramid, t is the time step corresponding to each column, and Λ_t is the index of the highest pyramid level that contains an image for time step t

mid levels as Q^0 to Q^L and will discuss how to compute them in the next section.

10.3.2 Relationships Between Pyramid Levels

Figure 10.7 shows the relationships between different levels of the multiresolution pyramid and the way in which the upsampled versions $Q^0, \ldots, Q^L$ of the pyramid levels $P^0, \ldots, P^L$ are obtained.

$P^{l+1}(t)$ is obtained by low-pass filtering from images in P^l:

$$P^{l+1}(n) = \sum_{i=-c}^{c} w_i \cdot P^l(2n-i) \Big/ \sum_{i=-c}^{c} w_i \ .$$

The $w_{-c}, \ldots, w_c$ are the kernel coefficients. We use a binomial filter with $c = 2$ and $w_0 = 6$, $w_1 = w_{-1} = 4$ and $w_2 = w_{-2} = 1$.

Q^l is obtained from P^l by performing l upsampling steps. The intermediate results of these operations are denoted by Q_l^l to Q_0^l, where $Q_l^l = P^l$ and $Q_0^l = Q^l$. Conceptually, Q_k^l is obtained from Q_{k+1}^l by inserting zeros to upsample by a factor of 2 and then performing a low-pass filtering operation. In practice, these two steps are combined into one, as expressed in the following formula:

$$Q_k^l(n) = \sum_{\substack{i=-c \\ (n-i)\,\mathrm{mod}\,2=0}}^{c} w_i \cdot Q_{k+1}^l\left(\frac{n-i}{2}\right) \Big/ \sum_{\substack{i--c \\ (n-i)\,\mathrm{mod}\,2=0}}^{c} w_i$$

10.3.3 Sliding Window Boundaries

As described above, we wish to compute the P^l and Q_k^l on the fly, while the video is being displayed. To minimize memory requirements, we will keep only

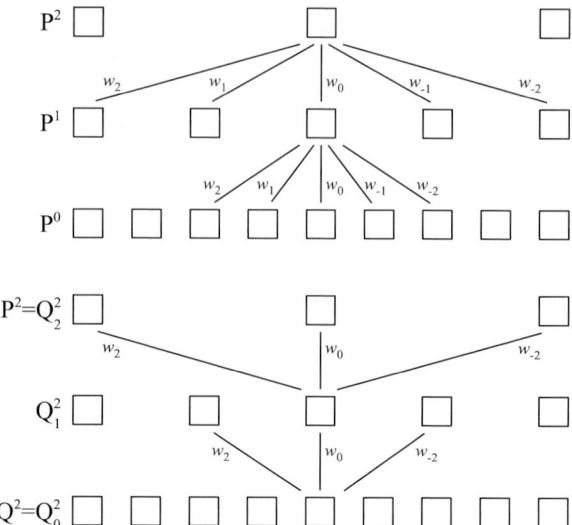

Fig. 10.7. *Top*: Relationships between different levels of the multiresolution pyramid. Each level of the pyramid is computed from the level below by filtering with the kernel $w_{-c}, \ldots, w_c$ (in this case, $c = 2$) and subsampling by a factor of 2. *Bottom*: Scheme for computing upsampled versions Q^l of pyramid levels P^l. Q^l is computed from P^l by repeatedly inserting zeros and filtering with the kernel $w_{-c}, \ldots, w_c$. In practice, only the non-zero frames are included in the convolution with the kernel

those frames of the P^l and Q_k^l in memory that are needed at any given time. This is achieved by using a circular buffer for each of the P^l and Q_k^l to implement a sliding window that contains the required frames. We will proceed in this section to derive the appropriate front and rear window boundaries, relative to the current frame. For example, a front and rear boundary for P^l of 4 and -2, respectively, would mean that, at time t, the sliding window would contain frames $P^l(\frac{t}{2^l} - 2)$ to $P^l(\frac{t}{2^l} + 4)$.

Because the filter used in the multiresolution pyramid is non-causal, the front boundaries have to extend into the future by different amounts (we also refer to this distance as the *lookahead*). For this reason, our algorithm is only suitable for pre-recorded video sequences, as noted in the introduction.

To derive the boundaries of the sliding windows, we start by noting that we need $Q^0(t)$ to $Q^L(t)$ in each time step to compute the final blended output image. We will work backwards from this to find the window sizes required in the preceding processing steps.

From the equation for Q^l (in Sect. 10.3.2), we see that to produce $Q^l(t) = Q_0^l(t)$, we need images $Q_1^l(i)$ with $\frac{t-c}{2} \leq i \leq \frac{t+c}{2}$ and $i \in \mathbf{N}$ (where the constant c depends on the width of the filter kernel). The front and rear boundaries for Q_1^l are thus $\frac{c}{2}$ and $-\frac{c}{2}$, respectively, relative to the current image $Q_1^l(\frac{t}{2})$. Because Q_1^l thus needs a lookahead of $\frac{c}{2}$ images, we need to produce $Q_1^l(\frac{t}{2} + \frac{c}{2})$ at time t (assuming that both t and c are even). We could now repeat this argument to compute the window boundaries required for all Q_k^l. The argument is simplified, however, by noting that we also obtain a valid result if we set a front boundary of c and a rear boundary of 0 for all k. These boundaries are sufficient for the following reason: At time t, we need to produce $Q_k^l(\frac{t}{2^k} + c)$ (assuming that t is a multiple of 2^k). This requires the images $Q_{k+1}^l(i)$ with $(\frac{t}{2^k} + c - c)/2 = \frac{t}{2^{k+1}} \leq i \leq (\frac{t}{2^k} + c + c)/2 = \frac{t}{2^{k+1}} + c$, i.e. a front boundary of c and a rear boundary of 0 is also sufficient for Q_{k+1}^l and by induction for all k.

Turning now to the downsampling side of the pyramid, we note that for all l, the sliding window for P^l needs to contain at least the image $P^l(\frac{t}{2^l} + c)$, because this image is needed for Q^l, which has a lookahead of c. Additional requirements are imposed by the dependencies between the P^l. We start with P^L, which no other pyramid level depends on, and assign it a front and rear boundary of c. Hence, at time t, we need to compute $P^L(\frac{t}{2^L} + c)$, for which we need the images $P^{L-1}(\frac{t}{2^{L-1}} + 2c - c), \dots, P^{L-1}(\frac{t}{2^{L-1}} + 2c + c)$. This means that pyramid level P^{L-1} needs to have a rear boundary of c and a front boundary of $3c$. By continuing in this fashion, we find that level P^l requires a front boundary of $(2^{L-l+1} - 1) \cdot c$. The rear boundary for all P^l is c because of the requirements imposed by the Q^l. We will refer to the front boundary or lookahead of P^l as $\lambda_l := (2^{L-l+1} - 1) \cdot c$. Because the input images are fed into P^0, the total latency of the pyramid is λ_0.

10.3.4 Pyramid Update Algorithm

We are now ready to present the algorithm that updates the multiresolution pyramid in each time step (see Alg. 1).

This pyramid update is carried out in each time step before blending the pyramid levels together to obtain the final output image (see next section). Because of the latency in the pyramid, a certain number of images at the beginning of P^l and Q^l are never computed; specifically, these are the $P^l(n)$ with $0 \leq n < \lambda_l$ and $Q^l(n)$ with $0 \leq n < c$. These images are assumed to be initialized to some suitable value (e.g. all black or equal to the first image in the video).

10.3.5 Gaze-Contingent Temporal Filtering Algorithm

The desired temporal resolution at each pixel is specified by a resolution map $R(x, y)$, where (x, y) is measured relative to the point of regard, $-(w - 1) \leq x \leq w - 1$, $-(h - 1) \leq y \leq h - 1$, and $0 \leq R(x, y) \leq 1$. The values contained in the map specify the temporal resolution relative to the resolution of the original video, i.e. a value of 1 corresponds to pyramid

---

Algorithm 1 Temporal pyramid update step

Input: t Time step to update the pyramid for
Globals: $P^0, \dots, P^L$ Pyramid levels
         $Q^0, \dots, Q^L$ Upsampled pyramid levels
         Q^l_k Intermediate results for upsampled pyramid levels
                            $(0 \le l \le L, 0 \le k \le l)$
Downsampling phase
$P^0(t + \lambda_0) = I(t + \lambda_0)$
$\Lambda_t = \max(\{\Lambda \in \mathbf{N} \mid 0 \le \Lambda \le L \wedge t \bmod 2^\Lambda = 0\})$
for $l = 1, \dots, \Lambda_t$ **do**

$$P^l\left(\tfrac{t}{2^l} + \lambda_l\right) = \sum_{i=-c}^{c} w_i \cdot P^{l-1}\left(\tfrac{t}{2^{l-1}} + 2\lambda_l - i\right) \Big/ \sum_{i=-c}^{c} w_i$$

end for
Upsampling phase
for $l = 0, \dots, L$ **do**
    if $\Lambda_t \ge l$ **then**
        $Q^l_l\left(\tfrac{t}{2^l} + c\right) = P^l\left(\tfrac{t}{2^l} + c\right)$
        $\hat{\Lambda} = l - 1$
    else
        $\hat{\Lambda} = \Lambda_t$
    end if
    for $k = \hat{\Lambda}, \hat{\Lambda} - 1, \dots, 0$ **do**
        $Q^l_k\left(\tfrac{t}{2^k} + c\right) =$

$$\sum_{\substack{i=-c \\ (\frac{t}{2^k}+c-i) \bmod 2 = 0}}^{c} w_i \cdot Q^l_{k+1}\left(\left(\tfrac{t}{2^k} + c - i\right)/2\right) \Big/ \sum_{\substack{i=-c \\ (\frac{t}{2^k}+c-i) \bmod 2 = 0}}^{c} w_i$$

    end for
end for

---

level P^0, 0.5 corresponds to pyramid level P^1, and in general, 2^{-l} corresponds to pyramid level P^l. Intermediate values are handled by interpolating between the two pyramid levels whose resolutions bracket the desired resolution, as described below. Values less than 2^{-L} are treated as referring to pyramid level P^L, since no lower-resolution versions of the image sequence are available.

Note that interpolating between two pyramid levels delivers only an approximation to the desired intermediate resolution. There are methods that deliver more accurate results (e.g. [334]), but they are computationally more expensive, and we believe that the approximation used here is sufficient for our purposes.

We compute blending functions that are used to blend between adjacent levels of the pyramid from the effective transfer functions for the pyramid levels (see [492] for details). The transfer function for pyramid level P^l can be approximated by the Gaussian

$$T_l(r) = e^{-r^2/(2\sigma_l^2)} ,$$

where $\sigma_l^2 = 1/(2^{2l+1} \ln 2)$, and r is the relative resolution. These transfer functions can now be used to define the blending functions. The blending function that is used to blend between levels P^l and P^{l+1} is given by

$$B_l(x,y) = \begin{cases} \frac{\frac{1}{2} - T_{l+1}(R(x,y))}{T_l(R(x,y)) - T_{l+1}(R(x,y))} & 2^{-(l+1)} < R(x,y) < 2^{-l} \\ 0 & R(x,y) \leq 2^{-(l+1)} \\ 1 & R(x,y) \geq 2^{-l} \end{cases} .$$

The algorithm for computing the output image $O(t)$ at time t for a gaze position $(g_x(t), g_y(t))$ is now as follows:

Algorithm 2 Pyramid level blending

Input: t Current time step
         $(g_x(t), g_y(t))$ Gaze position
         $Q^0(t), \ldots, Q^L(t)$ Upsampled pyramid levels
Output: $O(t)$ Output image
  $O(t) = Q^L(t)$
  for $l = L - 1, L - 2, \ldots, 0$ **do**
      for $x = 0, \ldots, w, y = 0, \ldots, h$ **do**
          $b = B_l(x - g_x(t), y - g_y(t))$
          $O(t)(x,y) = (1 - b)\, O(t)(x,y) + b\, Q^l(t)(x,y)$
      end for
  end for

To process an entire video, we now execute Algorithm 1 and Algorithm 2 for each video frame, as follows:

Algorithm 3 Gaze-contingent temporal filtering

---

  for $t = 0, 1, \ldots$ **do**
      Update pyramid for time step t (Algorithm 1)
      Get current gaze position $(g_x(t), g_y(t))$
      Compute output image $O(t)$ (Algorithm 2)
      Display image $O(t)$
  end for

Note that the gaze position is not needed for the pyramid update (Algorithm 1), so its measurement can be deferred until directly before the blending step (Algorithm 2).

10.3.6 Miscellaneous Considerations

As remarked above, colour images can be filtered simply by filtering the colour planes separately. In our implementation, we operate directly on the three colour planes of the YUV420 images read from the digital video files (MPEG-2 format). This has two advantages: First, we avoid having to transform the images to the RGB colour space before processing them; the processed images in YUV420 format can be output directly to the graphics card and are converted to RGB in hardware. Second, in the YUV420 image format, the chroma channels U and V are subsampled by a factor of two in horizontal and vertical direction. This reduces the number of pixels that have to be stored and processed in these channels by a factor of four and halves the total number of pixels compared to the RGB format.

The total memory requirements of the algorithm are quite large. The number of images that have to be stored on the downsampling side of the pyramid (i.e. in the pyramid levels $P^0, \ldots, P^l$) is $c \cdot (2^{L+2} - 2(L+2)) + L + 1$ (we omit the derivation). The number of images that have to be stored on the upsampling side (in the Q_k^l, $0 \leq l \leq L$, $0 \leq k \leq l$) is $(c+1) \cdot \frac{(L+1)(L+2)}{2}$. By taking advantage of the fact that the levels Q_l^l are identical to P^l, $0 \leq l \leq L$, we can avoid having to store these levels explicitly and thus reduce the total number of images in the Q_k^l levels to $(c+1) \cdot \frac{L(L+1)}{2}$.

Let us illustrate the memory requirements with an example, based on the parameters used in our experiments (see Sect. 10.4). We have $L = 5$ downsampled levels in the pyramid and use a filter kernel with $c = 2$. The total number of images that have to be stored is then 279. At a video resolution of 1024×576 pixels, as used in our experiments, one image in YUV420 format requires 884,736 bytes. The total amount of memory required is thus around 235 MB.

Another noteworthy point is that the processing time required by Algorithm 1 depends on Λ_t, the index of the highest pyramid level that is updated in time step t, and thus varies with each time step. Furthermore, the difference between the average and maximum processing time required is quite large. We omit the exact analysis here, but the average processing time is on the order of $O(2L)$, while the maximum time is on the order of $O(\frac{L^2}{2})$. Hence, if we choose the video resolution such that the maximum processing time will fit within the inter-frame interval, then, on most other frames, the CPU will be idle for a substantial proportion of the time. This can be avoided by introducing a variable-length buffer of several frames length to buffer the output of the multiresolution pyramid and compensate for the variation in processing time between frames.

It should be possible to implement our algorithm on the graphics processing unit (GPU) in the way that this has been done for spatial filtering [147, 447]. However, since the memory requirements for temporal filtering are quite large, as pointed out above, it should be ensured that the graphics card has a sufficient amount of video memory.

10.4 Implementation and Results

We implemented the algorithms in C++, with performance-critical parts written in Intel x86 assembly language using the SSE2 vector instructions. The software ran under the Linux operating system on a PC with a 2.8 GHz Pentium D dual-core processor and 1 GB of RAM. For gaze tracking, we used a SensoMotoric Instruments iView X Hi-Speed eye tracker that provides gaze samples at a rate of 240 Hz. The passive eye monitoring software ran under Windows 2000 on a separate PC, and the gaze samples were sent to the Linux PC using a UDP network connection.

The video sequences used in the experiments were taken with a JVC JY-HD10 camera and had a resolution of 1280×720 pixels at 30 frames per second (progressive scan). These video sequences were then scaled down to a resolution of 1024×576 pixels; at this resolution, with a multiresolution pyramid of six levels ($L = 5$ downsampled levels plus the original video sequence), the gaze-contingent filtering algorithm was able to process the video at the full rate of 30 frames per second.

Figure 10.1 shows a still frame from one of the video sequences (top) along with the image produced by the algorithm (bottom). The observer's point of regard (centre left of the image) is marked by a white square. The resolution map used in this example specified full resolution at the point of regard, with a smooth reduction towards the periphery (see Fig. 10.8). Accordingly, the area around the point of regard is unchanged, and increasing amounts of temporal filtering (causing moving objects to blur or even vanish) are visible towards the periphery. Note, for example, that the two men walking at the left of the image are slightly blurred and that the person who is about to leave the image at the right edge has almost vanished completely.

We have performed experiments on the detectability of the gaze-contingent temporal filtering and its effect on eye movements. These results have already been published [135, 136], and we will only summarize them briefly.

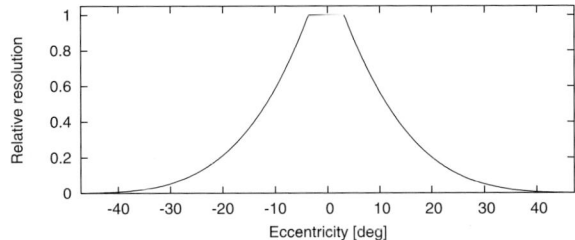

Fig. 10.8. A slice through the radially symmetric resolution map used to generate the filtered image in Fig. 10.1 and in the experiments on the effect of gaze-contingent temporal filtering on eye movements. The *horizontal axis* (radial eccentricity from the point of regard) has been scaled to give the visual angle in degrees, taking into account the monitor size and viewing distance used in the experiments

In one study [135], we investigated the detection threshold of the gaze-contingent temporal filtering effect as a function of retinal eccentricity. We used a resolution map that retained the temporal resolution of the original video across the whole of the visual field except for a ring-shaped region at a certain eccentricity from the point of regard, which was set to a reduced resolution R_r.

For various eccentricities of the reduced-resolution region, we measured the threshold for R_r beyond which the manipulation was no longer detectable. To determine this detection threshold, we used an interleaved staircase procedure, a common method for estimating thresholds in psychophysical 2AFC (two alternatives, forced choice) tasks [197].

Detection thresholds were measured for two subjects on four video sequences of twenty seconds each (a beach scene, a traffic scene at a roundabout and two scenes of pedestrian areas). Because this experiment was run on a slightly slower machine (Pentium 4, 3.2 GHz), the videos were scaled down to a resolution of 960×540 pixels to be able to process them at 30 frames per second. Subjects were seated 55 cm from a 22-inch CRT screen; this resulted in a field of view of 40.0×22.5 degrees (45.9 degrees diagonal).

Figure 10.9 shows the detection thresholds that were measured for one subject. (The results for the other subject were similar, and later measurements with more subjects [137] have confirmed these results.) It is apparent that the amount of temporal filtering that can be performed without being detected increases with eccentricity. Indeed, at high eccentricities, we can

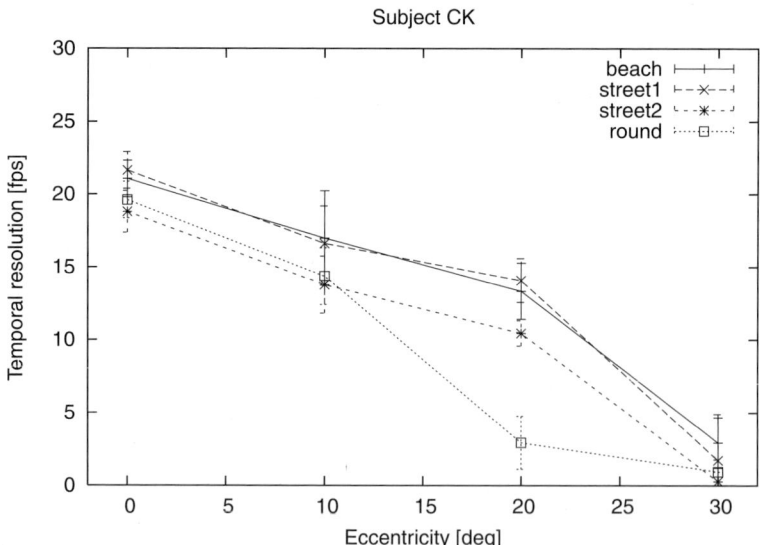

Fig. 10.9. Detection threshold for temporal filtering as a function of eccentricity. Each line indicates results for one image sequence

filter out almost all of the dynamic content of the video without the observer becoming aware of this manipulation. Note that we are not measuring whether the observer can detect the presence of certain temporal frequencies but whether the observer will notice that something is missing if we remove these temporal frequencies from the video. For example, in Fig. 10.1, the person who is about to leave the scene at the right edge is clearly visible in the top image; but the absence of this person from the bottom image is not noticed if one does not know the original image.

In a second study [136], we investigated the effect that gaze-contingent temporal filtering has on eye movements. Here, we used the same resolution map that was used to generate the sample image in Fig. 10.1, i.e. full resolution around the point of regard with a gradual falloff towards the periphery (see Fig. 10.8).

Eight subjects watched four video sequences of twenty seconds each; two of these sequences were identical to ones in the first study, while the other two were different but showed similar material. The observers were instructed to watch the videos attentively, but no other specific task was given. We then compared the results to measurements made for 54 subjects on the unmanipulated original videos. The results (not plotted here) indicated that, with temporal filtering, observers made fewer saccades to the periphery of the visual field, where the temporal filtering was strongest.

This is a plausible effect, since the temporal filtering removes high-frequency content in the periphery, and thereby for example moving objects, which is usually a strong bottom-up cue for saccades. However, further experiments are required to prove the effect to be statistically significant.

10.5 Discussion and Future Work

We have presented a new type of gaze-contingent display that manipulates the temporal resolution of an image sequence, and we have demonstrated that the temporal filtering effect can remain undetected by the observer if the cutoff frequency lies above an eccentricity-dependent threshold. We also have evidence indicating that gaze-contingent temporal filtering can reduce the number of saccades to the periphery of the visual field.

We see this type of display as a valuable tool for psychophysical research on the spatio-temporal characteristics of the human visual system when presented with natural scenes. Traditionally, this type of work has been done using synthetic stimuli such as dots and gratings (see e.g. [480]). Synthetic stimuli are attractive because they are easy to create and because their spatial and temporal frequency content is easy to manipulate. However, there is increasing evidence that many phenomena may be studied more profitably on natural scenes than on synthetic stimuli. For instance, the effects of crowding and masking can mean that phenomena observed on isolated synthetic stimuli do not transfer readily to natural scenes, which typically contain a great number of stimuli that are adjacent in space or time. (Crowding or masking is said to occur when one stimulus affects the perception of another stimulus that is close to it in space or time [490, 589].) Additionally, many studies that use synthetic stimuli present them under steady fixation to control the retinal eccentricity at which they are observed. There is evidence, however, that contrast thresholds, for instance, are higher during normal saccadic viewing than under steady fixation [196]. Gaze-contingent displays have the advantage of allowing natural saccadic viewing while at the same time allowing stimuli to be positioned relative to the retinal coordinate system. Apart from the possibility of differences in perception, some experiment designs, e.g. the study of perception-action cycles [189], require the subject to be able to make saccades. A gaze-contingent display is obviously necessary for these types of experiments if one wishes to manipulate some aspect of the scene with respect to retinal coordinates. For these reasons, we believe that gaze-contingent manipulation of images and image sequences will be an increasingly important tool for psychophysical research.

The gaze-contingent temporal filtering algorithm presented in this chapter can be combined with Perry and Geisler's algorithm [492] for gaze-contingent spatial filtering. To do this, one would simply run Perry and Geisler's algorithm on the output image produced by the temporal filtering algorithm. This would allow full control over both the spatial and the temporal resolution at each pixel, allowing very specific experiments to be carried out on the spatio-temporal characteristics of the visual system.

In a wider context, our work is motivated by applications that involve the guidance of eye movements [279]. In previous work, we showed that certain visual stimuli can trigger saccades to a chosen location [138]. The effectiveness of this stimulation, however, may be reduced if the scene contains other salient locations that can themselves trigger a saccade. We have shown that a saliency map can be used to predict a small number of such locations that are likely candidates for saccade targets [52]. The results of our experiments indicate that temporal filtering could be used to reduce the likelihood that the visual system will choose one of these locations as a saccade target; this should increase the probability that stimulation will then trigger a saccade to the intended location. For this application, it would be desirable to perform temporal filtering only at the specific locations that are likely saccade targets, instead of filtering the whole of the visual periphery, as in this work.

As mentioned in the introduction, gaze-contingent spatial filtering (foveation) can be used to improve the compressibility of video. We would expect that temporal filtering also improves compressibility, but we have not investigated this since compression is not the motivating application for us. However, we would like to point out another, perhaps less obvious way in which our work is relevant to video compression. If gaze-contingent filtering (either spatial or temporal) is to be used to improve compressibility, the short latencies required by the gaze-contingent effect place strong demands on the communcations channel between the sender and the receiver. Furthermore, the technique cannot be used to reduce the offline storage requirements of video.

Several researchers have therefore investigated the foveation of video using a fixed, observer-independent resolution map containing one or (usually) several foveation centres. The location of these centres may be determined either empirically by recording the eye movements of a set of test subjects [585] or by using an algorithm that is designed to identify the most relevant parts of an image [280, 663]. Assuming that the parts of an image containing the relevant information can be identified reliably, this procedure should lead to no loss of (high-level) information. However, if the observer happens to look at a part of the image where resolution has been reduced, the manipulation will become obvious, and the perceived quality of the video will suffer.

We therefore propose to combine this compression strategy with gaze guidance techniques so as to guide the user's gaze to the unfiltered areas of the image (which, after all, are the areas that have been identified as relevant) and to simultaneously use suitable temporal filtering to reduce the likelihood that the observer will look at the reduced-resolution parts of the image. This should improve the perceived quality of the video while, at the same time, communicating the information contained in the video more effectively.

Expanding further on this idea, we note that the use of foveation in video compression has exploited the fact that image content that cannot be detected by the retina may be omitted from the video. Later processing stages in the brain (i.e. cognitive processes) reduce the amount of information that is actually perceived much further. We propose that if gaze and attention can be guided reliably, the information that is filtered out by cognitive processes could also be omitted from the video, improving compressibility considerably.

Part IV

Eye Monitoring in Military
and Security Realms

11

Measuring the Eyes for Military Applications

Sandra P. Marshall

11.1 Introduction

In recent years, *Passive Eye Monitoring* technology has been used increasingly to understand key aspects of military operations. Advances in hardware design and implementation together with increased sophistication in software algorithms have led to robust systems that can be used in operational and near-operational settings as well as in conventional laboratory studies. Researchers from diverse fields such as cognitive science, cognitive psychology, engineering, and psychophysiology have come together to build research programs whose goal is to better understand human perception, attention, and mental effort in the context of performing real jobs. The U.S. Department of Defense has been a major source of support for these efforts. The Office of Naval Research (ONR), the Air Force Office of Scientific Research(AFOSR), the Army Research Institute (ARI), and the Defense Advanced Research Program Agency (DARPA) have all made important contributions in this area.

Three areas in which eye monitoring technology has shown enormous potential for military application are interface design and evaluation, performance appraisal, and workload assessment. This chapter looks at these three broad areas and gives specific examples of eye monitoring contributions in each.

11.2 Background

The research described in this chapter was carried out under support from the Office of Naval Research under several different grants (including $N000149310525$, $N000140010353$, $N000140210349$, and $N000140610162$). Two projects discussed here were large multi-disciplinary research efforts involving researchers across the country. A third project was a collaboration with a local research group in San Diego.

11.2.1 The TADMUS Program

During the 1990s, the Office of Naval Research carried out a seven-year project called the Tactical Decision Making Under Stress (TADMUS) Program. This large multidisciplinary research effort involved the major Navy labs as well as many defense contractors and university researchers. One goal of the TADMUS Program was to understand the factors that contribute to errors in tactical decisions. Two ancillary goals were to create new training techniques as well as new computer displays for the decision maker that would reduce or lessen the impact of such errors. As part of this large collaborative effort, I pursued my own research responsibilities within the program and also contributed eye monitoring analyses and evaluations to other team members. The work described here comes from both aspects of this research and additional features are detailed in [396, 397].

Much of what was studied under the TADMUS Program falls in the broad category of situation awareness or SA. Situation awareness is an essential part of tactical decision making, requiring an operator to perceive and understand the various elements in his immediate sphere of operations. SA also contributes to future performance of the operator as he links details of the current situation to other situations he has experienced in the past. To study situation awareness, the TADMUS Program established two primary research venues, each of which allowed testing with qualified operators engaged in high fidelity simulations of tactical situations. Several scenarios were developed to simulate many different ambiguous events. Each scenario typically lasted about 30–45 minutes.

The first environment was a simulated Combat Information Center of an Aegis Cruiser, located at the Space and Naval Warfare Systems Center (SPAWARSYSCEN) in San Diego, CA. This study focused on the development and manifestation of situation awareness in teams of Navy officers as they worked together to solve tactical problems [396, 397, 424]. The officers were Lieutenants in the U.S. Navy and were shipboard qualified as Tactical Action Officer (TAO). Many had practical experience as officers in the Persian Gulf. They participated in pairs, with one assuming the role of Commanding Officer and the other assuming the role of Tactical Action Officer for the study. The simulation was elaborate. It included numerous technicians who were confederates of the experimenters and fed prescribed information to the officers. Other confederates simulated inter-aircraft chatter as commercial pilots speaking English with heavy accents.

In the second environment, we studied groups of Navy officers working together at the Surface Officers Warfare School in Newport, RI. One member of each team was selected to serve as TAO, and he or she was the focus of the eye monitoring analysis. Again, the team worked in a simulated Combat Information Center and solved a series of tactical problems. A central objective of the TADMUS Program was to assess a new decision support system which was implemented on computer monitors available to the Commanding Officer (CO) and the Tactical Action Officer (TAO).

In most studies involving eye monitoring, the TAO had one or two computer screens available to him. Sometimes he had access to two different systems, the current screen he was accustomed to using onboard ship and a second screen that displayed the decision support system. At other times, he used a large two-screen display containing only the new decision support system. As part of the study, we used eye movements to evaluate how often and when the TAO used the new display.

In both environments, the officers serving as TAO wore head-mounted eye monitoring hardware, together with their usual ear phones for group communication. In the first study, the equipment was ASL's 4000 eye monitoring system, a monocular system that samples at 60 Hz. In the second study, the equipment was SR Research's EyeLink system, a binocular system that samples at 250 Hz.

11.2.2 The $A2C2$ Program

A second large collaborative effort at ONR is the ongoing Adaptive Architectures for Command and Control ($A2C2$) Program under the leadership of Daniel Serfaty of Aptima, Inc. Its goals are to examine and model important aspects of organization and decision making through carefully designed simulations. This long-running program has carried out a series of interdisciplinary experiments at the Naval Postgraduate School in Monterey, CA, using a simulation environment called the Distributed Dynamic Decision Making (DDD) Simulation [326]. In this setting, a team of decision makers is given a mission, a set of predefined mission requirements, and information about the team members' own assets. As a team, they formulate plans of action and execute those plans to accomplish their overall mission objective.

A key aspect of the simulations under the $A2C2$ program is that each team member shares a visual representation of the world through a coordinated computer system. Each team member has unique access to computer files and displays about his or her own areas of responsibility but the main display is a common tactical display available to all team members. Hence, as elements in the situation change, all members see the changes simultaneously. Throughout the simulation, the team members interact with each other continuously, giving each other status reports and making requests of other team members when help or support is required. Each simulation lasts about 45 minutes. $A2C2$ provides an extremely rich and fast-paced environment in which to collect data.

Eye monitoring in the $A2C2$ Program was done using SR Research's Eye-Link II System. One member of each six-person team was selected to be monitored, and he or she was tracked throughout the team's engagement with all simulations. All audio communications as well as all eye data were recorded. Detailed descriptions of the entire experiment may be found in [130, 168].

11.2.3 Other ONR Collaborative Research

A final source of eye monitoring research discussed in this chapter is work initially carried out with Mark St. John of Pacific Science & Engineering, in San Diego, CA, and extended in a study with David Kieras of the University of Michigan [301, 319]. The overall goal was to help operators manage their attention as they view a computer display so that they can concentrate on the most important or threatening features and spend less attention on the less important features. Related objectives were to study the ease and speed with which military operators detect specific aircraft or ships (tracks) on the display and to examine the extent to which highly cluttered screens delay the operator's response. St. John and his colleagues developed a test program in which to evaluate symbols on geographical displays used by military operators. Three symbol systems were evaluated under varying screen conditions. Although the test materials for these studies were based on actual military symbols, the test participants were university students. The task was created as a laboratory task that could be performed by trained or untrained individuals.

Eye monitoring in this study was done using *SR Research's EyeLink II* system. Each individual was tracked as he or she completed all parts of the test program.

11.3 Interface Design and Evaluation

11.3.1 Example 1

The TADMUS Program was an excellent opportunity in which to demonstrate the value of eye monitoring in military applications. eye monitoring was used in the study described here to understand situation awareness through a repeated series of six different questions in which the TAO responded while looking at the new tactical display. Each question was asked five separate times, focusing each time on a different track (i.e., aircraft or ship) in the immediate vicinity. The display showed all current information available to the TAO about the particular track of interest, and it changed with each question. The officer scanned the display when the question was given, searching for the necessary information to make a decision. The questions were randomly ordered under the constraint that the same question never appeared twice in a row. Officers were asked to search the display as quickly as possible and to make rapid responses.

The display is shown in Fig. 11.1. In its implemented form, it had a black background with blue text and figures. Important features were often in red or yellow. The display shows a great deal of potentially useful information about a specific track, and the question being answered at the moment focused only on that track. The different areas in which information could be located are

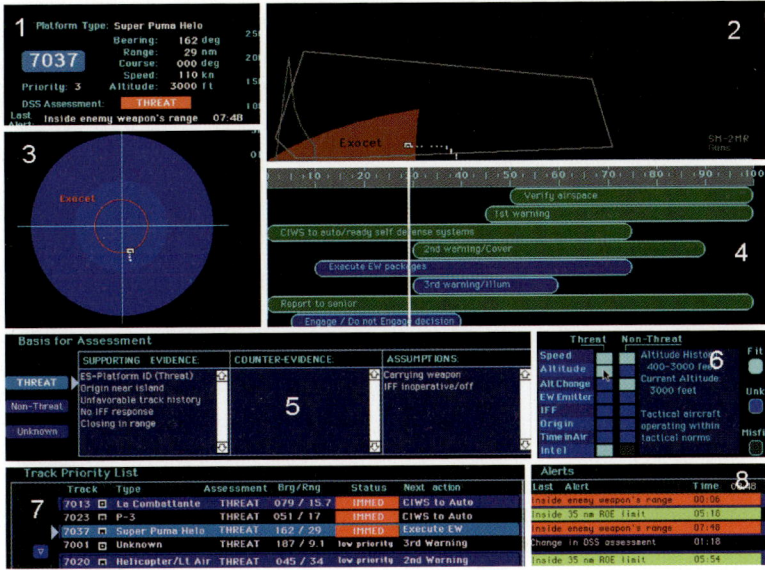

Fig. 11.1. The Decision Support System, developed by Dr. Jeffrey Morrison and colleagues, SPAWARSYSEN, San Diego, CA

indicated in the figure by the lighter rectangles in Fig. 11.1 (which are added here for emphasis and were not part of the display). For example, the track's kinetic information (speed, course, etc.) is given in the upper left corner, and its weapons capability is given in the upper right. For each question, we calculated the gaze pattern made by each officer, looking separately at each repetition of the question. Two examples from one officer are shown in Fig. 11.2.

The diagram on the left side of Fig. 11.2 shows the officer's responses to the question: What priority would you assign to Track X? where X varied from an unknown helicopter to a known commercial airliner. In all instances, the officer sought information from roughly the same parts of the display, namely the upper four regions numbered 1 − 4 in Fig. 11.1. In three of the five questions, he moved between regions 1 and 2. In four of the five questions, his eye traveled diagonally between regions 2 and 3. Information about viewing patterns such as these, when collected from actual operators or users, are extremely valuable for designers of interfaces. For instance, in this case the designers were concerned about the observed diagonal pattern because it is time consuming. It takes valuable time to move the eye from the upper right to the lower left. If all officers tested showed a similar pattern, the designers might decide to relocate some of the information in an attempt to optimize efficiency for the operator.

No such patterns are evident in the diagram on the right side of Fig. 11.2. Here the question was: Why might you be wrong in your assessment of

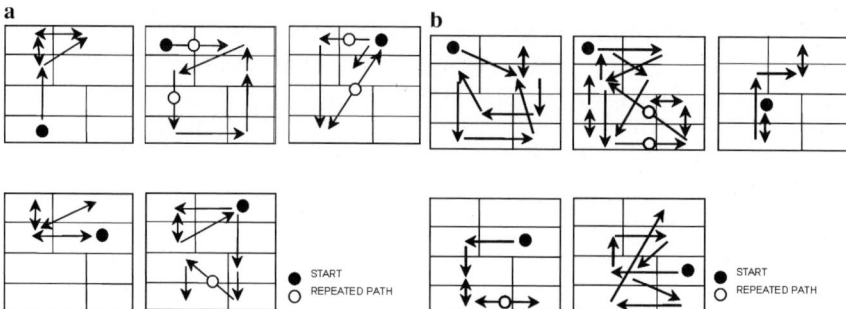

Fig. 11.2a,b. Patterns of eye responses to two questions about potentially hostile aircraft. Regions correspond to those in Fig. 11.1

Track X? Every region of the display is useful at some point, and the transitions are not predictable. Different tracks require different types of information, and there was no single, typical way that the officers approached this question.

Thus, eye monitoring analyses provided two important pieces of information: we were able to determine where the operator looked as he sought answers to typical questions faced in the Combat Information Center and we identified the regions that were most used as he did so. The first helps us understand which features of the display are considered, which are ignored, and which receive the most attention. Patterns of use give valuable feedback to the design of the display. The second gives us valuable input for cognitive models of tactical decision making. Knowing where and when the operator sought specific information enhances our understanding of his decision making process.

11.3.2 Example 2

A second example of using eye monitoring in military environments shows how eye monitoring can be used to examine specific components of computer displays. In this case, we focused on the symbols that appear on many computer displays. Our goal was to determine the extent to which different shapes, colors, and levels of degradation influence patterns of visual search. The task itself was simple. The operator was shown a display containing a large number of fully visible and or somewhat degraded symbols and was asked to find two fully visible symbols that matched exactly a target symbol. The task was repeated multiple times, and each operator saw multiple research conditions for a total of 172 trials. Full details of the study can be found elsewhere [301].

Three types of degraded symbols were used, and they varied according to the amount of information they conveyed:

- **High Information**: the standard military symbology $MIL-STD2525B$ which has color, shape, and other spatial features that clearly identify an aircraft or ship [126] ,
- **Intermediate Information**: symbicons which are simplified outlines of an aircraft or ship, intended to convey quickly the general features of a track [570],
- **Low Information**: dots which were essentially placeholders indicating that something was located in a specific place but giving no information about it.

Targets were always presented in MIL-STD form as were all fully visible symbols. Degraded symbols could be any of the three types given above. For blocks of stimuli, the level of degradation varied from 0% (no degradation and all symbols fully visible) to 75% (75% degraded and only 25% fully visible). The research design varied systematically with three types of degraded symbols, four levels of degradation (0%, 25%, 50%, 75%), and two colors (gray versus color). Figure 11.3 shows an example screen with 0% degradation.

We looked at two types of eye monitoring data. First, we analyzed the amount of time that individuals spent looking at the different types of symbols across the research design. Second, we analyzed cognitive effort by looking at the changes in pupil diameter that resulted across the different conditions of the task. Both are described briefly below.

One major question in this research was whether participants would notice the degraded symbols and if so, whether they spent more time looking at one type of symbol than another. This is an important question because it can influence the choice of symbol used in a display. It is advantageous to degrade symbols that represent non-threatening or irrelevant tracks. However, it is likewise important to convey enough information that situation awareness is not compromised.

We found that participants rarely looked at the degraded symbols. They were very skilled at ignoring those symbols that were not fully visible. Most of the time, their search patterns involved only the fully visible symbols. However, they did notice some of the degraded symbols. One interesting finding that emerged from this research was that individuals did not distribute their viewing time equally across the screen. Figure 11.4 shows the aggregate percentage of viewing time when individuals looked at the degraded symbols. While they rarely did look at the degraded symbols, they looked primarily at the left central part of the display when they did so.

A second question focused on the nature of the visual search. How often did an individual move around the display searching for target matches? We answered this question by looking at the number of transitions that individuals made from one region of the screen to the other (with each location of a symbol as shown in Fig. 11.3 being a region). The mean numbers of transitions per trial across the four proportions of degradation were: 13.84, 11.81, 10.76 and 8.53, with the highest number occurring with no degradation and

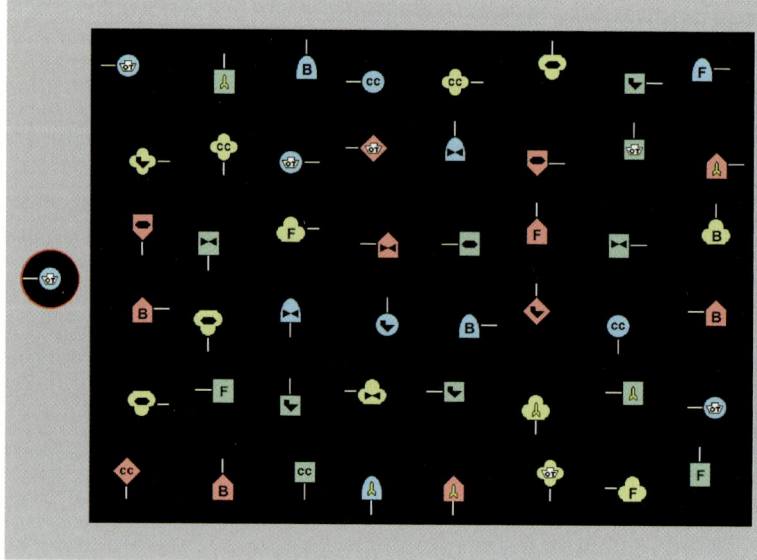

Fig. 11.3. The visual search task with no degradation. Target is small box at far left

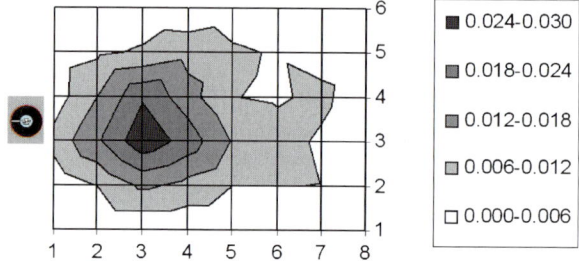

Fig. 11.4. Proportion of time spent viewing degraded symbols when they appeared in different regions of the display. Target is box at far left

the lowest occurring with the most. A multivariate repeated measures analysis of variance was significant across the conditions, $F = 16.19$, $df = 3, 17$, $p \leq .001$, and the test of linear trend was also significant, $F = 7.69$, $df = 1, 19$, $p \leq .001$. As the screen grew less cluttered, the participants looked at fewer regions on the display.

In addition to eye movements, we analyzed the changes in pupil dilation using a metric called the Index of Cognitive Activity. The Index was developed to assess the level of cognitive workload experienced by an individual doing various types of cognitive tasks such as reading, solving problems, or making a visual search. Details of the Index are given elsewhere [394, 395]. Basically, the Index detects small variations in pupil dilation and determines how often these variations occur. The Index is scaled by second, with values indicating the frequency with which unusual pupil activity is detected per second. As cognitive effort increases, so too does the number of measurable changes in pupilsize.

In this study, the Index of Cognitive Activity was measured for each trial for each individual and then averaged across the conditions of the study. The mean values for the four levels of degradation (0%, 25%, 50%, and 75%) were 4.28, 4.15, 3.93, and 3.79. The overall multivariate repeated measures analysis of variance for these data was significant, $F = 4.15$, $df = 3, 17$, $p \leq .01$, with reduced cognitive effort accompanying the degrading of symbols. A test of linear trend was computed for the overall Index scores across the four levels of degradation, and the repeated measures linear trend was statistically significant ($F = 12.15$; $df = 1, 19$; $p \leq .01$). As more symbols were degraded, the cognitive effort as measured by the Index went down.

Several conclusions may be drawn from this study. First, it does not seem to be necessary to greatly degrade the symbols in order to reduce the clutter on a display. Slightly faded MIL-STD symbols carrying full iconic information were ignored as easily as gray dots carrying little information. We have evidence that these different symbols did not differentially attract the eyes of the participants. This finding means that symbols carrying their full set of information can still be presented so that at all times the operator would have access to all the information. Situation awareness is not compromised by such degradation. Moreover, we also have evidence that cognitive workload drops when the display is less cluttered. If the goal is to lessen the workload for operators, one way to achieve it is to declutter the display by degrading the symbols of some objects on the display.

A subsequent collaboration using the data from the original research was undertaken with David Kieras at the University of Michigan. Using his well-known EPIC Cognitive Architecture [320], we modeled the visual search patterns as individuals performed this task. The EPIC architecture was developed to model humans in high-performance tasks and was the first computational cognitive architecture to explicitly represent visual availability and the time course of programming and executing eye movements. The details of the EPIC analysis are complex and too lengthy to include here. They may be found elsewhere [319].The importance of the study is the direct use of eye movements in a complex cognitive model. The results demonstrate that eye data are sufficiently sensitive to serve as the basis for important tests of theories of visual search.

11.4 Performance Appraisal

In addition to providing useful evaluations of the displays that individuals view, eye data also provide important information about aspects of individuals' performance. This use is described here by contrasting the gaze patterns of two officers in a tactical simulation. The study was part of the TADMUS Program described above. Situation awareness clearly plays a role in how quickly an officer responds to a developing tactical problem. Consider that a typical display might have 30 to 50 different tracks displayed on the TAO's screen, which depicts military and commercial traffic in the area. A frequent condition arises when one track on the display exhibits unusual behavior or cannot be clearly identified as friend or foe. In such situations, the officer must simultaneously address the problematic track while maintaining awareness of everything else occurring in the vicinity. His or her ability to do so develops over time with experience, with the result that highly experienced TAOs are able to make very rapid decisions about individual tracks while clearly keeping watch over all possible problems. Less experienced officers are more likely to develop ¡tunnel vision' with respect to the individual track and lose contact with the rest of the display.

The following is a brief example of using eye movement analysis as a tool for monitoring the development of situation awareness. In this example, the eye movements of a highly experienced officer are compared with those of a less experienced officer. The analysis involves three aspects: the gaze traces of both individuals during two critical time periods, a comparison of the amount of time spent looking at any specific region of the display during those times, and an analysis of the number of movements or transitions between these regions. Figures 11.5, 11.6, and 11.7 show these analyses.

In this study, officers were working with the two-screen decision support system developed in San Diego by Dr. Jeffrey Morrison and his colleagues at SPAWARSYSCEN. Four regions of the display are especially important. The large rectangular area on the left contains a geoplot, essentially a radar screen showing the location of all ships and planes in the vicinity. This region is generally considered to be the most important by the officers because it gives them an immediate view of the entire situation. The rectangle in the extreme upper right corner contains important kinetic information about a specific track of interest. To obtain this information, the officer must first select the track on the geoplot using a selection tool such as a track ball or mouse. Immediately to the left of the kinetic information is a region that shows the weapons capabilities of the selected track and also shows the officer's own weapons' range with respect to that track. Finally, the two wide rectangles at the very bottom of both screens show abbreviated kinetic information window for approximately 10 other tracks that are in the immediate vicinity. This information is provided automatically to the officer without his asking for it.

Figure 11.5 shows eight panels, each containing all eye movements (right eye only) recorded for one minute during two stressful episodes of one simulation. The left four panels are observations from the experienced officer; the right four panels are from the less experienced officer at precisely the same time in the simulation. Different patterns of eye movements are clear.

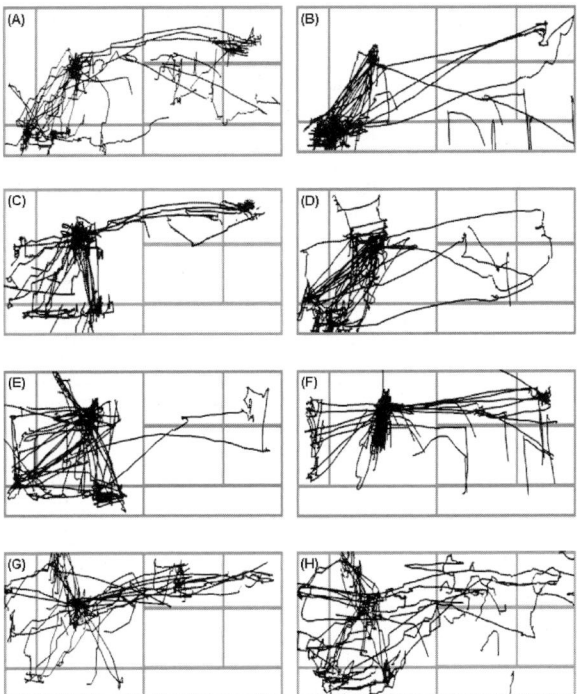

Fig. 11.5. Gaze trace patterns for two officers. Each panel shows one minute of data. Data from the more experienced officer are on the *left*; data from the less experienced officer are on the *right*

In general, the more experienced officer used more information on the display, referring often to all areas containing kinetic information as well as the areas showing his weapons capabilities. The less experienced officer tended to keep his point of gaze focused on a single region rather than moving around the display to gather information from all available sources (see panels B and E). The patterns on the left also appear to be more purposeful; that is, the transitions back and forth across the entire display are relatively short and deliberate. Those on the right tend to drift across other areas (panel H).

Figure 11.6 contains gaze statistics for these same two officers at the same time periods shown in Fig. 11.5. The differences in the gaze patterns are highlighted by comparisons of the number of seconds each officer spent looking at each region. Chi square tests show significant statistical differences in the patterns. Three key areas were contrasted in the chi square tests: (1) the geoplot, (2) the combination of all rectangles on the right side excluding the bottommost one, and (3) the two rectangles across the bottom. The combinations were necessary because the small number of seconds in many of them made it impossible to evaluate them individually. Four chi square tests were run, one for each minute of data shown in Fig. 11.5. Each test

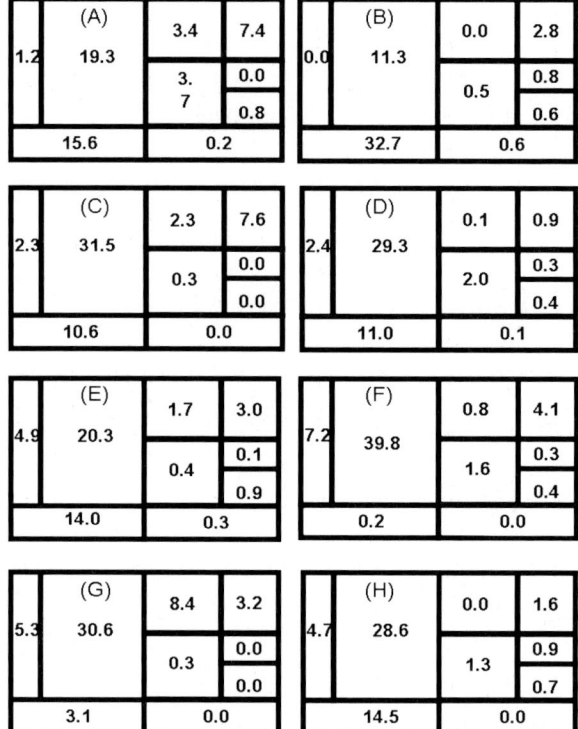

Fig. 11.6. Number of seconds spent in each region of the display of Fig. 11.5. Data from the more experienced officer are on the *left* and those from the less experienced officer are on the *right*

compared the number of seconds the less experienced officer spent in each of the three areas with the number of seconds the more experienced officer spent in each area. For example, the times in panels A were contrasted with B, those in C with D, and so on.

Three of the tests were significant. The second test (minute two) showed no significant difference. In the first minute displayed in Fig. 11.6 (panels A vs B), $\chi^2 (2) = 13.3$, $p < .01$. As can be seen from Fig. 11.5 , the more experienced officer spent a considerable amount of time on the right side of the display while the less experienced officer was focused on the lower rectangles, especially the left one.

In the third minute (panels E vs F), the largest difference is in the amount of time the less experienced officer spent on the geoplot. In contrast, the more experienced officer is using the kinetic information in the bottom rectangles. The patterns are statistically different, with $\chi^2(2) = 26.8$, $p < .01$.

In the fourth minute (panels G vs H), the two officers are using the geoplot to roughly the same extent. However, the more experienced officer is also us-

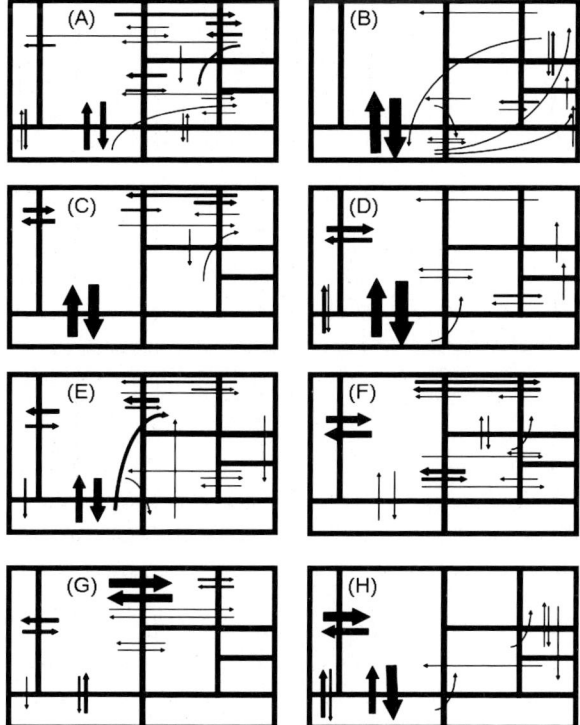

Fig. 11.7. Transition patterns for the two officers during each minute as shown in Fig. 11.5. Heavier lines indicate multiple transitions

ing the information on the right side of the display while the less experienced officer is focused on the kinetic information at the bottom. The patterns are statistically different, with $\chi^2(2) = 10.8$, $p < .01$.

Finally, Fig. 11.7 shows the patterns of transitions between the regions. As the officers sought specific information, they often went back and forth between two or more regions. These movements are captured by the arrows in Fig. 11.8, with the wider lines indicating multiple transitions. One noticeable difference between the two officers is the number of transitions made from the geoplot (large rectangle on left) and the tall rectangle immediately to its left. This leftmost area contained the controls for changing parameters of the display, such as zooming in or out and showing text. The less experienced officer spent much more time adjusting his display during these critical time periods than did the experienced officer.

What do these analyses tell us? That is, what do we gain from having the eye data? Most importantly, we see clearly the differences between the more experienced and less experienced officer. Observing their performance without knowing anything about their points of gaze, their focal attention, or their shifts in attention to regions, we could only say that the less experienced officer took more time to respond. With the eye information, we can begin to understand why he needed more time. We see the elements to which he allocated his attention, and we also see the elements that he failed to notice. This information is important for multiple reasons.

First, it has instructional value. It may be possible to improve the less experienced officer's performance by showing him the differences between himself and his experienced counterpart. We may assume that he will develop expertise over time as he gains more experience. Using eye movement analyses may shorten the time to develop this expertise.

Second, eye-movement analysis has design value. Understanding which regions of the display the officers access at critical times will give crucial information to the developers of such systems. Some elements were never or rarely viewed. Perhaps they should be modified or reduced, giving more room for the areas that are used more often. Similarly, looking at the patterns of transitions over multiple officers will show which regions are likely to be accessed contiguously. It may be advantageous to place information so that individuals do not spend large amounts of time in saccadic movement from one side of the display to the other.

Third, eye-movement data once again have cognitive modeling value. As noted above, situation awareness is an essential ingredient for an officer making decisions in tactical settings. Understanding how that awareness develops in any specific situation gives us, the cognitive researchers, more insight into how the officer is thinking about the situation and does so without our intruding into his decision making by asking him questions while he is carrying out his duties. Knowing when, where, and for how long the officer considered information gives us an estimate of the value of that information to him and its importance in his decision-making process.

11.5 Cognitive Workload

Cognitive workload was mentioned earlier in this chapter in the second example of interface design and evaluation. The interface example involved a relatively simple task of visual search. The current example looks at a more complex case in which operators are engaged in stressful situations and need to make accurate and timely responses. The environment is the DDD Simulation under the $A2C2$ Program as described above. Eight teams of six Navy personnel participated in the study, with each team member assigned to specific roles and responsibilities. This discussion focuses on two operators who were members of two different teams to illustrate the measurement and use of the Index of Cognitive Activity.

As part of this study, each team was given a mission and each team member was given explicit responsibilities. These responsibilities were based on two different organizational structures: divisional and functional. Divisional organization allocates responsibilities based on platform (such as an aircraft carrier). A team member would be responsible for all weapons, assets, and other properties of the platform. Functional organization allows each team member to specialize in one or two functions (such as air warfare or search and rescue) and to be responsible for assets that may reside in many locations or on many platforms. These two organizations are quite dissimilar and require different strategies and decision making by the team members as they work together to complete the mission. The mission demanded that they coordinate their assets and work closely with each other.

The design of the study called for each team to be trained and tested under one organization and then to be suddenly switched to the other, producing a mismatch between original training and testing. They were evaluated in two test conditions: (1) the organization matched original training and (2) the organization did not match original training. To assess the impact of changing from one team organization to another, the cognitive workload of the operators was measured under each organization using the Index of Cognitive Activity.

Recall that the Index is based purely on changes in pupil dilation as the individual performs a task or series of tasks. In this case, the task was a 35 minutes scenario in which all six team members were active participants. Only one team member was selected for observation and measurement using the eye monitoring equipment, and this team member participated fully in all aspects of planning, decision making, and performance.

Average Index values were computed for each operator across each minute of each scenario. Figures 11.8 and 11.9 show the measured Index for the two operators. Both figures have two lines, one showing the changes in cognitive activity during the operator's participation in the divisional scenario and the

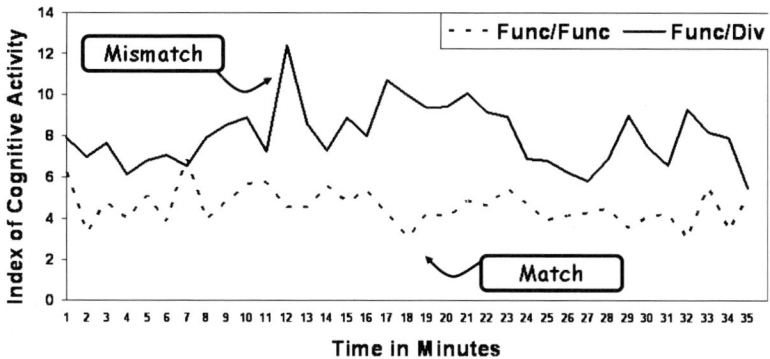

Fig. 11.8. Index of Cognitive Activityfor operator trained under Functional Organization

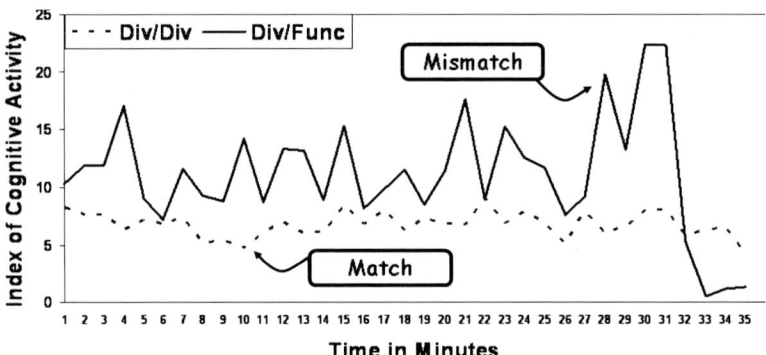

Fig. 11.9. Index of Cognitive Activity for operator trained under divisional organization

other showing the operator's participation in the functional scenario. Each figure depicts the matched condition (i.e., training and testing under the same organization structure) with dotted lines and the mismatched condition (i.e., training and testing under different organization structure) with solid lines.

Figure 11.8 shows results from the operator who trained first under the functional organization, tested under that organization, and then switched to the divisional organization. The Index for the matched condition (Func/Func) is relatively low and stable across the scenario, suggesting that the operator was maintaining a fairly constant state of cognitive workload. In contrast, the Index in the mismatched condition (Func/Div) is high and has several large spikes.

Figure 11.9 shows the opposite pairing: the operator was trained first under divisional and then later switched to functional. The same pattern is apparent. The matched condition (Div/Div) produced low values for the Index and they are relatively flat across the entire duration of the scenario. In contrast, the mismatched condition (Div/Func) produced very large Index values that repeatedly spiked over time. Once again, the mismatched condition produced values that were always larger than those observed in the matched condition.

Using the Index of Cognitive Activity in this way produced several important results. First, it was the mismatch, not a specific type of training, which produced high workload estimates. No matter which organization was used in training, switching to the other organization produced very high workload. Second, these estimates were obtained as non-intrusive psychophysiological measurements. We did not interrupt the operators as they performed their duties to inquire about their perceived cognitive workload. And third, the workload estimates were important pieces of information that could be correlated with other metrics from the study such as errors in performance, level of communication, and team structure.

11.6 Conclusion and Future Work

This chapter has presented several examples in which eye movements and pupil dilation data have been studied in military settings. The findings are strong and have been well received by the organizations that supported the research. These examples, of course, are only a few of the many studies that are now underway in military applications.

As research proceeds, the next step is the operational use of eye monitoring. For this step to occur, it has been first necessary to demonstrate the value of eye monitoring in general, the benefits of eye tracking to specific applications, and the feasibility of using the technology in laboratory settings. The methodology, software development, and analysis techniques exist. Transition to the field now requires hardware adaptations such as miniature cameras, wireless connections, and integration with other psychophysiological monitors. These adaptations are currently under development and show great promise. As they come into wide use, we can anticipate many more applications of this technology in the military arena.

Passive Eye Monitoring in Information Security

Bogdan Hoanca and Kenrick Mock

12.1 Introduction

In the post-September 11 era, security is becoming more and more critical. An important component of information security is user authentication, the ability of an information system to certify that a user is who she claims she is. Authentication can involve one of two processes: identification or verification. For identification, information about an unknown user must be compared against similar information for all possible users. The best match is returned within a confidence level. For verification, a user identity (entered as a user name for example) must be compared only against an existing signature (usually a password) stored for that user. While identification is important for database searches, for example to locate a person based on fingerprints left at a crime scene, most information systems implement authentication as verification. A user types in his or her user name or scans an identification card, then enters a password to verify the identity. Authentication as verification is used for both physical access (for example to secure areas) and for online access (for example to log in to a computer terminal). Secure user authentication requires that users be endowed with credentials which are i) unique for each user, ii) not easily stolen or lost and iii) reasonably affordable and iv) convenient to use. The order above is not an indication of importance of the various requirements.

It might not be obvious that the four requirements listed above are conflicting. No one technology has been demonstrated that can offer the perfect solution, meeting all of the four requirements simultaneously. The challenge is compounded today, with the proliferation of mobile networks that expose users to an increased volume of external threats. More and more users need to access several different information systems, often from public or unsecured computers, and sometimes over unsecured network connections. As users need to access multiple information systems, each with different authentication mechanisms, this increases the number of credentials the user must carry (and the likelihood that some of the credentials will get lost). Finally, as users need to carry an ever larger number of credentials, the cost increases and the convenience decreases.

Recent developments in eye monitoring technologies make them a suitable candidate for use in strengthening the security of information systems. Very little work has been reported on the use of eye monitoring technologies for information security. This chapter is the first to summarize the existing research and to propose directions for future applications. The chapter starts by introducing the basics of user authentication. We then review reported results of authentication using eye monitoring hardware. We discuss on one hand the use of eye monitoring as a replacement for computer mice, which has implications for users of public computers. We then discuss the use of eye monitoring for biometric authentication [10, 219], which could revolutionize human–computer interaction in any setting. We discuss both iris scanning solutions as well as solutions based on feature extraction from eye movements. Combining iris scanning and eye based biometrics can lead to even more secure and user-friendly interfaces. For each of the approaches we assess the usability and the limitations of eye monitoring based systems, given the current development stage of eye monitoring technologies. We conclude the chapter with comments on the future capabilities and the potential of the technologies based on current development trends.

12.2 Authentication in Information Systems

There are three well-established forms of user authentication currently in use. They include: i) something the user knows and enters at a computer keyboard or with a mouse on a graphical display (passwords), ii) something the user carries encoded on her body as a biometric trait (biometrics) or iii) something the user carries with her, for example, information encoded on a magnetic access card, chip or other physical device (hardware tokens). Often, several of these techniques are combined to increase the security of a system, in what is known as multiple-factor authentication. In particular, physical access cards are almost always used with an associated personal identification number (PIN); otherwise, a lost or stolen card would give full unrestricted access to an unauthorized bearer.

Among the three authentication options above, passwords are the most commonly used method because of their low cost and ease of use. They do not require any special hardware, are easily changed and always available (unless the user forgets his password, which is one of the drawbacks). On the flip side, passwords can be insecure if users select weak, easy to guess passwords, if users write down passwords in places where attackers can find them, or if the users are not careful about when and how they enter the password. Also, a malevolent observer who watches or records the user entering a password without the user's knowledge will be able to obtain even the strongest password. The danger of this information leak, alternately called "shoulder surfing" or "peeping attack" is highest in the case of authentication from crowded public areas, where multiple observers may be watching or even recording several access sessions, then using this information to recover passwords.

To quantify how strong passwords are, the usual metric is the number of possible passwords (also called the size of the password space), an indication of how difficult to guess a password may be. Intuitively, if a password is one of a large number of possibilities, it is more difficult to guess by an attacker who tries each possible password in turn (a so-called brute force attack). For example, for a password which may include up to eight printable characters (of a set of 96 possible values – uppercase, lowercase, digits etc), the size of the password space is higher than if the password is a dictionary word, which might be shorter than eight letters and include only characters a–z. Sometimes, the size of the password space is also known as cryptographic complexity. For the first example above, the space includes $96^8 \approx 7 \times 10^{15}$ possible combinations, while the eight letter words amount to fewer than 0.01% of this number of combinations. The total number of possible passwords is only relevant if the attacker needs to guess the password, but it is not relevant for shoulder surfing attacks. No matter how large the password space, if an attacker is able to record the user when she is entering the password, the system security is compromised.

Among the other two authentication options, physical cards are only moderately expensive and relatively convenient to use, but are more likely to be lost or misplaced. The price of smart cards, USB tokens and RFID cards has been decreasing steadily, making such authentication options ever more affordable. On the other hand, hardware tokens can quickly become inconvenient if several such devices are needed in order to access the various information systems a user might need to use. Additionally, the process of assigning and maintaining a large number of tokens is also impractical for the administrator of the information system. Single sign-on systems allow users to authenticate once and then use any number of linked information systems. Such settings are only practical for information systems managed within a single organization or among closely cooperating entities.

Similar considerations apply to the management of systems requiring large numbers of passwords. Single sign-on systems are preferred, but not always available. When faced with the need to memorize several different passwords for different information systems, users either write them down (an insecure practice already mentioned), they confuse passwords across different information systems, or they forget their passwords. In the last two situations mentioned, the systems administrator must provide assistance with resetting passwords, a costly process that also reduces the security of the system [71]. Any time a password must be reset, it is either set to a random value that needs to be sent to the user (with risks associated with information loss while the password is in transit) or it is set to a default value (which is more easily guessed by attackers).

Biometric identification techniques such as face scanners [219, 556] and particularly fingerprint readers [414] are convenient, because the user always carries the authentication information with them. Biometric credentials do not require users to memorize anything, or to carry several objects as credentials for authentication. For this convenience, biometric techniques for authentication have been steadily gaining in popularity. However, some of the biometric identification methods are still prone to false positive and false negative identification [632]. Attacks against biometrics include synthetic data (a fake biometric, for example made out of plastic), recording data from a real authentication session and playing it back later on (this is known as replay attacks), or attacking the database of biometric data to give access to a biometric of the attacker [336]. Moreover, some users do not possess certain biometrics (for example people with missing fingers etc). Most problematically, several privacy groups have also expressed concerns over the potential theft of biometrics, or the gathering and sharing of personal biometric data. This is even more troubling given the persistence of biometrics: if a biometric trait is compromised, it cannot be replaced or reset. Some of the problems described above for biometrics are also problems with the other authentication approaches: replay attacks or attacks on the database of user signatures are not limited to biometric authentication techniques.

12.3 Passive Eye Monitoring for Password Entry

Passwords entered as text from a computer keyboard were the first authentication method used. They continue to be the preferred method because of the low cost and ease of use. More recently, graphical passwords have also been proposed, as being easier to recall, less likely to be written down and also potentially able to provide a richer space (larger number of possible passwords) than the limited letters and symbols of text-based passwords [51]. For example, the user might authenticate by selecting with mouse clicks a series of points on an image, or a series of tiles in a tiled image. An example is the Real User Corporation's PassFaces[TM] system [483]. A variation of this is to require the user to choose several images among the set presented on each screen [491]. These schemes are illustrated in Fig. 12.1. One of the most studied approaches to using graphical passwords is the "Draw a Secret" scheme in which the user draws lines or shapes that constitute the password [118]. The strength of the graphical passwords is strongly dependent on the number of points to select or the number of strokes in the drawing (and most dependent on the number of short strokes in the drawing) [613].

Both the text and the graphical password entry systems suffer from a common weakness. The user needs to be aware of her surroundings when entering

Fig. 12.1. Examples of Graphical Password Schemes: *Left*: multiple tile image. *Right*: points on a single image. The multiple tile idea is similar to the one in the PassFaces™ scheme. The user selects a face out of a series of faces for their password. To enter a password in the single image approach, the user clicks in a prescribed order on a series of predetermined points in the image. The password points, *white circles* in the figure on the *right*, are not visible in an actual authentication scenario

a password, in case somebody is watching and recording the keystrokes or the images selected. This type of attack has been called "shoulder surfing" or "peeping attack." As mentioned above, if somebody is able to record the user when she is entering the password, even the most complex password scheme is easy to break. Graphical passwords may be even more susceptible to shoulder surfing, because images are likely to be larger than the size of the typical text characters on a screen or on a keyboard. Most often, authors assume that graphical passwords would be entered on a small screen, with a reduced observation angle [293, 295], and they dismiss the likelihood of shoulder surfing. In practice, this assumption is rather limiting.

Clearly, the danger of shoulder surfing is highest in high-traffic areas, for example, in airports where an authorized user needs to enter a code to open an access door to a secure area, next to a crowd of people passing by or waiting near the door. Another critical environment is at an automated teller machine (ATM) where many customers could be lined up awaiting their turn to access the machine. As the user is entering the password, multiple assailants closely located might be observing the password entry simultaneously, or may even record the password entry process using hand-held cameras and then later play it back at slow motion to observe the password entry sequence.

Several authentication techniques have been proposed that can reduce the danger of shoulder surfing. Some involve the use of special screen filters and guards, the use of screens with narrow viewing angles, or the use of special hardware. Other techniques require the user to remember information across multiple areas on the screen and to make an entry based on the aggregate information. For example, the user can either enter a number [261] (e.g., the sum of numbers associated with the letters in the password) or click on a graphical symbol [255] or area of the screen [574] that is somehow related to symbols that are part of the password. If the number of possible letters or symbols that would lead to a particular sum or to a particular click action is large enough, the danger of shoulder surfing could be greatly reduced. An observer who has full access to the information the user entered would still have to exhaustively explore all possible combinations that could have led to the user entering such information. This would make the attack much less practical. The attacker would have to complete a comprehensive search or to observe several authentication sessions to obtain the user's password.

Eye monitoring hardware can offer a much simpler solution to the problem of shoulder surfing. Such an approach was proposed by Hoanca and Mock [256], namely to use eye monitoring hardware without on-screen feedback. This way, the user can look at various areas of the screen to select symbols or points on graphical areas and the computer would be able to "see" what the user is selecting. An observer trying to get the same information would have to do one of two things. On one hand, she might have to mount equipment equivalent to the eye monitoring hardware, to track the user's gaze. It is unlikely that the user would not notice such equipment, especially if it comes in addition to the eye monitoring hardware already on the computer. Alternatively, the attacker could gain administrator access to operating system or to the eye monitoring hardware on the computer. If the computer has been compromised to such a degree, the attacker can easily install a key logger or screen recorder that would capture any kind of information entered by the user; shoulder surfing is only a minor concern in comparison with this. In the more distant future, the eye monitoring camera may also allow some real time monitoring of the user area, to warn about suspicious people looking at the user's screen, or even to detect whether the user was under coercion or was willingly entering the password on the screen. Thus, eye monitoring devices would allow access from a crowded public area, even with numerous potential attackers located at various incidence angles to the user, without allowing an attacker to see what the user is selecting on screen as a password.

One of the main complaints against using eye monitoring devices as a user interface has been the so-called "Midas touch," after the character from the Greek mythology who was cursed to turn into gold anything he touched. When using eyes to control objects on the computer screen, the danger is that any button or control the user will look at might be triggered by the gaze, whether this was the intention of the user or not. As such, the eye monitoring interface must be used as a pointing tool and should be assisted by an additional mechanism for selecting objects, either via a physical button click, voice command, or through eye fixation time upon the item to be selected. Eye fixation for selection might be the most appropriate to integrate with eye monitoring, but may require established user profiles. The threshold for fixation may have to range from 800–2000 ms [273] to accurately convert fixation to object selection. As an added benefit, an eye gaze tracking system could be equally accessible to people with various levels of typing abilities, with impaired control of their fingers and hands, or simply to people carrying something through the secured access door (weapons, luggage etc) that limits their freedom of hand movement.

Ideally, the user would be able to simply walk to a gaze tracking enabled computer and start to use the system. In practice, calibration of the eye monitoring hardware is required, but this could potentially be integrated in the log on procedure. The user would perform the calibration, use the eye monitoring equipment to log on, and finally go ahead with using the computer with eye monitoring capabilities.

The drawbacks of the proposed systems today are the relatively high cost of the eye monitoring hardware, and also the relatively high probability of error in selecting desired screen areas, especially if the eye monitoring hardware needs to operate without on-screen feedback. The cost of the hardware is expected to decrease, while the accuracy and performance of the technology will continue to increase [369]. It is even possible that computers in the future might have low cost eye monitoring hardware as a standard type of user interface, potentially as a replacement for the current computer mouse. As the accuracy of eye monitoring hardware continues to increase, it is likely that in the future eye monitoring will be able to operate accurately without on-screen feedback, allowing users to authenticate with minimal danger of shoulder surfing. The solution of using eye monitoring might already be cost effective in cases where hands-free and non-contact options are required (for example at high security access gates).

In the next two sections we review techniques that can bridge the gap between the current technological limitations of eye monitoring hardware and the capabilities needed for using eye monitoring for authentication in graphical user interfaces. We review separately the case for graphical interfaces where the user authentication requires i) selecting specific locations from a single image or ii) selecting tiles from a composite image.

12.3.1 Analysis of the Single Image Approach

If the gaze tracking system is used to select specific location(s) from a specific picture, then the simplest technique is to merely allow the user's gaze to control the active area on the screen, the way a traditional mouse cursor would. For security purposes, no cursor will be displayed on the screen in the active area, but the focus point of the user's gaze would act as an "invisible cursor."

Due to the lack of on-screen feedback as to the location of the cursor, calibration error with the eye monitoring hardware, and the inherent difficulty remembering a password location down to a specific pixel, research shows that a reasonable activation cell size for mouse based graphical password interfaces is a 10×10 pixel cell [678]. This resolution is easily within the error of existing eye monitoring hardware.

If the image has a resolution of 600 by 400 pixels, a 10×10 pixel cell yields 2,400 selectable locations. This is a rather modest number of possibilities. To increase the size of the password space, a user must be required to select multiple locations, in a prescribed sequence. If the user is required to select 5 locations in sequence then the password space is $(2400)^5$ or 7.9×10^{16}, which is an order of magnitude stronger than an 8-character text password using any of 96 printable characters.

In practice, having even 2,400 selectable locations per image is highly unrealistic. In typical images many of these locations will be non-distinguishable. Instead, users are likely to select key locations or "points of interest" along the edges of objects or items that stand out. We estimate a typical image may have between 30–50 distinguishable locations that make suitable passwords. This significantly reduces the likely password space to somewhere around 30^n where n is the number of locations the user must select. For a password space of size comparable with that of eight printable characters, the number of click points would need to be $8 \times \frac{\log(95)}{\log(30)} \approx 11$ click points, which is not unreasonably high.

Although the preceding analysis assumed fixed cell sizes of 10×10 pixels, the "interesting" points would be determined by the user, rather than by the system imposing a rigid grid on the image. Thus, several points might be clustered together, separated by large areas with no "points of interest." This should not affect the cryptographic strength of the password, as long as each image includes sufficient points of interest.

12.3.2 Analysis of the Multiple Tile Approach

Another approach to graphical password selection would be to successively select individual tile images from a set of image tiles, rather than points on a larger image. This is the approach in the PassFaces^{TM} case [483]. As in the previous section, the direction of the user's gaze points to the image to be selected, and no cursor (feedback) is displayed on the screen. For a tiled image, a practical number of tiles would be 9 to 36, although up to several hundred have been used for database search applications [91]. For small numbers of images (up to 36), the lack of feedback and calibration errors are not likely to be significant. A 6×6 grid of tiled images on an 800×600 pixel screen allows 130×100 pixel icons, much larger than the likely margin of error.

However, an issue that should be addressed is the selection of the images to be displayed. Consider a scheme where nine images are shown per screen and the user must select the correct image out of three screens. This means that one of the nine images per screen must always be the same – the password image. What about the remaining eight images?

On one hand, it is desirable to select the remaining eight images randomly out of a set of N images, where $N \gg 8$. However, an attacker who is able to record multiple login sessions will easily be able to determine the password by merely noting which images reappear upon a repeat login sequence. This problem can be alleviated by randomly selecting the eight images only once, and then displaying the same set of images upon a repeat login sequence. An attacker will no longer be able to determine the password by comparing previous login images, but the attacker is easily able to determine which of the N images are always presented on each screen, which reduces the effective password space. In practice, displaying the entire set of image tiles on every screen will make the most sense. As above, if the number of image tiles is 30, the user will need to go through 11 screens of selection to achieve a size of the password space equivalent to that of a text password with 8 printable characters.

12.3.3 Eye Monitoring Aware Authentication Schemes

The discussion in the previous section centered around using eye monitoring hardware for authentication, but used a relatively naïve approach, replacing the use of a hand-driven mouse with that of the eye-driven eye monitoring hardware. In an ideal world, if the eye monitoring hardware would work perfectly, the solution would be simple to use and would not require any modifications. In practice, the limited resolution of current eye monitoring hardware introduces errors, sometimes making the authentication scheme above rather impractical. Solving this problem requires substantial modifications to the traditional authentication approach, in particular making the authentication software aware of the potential limitations and sources of error of the eye monitoring hardware.

There are two major sources of error that can arise in the use of eye monitoring hardware: random errors and systematic errors. Both of these types of errors can be accounted for, and, to some extent, both types of errors can be corrected. Random errors occur because of the way the eye perceives an image. Even if a subject fixates an image (to select a particular tile or point of an image), the eye is constantly moving. In fact, the movement is required to get a good response of the visual organ. As the eye monitoring hardware samples the direction of the eye gaze, the estimated gaze point will be slightly different for each sampling point, even if the user makes the conscious effort of fixating a single point. Moreover, even if the eye would be stationary as it fixates the target point, the eye has a relatively wide angle of focus for the central vision (as opposed to peripheral vision). For people with a wider angle of the central vision range the uncertainty in the estimating their gaze point will be higher.

The second type of eye monitoring errors is due to improper calibration. Systematic errors occur when the calibration is faulty and the estimated gaze point is consistently different from the actual gaze point (for example if the gaze point is always below and to the left of the estimated position). Sometimes, such errors can arise not because of a faulty calibration procedure per se, but because the user moved his head or changed the angle of tilt of his head since the last calibration session.

12.3.3.1 Authentication Schemes that Can Handle Random Eye Monitoring Errors

To include the effects of random errors, we model the errors as randomly and uniformly distributed Fig. 12.2. When the user is fixating the center of a tile, the probability distribution function (PDF) of random error includes a central area where the PDF is of value p_1, and a surrounding area where the PDF is of value p_2. For simplicity, we consider a rectangular grid as shown in Fig. 12.2, although a circular geometry would more accurately describe the situation. We assume that the random errors are small enough that the PDF is zero outside the immediate neighbors of the fixated cell. The model can easily be extended to remove this assumption, although the practical utility of a system with such large errors would be questionable. We assume the tiles to have unit areas, for simplicity.

With the PDF above, the probability for the estimated gaze point to be on the same tile as the actual gaze point is p_1. The probability that the estimate is on one of the eight neighbors of the actual gaze point is $8p_2$. To normalize the PDF, we require that $p_1 + 8p_2 = 1$. We assume that the authentication requires selecting points or tiles on S successive screens of data. In this case, we call event N the situation when the estimate falls

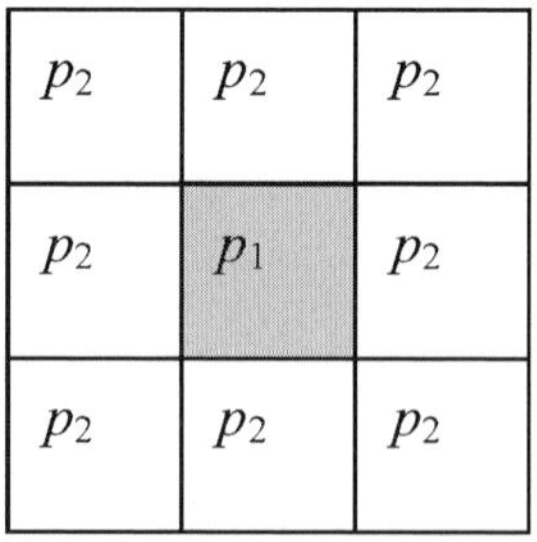

p_2	p_2	p_2
p_2	p_1	p_2
p_2	p_2	p_2

Fig. 12.2. Estimated gaze point PDF: Probability distribution function (PDF) for the estimated gaze point from the eye monitoring hardware when the user is fixating the center (*grey*) tile. The probabilities are as indicated by the p_1 or p_2 numbers for the tiles shown, and are zero for tiles more remote from the grey tile

not on the actual fixated tile, but on neighboring tiles. The complementary event A, is that the estimate falls on the actual tile fixated. If the errors are independent and randomly distributed, over the S screens required for one authentication session, the probability to have k exact matches of the estimated and the fixated tile will be given by the well-known binomial distribution:

$$p_k = \binom{n}{k} p_1^k (1 - p_1)^{S-k} \tag{12.1}$$

This information could be used to optimize the authentication process in the presence of random errors in two different ways.

First, if the estimate gaze points follow a pattern $AA...A$ (the estimate matches the actual expected tile for all screens of the authentication session), the password would be accepted, just as in traditional authentication. On the other hand, if the pattern includes only $k < S$ occurrences of event A and also $S-k$ occurrences of event N (estimate falls on a neighbor of the expected tile), the authentication could be deemed successful with a normalized probability given by:

$$p_{A(k)} = \frac{p_k}{p_S} \tag{12.2}$$

Consequently, even if the estimate of the user gaze always falls on neighboring tiles and not on the expected ones, there is a probability, albeit small, that the session would be successful.

This "soft reject" relaxes the authentication requirements to account for some of the more likely failures due to random errors. At the same time, any estimate that falls outside the set of immediate neighbors of the expected tile will deem the session as a failed authentication attempt. In fact, to make this "soft reject" process more robust, any attempt where $p_{A(k)} \leq p_{rand}$ (where p_{rand} is the probability of occurrence of the pattern due to a random try) should be automatically deemed as a failed attempt.

The process could be further optimized by considering the errors across successive authentication attempts. If an authentication session does not succeed, the user is likely to attempt a second, and even a third session within a short time. The idea is to evaluate the probability over successive authentication sessions that a given pattern of events A and N is due to chance or due to a legitimate user encountering random errors. If the likelihood of the pattern being caused by errors is high enough, the system would grant access to the user, even though none of the authentication sessions taken individually would be sufficient to grant access.

The impact of the proposed "soft reject" would be to relax the authentication rules, increasing the likelihood that a user would authenticate successfully even in the presence of errors. This will have the undesired effect of increasing false positives, allowing some sessions to succeed even when the user did not know and did not select the correct password.

12.3.3.2 Authentication Schemes that Can Handle Systematic Eye Monitoring Errors

The other type of errors likely to happen in eye monitoring systems are systematic errors. This type of error is when the estimated gaze position and the actual gaze position differ by a constant translation vector. For example the estimate could always be lower and to the right of the actual gaze position (Fig. 12.3). The cause for such systematic errors could be a change in head position or angle since the last calibration, or a faulty calibration itself.

The authentication system could account for such types of errors and automatically correct for them. As described above, the correction could be done when the estimate falls in neighboring tiles to the expected tile for correct authentication, or might even extend for systematic errors that take the estimate beyond the neighboring tiles. Because the translation vector can be random, accepting truly arbitrary translation vectors would lead to a loss of cryptographic complexity equal to one screen worth of data. In other words, if accepting any magnitude and orientation for the translation vector (difference between the estimated gaze position and the expected position for authentication), a session that would require S screens of selecting data would have the equivalent complexity of a session of $S - 1$ screens with no tolerance for systematic errors.

12.3.4 Experimental Investigation

The system used by Hoanca and Mock [256] to investigate eye monitoring based authentication is the ERICA system manufactured by Eye Response

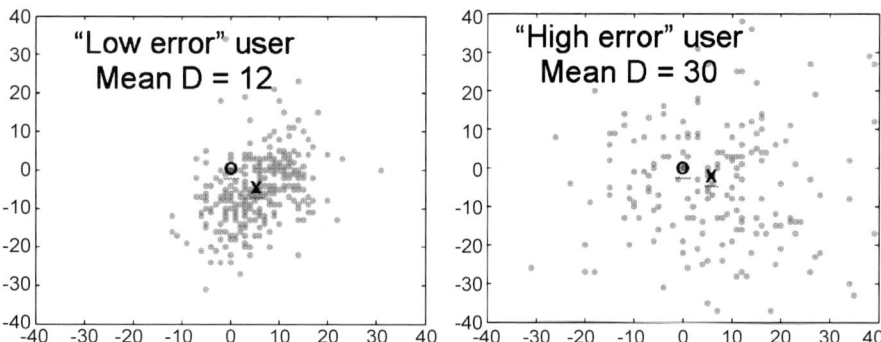

Fig. 12.3. Scatter plots of gaze points for low error and high error users: Coordinates are in pixels, relative to the target location (the exact password point, indicated by "*o*"). Even thought the absolute value of the error in the case of the "high error" user is three times larger than for the "low error" user, the center of gravity of the points (indicated by "*x*"), is approximately within the same distance in both cases. This distance is what the authors call the systematic error [256]

Technologies. The system operates on a notebook PC, and has a resolution of 1024×768 pixels. With a screen tiled in a 8×8 array, the probabilities in Sect. 12.3.3 above would be approximately $p_1 = 0.9$, $8p_2 = 0.1$. For a session requiring three screens for authentication, $S = 3$, the probability that a user would authenticate successfully with the basic technique in Sect. 12.3.1 would be approximately 0.73. In contrast, if using the eye monitoring aware approach in this section the probability would be increased to 0.81. The cryptographic complexity is reduced by a similar amount, $\frac{0.81}{0.73} = 1.11$ (less than 0.15 bits of equivalent complexity).

Using one of the low-cost models available on the market, Hoanca and Mock [256] mention the advantage of using a chin rest, but did not use one in the experiment. The participants reported taking 3-5 seconds per authentication point, which is considerably slower than the traditional password entry, but comparable with the time necessary for some of the techniques proposed for reducing the danger of shoulder surfing [677–679]. Finally, the recognition rates were as high as 90% (after correcting for systematic errors) when the users were required to click within 20 pixels of the actual password (Fig. 12.4). In most cases, the recognition rate was much lower, even with larger tolerances, especially for users wearing eye glasses (83% for clicking within 80 pixels). Clearly, there is a tradeoff between increasing the recognition rate and decreasing the cryptographic complexity, since both of these happen with increasing the tolerance. Using averaging to remove the systematic errors was reported to increase the recognition rate from 20% to 90% for an accuracy of 20 pixels, but this increase is only for one of the best cases, and is highly user dependent. In fact, for a user wearing eye glasses, the same averaging did not result in any change in the recognition rate and might even decrease the rate in some cases.

12.4 Eye Monitoring for Biometric Authentication

For many years now, eye-related biometric data has been used for user authentication. The approaches in use include iris recognition [116] and retinal scan [110]. Iris recognition relies on the fact that the human iris exhibits unique patterns that can be used to identify the user (and even to identify the left eye from the right eye of the same person [110]). A high resolution camera can capture an image of the iris and compare that image with a reference signature. A similar approach can be based on retinal scanning, where the unique pattern is that of the blood vessels on the back of the eye. According to a paper [117] by John Daugman, one of the inventors of the technique, iris scanning has not exhibited any false positives in 9.1 million comparison tests, even when used in identification mode. Iris scanning is susceptible to attacks, because a high resolution photograph of the iris can be substituted for a live eye [336]. The retinal scan is not susceptible to such attacks, but requires much more complex imaging devices, and even requires that users

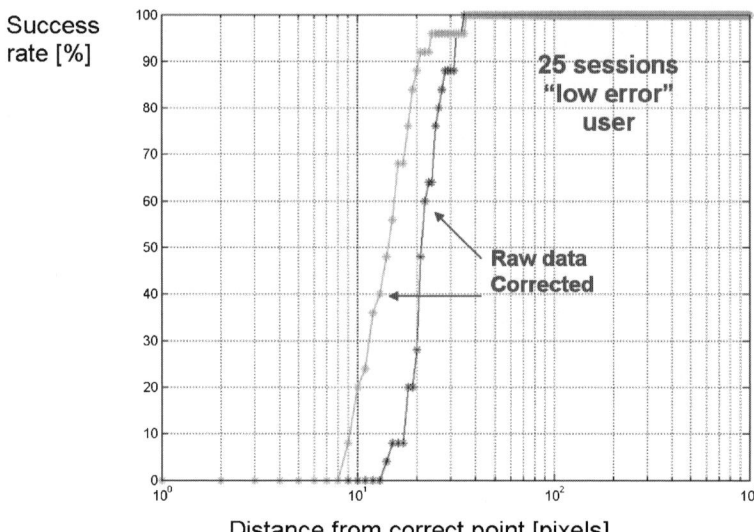

Fig. 12.4. Success rate as a function of acceptable distance from the correct password point: Passwords where click points are farther away from the correct point than the acceptable distance result in failed authentication sessions. The raw data is as captured by the eye monitoring hardware. In the corrected data, the position of the center of gravity of multiple gaze points was calculated, then the systematic error between this position and that of the correct point was subtracted offline to plot the success rate

place their eye on a scanning device, which could cause discomfort to some users.

Another approach to using eye monitoring technologies for authentication would be to extract user defining information from the eye movement patterns themselves. This would in essence be a biometric approach, where statistical properties of the eye monitoring data and not the actual eye gaze locations would identify the user. Ideally, the user might not even need to memorize a password, but would simply gaze at an image and allow the system to collect and process the gaze patterns.

To date, the authors are aware of only two published studies that use eye monitoring for biometrics-based authentication [32, 312]. The only other attempt even close to this idea was to classify users into broad categories, but the work was at an early stage of defining even the concept [244]. Bednarik et al. report on using 12 users to evaluate biometrics-based eye monitoring [32]. They measured both static and dynamic characteristics of eye gaze data. Measurements were taken for pupil diameters and eye gaze direction, as well as for time variations of these quantities (variations in pupil diameters and eye gaze velocities). Various combinations of Fast Fourier Transforms (FFTs) and Principal Component Analysis were used for feature extraction.

Even with one of the high-end eye-tracker models on the market, researchers found it advantageous to use a chin rest. Additionally, data collection took up to 5 minutes per participant. Finally, the false negative rates were in the range of 10–60%, unacceptably high for a realistic authentication application. Moreover, among the features the authors considered, a static metric, the distance between the eyes of the user, was the most accurate. The authors concede that this could be measured with a regular camera, without need for eye monitoring hardware. With the 12 users in the study, the authors dismissed fusion metrics based on combining distance between eyes with other features of the eye gaze pattern. Such fused metrics appeared no more accurate than the distance between eyes alone.

Clearly, if using more than the 12 users, the distance between eyes would quickly degenerate and become unacceptable as a discriminating feature; users would fall along a continuum of values for the distance between the eyes, making it impossible to differentiate among them. For such larger numbers of users, fused metrics would be able to remove the degeneracy and to separate users into groups or even into individual biometric sets.

Kasprowski et al. report somewhat more promising results in their study that extracted only dynamic characteristics of eye gaze data [312]. Subjects were stimulated by a point that jumped between nine positions in a 3×3 matrix over a period of eight seconds. The raw data consisted of eye displacement vs. time and cepstral coefficients were extracted from the waveform to use as features for classification. Nine participants provided 30 training cases and 30 test cases. Naive bayes, k-nearest neighbor, decision trees, and support vector machines were used to train classifiers to predict the class of the test cases. While the average false positive rate was as low as 1.36%, the average false negative rate ranged from 12% to 36%.

Although results reported to date are not sufficient to prove the viability of biometric approaches to eye monitoring based authentication, the approach is worth investigating further. Neither the Bednarik et al. study [32], nor the Kasprowski et al. work [312] required users to memorize any secret type of data. In fact, users were not even told that they were providing biometric data, but rather, that they were helping to calibrate the eye monitoring hardware [32] or were asked to follow a "jumping point" [312] a technique often used to calibrate eye monitoring equipment. All users were presented with the same screens for collecting the eye monitoring data. If more accurate features can be extracted from eye gaze data, users will be able to authenticate while calibrating the eye monitoring equipment, which could one day be available as standard equipment for the current computer mouse. There would be no passwords to remember or to be stolen by attackers.

Another strength of the biometric approach would be that theft of this type of biometric data would be unlikely. Fear that fingerprints might be lifted from drinking glasses or even from biometric check stations [336], or fear that iris patterns could be reproduced on artificial eyes is currently one of the main reasons why biometrics have not been used more widely [705]. The fear of biometrics theft is compounded with the fear that attackers might cut off fingers or remove eyes from victims to use these body parts for biometric based authentication [318]. If biometric devices would be able to detect whether the finger or eye is alive, this would help to alleviate fears of these last types of attacks. As with any security application, even one success of using a fake biometric (as for example reported by Thalheim et al. [336]) overrules all the claims about the ability to detect fakes.

To date, no published data exist on the ability to fake eye monitoring as a biometric. Of course, the history of the technique is too recent to allow for definitive conclusions. Eye gaze data on the other hand is clearly a dynamic quantity, which would signal to both attackers and fearful users that only a live eye used by the rightful owner can be used to generate the appropriate dynamic associated with the owner's eye gaze signature. The likelihood that eye monitoring could be faked is expected to be very low.

12.5 Multiple-Factor Authentication Using Passive Eye Monitoring

A simple extension of the techniques described in the previous two sections would be to combine eye gaze based password entry with a biometric approach. Users would use their eyes to select points as in the graphical password scenario. The selection points would factor in the authentication process, along with biometric information extracted from the gaze pattern dynamics. To some extent, this would remove one of the main advantages of biometric authentication, because users would still be required to memorize passwords. On the other hand, like with any multiple-factor authentication approach, users would be less concerned about a compromised password, because an attacker who stole a password would not be able to fake the biometric component to authenticate.

An even more powerful combination would be to combine two types of biometrics. Together, iris scanning and eye gaze dynamics could increase the accuracy of the authentication [312], while removing user fears that their eyes might be copied or forcefully removed by an attacker. Even if iris scanning alone is vulnerable to attackers using high-resolution images of the user's iris, this type of attack cannot replicate the behavior of a live eye. The challenge with such a combined approach rests on the different resolutions required for eye monitoring (medium) and for iris scanning (medium-high, 70 pixels for the iris radius, or 100–140 pixels preferred [110]). As the price of high resolution digital cameras continues to decrease, the resolution needed for iris scanning may be available and cost effective for eye monitoring cameras. The illumination scheme for both eye monitoring and iris scanning uses near infrared radiation, making it easy to combine the two types of authentication.

Combining eye monitoring with retinal scanning does not lend itself to easy implementations, but might become a possibility as technology matures.

Another multiple-factor authentication approach would be to use three factors. The user could enter a user name and a password using eye monitoring hardware as a hands-free mouse on an on-screen keyboard without on-screen feedback. This will minimize the danger of shoulder surfing and would provide the first authentication factor. While the user is entering the information above, the eye monitoring hardware could collect statistical data about the gaze dynamics, as a second, biometric based authentication factor. Finally, the same eye monitoring hardware could simultaneously perform an iris scan. The user id could be used to identify the user in a database, and the password, the gaze dynamics, and iris the scan data could be used to verify the user's identity.

Combining multiple authentication mechanisms in this way leads to an increase in the false negative probability of the overall scheme [114] (the probability that a legitimate user might be rejected). Because the probability of false negatives is zero for the password entry and extremely low for iris scanning, the overall false negative probability should be sufficiently low. To date, no data have been reported on the performance or feasibility of such an multiple factor scheme .

12.6 Conclusion and Future Work

Eye monitoring has the potential to offer new options for secure authentication. Using eye monitoring hardware as a "hands-free mouse" without on-screen feedback allows a user to enter information (not limited to passwords) without disclosing it to shoulder surfers. Alternatively, by extracting biometric features from the eye monitoring data, a user can be uniquely identified in a manner that is extremely difficult to replicate. Because the eye-tracker requires a real live eye to gather data, an attacker will not be able to record the user and play back the biometric at a later time. Even if a prosthetic eye is built to be used for such purposes, it is unlikely that the fake eye would be able to correctly reproduce the dynamic behavior of the real eye of the user. When combining the biometric of eye dynamics with other biometrics like iris scanning or retinal scanning, the security of the authentication is further increased. Finally, by combining eye monitoring dynamics, static eye biometrics (iris or retinal scan) and user password entry via eye monitoring without on-screen feedback, the potential is there for a multiple-factor authentication with high security and flexibility.

Eye monitoring hardware will soon be available at a cost and with accuracy that would allow wide deployment of the techniques described in this chapter. It is highly likely that solutions involving combined use of eye monitoring for data entry and authentication will emerge or become standard within the coming decade.

Part V

Eye Monitoring in Automotive and Medicine

Driver Cognitive Distraction Detection Using Eye Movements

Yulan Liang and John D. Lee

13.1 Introduction

Driver distraction is an important safety problem [518]. The results of a study that tracked 100 vehicles for one year indicated that nearly 80% of crashes and 65% of near-crashes involved some form of driver inattention within three seconds of the events. The most common forms of inattention included secondary tasks, driving-related inattention, fatigue, and combinations of these [325]. In-vehicle information systems (IVIS), such as navigation systems and internet services, introduce various secondary tasks into the driving environment that can increase crash risk [5,516]. Thus, in future it would be very beneficial if IVIS could monitor driver distraction so that the system could adapt and mitigate the distraction. To not disturb driving, non-intrusive and real-time monitoring of distraction is essential.

This chapter begins to address this issue by describing techniques that draw upon the data from a video-based eye monitoring system to estimate the level of driver cognitive distraction in real time. First, three types of distractions and the detection techniques are briefly described. Then, the chapter focuses on the cognitive distraction and describes a general procedure for implementing a detection system for such distraction. Data mining methods are proposed to be promising techniques to infer drivers' cognitive state from their behavior. Next, Support Vector Machines (SVMs) and Bayesian Networks (BNs) are used and compared in this application. Finally, several issues associated with the detection of cognitive distractions are discussed.

13.2 Types of Distraction

Three major types of distraction have been widely studied: visual, manual, and cognitive [518]. These distractions deflect drivers' visual and cognitive resources away from the driving control task. These distractions can degrade driving performance and even can cause fatal accidents. Visual distraction and manual distraction can be directly observed through the external behaviors of drivers, such as glancing at billboards or releasing the steering wheel to adjust the radio. Visual distraction usually coexists with manual distraction because visual cues provide necessary feedback when people perform manual tasks. Visual and manual distractions interrupt continuous visual perception and manual operation essential for driving and results in the absence of visual attention on safety-critical events. In the 100-vehicle study, visual inattention contributed to 93% of rear-end-striking crashes. Interestingly, in 86% of the rear-end-striking crashes, the headway at the onset of the event the led to the crash was greater than 2.0 seconds [325]. These facts show that visual distraction dramatically increases real-end-striking crashes because two seconds are long enough for an attentive driver to avoid a collision. Moreover, the degree of visual distraction is proportional to the eccentricity of visual-manual tasks to the normal line of sight [354].

To identify how visual distraction delays reaction time, several researchers have created predictive models that quantify the risks associated with visual or manual distraction from drivers' glance behavior [133,713]. Frequent glances to a peripheral display caused drivers to respond slowly to breaking vehicles ahead of them. The reaction time, the time from when the lead vehicle began to brake until the driver released the accelerator, could be predicted by the proportion of off-road glances in this reaction period using a linear equation: *(accelerator release reaction time)* $= 1.65 + 1.58$ *(proportion of off-road glances)* [713]. This relationship accounts for 50% of reaction time variance.

Another study took historic performance into account. It used a linear function of current glance duration away from road, β_1, and total glance duration away from road during the last three seconds, β_2, to calculate warning threshold, γ, as: $\gamma = \alpha\beta_1 + (1 - \alpha)\beta_2$. γ influenced the frequency of alarms for reminding drivers when they were too engaged in a visual-manual task, and α presented the weights of the two glance durations on γ [133]. Using this equation, those drivers who had been identified as risky drivers defined by their long accelerator release reaction time received more warnings per minute than non-risky drivers for a broad range of α and γ. These studies found diagnostic measures, such as frequency and duration of off-road glances, and used these measures to predict the degree to which visual and manual distractions delayed drivers' reaction time to critical roadway events. At the same time, these predictive models can be used to monitor drivers' visual and manual distraction non-intrusively in real time because remote eye tracker cameras can collect eye movement data without disturbing driving.

However, unlike visual and manual distraction cognitive distraction is internal and impossible to observe directly from external behavior. With visual distraction, it is possible to detect when the eyes are off the road, but with cognitive distraction there is no direct indicator as to when the mind is off the road. Nonetheless, the effects of cognitive distraction on driving performance may be as negative as those of visual distraction. A meta-analysis of 23 studies found that cognitive distraction delayed driver response to hazards [265]. For example, drivers reacted more slowly to brake events [354,359] and missed more traffic signals [591] when they performed mental tasks while driving, such as using auditory e-mail systems, performing math calculation, or holding hand-free cell phone conversations.

Identifying cognitive distraction is much more complex and less straightforward than visual and manual distraction. There is no clearly diagnostic indicator of cognitive distraction. In controlled situations, four categories of measures are used to assess workload and cognitive distraction: subjective measures, secondary task performance, primary task performance, and physiological measures [676, 714].

Subjective measures and secondary task performance can not be used to identify cognitive distraction for future IVIS. Commonly-used subjective measures include the NASA Task Load Index (NASA-TLX) and the subjective workload assessment technique (SWAT). Collecting subjective measures disturbs normal driving and cannot provide an unobtrusive, real-time indicator. The secondary task method for assessing workload will require drivers to perform a task in addition to driving and so is clearly inappropriate for measuring distraction as an unobtrusive, real-time indicator of distraction because the secondary task will confound the estimation by adding extra workload.

Primary task performance and physiological measures represent promising approaches for real-time estimation of cognitive distraction. The primary task refers to driving control task. The commonly used driving performance measures include lane position variability, steering error, speed variability, and so on. These measures can be collected non-intrusively using driving simulators or sensors on instrumented vehicles in real time. Physiological measures, such as heart-rate variability and pupil diameter, represent promising sources of information regarding drivers' state. But, there are substantial challenges in developing robust unobtrusive and inexpensive sensors for production vehicles. Although monitoring heart rate and pupil diameter on road is difficult, it is feasible to track eye movements of drivers in real time with advanced eye monitoring systems.

Moreover, eye movement patterns change with different levels of cognitive distraction because cognitive distraction disrupts the allocation of visual attention across the driving scene and impairs information processing. Some studies show that cognitive distraction impaired the ability of drivers to detect targets across the entire visual scene [514, 649], and reduced implicit perceptual memory and explicit recognition memory for items that drivers fixated [590]. Eye movement patterns reflect increased cognitive load through longer fixations, gaze concentration in the center of the driving scene, and less frequent glances to mirrors and the speedometer [513]. One study that systematically examined the sensitivity of various eye movement measures to the complexity of in-vehicle cognitive tasks found that standard deviation of gaze was the most sensitive indicator for the level of complexity [649]. Of the four categories of potential measures of cognitive distraction, eye movements and driving performance are the most suitable measures for estimating cognitive distraction [373, 374, 714].

Although cognitive, visual, and manual distractions are described separately, they coexist in most situations. For example, when entering an address into a GPS while driving, drivers need to recall the address and then glance at the system to enter it. This leads to both a visual and a cognitive distraction. Ultimately algorithms that detect visual distraction and cognitive distraction will need to work together to provide comprehensive prediction of driver distraction. However, the balance of this chapter focuses on the challenging task of estimating cognitive distraction using eye movements and driving performance measures.

13.3 The Process of Identifying Cognitive Distraction

Detecting cognitive distraction is complex procedure and requires a robust data fusion system. Unlike visual and manual distraction, the challenge of detecting cognitive distraction is to integrate multiple data streams, including eye movements and driving performance, in a logical manner to infer the driver's cognitive state. One way to address this challenge is by using data fusion. Data fusion systems can align data sets, correlate relative variables, and combine the data to make detection or classification decision [655]. One benefit of using a data fusion perspective to detect cognitive distraction is that data fusion can occur at different levels of abstraction. For instance, sensor data are aggregated to measure drivers' performance at the most concrete level, and then these performance measures can be used to characterize drivers' behavior at higher levels of abstraction. This hierarchical structure can logically organize the data and inferences and reduce parameter space in the detection procedure. The fusion systems also can continuously refine the estimates made at each level across time, which enables a real-time estimation of cognitive distraction.

To implement a data fusion system, there are two general approaches: top-down and bottom-up (see in Table 13.1). The top-down approach identifies the targets based on the known characteristics, such as shape and kinematic behavior of the targets. In detecting cognitive distraction, the top-down approach uses drivers' behavioral response to cognitive load that reflect existing theories of human cognition, such as Multiple Resource Theory [676] and ACT-R [550]. The limitation of the top-down approach is that it is impossible to implement data fusion without a complete understanding of the underlying process, which is lacking in the area of driver distraction.

Table 13.1. Matrix of data fusion strategies and the availability of domain knowledge

Domain knowledge	Data fusion strategies		
	Top-down approach	Mixed approach	Bottom-up approach
Available	√	√	√
Partially available	-	√	√
Not available	-	-	√

The bottom-up approach overcomes this limitation and uses data mining methods to extract characteristics of the targets from data directly, see in the bottom-up approach column of Table 13.1. Data mining includes a broad range of algorithms that can search large volumes of data for unknown patterns, such as decision trees, evolutionary algorithms, Support Vector Machines (SVMs), and Bayesian Networks (BNs). These methods are associated with multiple disciplines (e.g., statistics, information retrieval, machine learning, and pattern recognition) and have been successfully applied in business and health care domains [19, 604]. In driving domain, decision tree, SVMs, and BNs have successfully captured the differences in behavior between distracted and undistracted drivers [373, 374, 714].

The strategies of constructing data fusion systems include using the top-down approach alone, the bottom-up approach alone, or the mixed approach that combines top-down and bottom-up. The choice of the strategies depends on the availability of domain knowledge, as shown in Table 13.1. When the targets are understood very well, the data fusion system can be constructed using only the top-down approach. Currently, most data fusion systems use this strategy. Nevertheless, the lack of domain knowledge presents an important constraint of this top-down-alone strategy in some domains, such as the detection of cognitive distraction. The bottom-up-alone and mixed strategies overcome the limitation. Oliver and Horvitz [469] have demonstrated the effectiveness of these two strategies. They successfully used Hidden Markov Models (HMMs) and DBNs to construct the layered data fusion system for recognizing office activities by learning from sound and video data.

Detecting cognitive distraction requires a bottom-up data mining strategy because the effects of cognitive demand on driving have not been clearly understood. Although some theories of human cognition can help explain drivers' behavior, most theories only aim to describe, rather than predict, human performance and cannot be used to detect cognitive distraction. Some theories, like ACT-R, represent promising approaches that are beginning to make predictions regarding distraction and driver behavior [550]. Nonetheless, there are still substantial challenges to use these theories to describe complex behavior in driving. On the other hand, various data mining methods have been used to detect cognitive distraction. Zhang et al. [714] used

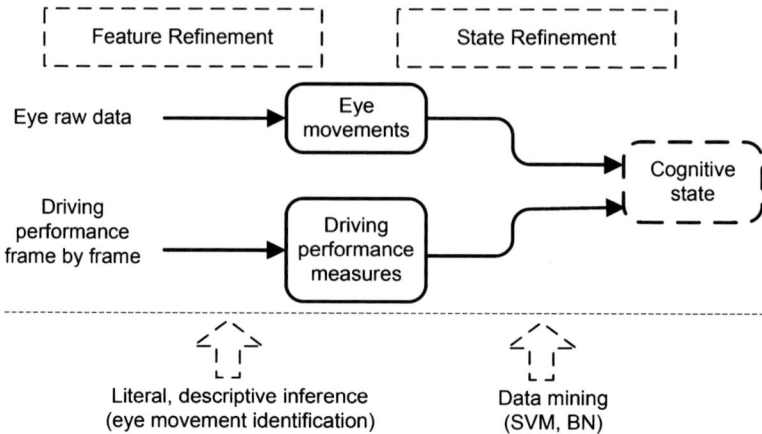

Fig. 13.1. Data fusion that transforms raw driving and eyemovement data into estimates of drivers cognitive distraction

a decision tree to estimate driver cognitive workload from glances and driving performance. In two other studies, Support Vector Machines (SVMs) and Bayesian Networks (BNs), successfully identified the presence of cognitive distraction from eye movements and driving performance [373, 374]. Thus, the strategies using bottom-up and mixed approaches are suitable for data fusion to detect cognitive distraction.

The procedure for detecting driver cognitive distraction can be formulized in two sequential stages, feature refinement and state refinement, as shown in Fig. 13.1. Feature refinement uses the top-down approach. This refinement transforms sensor data (such as eye and driving performance raw data) into performance measures based on an existing understanding of what measures may be the most sensitive to distraction. The sensor data are collected at a high frequency (e.g. 60 Hz) and include outputs from eye monitoring systems, vehicle lane position, and driver steering inputs. These raw data are transformed into eye movements described as fixations, saccades, and smooth pursuits according to the speed and dispersion characteristics of the movements. Various eye movement measures (such as fixation duration and saccade distance) are then calculated to describe drivers' scanning activity. Indicators of cognitive distraction, such as the standard deviation of gaze, are used as inputs for state refinement. State refinement then fuses these measures to infer a driver's cognitive state. In this stage, data mining methods are applied to train detection models from the data.

13.4 Data Mining Techniques to Assess Cognitive State

Different data mining methods produce different models. Numerical and graphical models represent two classes of models for assessing cognitive state.

Typical numerical models include SVMs, linear regression, and polynomial fitting, and typical graphical models include decision trees, neutral networks, BNs. Each of these classes of models has advantages in detecting cognitive distraction. Most numerical models have mature techniques for training and testing and have fewer computational difficulties compared to graphical models, which may cause fewer computational delays for detecting distraction in real time. Nonetheless, graphical models explicitly represent the relationships between drivers' cognitive state and performance and helps summarize knowledge from resulted models. SVMs and BNs are the representatives of numerical and graphical models, respectively.

13.4.1 Support Vector Machines (SVMs)

Originally proposed by Vapnik [642], Support Vector Machines are based on the statistical learning theory and can be used for pattern classification and inference of non-linear relationships between variables [112,642]. This method presents several advantages when identifying cognitive distraction. First, the relationship between human cognition and visual behavior can seldom be represented by a purely linear model. With various kernel functions, SVMs can generate both linear and nonlinear models with equal efficiency [6]. Second, SVMs can extract information from noisy data [78] and avoid overfitting by minimizing the upper bound of the generalization error [6] to produce more robust models compared to those that minimize the mean square error. SVMs have outperformed the linear logistic method in detecting cognitive distraction [374]. Nonetheless, SVMs can not explicitly present the relationships learned from data.

13.4.2 Bayesian Networks (BNs)

Bayesian Networks (BNs) is a graphical approach that represents conditional dependencies between random variables. A BN includes circular nodes depicting random variables, and arrows depicting conditional relationships. As shown in the left graph in Fig. 13.2, the arrow between variable nodes E_2 and S indicates that E_2 is independent of any other variable given S. BNs are either Static (SBNs) or Dynamic (DBNs). SBNs (Fig. 13.2, left) describe the situation that does not change in time. DBNs (Fig. 13.2, right) connect two identical SBNs at successive time steps and model a time-series of events according to a Markov process. The state of a variable at time t depends on either variables at time t or time $(t-1)$ or both. For example, the state of St depends on both S^{t-1} and H^t.

Like SVMs, BNs can model complex, non-linear relationships. However, BNs aim to identify relationships between variables, and to use these relationships to generate predictions. BNs can explicitly present the relationships learned from data. As a consequence, studying a trained BN helps identify

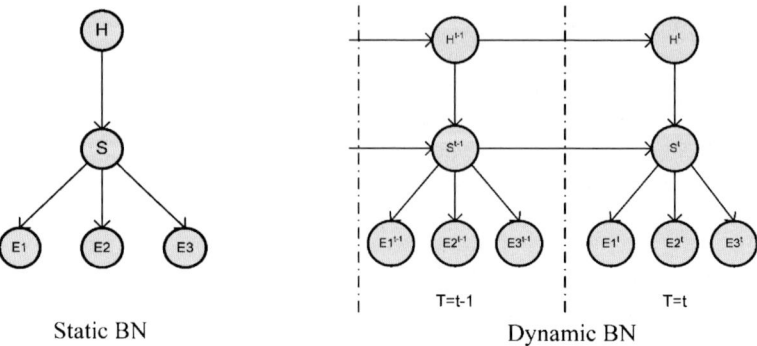

Static BN   Dynamic BN

Fig. 13.2. Examples of a SBN and a DBN

cause-effect links between variables, and the hierarchical structure of BN provides a systematic representation of these relationships. BNs are applicable to human-behavior modeling and have been used to detect affective state [371], fatigue [296], lane change intent during driving [344], pilot workload [211], and driver cognitive distraction [373]. One disadvantage of BNs is that they become computationally inefficient when models have a large number of variables (i.e., 20). Moreover, compared to SVMs training techniques for BN are less robust than those available for SVMs.

13.5 Comparisons of SVMs and BNs

Although both SVMs and BNs have been successfully applied to detect cognitive distraction, it is not clear which might be more effective and how they might differ from each other when detecting cognitive distraction. Here, we compare the effectiveness of SVMs, SBNs and DBNs methods in detecting driver cognitive distraction using eye movements and driving performance. Nineteen performance measures were used to detect cognitive distraction. The training data were randomly selected from experimental data collected in a simulated driving environment. The models were evaluated using testing accuracy and two signal detection theory measures including sensitivity and response bias. DBNs, which consider time-dependent relationships, were expected to have the best performance. Of the two methods that only consider the relationship at single time point, it was expected that SVMs would perform better than SBNs because SVMs have fewer computational difficulties in terms of training and testing.

13.5.1 Data Source

Data were collected using a fixed-based, medium-fidelity driving simulator. The driving scenario consisted of a straight, five-lane suburban road (two

lanes of traffic in each direction divided by a turning lane). It was displayed
on a rear-projection screen at 768×1024 resolution, producing approximately
50 degrees of visual field. During the experiment, participants, 35–55 years
old, were required to drive six 15-minute drives: four IVIS drives and two
baseline drives. During each IVIS drive, participants completed four separate
interactions with an auditory stock ticker system. The interactions involved
participants continuously tracking the price changes of two different stocks
and reporting the overall trend of the changes at the end of the interaction.
In baseline drives, participants did not perform the task. During all drives,
participants were instructed to maintain vehicle position as close to the center
of the lane as possible, to respond to the intermittent braking of the lead
vehicle, and to report the appearance of bicyclists in the driving scene. Nine
participants' eye and driving performance data was collected using a Seeing
Machines faceLAB^{TM} eye tracker and the driving simulator at a rate of 60 Hz.
Further details can be found in a companion study [374].

13.5.2 Feature Refinement

The raw eye data describing dynamic change of gaze vector-screen intersection coordinates were then translated into sequences of three types of eye
movements: fixations, saccades, and smooth pursuits. Segments of the raw
data were categorized based on two characteristics: dispersion and velocity
(see Table 13.2). Dispersion describes the span (in radians of visual angle)
that the gaze vector covered, and velocity describes the speed (in radians of
visual angle per second) and direction of the gaze vector (in radians) during
a movement. The identification process began with a segment of six frames;
based on the characteristics demonstrated in these frames, the posterior probabilities of the eye movements were calculated for the segment (see Fig. 13.3).
If the highest probability was greater than a certain threshold, the segment
was identified as that eye movement. The segment was then increased by one
frame, and the process was repeated to check if the eye movement continued
in the new frame. If no movement could be identified, the segment was decreased by one frame, and the posterior probabilities were calculated again. If
only three frames remained in the segment, the eye movement was identified
using only the speed characteristic. When speed was high, the movement was
labeled as a saccade; when the speed was low, it was labeled as a smooth

Table 13.2. The characteristics of fixations, saccades, and smooth pursuits

Types	Dispersion	Velocity
Fixation	Small ($\leq 1°$)	Low, random direaction
Saccade	Small ($> 1°$)	400–600°/sec, straight
Smooth pursuit	Target decided ($> 1°$)	1–30°/sec, target trajectory

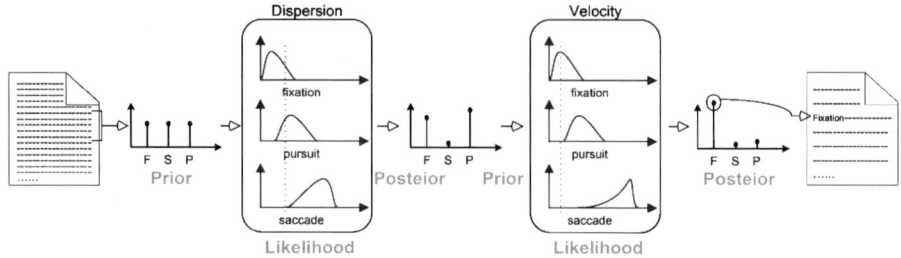

Fig. 13.3. Illustration of the algorithm used to identify eye movements

pursuit. After each eye movement was completely identified, the identification process began again with a new six-frame segment. The likelihood of eye movements and the threshold of posterior probability were chosen according to the previous studies [283], and adjusted according to the particular characteristics of the data.

After identification, measures of eye movements were summarized over various windows to create instances that became the model inputs. Sixteen eye-movement measures included the mean and standard deviation of fixation duration, horizontal and vertical fixation location coordinates, pursuit duration, pursuit distance, pursuit direction and pursuit speed, mean blink frequency, and percentage of time spent performing pursuit movements. Fixation and pursuit duration and the percentage of time spent in pursuit movements represent the temporal characteristics of eye movements, horizontal and vertical position of fixation relates to spatial distribution of gaze, and standard deviation explicitly represents gaze variability. The pursuit distance, direction and speed capture the characteristics of smooth pursuits. Saccade measures were not included because the study was interested in how cognitive distraction may interfere with drivers' acquisition of visual information, which is greatly suppressed during saccades [283].

The driving simulation generated steering wheel position and lane position at 60 Hz. Steering error was calculated at 5 Hz, which describes the difference between the actual steering wheel position and the steering wheel position predicted by a second-order Taylor expansion [431]. Driving measures were summarized across the same window as eye movement measures. The driving performance measures consisted of the standard deviation the of steering wheel position, mean of steering error, and the standard deviation of lane position.

13.5.3 Model Training

For each participant, three kinds of models were trained with randomly-selected data and the best model settings obtained from the previous studies [373, 374]. Testing accuracy and two signal detection theory measures including sensitivity and response bias were used to assess the models.

The best parameter settings for each kind of model were selected. IVIS drives and baseline drives were used to define each driver's cognitive state as "distracted" or "not distracted", which became the prediction targets. The 19 performance measures – including 16 eye movement measures and 3 driving performance measures that were summarized across a window (5, 10, 15, or 30 seconds long) – were used as predictive evidence. SVM models used a continuous form of the measures, while BN models used a discrete form.

To train SVM models, the Radial Basis Function (RBF), $K(x_i, x_j) = e^{-\gamma|x_i - x_j|^2}$, was chosen as the kernel function, where x_i and x_j represent two data points and γ is a pre-defined positive parameter. The RBF is a very robust kernel function. Using the RBF, it is possible to implement both non-linear and linear mapping by manipulating the values of γ and the penalty parameter C. C is a pre-defined positive parameter used in the training calculation [268]. In training, we searched for C and γ in the exponentially growing sequences ranging from 2^{-5} to 2^5, using 10-fold-cross-validation to obtain good parameter settings [87]. "IBSVM" Matlab toolbox [87] was used to train and test the SVM models.

BN training included structure learning, which tested the existence of conditional dependencies between variables, and parameter estimation, which decided the parameters of the existing relationships. With 19 measures, the structures of the BNs were constrained so that the training procedure was computationally feasible. For SBNs, the direction of the arrows was from the target node ("distraction" or "not distraction") to performance nodes. The performance nodes could connect with one another. For DBN models (see Fig. 13.4), the arrows within a time step were present only in the first time step and constrained as SBNs. After the first step, the arrows were only from the nodes in the previous time step to the current one. The BN models were trained using a Matlab toolbox [427] and an accompanying structural learning package [362].

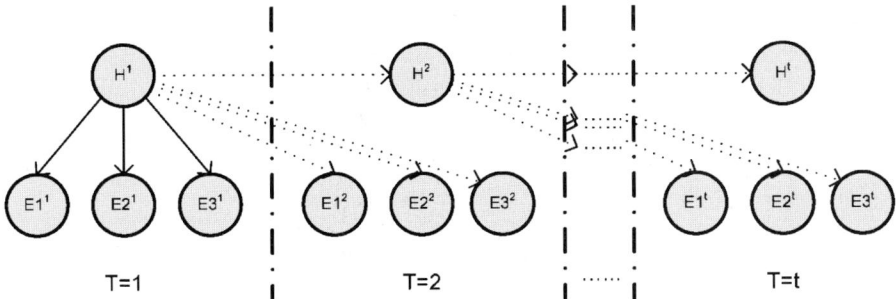

Fig. 13.4. Constrained DBN Structure, where *solid arrows* represent relationships within the first time step, *dotted arrows* represent relationships across time steps, and H and E presents the predictive target and performance measures, respectively

Three types of models, SVM, SBN, and DBN models, were trained with randomly selected data for each participant. The training data represented about 2/3 of the total data. The other 1/3 was used as testing data. The training data for SVM and SBN models were time-independent instances, and the training data for DBN models were 120-second sequences. In total, there were 108 models trained, 36 models (9 participants × 4 window size) in each of DBNs, SBNs, and SVMs.

13.5.4 Model Evaluation

Model performance was evaluated using three different measures. The first was testing accuracy, the ratio of the number of instances correctly identified by the model to the total number of testing instances. The other two measures were associated with signal detection theory: sensitivity (d'), and response bias (β), which were calculated according to (1) and (2).

$$d' = \Phi^{-1}(HIT) - \Phi^{-1}(FA) \tag{13.1}$$

$$\ln(\beta) = \frac{|\Phi^{-1}(FA)|^2 - |\Phi^{-1}(HIT)|^2}{2} \tag{13.2}$$

where HIT is the rate of correct recognition of distracted cases, FA is the rate of incorrect identification of distraction among not-distracted cases, and Φ^{-1} presents the function of calculating the z-score. d' represents the ability of the model to detect driver distraction. The larger d' value, the more sensitive the model. β signifies the bias of the model. When β equals 0, the model classifies segments as "distracted" or "not distracted" at equal rates, and false alarms and misses (not detecting distraction when it is present) tend to occur at similar rates. When $\beta < 0$, cases are classified more liberally, and there is a tendency to overestimate driver distraction and to have higher false alarm rates than miss rates ($1 - HIT$). When $\beta > 0$, cases are classified more conservatively, and there is a tendency to underestimate driver distraction and have more misses than false alarms. Both d' and β can affect testing accuracy. Separating sensitivity to distraction from model bias makes for a more refined evaluation of the detection models than a simple measure of accuracy with which the cases are identified [580].

13.5.5 Model Comparison

We conducted a 3×4 (model types: DBN, SBN, and SVM by window size: 5, 10, 15, and 30 seconds) factorial analysis on three model-performance measures using a mixed linear model with participants as a repeated measure. We then performed post hoc comparisons using the Tukey-Kramer method with SAS 9.0.

DBN, SBN and SVM models were significantly different for testing accuracy and sensitivity (testing accuracy: $F_{2,16} = 6.6$, $p = 0.008$; sensitivity:

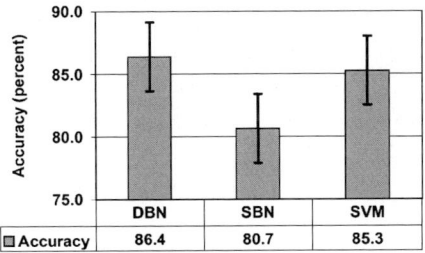 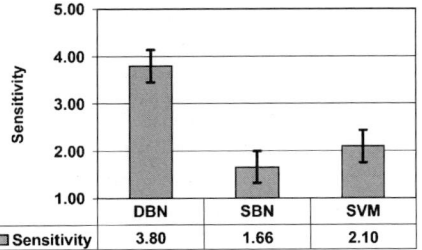

Fig. 13.5. Comparisons of testing accuracy and sensitivity

$F_{2,16} = 32.5$, $p < 0.0001$). Shown on the left in Fig. 13.5, DBNs and SVMs produced more accurate models than SBNs (DBNs: $t_{16} = 3.4$, $p = 0.003$; SVMs: $t_{16} = 2.77$, $p = 0.01$). The DBN and SVM models here had similar accuracy ($t_{16} = 0.66$, $p = 0.5$). On the right side of Fig. 13.5, the DBN models are significantly more sensitive than the SVM and SBN models (SBN: $t_{16} = 7.7$, $p < 0.0001$; SVM: $t_{16} = 6.1$, $p < 0.0001$). Here, however, the SVM and SBN models have similar sensitivity ($t_{16} = 1.6$, $p = 0.13$). These comparisons indicate that using similar training data, DBNs can capture more differences in driver distraction and generate more sensitive models than the SBNs and SVMs. Although the SVM and SBN models showed similar sensitivity, the SVM models had an advantage in testing accuracy, perhaps due to their robust learning technique.

The decision bias was marginally different for the three types of models ($F_{2,16} = 2.8$, $p = 0.09$). The DBN models were more liberal than the SBN and SVM models (DBN: -1.85; SBN: -0.47; SVM: -0.55) with marginal significance (SBN: $t_{16} = 2.1$, $p = 0.051$; SVM: $t_{16} = 2.0$, $p = 0.06$). The SBN and SVM models had similar response biases ($t_{16} = 0.1$, $p = 0.9$), which were not different from the neutral model that is characterized by zero (SBN: $t_{16} = 1.1$, $p = 0.3$; DBN: $t_{16} = 1.3$, $p = 0.2$). These results can be used to explain the discrepancy in the comparisons of the DBNs' and SVMs' testing accuracy and sensitivity. Although being less sensitive than the DBN models, the SVM models reached a similar level of similar accuracy as the DBN models by using a more neutral strategy. Nevertheless, to explain how SBN and SVM models resulted in different accuracy given their similar sensitivity and response bias, the analyses of hit and false alarm rates were needed.

As be seen in Fig. 13.6, the comparisons of false alarm and hit rates show that DBNs, and SVMs marginally, had higher hit rates compared to SBNs (overall: $F_{2,16} = 4.8$, $p = 0.02$; DBN: $t_{16} = 3.1$, $p = 0.008$; SVM: $t_{16} = 2.0$, $p = 0.06$). False alarm rates of the three types of models were similar ($F_{2,16} = 1.1$, $p = 0.4$). This indicates that the DBN and SVM models reached higher accuracy than the SBN models by elevating hit rate.

The effect of window size interacts with model type to affect the false alarm rate (window size: $F_{3,24} = 3.0$, $p = 0.052$; interaction: $F_{6,46} = 2.0$,

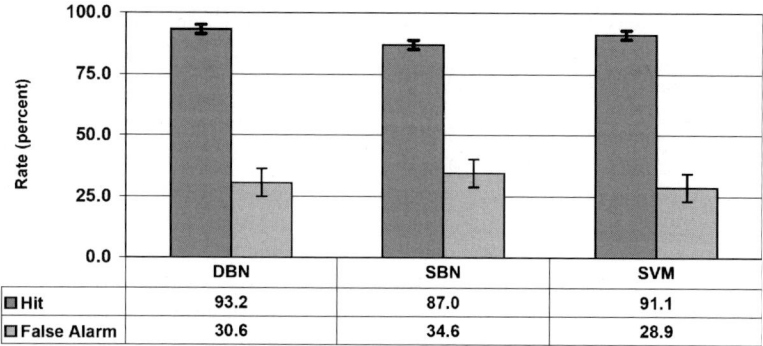

Fig. 13.6. The comparisons of hit and false alarm rates

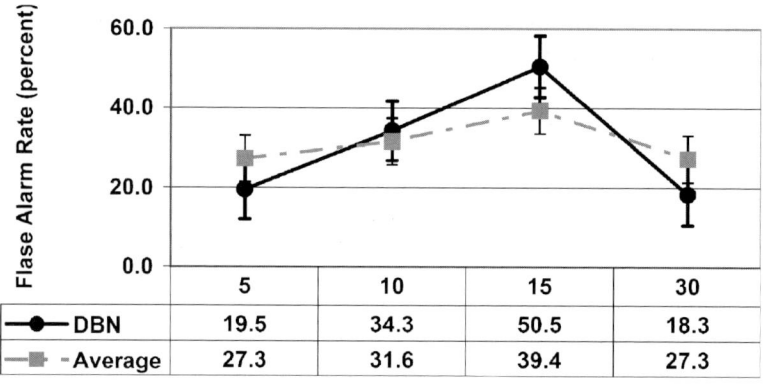

Fig. 13.7. The comparison of false alarm rate for DBNs and the average

$p = 0.08$). No effect or interaction was found for testing accuracy, sensitivity, response bias, or hit rate. False alarm rate increased with window size from 5 to 15 seconds, and then decreased at 30 seconds (see dotted line in Fig. 13.7). The DBN models followed this trend, but the magnitude of the change was more dramatic than that seen with the SBN and SVM models.

In summary, DBNs produced more sensitive models than SBNs and SVMs. The DBN and SVM models were more accurate and had higher hit rates than the SBN models. However, the effects of response bias on the three types of models were only marginally significant. Window size and its interaction with model type did not affect testing accuracy, sensitivity, response bias, or hit rate, but marginally affected false alarm rate.

13.6 Discussion and Conclusions

Compared to SBN and SVM, the DBNs, which model time-dependent relationships between drivers' behavior and cognitive state, produced the most accurate and sensitive models. This indicates that changes in drivers' eye movements and driving performance over time are important predictors of cognitive distraction. At the same time, the SVM models detected driver cognitive distraction more accurately than the SBN models. This suggests that SVMs have advantages over BNs. The cross-validation seems to have resulted in better parameters for the SVMs. We used a 10-fold cross-validation to search for the best C and γ values in the range of 2^{-5} to 2^5. One possible result of selecting good parameters may be evident in the marginally increased hit rates and relatively lower false alarm rates of the SVM models, although the difference in false alarm rate was not significant. In addition, SVMs have less computational difficulties than BNs. It took less than one minute for SVMs to train a model, but approximately a half hour for SBNs and even longer for DBNs using an equal amount of training data.

A good real-world detection system would need to both accurately detect driver cognitive distraction and minimize the false alarm rate to promote acceptance of the system. An interesting finding of this paper is that window size had a marginal effect on false alarm rate though window size did not affect the other evaluation measures. This effect was particularly salient for DBNs. It also means that either a small (5 s) or large (30 s) window size used to summarize drivers' performance measures will reduce the false alarm rate without affecting overall model performance. This relationship can be used in practice to improve system acceptance. However, as shown in Fig. 13.7, the false alarm rates are still relatively high. Reducing false alarm rate is an important issue that future research.

Based on these comparisons, a hybrid learning algorithm that combines the time-dependent relationship and SVM learning technique improve model performance. Possible ways to integrate SVMs and DBNs include bagging or paralleling. Bagging describes using multiple models to make a prediction of one target. The procedure of bagging first involves training multiple (an odd-number of) SVM and DBN models with different training datasets to form a set of models. Then, all models make their own predictions for the same datum (or case). The final decision of cognitive state for this datum ("distracted" or "not distracted") depends on the vote from all models in the set. This method can reduce the variance of prediction and avoid overfitting.

Paralleling involves connecting two models sequentially. For example, if some aggregated descriptions of drivers' behavior, such as concentrated eye scanning pattern, are demonstrated to be essential for identifying cognitive distraction. We can first use SVMs to build the models that can identify such an eye scanning patterns from eye movement measures and then use DBN models to infer cognitive states with the description of scanning patterns identified from the SVM models. One study combined a Bayesian Clustering by Dynamics and SVM model in a parallel matter to forecast electricity demand [175]. The hybrid learning algorithm combing SVMs and DBNs can be used in the state refinement shown in Fig. 13.1 to implement detection of cognitive distraction.

Such bottom-up approach using data mining methods can complement top-down approaches. For example, the relationship identified from BNs may help discover evidence to support or extend theories regarding the cognitive processes that underlie distraction. The theories related to human cognition, in turn, can provide some pre-knowledge that can be integrated into bottom-up models, such as the initial structure and structural constraints of BN models. Data mining and cognitive theories can also directly cooperate to identify cognitive distraction. For instance, data mining methods can summarize driver's performance into an aggregated characteristic of driving behavior, such as seriously-impaired driving behavior or diminished visual scanning. Then, a model based on a cognitive theory can take the characteristic as an input to identify a driver's cognitive state. Thus, combination of top-down and bottom-up models may provide more comprehensive prediction for cognitive distraction than either alone.

However, some practical limitations exist in implementing detection systems. The first is to the problem of obtaining consistent and reliable sensor data, such as eye data [649]. For example, eye trackers may lose tracking accuracy when vehicles are travelling on rough roads or when the lighting conditions are variable. More robust eye monitoring techniques are needed to make these detection systems truly reliable. Second, it is necessary to identify the delay from the time when driver's cognitive state changes to when "distraction" state is detected. Three sources of delay have been identified [374]. They are the delay caused by response time of sensors, the time used to reduce sensor data and make prediction, and window size used to summarizing performance measures. The magnitude of each delay was approximately estimated in a previous study and depends on factors such as the computational efficiency of the algorithm [374]. Depending on how the estimated distraction is used, such delays could lead to instabilities as a distraction countermeasure is initiated just as the driver returns to the task of driving and is no longer distracted.

We have discussed how to detect visual and cognitive distraction as if they occur in isolation. In real driving environment, cognitive distraction often occurs with visual and manual distraction. To obtain comprehensive evaluation of driver distraction, the detection needs to consider all kinds of distractions. One approach is to create an integrated model that provides a comprehensive assessment of driver state. Alternatively, two models, one that detects visual and one cognitive distraction, need to work together in future IVIS. To simplify the detection, the procedure can begin by checking the level of visual and manual distraction because detecting visual-manual distraction is much easier to detect than cognitive distraction. If visual distraction is detected then it may not be necessary to check cognitive distraction.

Alertometer: Detecting and Mitigating Driver Drowsiness and Fatigue Using an Integrated Human Factors and Computer Vision Approach

Riad I. Hammoud and Harry Zhang

14.1 Overview

A significant number of highway crashes are attributable to driver drowsiness and fatigue. Drowsiness-related crashes can often cause more serious occupant injuries than crashes that are not related to driver drowsiness [291]. In order to better understand and curtail this problem, human factors researchers and engineers have studied the problem of driver drowsiness and fatigue in a variety of driving environments (for example, driving simulators). They have determined the behavioral characteristics associated with driver drowsiness and investigated the effectiveness of different mitigation strategies.

Slow eyelid closure has been identified as a reliable and valid measure of driver drowsiness. It can be detected non-obtrusively in real time by using an in-vehicle drowsiness detection system consisting of a camera used for capturing driver face images and a microprocessor for detecting eyelid closures. In the past few years, computer vision scientists have made great strides in developing advanced machine learning algorithms to detect slow eyelid closures.

In this chapter, we describe the extend of the contribution of driver drowsiness and fatigue as factors highway crashes. Next, different approaches of driver drowsiness mitigation is discussed, and the development of driver assistance systems is identified as a very promising approach. Human factors research is also reviewed to identify and validate slow eyelid closures as a diagnostic measure of driver drowsiness. The assessment of slow eyelid closures using an automatic eye tracking system and an intelligent computer vision algorithm will be described. Finally, research on effective methods of mitigating driver drowsiness is discussed.

14.2 Driver Drowsiness and Fatigue as Risks to Highway Safety: Problem Size

In modern society, people are increasing the amount of work in their daily schedule and thus spend more hours at work. Global companies have offices all over the world and many employees experience jet lag as a result of traveling across different time zones and countries for business meetings and product demonstrations. When not traveling, many employees attend conference calls that take place during early mornings or late evenings in order to accommodate the needs of meeting participants from different time zones and countries. These schedule irregularities disrupt their circadian rhythms and may reduce the quality of their sleep. Similarly, our personal life is overwhelmed with household chores. Parents with school-age children often have to send their children to school in the morning, pick them up from school in the afternoon, and provide transportation to and from extracurricular activities. Parents with young children cannot always get a good night sleep without experiencing interruptions. Young people may attend parties extending into the late night and travel a long distance to visit friends or family members. Faced with the aforementioned work-related and personal responsibilities, many people are cutting back on sleep. They may not get a quality sleep and therefore feel exhausted. In addition, a large number of people take allergy or cold medicine that may cause drowsiness. A small percentage of the general population suffers from sleep disorders (e.g., sleep apnea) preventing quality sleep. Many of these people drive to work, school, sports venues, and vacation destinations even when feeling drowsy. The drowsy driver problem is exacerbated because driving is frequently monotonous, especially on straight and level roads in late evenings.

In a 2003 interview with 4010 drivers in the U.S.A., 37% of the drivers reported having nodded off while driving at some point in their lives and 29% of these drivers reported having experienced this problem within the past year [543]. In a 2006 survey with 750 drivers in the province of Ontario, Canada, nearly 60% of the drivers admitted driving while drowsy or fatigued at least sometimes, and 15% reported falling asleep while driving during the past year [640].

Safe driving requires an adequate level of attention allocated to the driving task. Drivers must be vigilant and alert to monitor roadway conditions, detect hazardous situations, and make judicious decisions in order to steer the vehicle out of harm's way. For drowsy drivers, the level of attention possible for driving can be drastically reduced. As illustrated in Fig. 14.1, drowsy drivers may experience an attention deficit that could lead to a loss of control of their vehicle. Because of attention lapses, they may fail to monitor the lane position and depart from the roadway to cause a single-vehicle road departure crash, or drive into opposing traffic causing a head-on collision. According to Royal et al. [543], an overwhelming majority (92%) of the drivers who have nodded off while driving during the previous six months reported

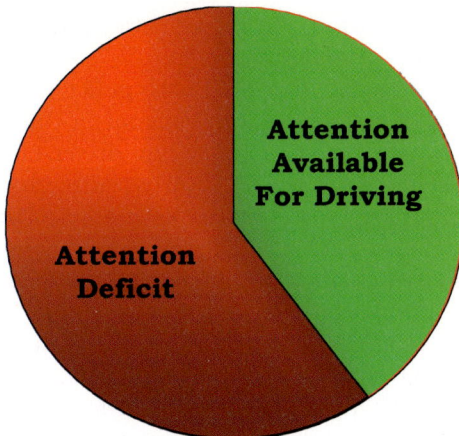

Fig. 14.1. The relationship between attention deficit and attention available to driving as experienced by a drowsy driver

that they were startled awake by a crash. One in three drivers (33%) drifted into another lane or onto the shoulder, 19% crossed the centerline, and 10% ran off the road.

The National Highway Transportation Safety Administration (NHTSA) has estimated that driver drowsiness is the primary cause in 100,000 police reported crashes, resulting in 76,000 injuries and 1500 deaths annually in the U.S.A. These numbers represent 1–3% of all police reported crashes and approximately 4% of fatalities [328, 658]. Other studies have reported higher statistics. Based on 2001 Crashworthiness Data System (CDS) data, Eby and Kostyniuk [156] reported that 4.4% of crashes are attributed to driver drowsiness. This statistic would be considerably higher if 55.6% of the "unknown" cases were excluded from the analysis. In Australia, approximately 6% of crashes may be attributable to driver drowsiness and fatigue [176]. In England, up to 16–20% of police reported vehicle crashes are related to driver drowsiness and fatigue [262]. The different statistics may be attributed to an inherent difficulty in determining the role of driver drowsiness and fatigue in highway crashes. Unlike metal fatigue, driver fatigue and drowsiness do not leave telltale signs after a crash occurrence. Typically, drowsy drivers will become awake and alert after a crash. Therefore, it is difficult for police officers to verify the involvement of driver drowsiness at the crash scene. Police officers may interrogate the drivers or use circumstantial evidence to infer the involvement of driver drowsiness. Drivers may not always recognize or report the involvement of drowsiness when interrogated by police [592]. The problem becomes even more complex when drugs and alcohol appear to be present. Because drowsiness and alcohol are inter-related, police officers may report the presence of alcohol (which can be verified) but not driver drowsiness (which is difficult to verify).

Because of the limitations associated with police report-based statistics, an alternative method is to install video cameras in vehicles to record roadway conditions and driver behavior (including driver face) and analyze the videos to determine the involvement of driver drowsiness in crashes. Recently, NHTSA has sponsored a 100-car Naturalistic Driving Study in order to gain a better understanding of circumstances surrounding crashes and near-crashes [324]. Based on the 100-car study, 22–24% of crashes and near-crashes can be attributable to driver drowsiness. Klauer et al. [324] revealed that driver drowsiness increases the likelihood of crashes or near crashes by 4–6 fold.

Drowsiness-related crash scenarios appear to be quite unique. Fatigue or drowsiness-related crashes tend to occur after midnight or in the afternoon with vehicles traveling at high speeds. Often time drivers do not perform any maneuvers to prevent a crash before the crash occurrence. The most typical crash scenario tends to be the "single vehicle run-off-the-road crash", in which a single vehicle with the driver alone in the vehicle leaves the roadway resulting in a serious crash. Campbell et al. [79] reported that drowsy drivers contribute to approximately 8–10% of single vehicle run-off-the road crashes. However, a small percentage of rear-end crashes and lane change crashes is attributable to drowsy drivers.

14.3 Approaches to Minimize Driver Drowsiness

Several approaches have been taken in order to reduce driver drowsiness and fatigue. One approach is the educational[1] approach [592]. With this approach, educational materials are made available to the general public in order to dismiss common misconceptions about the drowsy driver problem. For example, misconceptions such as "It does not matter if I am sleepy", "I do not need much sleep", and "I cannot take naps". In order to maximize the effectiveness of educational campaigns, they should be targeted towards at-risk groups such as young males and shift workers who are especially vulnerable to drowsiness related crashes. When possible, work schedules should be adjusted to minimize driver fatigue.

Another approach is the development of driver assistance systems [204, 291,712]. As depicted in Fig. 14.2, driver states such as drowsiness and fatigue are assessed in real time using a driver state sensor. When the drowsiness level exceeds a pre-defined threshold, drowsiness warnings are delivered to the driver. Heeding the warnings, the driver will take certain countermeasures (e.g., taking a short nap) to reduce the drowsiness level and restore alertness. The driver assistance system will continuously monitor the driver drowsiness

---

[1] http://www.nhtsa.dot.gov/people/injury/drowsy_driving1/drowsy.html

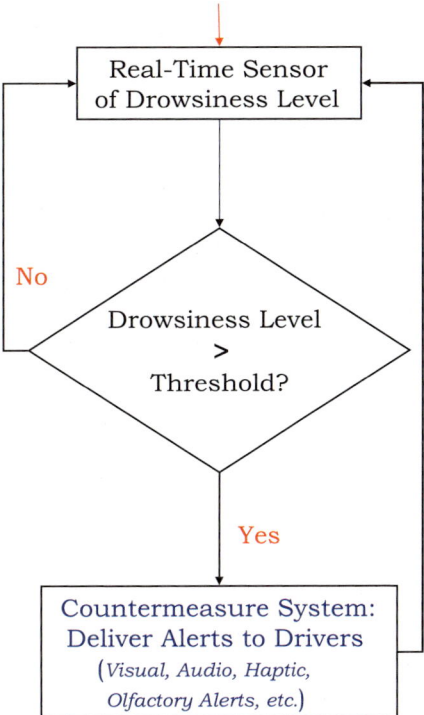

Fig. 14.2. A closed-loop driver assistance system for driver drowsiness mitigation

level. When a higher level of drowsiness is detected, more aggressive warnings will be delivered. This closed-loop feedback system has shown some promise in reducing driver drowsiness levels [204, 291, 712]. Because driver drowsiness is typically associated with single vehicle road departure crashes, the drowsiness information can also be fed into lane departure warning systems to enhance the lane departure system effectiveness [335]. When driver drowsiness leads to an unintentional lane departure, a warning may be issued to re-gain the driver's attention and steer the vehicle back to the road.

The driver assistance system includes two major components as presented in Fig. 14.2. The first component is a driver state sensor that gathers and processes information from multiple variables including driving performance and infers the drowsiness level from the information. The second component is a countermeasure system that determines whether or not the current drowsiness level warrants the delivery of an alert to the driver, and how the alert should be delivered (e.g., with visual, audio, haptic, or olfactory alerts).

Both components are critical to the success of driver assistance systems. Assistance would be impossible without a reliable and valid assessment of driver drowsiness levels from the driver state sensor. If the sensor has many false negatives (misses) and cannot sufficiently detect drowsiness epochs, drowsiness alerts will not be issued when needed. If the sensor system has many false positives (false alarms) and erroneously identifies non-drowsy periods as "drowsy", false drowsiness alerts will be issued to drivers who are in fact alert. In either case, drivers will not trust the system, disregard the drowsiness alerts, or even turn the system off. Therefore, it is imperative that the sensor component has a high level of reliability and validity.

The countermeasure component is also critical to the success of driver drowsiness mitigation. If the drowsiness alerts are confusing or annoying, the alerts will not be comprehended or accepted by the users. The driver acceptance level will be low and drivers may even turn the system off. The drowsiness problem cannot be mitigated if effective countermeasures are not provided to the driver.

14.4 Measures for Detecting Driver Drowsiness

It is virtually impossible for drivers to fall asleep without any behavioral indications or performance degradations. Certain behavioral indications invariably precede the onset of sleep, including slow eyelid closures, increased number and duration of eye blinks, reduced pupil diameter, excessive yawning, head nodding, slowdown of breathing and heart rate, decline of muscle tone and body temperature, electromyogram (EMG) shift to lower frequencies and higher amplitude, increase of electroencephalogram (EEG) alpha waves, and struggling to fight sleep and stay awake [131, 171, 174, 472, 568]. However, each driver's behavioral and physiological manifestations of drowsiness and fatigue are different. Because of the individual differences, several drowsiness measures are capable of detecting a high level of drowsiness in some individuals, but fail to do so in other individuals. The challenge for human factors researchers is to identify measures that can reliably detect driver drowsiness in a vast majority of individuals.

After reviewing many drowsiness-related measures, Erwin (1976) concluded that slow eyelid closure is the best measure of driver drowsiness [171]. When a driver becomes drowsy, the eyelids tend to droop and a smaller portion of the eyes becomes visible. When the eyes are fully open, the iris and the pupil are clearly visible from the face image. If the eyes are 80% closed, only 20% of the iris or the pupil will be visible from the face image. If the eyes are 100% closed, the iris or the pupil will not be visible from the face image. Several field operational studies and laboratory experiments [131, 174, 508, 680] have revealed that among all driving performance and physiological measures that have been tested, PERCLOS is the most reliable and valid measure of driver drowsiness. PERCLOS is defined as the percentage of time that the eyelids are 80–100% closed over a time window (e.g., over a 1-minute, 3-minute, or 20-minute time window). For example, if the eyelids are 80–100% closed for a total of 6 seconds within a one-minute time window, PERCLOS would be 6/60 or 10%.

The most definitive study for establishing the validity of PERCLOS as a driver drowsiness measure is the controlled laboratory experiment by Dinges, Mallis, Maislin, and Powell (1998) [131]. The psychomotor vigilance task (PVT) measures the latency between a visual stimulus and a motor response (e.g., pressing a button) and is a very sensitive measure of fatigue based on night work and sleep loss studies. In Dinges et al. (1998) [131], fourteen male participants stayed awake for 42 hours and the PVT performance was measured on a bout-by-bout basis. Each bout consists of 20 minutes. They derived nine drowsiness measures, including EEG measures, head position measures, eye blink measures, and PERCLOS measures that are based on manual coding of video images by trained human observers. A good measure of driver drowsiness should be highly correlated with the PVT performance on the bout-by-bout basis. The EEG measures are positively correlated with the PVT performance, but the Pearson correlation coefficients are moderate, averaging 0.58 from eight calculations. The correlation coefficients vary from a low value of 0.31 to a high value of 0.95. Head position measures are positively correlated with the PVT performance for most participants, but a negative correlation is shown for one participant. The correlation coefficients may be a negative value (-0.54), a low positive value (0.13), or a high positive value (0.91). Similarly, the eye blink measures may be positively or negatively correlated with the PVT performance. The correlation coefficients vary widely, from -0.48 to 0.90. These results indicate that EEG measures, head position measures, and eye blink measures are reasonable indications of driver drowsiness for some individuals, but poor indications for other individuals.

In contrast, PERCLOS measures are consistently correlated with the PVT performance for all individuals, with Pearson correlation coefficients ranging from 0.67 to 0.97. Using different time windows, it is found that PERCLOS over a 20-minute time window has a higher correlation with the PVT performance than does PERCLOS over a 1-minute time window. Regardless of the time window, however, PERCLOS is found to be more highly correlated with the PVT performance than other measures.

A recent study in [296] has provided the converging evidence for the PERCLOS measure. They employed a test of variables of attention (TOVA) as a vigilance measure to detect driver fatigue and drowsiness. In the TOVA test, participants are required to sustain attention over a time window (e.g., 20-minute interval) and press a button when a light randomly appears on a computer screen. The TOVA response time, defined as the time interval between the light onset and the button press, is a measure of driver drowsiness. When a driver becomes drowsy, the TOVA response time increases accordingly. As depicted in Fig. 14.3, as the TOVA response time increases, PERCLOS increases. This indicates that PERCLOS is closely related to driver drowsiness and fatigue levels.

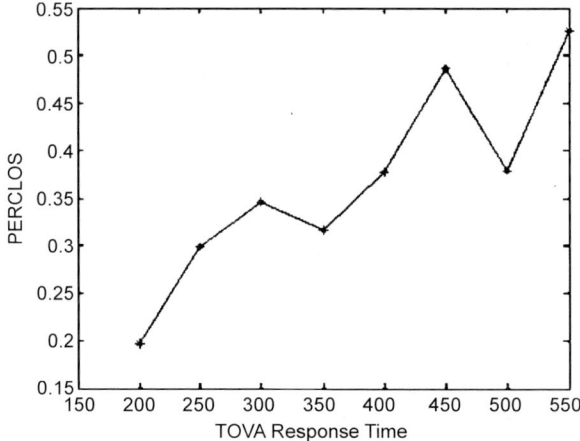

Fig. 14.3. Relationship between PERCLOS and TOVA response time. Courtesy of [296]

In order to calculate PERCLOS accurately, a distinction must be made between slow eyelid closures and eye blinks. The research on the relationship between eye blinks and driver drowsiness has mixed results. Some studies have found that more frequent and longer eye blinks are associated with driver drowsiness and fatigue [179, 472], but other studies have not demonstrated similar effects [131]. In any event, eye blinks have markedly shorter durations than do slow eyelid closures. In terms of video coding, eye closures with a duration of over $300 - 500$ milliseconds are typically coded as slow eyelid closures and entered in the PERCLOS calculation, and eye closures with a duration of under 300–500 milliseconds are coded as normal eye blinks and disregarded in the PERCLOS calculation. Human factors researchers may pick a threshold value anywhere between 300–500 milliseconds because the PERCLOS values do not seem to vary significantly with a different closure value in this range.

Wierwille et al. [680] have defined 8% and 15% as two PERCLOS thresholds for driver drowsiness. If PERCLOS is over 15%, the driver is declared as "drowsy", and if PERCLOS is between 8–15%, the driver is declared as "likely drowsy". These threshold values have been adopted in nearly all subsequent research [131]. A large-scale field operational test (FOT) has been conducted to evaluate a drowsiness mitigation system using these PERCLOS thresholds [508]. High PERCLOS values appear to be directly linked with crashes. As illustrated in Fig. 14.4, a driving simulator study revealed that PERC-LOS exceeded 15% approximately 60 seconds before a crash occurrence, and it peaked approximately 20–30 seconds before the crash occurrence [179].

In order to make PERCLOS a useful metric in driver drowsiness sensing systems, the determination of slow eyelid closures must be automatically performed by computer programs rather than by trained human ob-

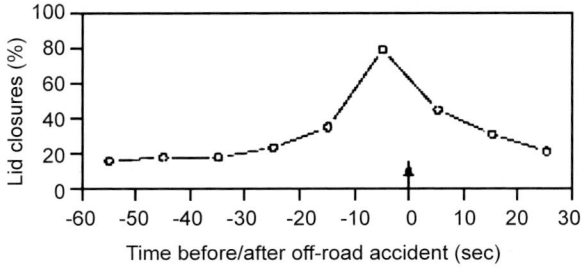

Fig. 14.4. PERCLOS values preceding a crash occurrence. Courtesy of Trucking Research Institute, etc. [179]

servers. Instead of generating eyelid closure information off-line with a long temporal delay in the manual coding scenario, the automatic system must produce eyelid closure information in real time (e.g., with minimal temporal delays) in order to deliver drowsiness alerts promptly. In order to be well accepted by drivers, the automatic system must be unobtrusive and make no physical contact with drivers. Furthermore, the system should work during both daytime and night time. The system should operate automatically without any driver's interaction. It should operate for drivers of diverse ages, genders, skin complexions, and eyeglass prescriptions. The reliability and validity of the system must be evaluated in the driving context. In the next section, we will describe some recent eyelid closure algorithms using computer vision and video image processing that satisfy these requirements.

14.5 Computer Vision Algorithms for Detecting Driver Drowsiness

As mentioned earlier in this chapter, PERCLOS is defined as the percent of time the driver's eyes are closed. That is the percent of time slow eyelid closures cover the pupil. More specifically, it is the percent of time the eyelids are closed 80% or more. Figure 14.5 depicts a progressive disappearance of the pupil as a slow eyelid closure event occurs. PERCLOS is computed for each monitored eye separately.

To calculate the value of PERCLOS measure at time t for one eye, the following steps are considered:

Fig. 14.5. Illustration of eye closure events in a driver fatigue scenario

1. Select a time window $[t - n, t]$ of a predefined length, n, in which the spatial eye position has been tracked (i.e. in each frame i the eye region is located and extracted).
2. Estimate the percentage of eyelid closure, in frame i, by performing a specialized eye closure determination routine (see next section); the estimated value lies within 0% (open eye) and 100% (complete closed eye) range.
3. Filter the generated outputs over n frames by eliminating those corresponding to normal (fast) eye blinks.
4. Evaluate the following equation (Eq. (14.1))

$$PERCLOS(t) = \frac{\sum_{i=1}^{n} \delta_i}{n} \times 100 \qquad (14.1)$$

where,

$$\delta_i = \begin{cases} 1, & \text{if closure percentage at the } i^{th} \text{ frame is greater than 80\%;} \\ 0, & \text{otherwise} \end{cases}.$$

$$(14.2)$$

For example, if 16 slow eyelid closures, each of 3 seconds, are detected over 60-s window, then the PERCLOS value is 80%, indicating a critical fatigue level. The evaluation time window of PERCLOS is usually considered constant (see previous sections for details). The eye data are usually gathered in eye fixation mode, with limited head movements and frontal head scenarios to limit the artifacts and noise contributions to the computation of PERCLOS.

The key step in the PERCLOS algorithm is measuring the percentage of eyelid closure per frame. To that end, several image processing techniques that have been investigated in the literature are briefly described in next section. The common approach has two phases: (1) identify key eye components like eyelids, pupil and iris – for a detailed description of the eye region model the reader may refer to Table 2.1 in Chap. 2 – and (2) derive statistics from these key eye components including the eyelids spacing and width-height pupil and iris ratio [43, 296].

14.5.1 Eyelid Spacing Estimation

The Eyelid spacing based technique for eye closure determination is achieved by monitoring the edge boundaries of the monitored eye and determining the spatial distance that separates the selected upper and lower eyelids points, that is indicative of the closure state (degrees of openness) of the eye. These points could be obtained from *selected eyelids lines* or *eye contour*, both of which are briefly described below. This distance can be used to compute the eyelid opening width, the eyelid opening and closing speeds, eyelid closure duration, and blinking frequency.

14.5.1.1 Eyelid Detection

Driver drowsiness can be determined from whether and how a driver's eyelids are moving or whether they are actually closed [291]. Eyelid modeling with lines is comprised of several steps. First, an edge detection function is applied for detecting edge boundaries of the eye in the captured image. The pixels at the edge boundaries exhibit a characteristic shading (e.g., dark gray contrasted with white, or vice versa) that identifies the edge boundaries. The edge map (set of edge pixels) is computed using standard edge detectors. Second, a Hough Transformation function is performed to transform plotted edge pixel points from Cartesian coordinates to polar coordinates to enable the generation of lines on the edge boundaries of the monitored eye. Finally, a line selection algorithm is applied for selecting a pair of lines that represent the upper and lower edge boundaries of the eye. Assuming the head is steady and frontal, from the Hough Transformation, lines that are substantially horizontal are extracted and, from the substantially horizontal lines, the upper and lower horizontal lines are selected.

The Hough Transformation transforms two-dimensional images with lines into a domain of possible linear parameters, where each line in the image provides a peak position at the corresponding line parameters. The Hough Transformation is generally known in the art of image processing and computer vision for detecting shapes and in particular straight lines. Other transformations such as a Radon Transformation, or Least-square type methods for line fitting may be employed to generate substantially horizontal lines and select the upper and lower lines. Fig. 14.6 depicts the results of eyelids fitting with straight lines. Applying this technique on low-resolution infrared images is challenging due to the potential loss of eye features and low contrast level. The selection of eye state thresholds is another challenge with low-resolution eye images. Bueker et al. [73] reported that even when using a large focal

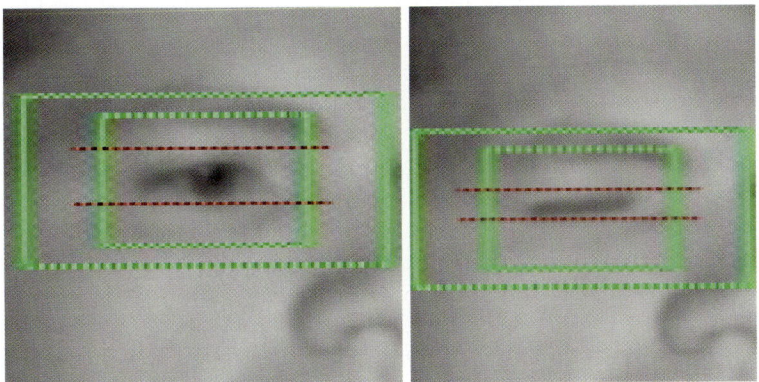

Fig. 14.6. Illustration of eyelids modeling with horizontal lines: open eye (*left*), closed eye (*right*)

length – adjusted to the size of a truck compartment to allow driver head monitoring over a wide range of different seating positions – the size of the eyes in the resulting image is very small. Depending on the seating position, in VGA resolution typically the diameter of the pupil is smaller than 15 pixels. This results in eyelid openings between 10 and 15 pixels, which makes it extremely difficult for the image processing algorithms to extract an accurate eyelid signal.

14.5.1.2 Eye Contour Detection

A deformable contour model is a powerful image analysis tool and one well-known method for continuous eye contour extraction [309,353,701,708]. Both continuous contour extraction and approximate contour detection methods require a segmentation of the eye region image. Edge-based detection techniques are usually employed at the segmentation phase. The spatial luminance gradients image (edge map) are used as the feature for the contour detection methods.

Deformable models need careful formulation of the energy term and close model initialization, otherwise, unexpected contour extraction result can be acquired. Frequently, a step-by-step coarse to fine shape estimation is used to lessen the number of degrees of freedom for the model, because active contour models tend to be attracted by local minima [309, 701] and move away from the real contour. Sometimes many internal constraints on interdependence between the different eye features are used [708] to achieve a stable detection result. Using luminance edges as the feature for the contour detection is risky, because eye areas may have many unwanted spurious edges and lack the informative ones.

Understanding the challenge in extracting complete continuous eye contours, some researchers have tried different approaches to approximate eyelid contours through the detection of several landmark points like eye corners and iris border points [177], [614]. This results in less accurate, but more stable detection. Vezhnevets and Degtiareva [648] have proposed a technique, which combines stability and accuracy. First, iris center and radius are detected by looking for a circle separating dark iris and bright sclera (the eye white). Then, upper eyelid points are found based on the observation that eye border pixels are significantly darker than the surrounding skin and sclera. The detected eye boundary points are filtered to remove outliers and a polynomial curve is fitted to the remaining boundary points. In order to reduce the effect of image noise, the image is preprocessed with median filtering and 1D horizontal low-pass filtering. Finally, the lower eyelid is estimated from the known iris and eye corners. The iris center and radius detection are performed in image's red channel, which emphasizes the iris border. This is due to the fact that iris usually exhibits low values in the red channel (both for dark and light eyes), while the surrounding pixels (sclera and skin) have significantly higher red values.

Figure 14.7 illustrates the detection of eye image segmentation and eye contour detection. The reader may refer to Chaps. 2 and 7 for mathematical details and visual illustrations using this type of eye contour extraction technique.

14.5.2 Pupil and Iris Detection

One would observe that, as the eye closes, the pupil and the iris become occluded by the eyelids and their shapes become more elliptical. Therefore, the degree of eye opening could be characterized by the shape of the pupil or the iris described by the ratio of pupil/iris ellipse axes. If this ratio exceeds a threshold (e.g., 20% of the nominal ratio), a closing eye action is detected. When the pupil ratio is above 80% of its nominal size a closed eye state is provoked, which means the eyes are closed. The system is in an opening state if the pupil ratio is below 20% of the nominal ratio. In [43] the eye closure duration measurement is calculated as the time that the system is in the closed state. The authors computed this parameter by measuring the percentage of the eye closure in 30-s window, excluding the time spent in normal eye blinks. The eye closure/opening speed measurements represent the amount of time period during which the pupil ratio passes from 20% to 80% or from 80% to 20% of its nominal size, respectively.

The most popular and effective way to detect pupils is the subtraction technique. It uses a structured infrared light, a dual on-axis and off-axis IR LEDs system that generates bright and dark pupil images (see Chaps. 5 and 6 for more details on this technology). The binary pupil blob is identified easily by subtracting the odd and even eye regions, and then performing a specific parameterization of a general shape finding algorithm like the Circular Hough transform on the obtained image to find circles of a certain radius. While this method is simple and effective it is difficult to maintain its performance level in outdoor conditions and on subjects with eyewear. Another issue lies in the packaging size of such an eye tracking system (two distributed LEDs), which might be problematic in some environments like small vehicle compartments. When a wide-angle single camera with only on-axis LEDs is employed,

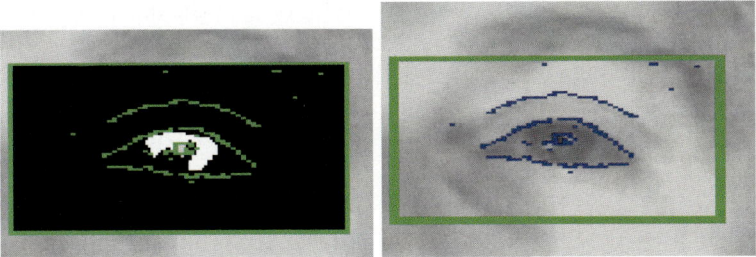

Fig. 14.7. Illustration of approximate eye contour detection: segmented eye region (*left*), overlaid contour (*right*)

a precise detection of dark-pupils becomes more challenging, due mainly to clutter presence and low contrast differences between the pupil and the iris, and between the pupil and the eyelashes. In addition pupil and iris intensities changes among people (dark, brown, and light irises). The general approach for dark pupil detection consists of a contrast enhancement pre-processing step, followed by a histogram-based thresholding and binary blob extraction, and finally filtering and blob selection logic (see Fig. 14.8). The selection rules are formulated based upon some prior knowledge like the shape of the pupil, position in the eye, intensity level, size, and the pupil's peripheries such as upper lids, the iris and eye corners. The pupil position could be used to initialize the iris localization technique. The Hough Transformation [149, 566] is generally known in the art of image processing and computer vision for detecting circular-like shapes, but more sophisticated and elegant approaches could be employed (see Chap. 2 for details). Figure 14.9 depicts the results of the luminance-based image segmentation step and the ellipse-fitting method for iris searching in low infrared images, while Fig. 7.4 (Chap. 7) shows detected limbuses (red ellipses) in good quality color eye images. Recently, Vezhnevets et al. [648] have proposed a stable technique for iris localization. Several other iris detection and tracking techniques, reviewed in Chaps. 2 and 7, have shown good performance and stability.

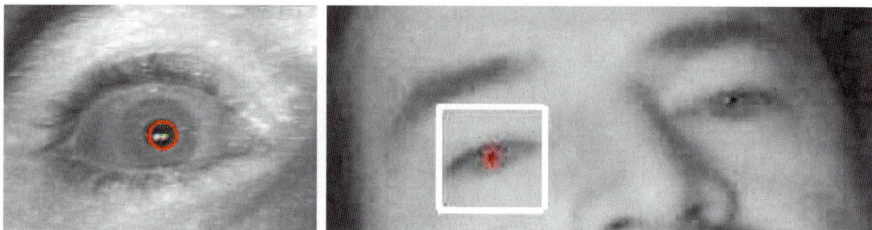

Fig. 14.8. Illustration of dark-pupil detection in a moderate (wide open, *left*) and low resolution infrared images (semi-open, *right*)

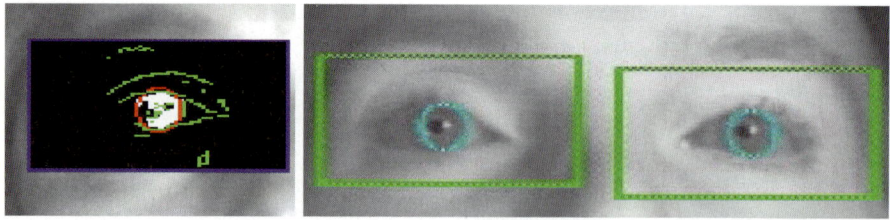

Fig. 14.9. Illustration of detection of the luminance-based image segmentation and dark blob extraction (*left*), and iris detection and modeling with an ellipse (*right*)

14.5.3 Eye Blink Detection

Eye blink detection allows measurements of blink frequency, blinking rate and the rate of eye closures. These measures are often used as indicators of driver awareness while operating a vehicle [430, 636].

An eye blink event occurs when the eye goes from an open state to a closed state, and back to the open state again. That is a consecutive transition among the three states, closing, closed, and opening. An eye blink has a duration and a velocity. Based upon these two elements, eye blink events are usually grouped into short, long, fast or slow categories. In case of driver fatigue scenarios, eye blinks are rather slow with long duration.

Eye blinks have been initially investigated in the computer vision community as a validation measure in the detection phase of the spatial eye location, in an attempt to reduce the detection rate of false positives. Numerous automatic spatial eye localization techniques have relied on detecting partial or complete fast eye blinking events [4, 49, 208, 384, 430, 573, 624, 636, 698]. In Yano et al. [698], frame differencing is used for blink detection. Al-Qayedi and Clark [4] track features about the eyes and infer blinks through detection of changes in the eye shape. Smith et al. [573] try to differentiate between occlusion of the eyes (due to rotation of the head) and blinking. The subject's sclera is detected using intensity information to indicate whether the eyes are open or closed (i.e., a blink is occurring). Black et al. [49] detect eye blinks using optical flow but the system restricts motion of the subject and needs "near frontal" views in order to be effective. Recently, Bhaskar et al. [46] detect eye blinks using optical flow. Motion regions are first detected using frame differentiation. Optical flow is computed within these regions. The direction and magnitude of the flow field are then used to determine whether a blink has occurred. The forementioned methods are applied on the entire frame and potential eye candidates are validated based upon an eye blink filter.

Slow eye closure detection, for fatigue assessment, requires a careful analysis of the change in the waveform (eye signal) that represents the monitored eye sequence. The *transition time* between two consecutive peak and valley points is used as indicative of the blink category.

To construct the waveform, a feature vector is selected first. This could take different forms including the eye-blob size, correlation scores and the aspect-ratio of the eye-shape. Figure 14.10 illustrates the obtained waveform of an infrared low-resolution tracked eye sequence. The grey-scale eye region is segmented first into foreground and background regions, then the foreground object in the middle of the binary image is extracted and its pixel count is employed as a feature. This figure depicts the two categories of fast and slow eyelid closures. To determine the state of a tracked eye, the eye-blob size is tracked over time. The upper and lower signal envelopes of the size are then found. The upper (maximum) and lower (minimum) envelope values are then used to compute a dynamic threshold. The current eye-blob size

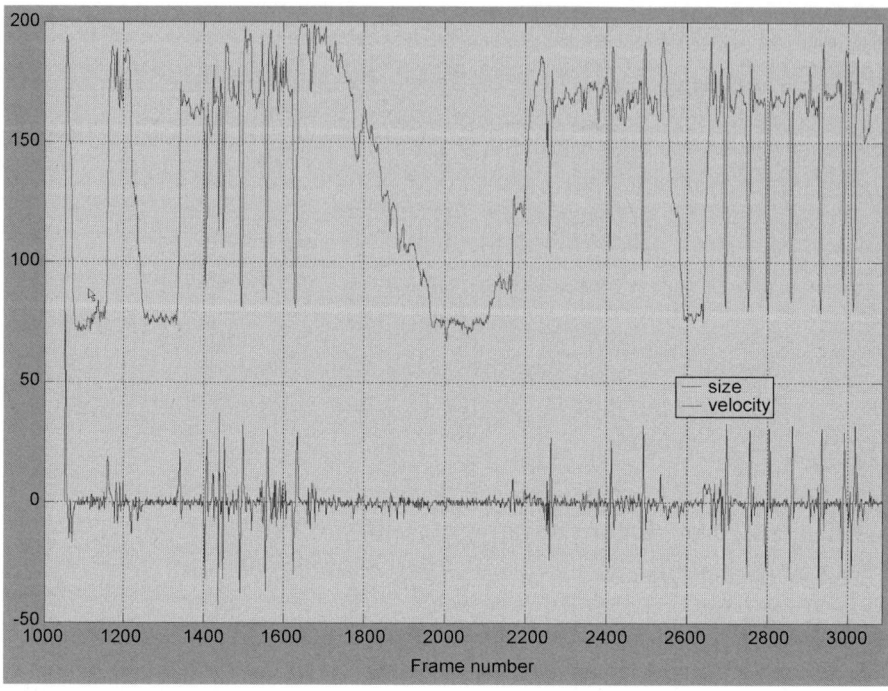

Fig. 14.10. Illustration of the eye signal: fast eyelid closures (frames 1700 to 2200) vs. slow eyelid closures (frames 2600 to 3000). The *upper curve* represents the size of the eye while the *lower curve* represents its corresponding velocity

may be compared against this threshold to classify the eye as open or closed. To determine which signal envelope to update, the current size is compared against the previous threshold. If the size is greater than the threshold, the maximum is updated; if less, the minimum is updated. Over intervals where a signal envelope is not updated, its value is allowed to decay (maximum decays to smaller values, minimum decays to larger values) to insure the adaptation does not become trapped in an incorrect state.

Another blink detection technique that avoids eye region segmentation, consists of correlating the eye images with an eye template. This latter could represent an open eye or a closed eye. It is either selected on-line at the initialization phase of the eye location tracker, or constructed off-line using machine learning tools like the eigen-picture method. Grauman et al. [208] update the eye template during tracking, and use the Normalized Cross Correlation (NCC) to compute the correlation scores. They reported that a close examination of the correlation scores over time for a number of different users of the system reveals rather clear boundaries that allow for the detection of the blinks. The waveforms representing degree of closure are distinct. When the user's eye is in the normal open state, very high correlation scores of about 0.85 to 1.0 are reported. As the user blinks, the scores fall to values of about 0.5 to 0.55. The drawback of this method is that it is prone to slight drift of the eye tracker as well as to out-of-plan head rotations. Grauman et al. [208] report a success rate of 96% in almost real-time. Here, frame differencing is initially performed to obtain motion regions. Then a pre-trained "open-eye" template is selected for the computation of the Mahalanobis distance measure which is used to retain the best pair of blobs as candidates for eyes. The eyes are tracked and correlation scores between the actual eye and the corresponding "closed-eye" template are used to detect blinks. However, it requires offline training for different depths from the camera for the computation of the distance. In addition, the system requires initialization in which a stored template for a particular depth is chosen. Once the template is chosen and the system is in operation, the subject will be restricted to be at the specified distance.

14.6 Effective Countermeasures for Mitigating Driver Drowsiness

Some researchers have argued that driver drowsiness mitigation may not be very effective because some drivers will continue to drive their vehicle despite their awareness of sleepiness. Therefore, motivation and awareness of the consequences of sleepiness are also factors that can contribute to the safety of drowsy drivers [712]. Education of at-risk drivers about the heightened risk of crashes associated with drowsiness should be a key component of drowsiness mitigation process [592].

A drowsiness mitigation system that delivers drowsiness warnings to drivers is a promising approach. Early studies found that over 90% of drivers felt that they can reverse the drowsiness process by taking actions to reduce their levels of drowsiness [618]. Recently, several automotive manufacturers have evaluated drowsiness mitigation systems with encouraging results [291, 335]. The feedback from test drivers who gave an assessment of the warning systems indicates that drivers welcome drowsiness mitigation systems. When being asked about whether they would accept a camera and infrared light sources in the vehicle to assess driver fatigue, over 80% of respondents answered affirmatively.

The most effective method of mitigating driver drowsiness is to take a short nap to increase the alertness level [142, 179, 554]. Trucking Research Institute [179] showed that an afternoon nap has a number of beneficial effects on driver alertness and performance. Performance on the computerized tests was generally faster, more accurate, and less variable in the nap condition than in the non-nap condition. Subjective sleepiness and fatigue scores were significantly lower in the nap than in the non-nap condition. Driskell and Mullen [142] reviewed the biomedical and behavioral research papers on the effectiveness of napping that have been published in the last 25 years.

A meta analysis from twelve studies revealed a small but significant effect of napping. As illustrated in Figure 5.11, participants improved their perform-ance in reaction time, visual vigilance, logical reasoning, and symbol digit substitution tasks after taking a nap, and the performance improvement was greater with a longer duration (e.g., a nap of 2–4 hours) than with a shorter duration (e.g., a nap of 15–30 minutes). Participants also reported feeling less fatigued after taking a nap and the duration of the nap (e.g., from 15-minutes to 4 hours) did not affect the reported fatigue level. Overall, it is evident that a nap as short as 15 minutes is very effective in combating driver drowsiness. Figure 14.11 also indicates that the performance improvement and reported fatigue reduction from the naps can last as long as for several hours. These studies suggest that the driver drowsiness problem can be alleviated by en-couraging drowsy drivers to take a short nap between trips.

Because taking a nap may not always be feasible, however, alternate meth-ods have been proposed to mitigate driver drowsiness. These methods include taking caffeine [432,554], exposing to bright lights (especially blue lights [66]), using biofeedback with visual and auditory indications [204], using rumple strips, blowing peppermint-scented air into the cabin [254], performing phys-ical exercise, lowering the cabin temperature, and changing airflow. Sanquist and McCallum [554] argued that some of these techniques (e.g., peppermint-scented air, auditory alerts, changing airflow or temperature, and physical exercise) have only a brief effect. Fairbanks et al. [174] asked drowsy drivers to rate the effectiveness of various alerting countermeasures. Drowsy drivers consistently rated male or female voice messages and peppermint scents as ef-fective countermeasures for mitigating driver drowsiness. In Grace and Stew-ard [204], up to three amber LED lights were illuminated and an audible advisory tone was sounded when the PERCLOS level exceeded 8%, and up to three additional red LED lights were illuminated and an audible advi-sory tone was sounded when the PERCLOS level exceeded 14%. Grace and Steward [204] found that the visual and auditory alerts had a high level of acceptance by the drivers.

Ho and Spence [254] investigated the differential effects of olfactory stim-ulation on dual-task performance under conditions of varying task difficulty. Participants detected visually presented target digits from amongst a stream of visually presented distractor letters in a rapid serial visual presentation task. At the same time, participants also made speeded discrimination re-sponses to vibrotactile stimuli presented on the front or back of their torso. The response mapping was either compatible or incompatible (i.e., lifting their toes for front vibrations and their heel for back vibrations, or vice versa, respectively). Synthetic peppermint odor or unscented air (control) was delivered periodically for 35 seconds once every 315 seconds. The results showed a significant performance improvement in the presence of peppermint odor (as compared to unscented air) when the response mapping was incom-patible (i.e., in the difficult task) but not in the compatible condition (i.e.,

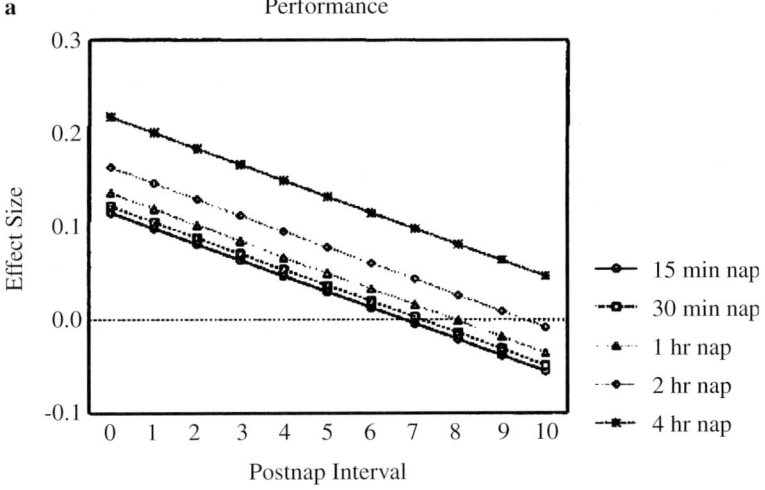

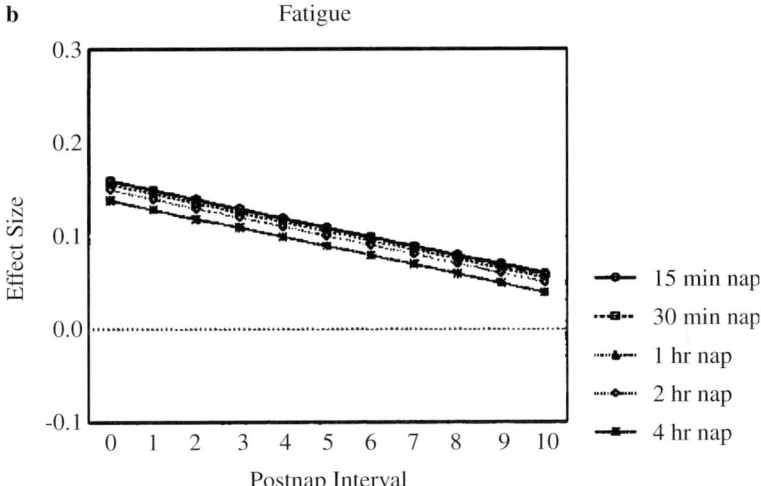

Fig. 14.11. Effects of nap duration and postnap interval on objective performance (*top panel*) and reported fatigue levels (*bottom panel*). Courtesy of [142]

in the easy task). These results provide the empirical demonstration that olfactory stimulation can facilitate tactile performance, and also highlight the potential modulatory role of task-difficulty in odor-induced task performance facilitation.

A potentially effective method of reducing crashes that are attributable to driver drowsiness is to fuse the driver drowsiness level with a lane departure warning system [335, 712]. Drive state information such as the driver drowsiness and distraction level is a key factor in determining whether a lane departure is intentional or unintended. If a driver is distracted or drowsy, unintended lane departures may occur. Lane departure warnings should be provided to these drivers to reorient their attention back to driving to avoid crashes. On the other hand, if a driver is alert and attentive, lane departures are likely intentional. Lane departure warnings may be suppressed for attentive drivers because the warnings would be annoying to an attentive driver who is making an intentional lane departure. Kozak et al. [335] found that steering reaction times for correcting lane departures were shortened by alerting drowsy drivers using a steering wheel torque, steering wheel torque and vibration, rumble strip sound with steering wheel torque, and head up display with steering wheel torque.

14.7 Discussion and Conclusions

In this chapter we have presented a summary of the state-of-the-art work in both human factors and computer vision fields to minimize the problem of driver fatigue and drowsiness. Today, driver fatigue and drowsiness are among the major contributing factors to automotive crashes. In the past few years, significant progress has been made in identifying and validating slow eyelid closures and PERCLOS as diagnostic measures of driver drowsiness. Furthermore, it has become feasible to detect eyelid closures in real time using video-based, automatic, unobtrusive eye tracking systems and intelligent computer vision algorithms.

This chapter presented also a summary of the considerable research that has been performed to determine effective methods of mitigating driver drowsiness using techniques such as making a suggestion for drowsy driver to take a short nap, and delivering visual, auditory, and olfactory alerts to drowsy drivers. The drowsiness state may also be fused with a lane departure warning system to deliver a timely warning when an unintended lane departure takes place and minimize false alarms.

Today, several challenges remain in detecting and mitigating driver drowsiness. One is related to the input image size and sharpness. When the size of the input eye region is small relative to the actual size of the eye or the input image is not sufficiently sharp, the structure of the eye may not be sufficiently visible. For instance, potential failures in the tracking of the iris are related to clutter presence where the iris model matches another dark portion in the eye region, such as shadow around the hollow between the inner corner of the eye and the root of the nose. An especially bright iris could contribute to this type of contrast error. Reported experiments (see Sect. 2.6 for details) have shown that iris localization techniques fail on Caucasian-like subjects who had bright irises with strong specular reflection and a thick and bold outer eye corner or even glasses pad. Improvements could be achieved through dynamic camera and light source controls as well as by fusing multiple eye closure indicatives.

The ultimate goal is to develop a low-cost effective system that can be used to predict the fatigue level for the entire driving population. One should notice that some drivers do not exhibit the type of behavior that indicates fatigue to a sufficient extent. Another issue to the algorithm developers is related to the eye nominal size estimation which varies depending on the driver. In [43], this value is obtained through forming a histogram of the eye opening degree for the last 2000 frames not exhibiting drowsiness. The most frequent mode value of the histogram is considered to be the nominal size. The PERCLOS is computed separately in both eyes and the final value is obtained as the mean of both [43].

Recent validation studies shown that the way drowsiness progressed under actual road conditions was different from what was observed in the simulator. Drowsiness came on significantly faster in the simulator if the assessment is based on the final fitness value of the prediction algorithm. While PERCLOS is the best human factors measure for fatigue detection, it appears challenging for computer vision researchers to provide a low-level image processing algorithm that would work for any subject, at any depth level, in presence of head movements, with any type of eyewear and in all operating driving scenarios.

Eye Monitoring Applications in Medicine

M. Stella Atkins, Marios Nicolaou, and Guang-Zhong Yang

15.1 Introduction

Eye trackers have been used for medical image perception in radiology since the 1960s, with pioneering work by Harold Kundel and Dr. Calvin Nodine from the Department of Radiology, University of Pennsylvania Health System, Philadelphia [348, 457]. Initial studies involved radiologists viewing 2D X-ray images of the chest or breast on photographic film displayed on large "light boxes". The aim of these early studies was to evaluate the use of image processing techniques, such as edge enhancement, to improve the visibility of lung tumours and hence reduce the error rates [351]. In order to make their experiments rigorous, Kundel and Nodine developed the technique of synthesising nodules photographically on otherwise normal chest images. The eye monitoring data provided understanding of the visual search process [346] and why errors occurred [349, 457]. These projects are motivated by the aim to provide eye-gaze driven decision support systems [123, 350, 456].

Modern studies using eye monitoring in radiology also determine the extent to which the user is distracted from visual search by using image navigation controls [13] to determine the effectiveness of different displays [341], and different interfaces for viewing 3D medical volume data [12, 164].

Eye trackers have been used more recently in surgery, especially for minimally invasive surgical training [357, 442], as well as for level of detail control for real-time computer graphics and virtual reality [475]. A number of exciting emerging applications are being developed to integrate eye monitoring information to provide gaze contingent control in surgery.

Models of Eye Trackers. Many different types of eyetrackers have been used for medical applications, as the technology has evolved over the decades. New devices such as the Tobii eye tracker offer fast calibration and user-friendly data interrogation. The system can be ported to busy medical professionals in their workplace. It works in the high luminance environment of operating rooms, as well as with users wearing glasses, so real medical studies can be performed. In this chapter, we will provide examples and illustrations in order to demonstrate the use of eye monitoring technologies in the medical field.

Chapter Organization. The remainder of this chapter is organised as follows. In Sect. 15.2 we provide a general overview of eye monitoring technologies in radiology with particular emphasis on the role of eye trackers as tools to enhance visual search (Sect. 15.2.2). Section 15.2.3 explores the visual search process when errors are made, classifying the errors according to the eye gaze fixations times. Eyetrackers have also been used to understand the nature of expertise in radiology (Sect. 15.2.4). All these results may be integrated to enable the future development of aids to image perception in radiology (Sect. 15.2.5).

Section 15.3 provides a general overview of eye monitoring in surgery, focussing on the role of eye monitoring in minimally invasive surgery (MIS) procedures. Robots can be used for MIS, introduced in Sect. 15.3.2. Simulators are commonly used for training MIS procedures in the laboratory (Sect. 15.3.4), where eye monitoring shows the differences between experts and novices. Eye Monitoring results may also be used to provide better visualisations for training. Section 15.3.5 shows how eye gaze tracking in the operating room provides insites into the surgeon's cognitive processes during different surgical steps. Section 15.3.6 provides other applications of eye gaze tracking in MIS. Section 15.3.7 presents future perspectives, including gaze contingent control in robotically-assisted surgery. Finally, Sect. 15.4 gives an overall summary of this chapter.

15.2 Eye Monitoring in Radiology

15.2.1 Overview

Radiologists' work often involves diagnosis, by searching for abnormalities in medical images. Traditionally, projection radiography images, formed by exposing an object such as the chest to X-rays and capturing the resultant "shadow" on 2D photographic films, were used. The films of medical images were displayed on large light boxes, but now filmless diagnosis using digital data displayed on computer monitors – softcopy diagnosis – is becoming more common. Most tomographic 3D image modalities such as Magnetic Resonance Imaging (MRI) and Computer Tomography (CT) now use digitised images displayed on computer monitors. However, most mammography screening still uses projection radiography onto films, although the films may be digitised and displayed on computer monitors, as there is evidence that digital viewing may enhance certain diagnoses [496]. There is also a proposal that chest X-rays for screening should be replaced with digitised computed tomography (CT) images viewed on computer monitors [241, 247].

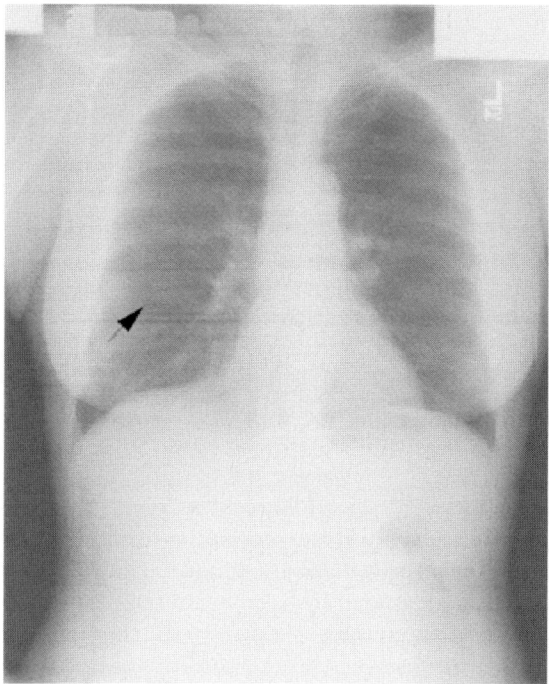

Fig. 15.1. Example of a subtle lesion on a chest X-ray. The *arrow* points to the subtle abnormality, which may be a lung nodule (image courtesy of H. Kundel)

Radiology training is designed to provide assumptions about normal medical images. Abnormalities usually exist as hyperintense regions in the images caused by tumours or nodules (solid lumps of tissue), and the process of diagnosis includes a visual search for these hyperintense regions, which we refer to as "lesions" . An example of a subtle lesion on a chest X-ray is given in Fig. 15.1. The arrow points to the lesion.

Yang et al. [696] proposed using eye monitoring in many applications in visual search tasks, of which radiology is an important example. In extension to this work, Atkins et al. showed that eye monitoring is a useful tool to help design and assess radiology workstation interaction techniques: different interfaces affect response time performance through disruption of the visual search process [13, 417]. Similarly, eye monitoring can help identify the amount of distraction using different image navigation tools for viewing 3D medical volume data such as CT and MRI [12].

Perception research can benefit clinical radiology by suggesting methods for reducing observer error, by providing objective standards for image quality, and by providing a scientific basis for image-technology evaluation. Krupinski from the Department of Radiology at the University of Arizona and others created the Medical Image Perception Society in 1997 to promote research in several aspects of modelling and evaluating computer-aided perception tools in radiology [339, 416].

The history of the use of eye trackers in radiology was summarised by Kundel in his keynote presentation at the 2004 SPIE medical imaging conference Reader error, object recognition, and visual search [345], and this chapter follows some of the important issues raised.

15.2.2 Visual Search

In early work, Kundel et al. observed that many lesions can be seen by peripheral vision and verified by foveal vision, but peripherally inconspicuous lesions must be found by scanning the fovea over the image [348]. Figure 15.2 shows two typical scan paths for the detection of lesions on chest X-rays, for the image with an inconspicuous lesion shown in Fig. 15.1. The resultant search pattern, as shown by eye-tracking recording, is influenced by both the patient's clinical history and the radiologist's experience [455].

Kundel and Nodine developed the concept of foveal fixation clusters of duration >100 ms, and used the centroid of a fixation cluster to describe the center of focal attention, surrounded by a 5° visual field for the region of interest (ROI) [457]. Using fixation clusters, the scan path can be alternatively viewed as a set of fixation clusters over the image. This is illustrated in Figs. 15.3 and 15.4, where the scan path for detection of tumours from mammogra-

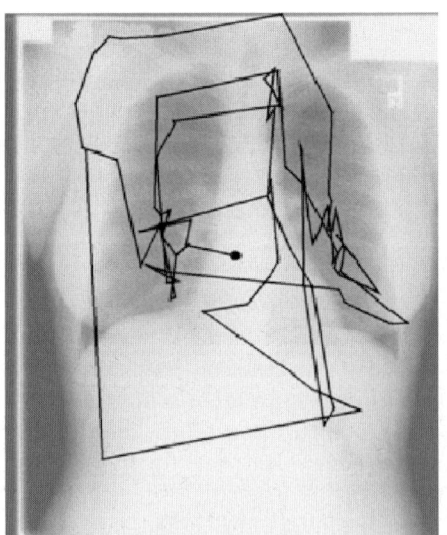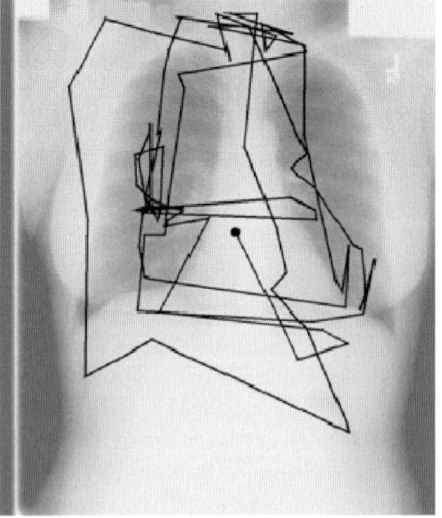

Fig. 15.2. Two radiologists scan paths for detection of lesions on chest X-ray with a subtle lesion in the lower lung corresponding to the arrow in Fig. 15.1 (image courtesy of H. Kundel)

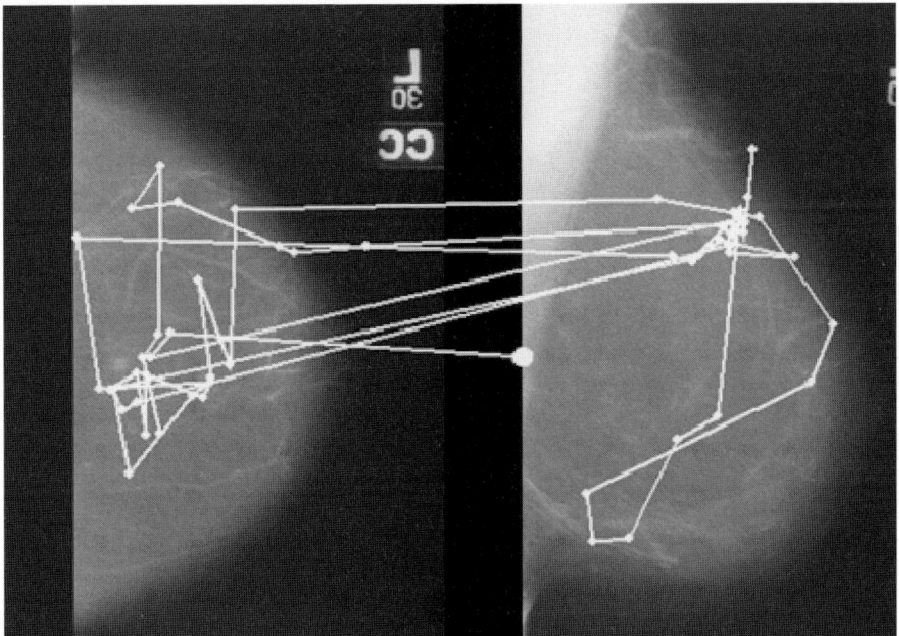

Fig. 15.3. Typical scan path for detection of tumours in mammography. One suspicious area is extensively fixated in each image. (image courtesy of H. Kundel)

phy images is shown in Fig. 15.3, and the corresponding fixation clusters are shown in 15.4. One suspicious area is extensively fixated in each image.

15.2.3 Errors and Visual Dwell Time

Unfortunately radiologists do make mistakes; some abnormalities may not be detected (false negative errors), and some hyper-intensity regions may be erroneously labelled as tumours (false positive errors). General estimates suggest that there is about a 20–30% miss or false negative rate in radiology, with a 2–15% false positive rate, a 10–20% inter-observer variation and 5–10% intra-observer variation [246]. Research in perception and visual search using eye tracking has led to important advances in understanding errors and revealing why some of these errors occur.

There are several types of errors, including technological, perceptual, and cognitive errors [345]. Technological errors arise when there is inadequate lesion conspicuity, resulting in a missed lesion. Perceptual errors arise when a suspicious object in the image is not recognized. Cognitive errors arise when a suspicious object has been identified, but has not been classified correctly i.e. a decision error has been made. Eye monitoring studies during the search for discrete lesions in projection chest X-ray images and mammograms have contributed to the understanding of perceptual and cognitive errors.

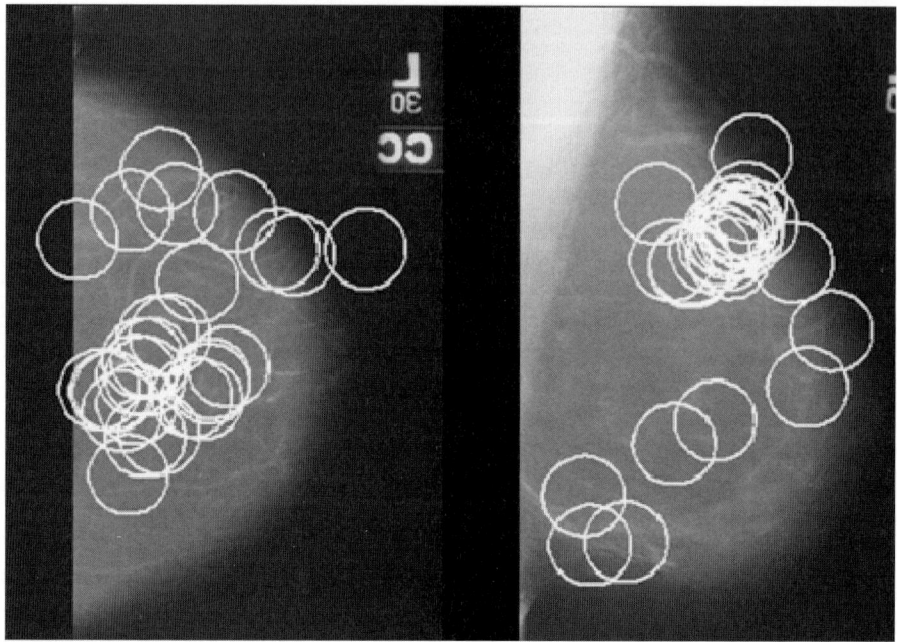

Fig. 15.4. Corresponding fixations and regions of interest from the scan path shown in Fig. 15.3. One suspicious area is extensively fixated in each image (image courtesy of H. Kundel)

15.2.3.1 Technological Errors

In addition to inadequate lesion conspicuity, noise in medical images can also limit the perception of contrast and detail. The higher luminance of the light boxes may permit enhanced perception of subtle abnormalities over computer monitor displays. Krupinski studied the accuracy of diagnoses using various computer monitors [337, 342]. She showed that detection performance was not affected significantly by display luminance, but the search behaviour was affected; total viewing time was longer with lower luminance displays, with more fixations on the normal images using lower luminance displays.

Models of human observers have been developed to determine the minimum conspicuity of lesions for the task specific detection of nodules in mammography and lungs [74]. Current research is extending the mathematical models that relate signal detection to physical descriptions of image signal and noise, to include realistic lesions and anatomical backgrounds. Recent research is focused on issues of conspicuity which may be used to determine the boundary between recognition and decision errors [347, 391, 412]. Tisdall and Atkins have developed a similar model for detection of simulated nodules in MR images [620]. All these models allow prediction of human performance given images of a certain quality, and aid greatly in suggesting performance limitations for detection of subtle lesions in noisy images, or on low contrast displays.

15.2.3.2 Perceptual Errors

Perceptual errors include visual search errors where the scanning path does not include the lesion or where there is a low peripheral conspicuity of lesion. Perceptual errors also include recognition errors which arise from lack of knowledge and experience with lesion and background properties.

15.2.3.3 Cognitive Errors

Cognitive errors arise when a suspicious object has been identified, but has not been classified correctly due to a decision error.

15.2.3.4 Classification of Errors: False Positive and False Negative

False positive errors can be explained by image noise and overlapping anatomic structures that often mimic disease entities. False negative errors arise when a lesion is missed. False negative errors are harder to understand, especially when missed lesions can be seen in retrospect [338].

Based on eye monitoring data of radiologists performing diagnoses, Kundel et al. [348, 454] proposed three types of false negative errors based on increasing fixation time on the missed lesion. These errors can be classified as perceptual errors (two kinds) or as cognitive errors. Perceptual errors arise as either a faulty visual search where there is failure to fixate the abnormal region (called a scanning error), or as faulty pattern recognition or failure to report a lesion that has been fixated for a short amount of time (called a recognition error). Cognitive errors arise as faulty decision making or failure to report a lesion that has been extensively fixated (called a decision error) [345]. These situations are depicted schematically in Fig. 15.5.

Perception research with eye monitoring shows that subjects typically spend fixations cumulating to about 1000 ms of dwell time on an object of interest [253]. This implies that radiologists might spend 1000 ms dwell time on a lesion to recognize it as an object of interest. This implies that if the subject has not fixated on the abnormality for 1000 ms, then there likely was a pattern recognition error (see Fig. 15.5(b)). Having recognized an object of interest, the radiologist then has to decide if the object is a lesion, or if it is benign. If an object of interest has been found with several fixations, and the wrong decision is made, we call it a decision error (see Fig. 15.5(c)).

Berbaum et al. reported [41] that the results of numerous studies indicated that the temporal threshold separating recognition and decision errors

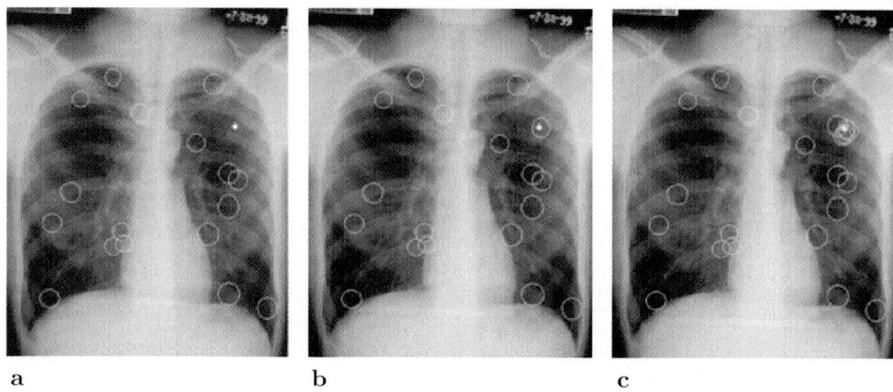

a   b   c

Fig. 15.5. a Scanning error where the target white dot is never foveated: **b** Recognition Error where the target is foveated but not recognized as being abnormal; and **c** Decision Error

in radiography should be about 1000 ms of dwell time for pulmonary nodules [348, 349, 454], for masses on mammograms [340, 458], for skeletal fractures [269] and for diverse chest abnormalities [42]. For these studies, the division between recognition error and decision error has been confirmed, based on the recorded dwell times and the decision outcome.

15.2.3.5 Comparative Visual Search

Comparative visual search involves two images, usually side-by-side, to complete a visual search task such as pattern matching two objects. This situation arises frequently in radiology when viewing a frontal and a side view of a chest X-ray, or in mammography when viewing a front and side view of a mammogram. Suspicious objects in mammography images will have different shapes depending on the view. Furthermore, features which may be hidden e.g. by the ribs, in one view, may be seen in another view. Only a few researchers, notably Pomplun et al. from the University of Toronto [278, 498–500], Atkins et al. from Simon Fraser University [13, 14] and ElHelw et al. from Imperial College [163] have studied eye scan path patterns during comparative visual search tasks, although picture matching tasks involving recognition of the same object within two pictures have been much studied [271]. Pomplun et al. were proposing answers to questions such as how are objects represented in memory, and how are objects recognised. Atkins et al. were discovering how an application design can benefit from the results of eye gaze tracking, and discuss how eye monitoring yields insights into the search and error processes [13]. All these researchers use eye scanpath and fixations patterns to examine the use of visual memory in visual perception during the search for a matching object (a target) in side-by-side images, and examine the impact of this on the visual search pattern and the errors.

Like other researchers such as [345] have reported for single images, these researchers found there are two phases of visual search – a first phase of search and comparison, and a second phase of detection and verification. The limitations of the working visual memory require that once a suspicious object has been seen, a second phase of verification is required. In general, their results suggest that oculomotor behaviour in comparative visual search is determined on the basis of cognitive economy, and that extra saccades are preferred to extra effort to remember multiple features in dense clusters across saccades.

The advantage of using scan paths of comparative visual search tasks rather than searching over a single image, is that the difference between recognition errors and decision errors can be much more easily observed.

A typical scan path involving a comparative visual search with a target is illustrated in Fig. 15.6. The subject has to view the two side-by-side images before deciding whether an object of interest (in this case, an artificial object drawn as a pair of circular discs one of which has a hollow fill) is a target or not. In this case, there is a target in the pentagon at the bottom center of each image. Full details of targets and the stimuli are given in [13]. The hypothesis was that the same assumptions about cumulated dwell times and decision errors for visual search in single images discussed above would

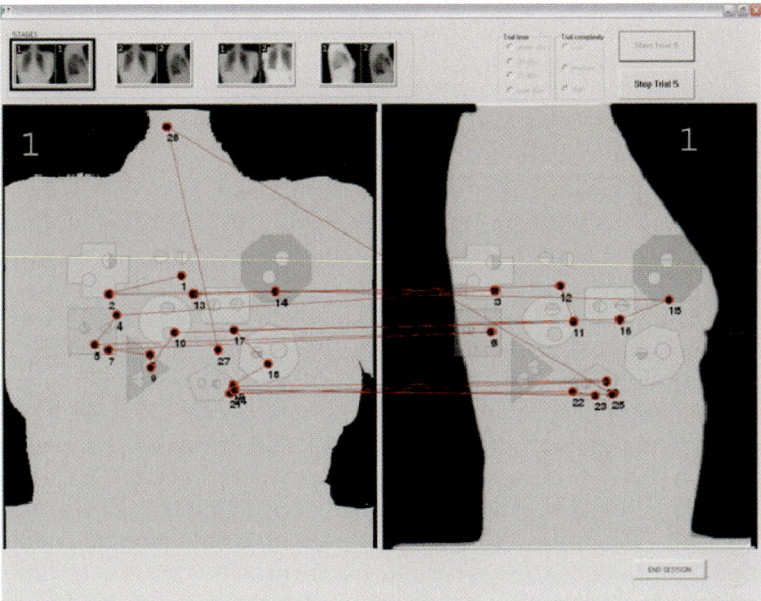

Fig. 15.6. Typical Scan path during comparative visual search with a target. The fixations show two phases corresponding to initial search (fixation 1–18) and to recognition (fixations 19–25)

hold for comparative visual search. The hypothesis was tested by performing several user studies with an artificial task searching for stimuli, both with radiologists and with laypersons. In these studies of comparative visual search where multiple fixations are required, a slightly longer dwell time on the targets of 1.25 secs was identified as the fixation duration to separate errors of recognition (dwell time <1250 msecs) from errors of decision-making (dwell time >1250 msecs) [14]. This likely arose because of the limitations of the human short term memory, and the nature of the targets which required checking several features before a decision could be made.

15.2.3.6 Satisfaction of Search

Another problem with diagnosis arises because of a phenomenon called satisfaction of search (SOS). This refers to the fact that detection of one radiographic abnormality may interfere with detection of other abnormalities on the same examination – in viewing radiographs there is a tendency to become satisfied after identifying the first abnormality, which may lead to failure to search for additional findings.

There are several theories that try to explain SOS. The theory of strategic termination of search is based on the minimization of the number of false-positive diagnoses. The adherence to a perceptual set promotes the inclination to discount findings of a different category from the ones found first. Another theory hypothesized a lack of attention to regions that did not contain contrast material on contrast-enhanced images, or that visual distractors, such as bright blood vessels, impaired the observer's ability to detect bright liver nodules on contrast enhanced spiral CT scans [673].

The traditional theories for SOS errors, indicating that the search is terminated after the discovery of an abnormality, have been discredited, after it was shown that observers continue to inspect images after an initial abnormality was reported. According to the Kundel-Nodine method of error classification, an unreported abnormality is assumed to have been recognized if the gaze dwell time on the abnormality exceeds some estimated value [348]. This theory is challenged by the fact the eyes can fall on an abnormality and the visual pattern may be analyzed as not corresponding to the abnormality. Using the same chest radiography cases, the "protocol analysis" method of collecting verbal protocols during the interpretation indicated a much higher proportion of search errors and much lower proportion of decision errors than with the Kundel-Nodine gaze dwell time method [42]. During the verbal protocol procedure, almost half of the observers acted like they created a kind of checklist on the spot to help them generate more systematic verbalization. This may explain the different indications for the source of errors the two methods provided, since some heuristic method of self-prompting, such as an automatic checklist might counteract SOS error.

Another experiment [41] demonstrated a SOS effect on test fractures with major, but not minor additional fractures. These experimental results suggest that detection of other fractures is inversely related to the severity of the detected fractures, thus eliminating the faulty scanning as a cause for SOS in musculoskeletal trauma. The average overall search time was significantly reduced by nine seconds when a fracture was added, regardless of the severity of the fracture.

To address SOS, the design of radiology workstations should evaluate different techniques, to allow for systematic viewing of images, like covering the study in a specific spatial manner (say starting in the upper left hand corner of the image and then proceeding in some set pattern over the remainder of the image) [530]. This simplistic solution may prove more like an impediment, since radiologists like to get a "gestalt" of the entire image/examination, and the radiologist's eye is often drawn from the very beginning to one or more findings (for reasons not always explicable, but very likely related to the context) [82]. A better solution may be to unobtrusively track the radiologist's eyes, providing active signalling of the unexplored image areas, and providing notification to prevent the completion of a report without the radiologist seeing all the images.

15.2.4 Understanding the Nature of Expertise in Radiology

The nature of expertise can be broadly divided into knowledge and experience. Perception research can look into clarifying situations when errors occur, differentiating between potential causes, such as ambiguous information rather than problems of perception, attention or decision making. Cognitive research, which draws on cognitive psychology, artificial intelligence, philosophy and linguistics, is primarily concerned with characterizing the knowledge of structures and cognitive processes underlying human performance. Using cognition the users of radiology workstations can be ranked as experts, intermediates and novices. Users can have different levels of medical competence, corresponding to different levels of training.

Some studies demonstrated specific and predictable differences between novices and experts in terms of perceptual search behaviours:

- experts tend to find lesions earlier in search than novices,
- experts have different fixation and dwell patterns,
- experts tend to have much more efficient search strategies than novices.

Thus is a radiologist somehow better at searching images than other clinicians or lay persons? Two separate studies compared radiologists and lay persons searching for hidden targets in complicated picture scenes [452]. One example of such tasks is finding Nina and Waldo in the "Where's Waldo" children's book. These tasks were similar to reading X-rays and searching for lesions because the targets of search were embedded in complicated backgrounds that also had to be searched and interpreted in order to understand the scene, much like tumours in chest radiographs. Overall, radiologists spent more time searching images for targets, and they also tended to fixate on the target much earlier in search than lay persons did. The radiologists scanning patterns suggested a more detailed visual search, which covered less of the image than the layperson's circumferential search pattern. The authors of this study conclude radiology expertise did not positively transfer to the limited art-testing experience. To support their conclusion, Nodine and Krupinski referred to the theories of Osgood, which relates the degree of transfer on the similarity of training and test situations, and Bass and Chiles, which suggest that performance on perceptual tests had little correlations with diagnostic accuracy in detecting pulmonary nodules [27, 473].

A similar experiment, this time involving lesion-detection, was conducted with 16 untrained and 16 trained subjects [246]. These results showed large intra-observer variability in both groups, with no significant difference in lesion detection between trained and untrained observers. Since the effect of training proved insignificant in this study, the author suggests that talent may be more important than training, and that detection may be a skill perhaps learned early in life.

15.2.5 Developing Aids to Image Perception

As mentioned in [349], about two thirds of the missed lesions receive prolonged perceptual attention and processing. One could improve these error rates by using perceptually-based feedback, where the eye-position of the radiologist is recorded when searching for lesions. The eye data could then be used to circle the areas associated with dwells longer than 1 second. Indeed, use of such visual feedback resulted in a 16% increase in observer performance for radiologists looking for pulmonary nodules, compared to just showing the image again without any dwell locations indicated [453]. It is possible that computer-aided detection schemes improve detection, for the same considerations perceptual feedback did: it focuses perceptual and attentional resources better than the unaided radiologist can do by himself.

Dempere-Marco et al. from Imperial College have shown how novices can be trained to follow experts visual assessment strategies in reading CT lung images, with significantly improved performance [123]. More recent work by Hu [270] shows how distributed lesions can be identified using hot spots of visual attention can be given as feedback, to improve nodule detection performance in CT lung images.

These results may influence the future design of decision support systems in radiology involving eye monitoring. Hence eye gaze tracking radiologists while they work, may lead to improved training of novices, and improved performance for experts.

15.3 Eye Gaze Tracking in Surgery

15.3.1 Introduction to Minimally Invasive Surgery (MIS)

Most eye monitoring work with respect to surgery has so far been carried out in the field of minimally invasive surgery (MIS). This type of surgery is typically performed using small incisions ($0.5 - 1.5$ cm long), tiny instruments and a video camera attached to fibre-optic viewing devices known as endoscopes. This specially adapted camera with complex optics relays video from the operative field onto a visual monitor which the surgeon can use to navigate the elongated instruments and perform the surgical procedure (see Fig. 15.7).

MIS has seen an exponential growth in popularity among patients and doctors in the last 20 years. With MIS, it is possible to perform major complex surgery through tiny incisions resulting in reduced post-operative complications such as pain or leg clots and enabling patients to recover faster. In fact, certain surgical procedures are currently being done almost exclusively using this technique such as the removal of the gallbladder (cholecystectomy) for gallstone disease.

From a technical point of view, the MIS endoscope can provide good visualisation and hence effective targeting of otherwise inaccessible areas. Minimised tissue handling significantly reduces local trauma and inflammation. Unfortunately, the biggest disadvantage of MIS is the sharp learning curve imposed on the surgeon. It is technically more demanding, requiring more concentration and operative time than open surgery. This increases the surgeon's fatigue and stress due to the remote interface of the laparoscopic technique [188]. The representation of the three-dimensional (3D) operative scene on a two-dimensional (2D) visual display can be extremely challenging especially for novices, yet there is a requirement to perform delicate manoeuvres in 3D using 2D visual cues. Furthermore, the misalignment of the visual and motor axes, as illustrated in Fig. 15.7, along with angular rotations and the increases the complexity of the task can result in a higher error rate and faster mental fatigue [227, 716].

The task is further complicated by the elongated instruments, the tip of which moves in an opposite direction to the handle, like a fulcrum . Tactile feedback is also significantly reduced [188,538,572]. In addition, magnification of the operative field by the endoscope requires that the surgeon mentally scales the instrument movements appropriately [228]. Research has shown that the learning curve for MIS is steeper than that of open surgery [593].

15.3.2 Robotically-assisted Surgery

Robotically-assisted surgery is a promising advance in MIS which allows stereoscopic visualisation of the operative field, with improved 3D accuracy,

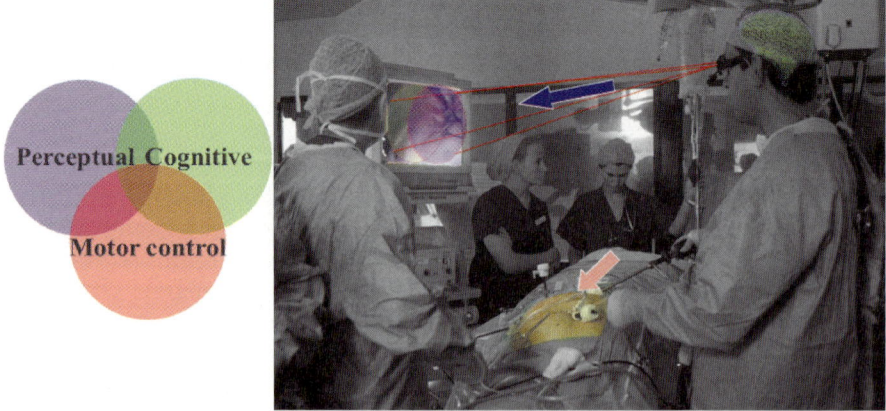

Fig. 15.7. A typical MIS setup. During the procedure, the surgeon is required to integrate knowledge of the anatomy and procedure (cognitive part) and dexterity (motor part) guided primarily by 2D visual cues from a display unit (perceptual part). The misalignment of the visual and motor axes (*arrows*) makes this process even more challenging

higher precision, tremor elimination, motion scaling and more natural instrument movement with up to seven degrees of freedom [85, 386]. By the end of the 90's, surgical robots were being used in several places throughout the world, and thousands of robotically-assisted procedures have now been performed [44]. Robotic assistance has empowered surgeons to perform otherwise impossible procedures such as in paediatrics or in-utero where instrument stability in limited space is vital. Robotic surgery is currently being increasingly used in a number of other MIS specialties such as orthopaedics, cardiothoracic and urology. The master-slave remote system has the additional advantage of allowing the remote manipulation of instruments, i.e., surgeon and patient being thousand miles apart (tele-surgery) [386].

15.3.3 Eye Monitoring in MIS

The visualisation of the operative field on a visual display unit either monoscopically (in the case of conventional MIS) or steroscopically (in the case of robotic surgery), permits the use of eye monitoring to study how the visual and motor axes are integrated, and offers insights on the subliminal processes that occur during training and how the lack of depth information is compensated. Furthermore, by using binocular eye tracking it is possible to extract 3D motion and tissue deformation from the surgeon's eye movements. This information could then be used to build motion stabilization systems (for beating organ surgery) or for registering and accurately displaying pre-operative data such as CT scans onto the operative field in order to supplement the surgeon's operative field [429].

15.3.4 Eye Monitoring in Lab-based MIS

The majority of the eye monitoring research in the field of MIS has focussed on identifying the differences in the visual behaviour between novices and experts performing the same laparoscopic task. Such knowledge is key to the understanding of how the motor learning process occurs and it elucidates the role of the human visual system on this process at various stages of training. Using a computer based virtual reality MIS simulator, Law et al., demonstrated a difference in the visual behaviour between experts and novices performing the same generic laparoscopic task . Novices required more frequent visualisation of the instrument that experts to perform the same task often tracking the instrument as it moved in the simulated operative field [357]. In a separate experiment, Kocak et al., used a saccadometer to assesses eye movement metrics in laparoscopists of varying experience performing the same simulated MIS tasks. They found a significant inverse relationship between the level of experience and eye movement metrics: saccadic rate and standardises peak velocity suggesting that experienced subjects required fewer fixations than less experienced surgeons to perform the same task [331]. The reason for these differences was not entirely clear from these experiments.

The research group at Imperial College, London, carried out a number of experiments using laparoscopic novices and experts performing inanimate and animate MIS. The aim of this work was to identify any differences in visual behaviour between these two groups that could explain the superior performance by the experts and shed light about how monoscopic depth perception develops with time. Using a desktop-based video eye tracker setup shown in Fig. 15.8, Nicolaou et al. observed the visual behaviour in the context of operative field visualised by the surgeon, to further delineate the differential behaviour between novices and experts.

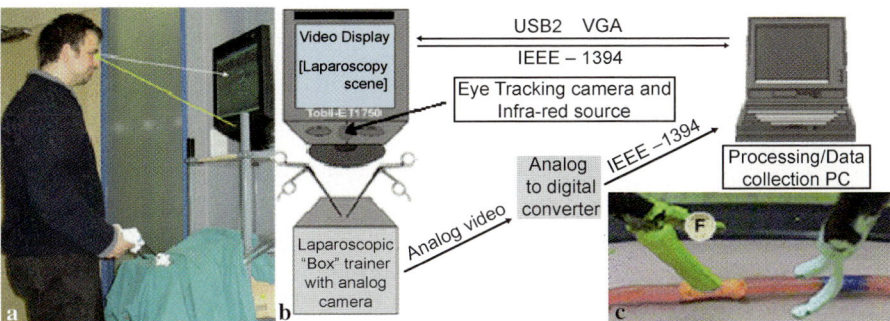

Fig. 15.8. a Experimental setup of the laparoscopic trainer used for the eye monitoring study, and **b** a schematic diagram showing the main data/command flow during real-time eye monitoring with a remote eye tracker. **c** Shows an example of the laparoscopic view of the setup with foveation points super-imposed

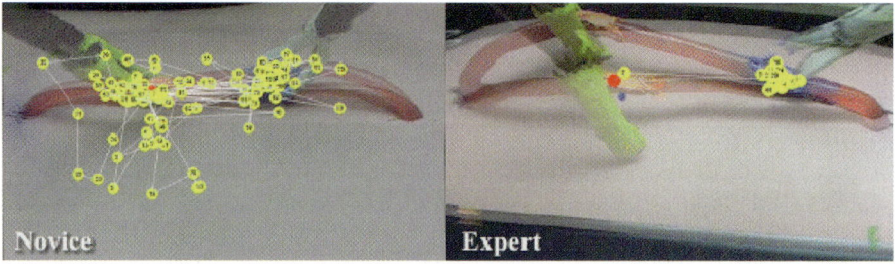

Fig. 15.9. Time-lapse gazeplot of a novice and an expert performing the laparo-scopic task of grasping and cutting a simulated blood vessel. Note the concentration of fixations on the two important targets as demonstrated by the expert

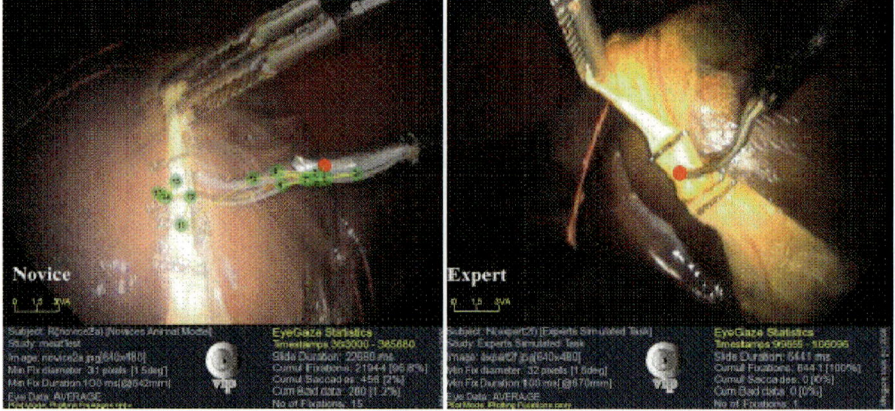

Fig. 15.10. A novice and an expert performing the same subtask of cutting the cystic duct in a porcine training model. Note the difference in visual behaviour with an "instrument tracking" in the novice and minimal fixational variation in the expert

As shown in Fig. 15.9, there is a striking difference between the visual behaviour of a novice and an expert laparoscopist performing a grasping and cutting a blood vessel exercise. The novice requires a far greater number of fixations and has to rely almost exclusively on a visual feedback mechanism to guide the instrument into position. In contrast, all of the expert's fixations are concentrated on the two important points on the blood vessel: the grasping point (left) and the point to be transacted on the right. There is no tool tracking behaviour during the task, yet the instruments are accurately positioned with a great economy of movement [442].

The tool-tracking behaviour demonstrated by the novices may represent a compensation mechanism for the lack of stereoscopic perception of the operative field as well as reduced tactile feedback and proprioception from the surgical instruments. It is suggested that the continuous visualisation of

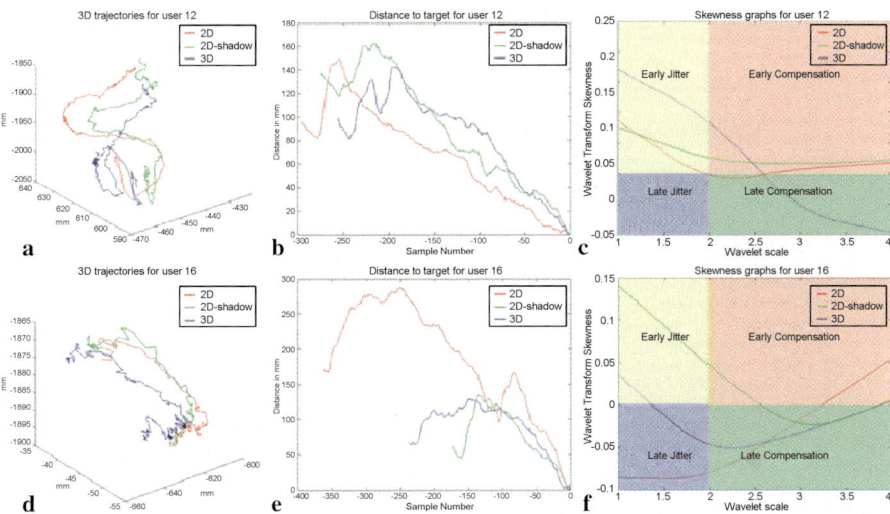

Fig. 15.11a–f. Graph to show how the instrument trajectory is affected by the method of visualisation. **a** and **d** show the 3D trajectory of two users performing an identical laparoscopic manoeuvre. **b** and **e** show a wavelet 2D representation of the trajectories when approaching the target with the three modes of visualisation. **c** and **f** show the wavelet extrema skewness graphs for these trajectories for each method of visualisation (*red* = 2D, *green* = 2D + shadow, *blue* = 3D)

the moving instrument acts a motor-visual feedback loop during task execution when input from other sensory modalities is minimal [440]. Nicolaou et al., were also able to further identify how the motor behaviour compensates from the depthless visualisation of the operative field. In an experiment using three different methods of visualisation, two distinct motor behaviours: instrument jitter (high frequency) and compensation (low frequency) were identified during a simple laparoscopic task (see Fig. 15.11). Figure 15.11(a and d) demonstrates the instrument trajectories for two users performing the same laparoscopic task under 3 different methods of visualisation: conventional laparoscopy (2D), laparoscopy with natural shadow (2D + shadow) and pseudo-stereoscopic visualisation (3D). The frequency of the trajectories are then represented in 2D using a wavelet framework (Fig. 15.11b and e) and the temporal motor behaviour can then be plotted in its component frequencies: jitter (high frequency) and compensation (low frequency) (Fig. 15.11c and f). The derived jittering and motor compensation during different stages of the task may be indicative of the user's ability for extracting depth from 2D and visual-motor coupling. With training and experience, depth extraction from monoscopic cues becomes more efficient, and the motor and sensory systems become more finely tuned. At this point the constant motor-visual feedback loop shown by the novices is no longer important [441].

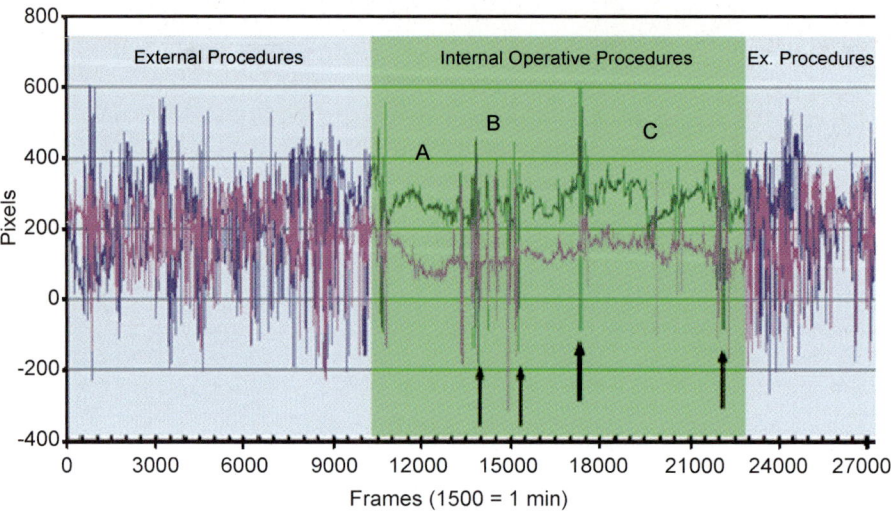

Fig. 15.12. A trace representing saccadic amplitudes during the various steps of a laparoscopic cholecystectomy procedure: (*A*) dissection of Callot's triangle, (*B*) cutting the cystic duct, (*C*) removal of gallbladder with diathermy. Note the dramatic variation in amplitude during the various steps of the procedure, and the formation of a catalogue of eye movements (Image courtesy of Mr Adam James)

15.3.5 Eye Monitoring in the Operating Room

Using wearable eye trackers, James et al. recorded a surgeon performing an actual MIS procedure in the operating room (see Fig. 15.7). By observing the saccadic amplitudes during a multiple step procedure such as the gallbladder removal (cholesystectomy) procedure they developed an eye movement catalogue for the various subtasks based on the variation in the eye metrics (see Fig. 15.12) [608]. Using the surgeon's fixational points, James et al. were also able to identify anatomical primitives during a procedure and suggested that their purpose is to serve as a feature map for the surgeon to guide the successful execution of each individual surgical step and subtask [290]. Figure 15.12 demonstrates the visual behaviour during the three principal steps of the laparoscopic cholecystectomy procedure; dissection of the Callot's triangle (A), cutting the cystic duct (B) and removal of the gall bladder with diathermy (C). Further to this, the arrows in Fig. 15.12 highlight instrument transition periods in the internal operative section of the gaze signal. It is apparent that basic characteristics of the eye movement are different, and episodes of focused attention with prolonged fixation and rapid pursuit movements are purposeful but may not be consciously registered by the surgeon.

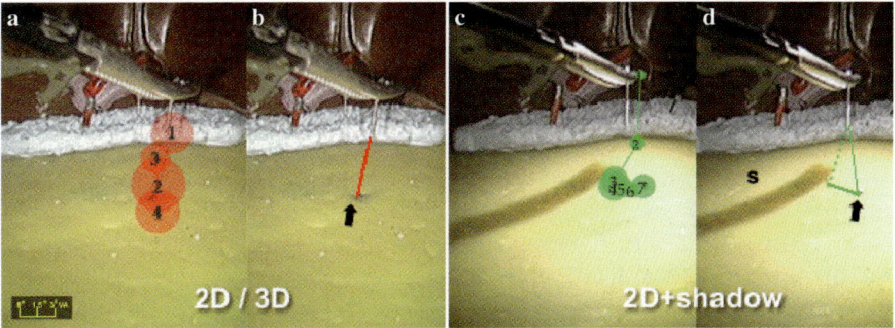

Fig. 15.13a–d. Eye monitoring used as a tool for assessing visual behaviour in three different methods of visualisation. Distinct visual behaviour can be seen from the gaze plots in the shadow casting method **c** as compared to conventional/pseudo3D ones. Unlike the direct gaze between the tooltip and the target (*arrow*) **b** a visual triangulation pattern between the instrument, its shadow (*S*) and the final target is observed in the shadow casting method **d**

15.3.6 Other Applications of Eye Monitoring in MIS

Eye monitoring has also been used to assess the effect of varying the method of MIS visualisation on visual behaviour. El Helw et al., using eye trackers, demonstrated the importance of specular highlights for improving the photorealism of graphic based MIS simulations [163].

Nicolaou et al. at Imperial College, assessed novices performing a laparoscopic task in a box trainer model using three different methods of MIS visualisation: conventional, a pseudo-stereoscopic method of visualisation based on the delivery of alternating disparate images using polarizing glasses and a method of visualisation that cast a shadow in the operative field. They found that task execution was significantly better with the shadow casting method. By considering the eye monitoring data recorded during the task, they demonstrated the differential visual behaviour between the shadow casting method and the other two methods of visualisation. In the former, subjects demonstrated a visual triangulation behaviour which was absent in the other two methods (see Fig. 15.13). They suggested that this behaviour enabled subjects to extract depth from monoscopic cues more efficiently than any of the other two methods leading to a more precise task execution [441].

15.3.7 Future Perspectives: Gaze Contingent Control in Robotically-assisted Surgery

As already discussed, robotically assisted surgery already offers further advancements in the field of MIS empowering surgeons to perform more complicated procedures with higher precision and greater control. By taking advantage of the surgeon's stereoscopic vision and by detecting vergence movements by binocular eye monitoring, Mylonas et al. were able to recover 3D information from the operative field [428]. Such 3D information could be used to either register medical images on important anatomical landmarks, thus enabling their superimposition of these scans on the appropriate anatomical points augmenting the surgeons field of view or to detect and track moving organs during a procedure. An example application for this is beating heart surgery where the surgeon fixates on the beating pericardial surface of the heart and tracks it with his eyes . By tracking the surgeon's eye movements, the 3D position of the fixation points can be accurately computed and used to mobilize the visualisation camera in synchrony with the heart thus creating an apparently stationary heart. Although much of this work is still in preliminary stages, important advances have already been made in this area showing that such an approach may become feasible in the future [429].

15.4 Discussion and Conclusions

The primary goal of a radiologist is to produce an accurate diagnostic report in the most efficient manner possible. Identifying various anatomic structures and pathological findings requires mental effort for the visual search, analysis and interpretation of increasingly complex radiological data. In order to correctly classify a pattern within an image as abnormal, the radiologists often times have to gather and register complementary information from several related images. The radiologist has a working memory filled by the task. If lesions are detected, their size must be measured. The information described above must be stored in the working memory, to be available for quick access. Accessing the workstation controls also requires access to the same cognitive resources, which usually flushes the diagnostic information from the short time memory. Such disruptions and interruption can not only increase the interpretation time, but can also affect the accuracy of interpretation. Eye gaze studies can aid in identifying distracting issues related to image navigation, and also to help determine a classification of errors.

Minimally invasive and robotically assisted surgery represents the greatest revolution of surgical technique in recent years. It empowers surgeons to perform complex procedures using small incisions with a high level of accuracy. Unfortunately, the foreign nature of MIS requires a very steep learning curve for the surgeons. The static frame of reference of the operative field on a computer monitor enables researchers to use eye monitoring to further study and understand this learning process. Eye monitoring research has shown important behavioral differences between novice and expert MIS surgeons and has shed light on the underlying perceptual processes that occur during extraction of depth from monoscopic cues. Eye tracking research has also illustrated how this type of monoscopic visual perception can affect the surgeon's motor behaviour during surgery. A repertoire of eye movements has also been identified during a multi-stage surgical procedure serving as "feature maps" guiding the surgeon to a successful execution of each step. Knowledge of such feature maps, and understanding of the perceptual challenges of MIS training, can aid the development of more efficient training programmes as well as help develop and evaluate newer technologies which can enhance the visualisation of the operative field. In the future, coupling eye trackers to surgical robots promises to further empower surgeons to perform more complex procedures such beating heart surgery with a higher accuracy and control.

Part VI

Eye Monitoring in Information Retrieval
and Interface Design Assessment

Eye Monitoring in Online Search

Laura Granka, Matthew Feusner, and Lori Lorigo

16.1 Introduction

The Web introduces both new opportunities and challenges for eyetracking research. Recent technological advances in hardware and software have greatly contributed to the use of eye monitoring (or eye gaze tracking) for the analysis of online applications. Eye Monitoring is now being used to offer insights into homepage marketing, advertising, reading of online news, and interpreting user behavior in online search environments. This chapter will assess some of the key issues surrounding data collection and analysis in online contexts. We will then specifically address the use of eyetracking for online search in more detail.

Eye Monitoring is becoming a popular tool for understanding user behavior in a number of computer and web based contexts, ranging from how viewers read online news stories [579], how net surfers respond to banner ads and other advertisements [40, 141], and how users interact with web displays and menus [200]. Eye Monitoring has been used only since 2003 to investigate online information retrieval, which is the context the rest of this chapter will focus [206, 545].

New technological advances allow us to more effectively answer questions about how a user scans and searches for online content, as well as combine this behavioral data with implicit forms of feedback, such as server log data. We will then discuss how eyetracking is an effective tool to augment standard analysis methods for studying information retrieval.

16.2 Methodology

16.2.1 Eye Monitoring Hardware

Eye Monitoring lets a researcher understand what a user is looking at while performing a task. There are many different methods for tracking eye movements, including video eye trackers (a type most suited for online and usability contexts), magnetic coil systems (placed directly on a subject's eye and more often used for medical research), and electro-oculography (EOG) recordings (based on muscular movements surrounding the eye). Despite some advantages in recording quality, coils and EOG systems can be very uncomfortable and invasive for the viewer. Video eye trackers can be slower (sampling at about 30-60 Hz) and less accurate due to the need to capture and process each frame, and lose tracking during blinks. However, they are more natural for the participant, and their accuracy is improving.

Ease of calibration also must be considered when selecting eye monitoring hardware, especially for use in online applications. If the process is too long or involved, it could be disruptive to a participant's perception of routine web browsing. Various automatic calibration systems are helpful for adding transparency to the use of eye monitoring in online contexts, but such options should be considered carefully before trading calibration accuracy for convenience.

Eye Monitoring experiments traditionally have been very costly and time consuming, though newer technology and improvements in hardware are opening new avenues and research directions. Hardware platforms such as the Tobii 1750 and x50 [623], and ASL R6 VHT [9] are now equipped with automated calibration, which is ideal for industry practitioners who use eye monitoring on a regular basis and need a quick and simple set-up.

16.2.2 Interpreting Eye Movements

Since its emergence as a popular research method, much progress has been made towards accurately interpreting the different eye movements that are captured by eyetrackers. There are now commonly accepted interpretation standards for ocular indices. A careful understanding of these is necessary to accurately interpret the collected eye monitoring data. Several key variables have emerged as significant indicators of ocular behaviors, namely fixations, saccades, pupil dilation, and scanpaths [405, 510].

Fixations/Fixation Duration

Though there are many different approaches to identifying fixations [549], a fixation is generally defined as a spatially stable gaze lasting for approximately 200–300 milliseconds, during which visual attention is directed to a specific area of the visual display. Fixations traditionally are understood to be indicative of where a viewer's attention is directed, and represent the instances in which information acquisition and processing are able to occur [510]. Based on existing literature, a very high correlation has been found between the display item being fixated on and that being thought about. Similarly, there is a close connection between the amount of time spent fixating on certain items and the degree of cognitive processing [308, 510]. Eye fixations are the most informative metric for evaluating information processing primarily because other indices, such as saccades, occur too quickly for the viewer to absorb new information [510].

At least three processes occur during an eye fixation: encoding of a visual stimulus, sampling of the peripheral field, and planning for the next saccade [653]. Research has shown that information complexity, task complexity, and familiarity of visual display will influence fixation duration [145]. The length of eye fixations is also largely dependent on a user's task. The average fixation duration during silent reading is approximately 225 milliseconds, while other tasks, including typing, scene perception, and music reading approach averages of 300–400 milliseconds.

From an eye monitoring perspective, information retrieval seems to encompass both a visual search scenario as well as reading, so it is expected that the average fixation duration will fall within the range of these two groups. The differences in fixation length can be attributed to the time required to absorb necessary information, and the speed at which new information should be absorbed. It is necessary for the eye to move rapidly during reading, while in visual search and scene viewing, it is less imperative that the eye quickly scans the entire scene, but rather that the user can absorb key information from certain regions.

Saccades

Saccades are the continuous and rapid movements of eye gazes between fixation points. They are extremely rapid, often only 40–50 milliseconds, and can have velocities approaching 500 degrees per second. No information is acquired by the viewer during a saccade due to the unstable image on the retina during eye movements and other biological factors. This lapse of information intake is traditionally referred to as saccadic suppression, but because saccades represent such short time intervals, individuals are unaware of these breaks in information perception [510]. Saccadic movement has been analyzed extensively in the context of reading. The research has identified saccadic behaviors including regressions (re-reading content) and word skipping, which are also important to consider in online contexts [510].

Pupil Dilation

Pupil dilation is a measure that is typically used to indicate an individual's arousal or interest in the viewed content matter, with a larger diameter reflecting greater arousal [145, 251, 510]. Studies can compare the average pupil dilation that occurs in a specific area of interest (AOI) with the average pupil dilation of the entire site to gain insight into how users might cognitively understand or process the various content matter [250].

Scanpath

A scanpath encompasses the entire sequence of fixations and saccades, which can present the pattern of eye movement across the visual scene. User scanpath behavior provides insight into how a user navigates through visual content. Studies analyzing properties specific to scanpath movement have enabled researchers to create a more comprehensive understanding of the entire behavioral processes during a visual or online search session [244, 307, 379]. Existing literature suggests that scanpath movement is not random, but is highly related to a viewer's frame of mind, expectations, and purpose [699]. Several researchers have explored the sequence of eye movements more closely using sequencing alignment algorithms, which will be discussed further on in the chapter [244, 489].

Other Eye Movements

Smooth pursuit is a type of eye movement where a fixating eye smoothly drifts in order to follow a moving target [361]. This type of viewing does not produce saccades, and generally is not analyzed in online eye monitoring for two reasons. First, web page content is almost entirely static; text and images generally are not animated. The emergence of embedded videos and Flash animations may prompt further study, but to date studies have focused on uncomplicated static scenes. Secondly, smooth pursuit tracking only occurs due to moving page elements, such as when a user decides to scroll. However, time spent scrolling is assumed to be low compared to total time viewing a static scene, so a user's time spent in smooth pursuit eye movements is assumed to be negligible and omitted. Vergence movements can occur when the eyes move inwards or outwards together in order to refocus at a new distance. However, viewing distance is almost always stable during computer use, so vergence movements are also assumed to be negligible.

Area of Interest (AOI)

Often, a researcher is interested in analyzing eye movements with respect to specific regions of a scene, or webpage, such as ads, images, and primary content areas. For this purpose, metrics such as number of fixations, fixation duration, or even pupil dilation, are often reported per each area of interest (AOI), also known as a "lookzone" in some software applications. Classifying unique regions of interest on a page lets a researcher make comparisons between, or even draw conclusions about specific types of content, such as whether there are differences in eye movements when viewing advertising versus standard content. Most eye monitoring software used in online applications comes with basic tools to map gaze coordinates to lookzones, though this functionality is often very cumbersome and ill equipped for extensive investigations on large numbers of pages. We discuss several software options that address some of these concerns later in this paper.

16.2.3 Eye Monitoring Methodology

Qualitative Analyses

In addition to formal controlled eye monitoring experiments, industry practitioners and user experience researchers are now developing qualitative approaches to using eye monitoring in day-to-day evaluations of web-based products [207]. In industry work, there is value in simply viewing the pattern and path that a user takes when interacting with a new product or design. For usability studies, Granka and Rodden [207] discussed the benefits of using the eyetracker, not necessarily for the in-depth data analysis, but because of the immense value in enabling product teams to view the real-time projection of the user's eye gaze during the completion of tasks in a usability study. Qualitative analysis in eye monitoring stands in stark contrast to tightly controlled experimental design where much care must be taken to conduct appropriate statistical analyses. Because eye monitoring data is so complex and multi-leveled, to fully account for all random and fixed effects in most eye monitoring experiments (e.g., assessing how task type and gender impact fixation duration and pupil dilation), and to generate appropriate estimates of error based on the nested data structure, three- and four-level linear mixed models are often the ideal solution to accurately analyze the data, particularly if the researcher is interested in differences between conditions, task types, and other related metrics [407].

Quantitative Analysis

For quantitative eye monitoring analysis, Goldberg and Kotval [199] summarize methods for analyzing eye monitoring data for computer-based usability studies, looking specifically at the use of menus and toolbars. Many of their suggested analyses, such as assessing the length of a scanpath, determining the fixation density in a region, and comparing fixation durations, are now regularly used metrics in recent web-based eye monitoring work. These numerical metrics (numbers, durations, and lengths of fixations and saccades) are analyzed using standard statistical methods.

Another type of data analysis that Goldberg and Kotval discuss [199] is fixation clustering. Fixation clustering is particularly important for identifying the division of attention between different segments of a web page. Such analysis can be done visually with a fixation map [685], but also analytically using spatial clustering algorithms [555].

Sequence Alignment Methods

Sequence alignment is an interesting technique for measuring the similarity between two scanpaths over the same stimulus. Scanpaths are interpreted as sequences of fixations, which can be compared using a well-known optimal pairwise alignment algorithm [363, 434]. The magnitude of dissimilarity between the scanpaths is computed by calculating the smallest combination of insertions, deletions, and replacements that need to be made in order to transform one sequence into the other. Each edit (insertion, deletion, substitution) has an associated cost, and the sum of costs is called the edit distance or Levenshtein distance. Josephson and Holmes [307] have used this pairwise matching method to compare scanpaths across different structural elements on a webpage. Another research effort used sequence alignment to group scanning behaviors on popular websites [481]. As an extension to sequence alignment, multiple sequence alignment is now being used to analyze and aggregate populations of viewers [244,672]. Multiple sequence alignment algorithms generally build on pairwise alignment by repeatedly aligning sequences until one sequence is left that contains elements of all scans in the group. Another estimate of aggregate vewing behavior is the "center star" sequence, the sequence already existing in the dataset that minimizes the distance to all other sequences. The advantage of a multiple sequence alignment is that it can actually combine elements of two or more individual sequences while the center star clearly cannot.

One unresolved issue regarding both pairwise and multiple sequence alignment algorithms is the choice of algorithm parameters. The first parameters needed for the Levenshtein distance are the insertion, deletion, and substitution costs. By default, unit values are used to equate total cost with total number of edits. However, in the two-dimensional space on a computer monitor, it makes sense to assign a substitution cost according to the distance in pixels. Consequently, insertion and deletion costs must be assigned in the same units. When substituting fixations based on their distance, it makes sense that an insertion or deletion edit would also involve an imaginary associated saccade. Therefore it also would make sense to use the average saccade length in pixels for both the insertion and deletion score. Other schemes for choosing parameters may be purely conceptual. For example, fixations can be labeled by the category of region where they occur, and sequence alignment parameters can be chosen based on similarity of categories [307].

Probabilistic Models

Probability values computed from eye monitoring data can be incorporated into many different descriptive models that represent eye movements. After configuring such a probabilistic model with recorded data like scanpaths, the model can be used to evaluate other scanpaths and eye movements. While use of these models is often more complex than sequence alignment methods [212] it can be more powerful as well. A transition matrix is one simple but effective way to model free-viewing eye movements as in visual search tasks [501]. Probabilities of movement between different locations are computed, and then interesting transitions are selected and evaluated. For example, horizontal, vertical, or diagonal movements between different elements indicate varying levels of effectiveness during reading or scanning tasks.

Markovian models compose one class of probabilistic models that incorporates varying levels of probabilities. For example, a first-order Markovian transition matrix would capture not only the probability of movement from each element to the next, but from the previous element or state as well. Higher order models capture transitions further and further into the past. A Hidden Markov Model can be used when states between transitions are not known or explicit. The complexity captured in the depth of order in a Hidden Markov Model used to model different eye movement strategies can indeed affect the results [546]. However, in print media, it has been shown that even a first-order Markov model can capture a great deal of interesting information about a scanpath [495].

Error Analysis

When conducting research in online environments, it is important to understand the degree of accuracy with which to interpret your findings. Currently the most popular monitor size and resolution is a 17″ monitor set to display 1024 × 768 pixels. At standard ergonomic viewing distances between 24 and 40 inches [474], eye monitoring hardware error can be surprisingly large. Many eye trackers claim to be accurate to within 1 degree of visual angle, which corresponds to an on-screen error of between 32 and 53 pixels (Table 16.1). A single line of text on the web is often around 10 pixels high, so in a normal viewing setting with a popular eye tracker it is unlikely that the researcher can be certain which line of text is being viewed. Some studies have compensated for this problem by using large monitors [639], but the ecological validity is a concern if the text or monitor size is drastically larger than what a user is typically sees.

The distribution of error also is not usually described for popular commercial eye trackers. What fraction of samples is within 0.1 degrees of the true value? In the absence of a complete study, it is worth noting that while one degree of error seems small, it can amount to a large number of pixels, so it is hard to draw conclusions about the specific words a user fixates. It also is worth noting that the number of pixels corresponding to 1 degree of visual angle on a flat monitor begins to increase dramatically around 30 degrees from center. However, even on a 21″ monitor at a 24″ viewing distance, the screen only spans $+/- 19.3$ degrees of the visual field.

Thus, while very robust, eye monitoring is not always the most suitable method for many questions about online analysis. While it can accurately portray a user's typical course of action on a search results page, it is costly and only possible for studies where the user is physically present in front of the equipment. Calibrating and monitoring the study also add overhead, meaning that analyzing and aggregating all user data from one study requires more work than even multi-session log analyses.

Table 16.1. Pixels per degree of visual angle for popular monitor sizes, screen resolutions, and viewing distances

Diagonal Screen Size	24″ viewing distance				40″ viewing distance			
	1024 × 768	1280 × 1024	1600 × 1200		1024 × 768	1280 × 1024	1600 × 1200
14″	38.98	48.73	60.91		64.24	80.31	100.38
15″	36.48	45.60	57.00		60.02	75.02	93.78
17″	32.37	40.46	50.57		53.07	66.33	82.92
19″	29.14	36.42	45.53		47.59	59.49	74.36
20″	27.77	34.72	43.40		45.27	56.59	70.73
21″	26.54	33.18	41.47		43.17	53.96	67.45

Software and Data Visualization

Software for visualizing and analyzing eye monitoring data is either custom-made or packaged with eye monitoring hardware. Heatmaps, such as those shown in Fig. 16.2, in the following section, are a popular and effective means to show aggregate viewing behavior on a given web page, and can be found in some of today's latest commercial eye monitoring software [622]. Some other commercial options, which are often bundled with hardware, include GazeTracker [355], and iView X [571], and each of these packages varies in the features they provide. There are efforts underway to standardize various eye monitoring data formats and protocols [28], and there are some more general analysis programs that are freely available, for example WinPhaser [260], many biological sequence alignment applications, and any number of free string edit tools and functions available on the web. However, the availability of, and functionality afforded by software for analysis should be considered before beginning an eye monitoring study.

One key limitation of current commercial eye monitoring software is that there is little support for analyzing and discerning patterns in the scanpaths themselves. At present, eye monitoring data is typically visualized either as an aggregate view of what users looked at, as in heatmaps, or as individual scanpaths, as in Figs. 16.3 and 16.4. While both of these techniques offer great insights over traditional log analysis, they lack the ability to convey what an "average" sequence looks like, how typical a given path is compared to another, or where common subsequences lie. One sequence analysis program, eyePatterns, announced its offering for free download [672], but is unavailable as of this writing. Pellacini et al. [489] also describe a work in progress for visualizing paths in the context of a collection of paths, and Hembrooke et al. [244] discuss methods for extracting such an "average" path. Scanpath analysis provides additional information about the process involved in a user's interaction with a search engine result page (SERP), though at present is often too cumbersome due to the scarcity of existing tools and the complex nature of the sequential data. Scanpath analysis is particularly useful in the context of information retrieval, where the path of eye movements offers direct insight into a user's decision making.

16.3 Eye Monitoring in Information Retrieval

The greatest strength of eye monitoring in the context of information retrieval is its ability to highlight what a user is looking at before selecting a document, depicting the process whereby individuals view the results presented to them on a SERP. In contrast to traditional methods that study information retrieval behavior, eye monitoring is better suited to assess the actual behaviors that users employ when reading and making decisions about which documents to select. Passive Eye Monitoring will enable us to detect

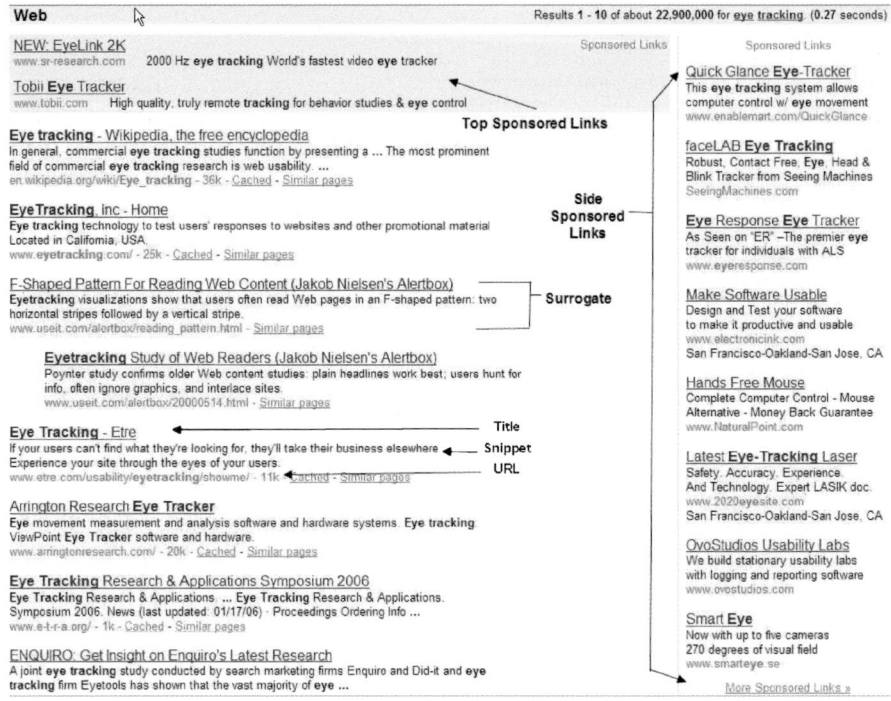

Fig. 16.1. Detailed breakdown of search engine result page (SERP) content

what the searcher is looking at and reading before actually selecting an online document, while other methods for information retrieval analysis, such as server log data, are all based on the single result that a user has selected.

Figure 16.1 shows a sample Google result page that illustrates the terminology that we will use throughout the rest of the chapter. Search engines typically return 10–20 results per page. We refer to each of these shortened results as the surrogate, which is comprised of a title, snippet, and URL. Later on in the chapter, we will discuss additional components to the page such as sponsored links, which are also highlighted in the example below.

16.3.1 Relevant Information Retrieval Research

The information retrieval literature can largely be split into two distinct classes of research: those that investigate the effectiveness and quality of the search engine, and those that evaluate the behaviors that users employ when engaging in an online search. While the eye monitoring research discussed here falls into the latter, the findings also provide direct recommendations for search quality, particularly those that use log data to measure result relevance [143, 298].

A majority of current information retrieval evaluations focus on evaluating the functionality of the actual search system in terms of its efficiency, precision, and ranking of results. These retrieval evaluations focus on assessing the precision and recall of the selected documents as a measure of the information retrieval system itself. Somewhat less focus has been placed on the user – towards addressing the cognitive behaviors involved in evaluating online search results [245, 267, 576]. Our primary objective with eye monitoring is to examine how searchers assess, evaluate, and selectively deem an online document to be relevant.

Capturing eye movements during search will address pre-click user behaviors and provide a comprehensive account of how a user views and selects an online document. Analyzing eye movements will enable us to recognize what a searcher reads, skips, or scans, and it is these metrics that offer insight into the use of current information retrieval systems.

Effectively using the ocular indices previously described will enable us to determine the content that searchers focus on, for how long, and in what order, enabling us to examine some of the assumptions that have limited the traditional evaluations of information retrieval systems [300]. As will be discussed in the last section of this paper, relating eye movement behavior with other measures, such as clickstream or mouse movement data, can provide an even more comprehensive picture of the process through which online information acquisition actually occurs.

In sum, the common trend throughout many of the existing user studies is that they look primarily at the outcome measures of user behavior, quantified in such ways as query wording, time spent searching, and the rank of the selected document. These measures are produced after the subject has selected a document. Eye Monitoring can be used to supplement these metrics, enabling us to understand the actual process whereby users reach a decision about which document to select.

16.3.2 Existing Research Findings

To date, a number of studies have investigated viewing behavior on SERPs, the first of which offered a primarily descriptive explanation of how users view the results presented to them [167, 205, 443]. Several researchers have subsequently gone beyond these descriptive measures to also assess what these eye movements mean in the context of interpreting server log data, as well as determining an average course of viewing on the SERP [300, 379]. This section will discuss some of the key findings in the eye monitoring and information retrieval literature, pointing out some of the slight differences in the data that are produced.

16.3.3 Overall Viewing Patterns

Several reports have likened the path of a user's eye on a search results page to an F shape, or a Golden Triangle, with the majority of attention being given to the top few results [167, 443]. Figure 16.2 presents a standard SERP with the golden triangle pattern of viewing. Aggregate analyses show that while users may read the first result in detail, they will rarely give this same degree of attention to the following results, thus each result is read successively less, tapering into the bottom of the triangle. Based on this finding, Jakob Nielsen again stressed the importance of effective web writing, indicating that users won't read your text word-by-word, and the most important information should be stated clearly in the beginning of the sentence or paragraph, ideally with bullet points or "information carrying words" [443].

While the golden triangle, or F-shape is certainly a relevant generalization, work has been done to suggest that the viewing behavior on SERPs is not quite so simple. In fact, viewing patterns highly depend on the user and task. In the first study produced by Enquiro, the authors called out attention to slight variations to the golden triangle produced by oneboxen and sponsored links [167]. They found that with onebox content and sponsored links present, the top portion of the triangle is elongated to compensate for the additional text.

Recent work goes even further to suggest how tasks and users can impact the way in which a results page is viewed [529]. Figures 16.3, 16.4, and 16.5 depict one given task that asked the user to find out who is the tallest active player in the NBA. Users were provided with the query term [tallest active player NBA], and were instructed to begin with that query to find the answer [529]. The heatmap presented in Fig. 16.3 is an aggregate of all 32 users who did this task, and shows some slight resemblance to the golden triangle and F-shape. The main similarity is that the first result is read the most completely, the second result is read fairly completely, and the following results have successively fewer fixations.

Individual User Differences. Upon closer inspection, however, it appears that users seem to be on average reading the snippets rather carefully, more thoroughly than the "golden triangle" generalization would suggest. An explanation for this behavior is that the answer to this task is explicitly stated in the snippet of the second surrogate. This offers evidence that slight differences in the results page and content presented, such as these information-bearing snippets, have the ability to produce different viewing behaviors. Thus, it is important to understand and account for the different variables on the page when interpreting aggregate plots and making assumptions about user behavior.

To identify individual user behaviors, two individual scan patterns are presented in Figs. 16.4 and 16.5 to emphasize the degree of variance between individuals, even when viewing the same exact page. In the Fig. 16.4, the user is quickly scanning only a few results before feeling satisfied and click-

Fig. 16.2. Golden triangle viewing: the aggregate plot of eye movements can resemble a triangle, with top results viewed more extensively than the lower

ing on one. In contrast, the user in Fig. 16.5 is evaluating the results much more carefully, reading snippets in depth, as well as exhausting more of the options available to her. Both users were presented with the same page, but approached the information gathering process very differently.

Aula et al. [15] also noticed these individual differences in scan activity in her own research. The authors used eye monitoring to assess how users evaluate search result pages, and subsequently classified two types of online searching styles – economic and exhaustive. Economic evaluators are similar to the first image shown, where users scan quickly and make a decision based on what seems to be the first relevant result. Exhaustive evaluators, on the other hand, are fairly similar to the second image depicted below, where the users prefer to assess multiple options before clicking on a result.

Much more work is needed to determine if scan behavior remains consistent on a per-user basis across a variety of tasks (e.g., does a given user always scan SERPs quickly?), or whether the task and presentation of results

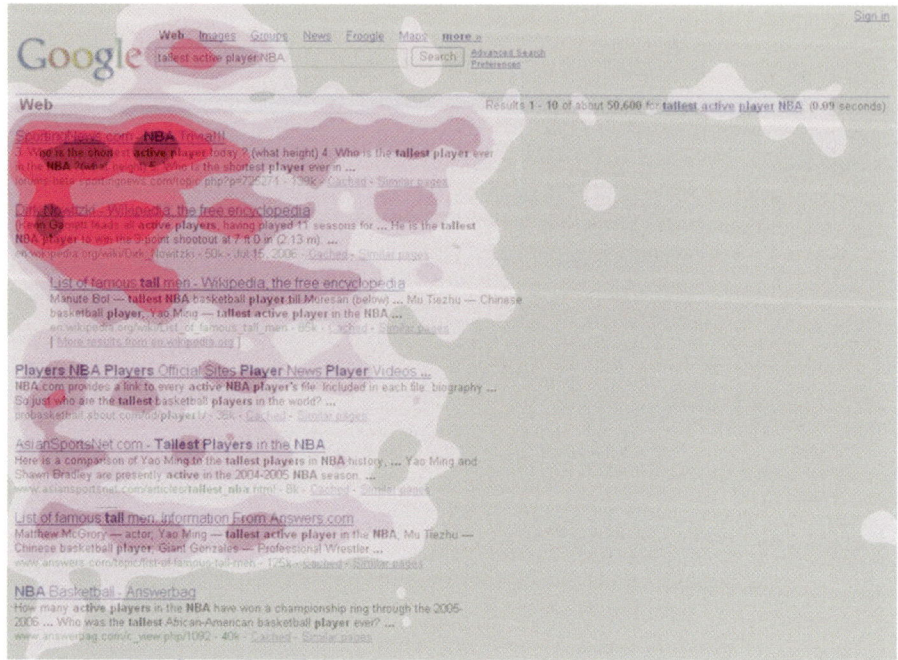

Fig. 16.3. Note the golden triangle is stretched a bit. Users are reading the snippets more thoroughly than the golden triangle would suggest

ultimately have the most impact in determining how the results are viewed. While research has found differences in viewing search results based on variables like task type and task difficulty [205,379], less user-centric analysis has been done.

16.3.4 Number of Results Viewed

While the previous example shows that users exhibit different behaviors in search, it is still helpful to know, on average, how exhaustively users attend to the information presented to them. On average, users only view about three to four surrogates on the SERP [167,205,300,379], and the number seems to vary based on the task and the expertise of the user. The research done by Granka, Joachims, and Lorigo [205, 300, 379] studied college students, with an average age of 20.5. They reported that users view about two to three surrogates before clicking on a result. Enquiro recruited participants in their 30s and 20s, with some over 40, and reported averages of 3.7 surrogates viewed for young users (under age 35), and an average of 3.8 for users over age 35. While there are slight differences based on age and education level, both studies consistently showed that individuals rarely visit a second result page. Lorigo et al. [379] reported that users look beyond the top 10 surrogates

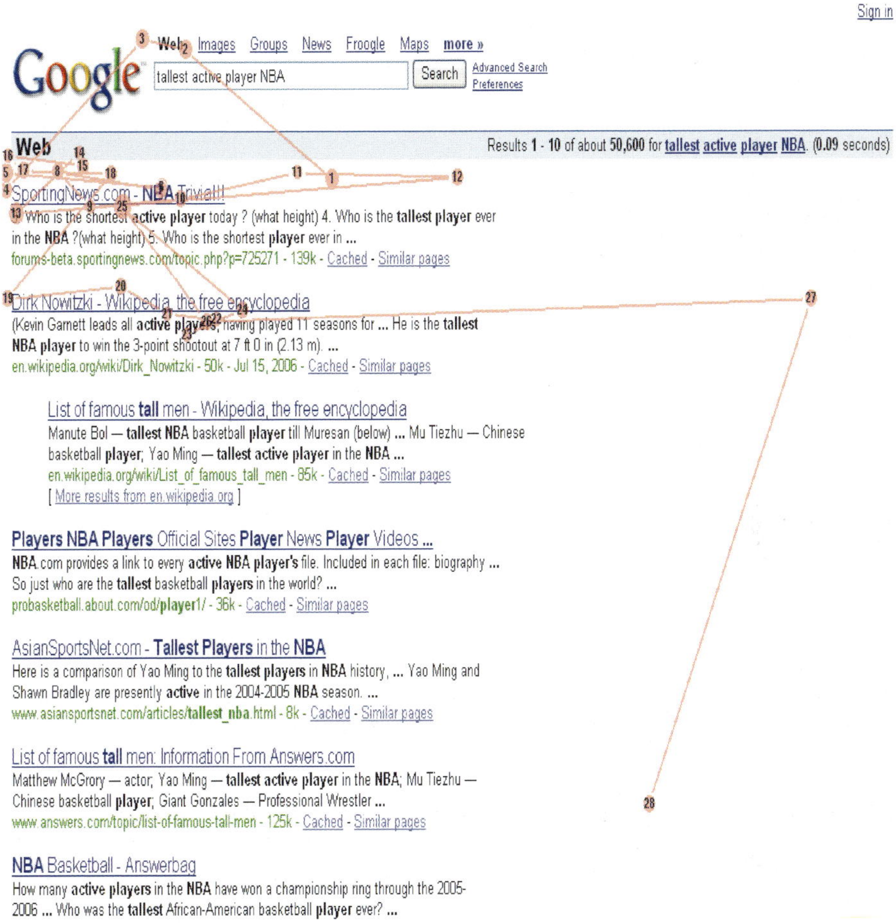

Fig. 16.4. "Efficient" style of scanning a SERP viewing less than three results

in less than 5% of the instances. Furthermore, users spend less than two seconds viewing each individual surrogate [167, 205, 300], indicating that the key components of the surrogate are parsed very quickly for a near immediate judgment.

Joachims and Granka also explored how viewing surrogates correlates with the search results that are clicked [299]. The graph in Fig. 16.6 depicts the relationship between the results that are clicked and the surrogates, or abstracts, that are viewed. Note the first two results are viewed nearly equally, while the first result is clicked upon disproportionately more often. A later section of this chapter will address this phenomenon in more detail.

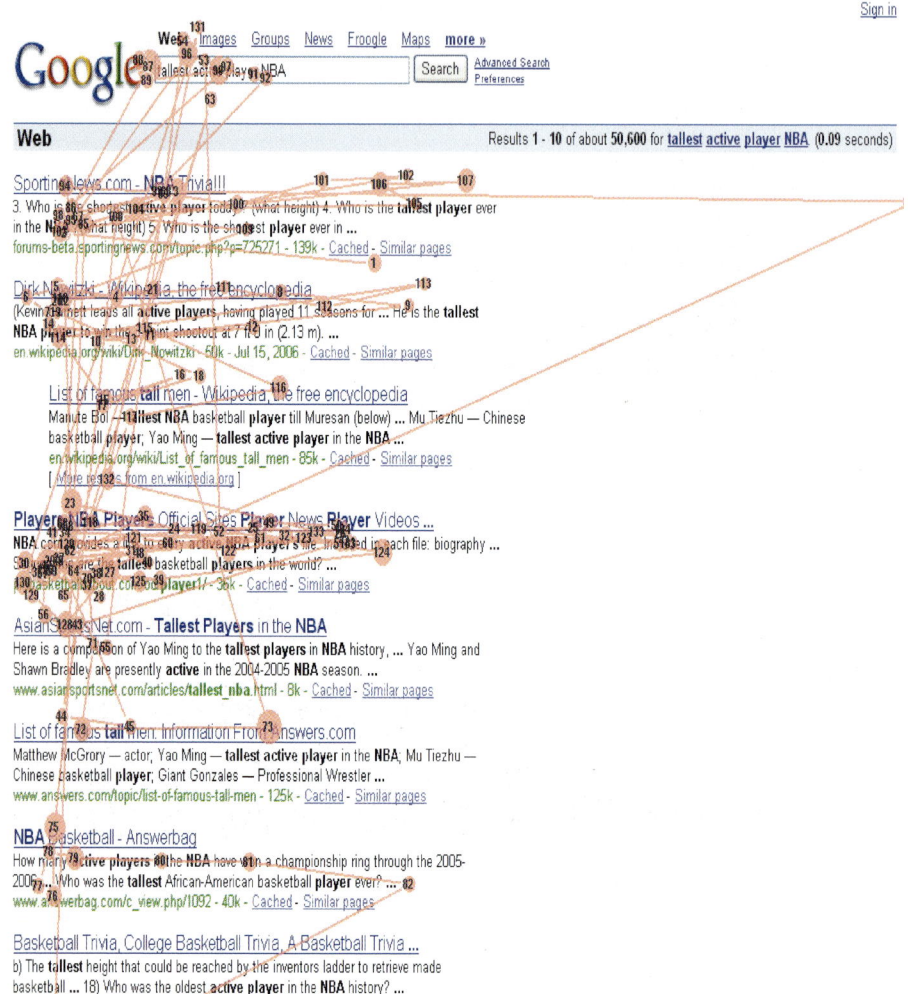

Fig. 16.5. This user view the page much more exhaustively, reading all of the surrogates in more detail, and taking the time to look at more than three

16.3.5 Viewing Sequence

While we know that in total, only about three to five surrogates are viewed on a search result page, how do users evaluate the options presented to them? More specifically, do users view the surrogates linearly in the order presented, skip around throughout, or backtrack and read a surrogate more than once.

Granka and Joachims et al. [206, 299] both looked at viewing sequence descriptively, recording the order a user viewed surrogates of a particular rank. For each surrogate, this was measured by the fixation at which a user first

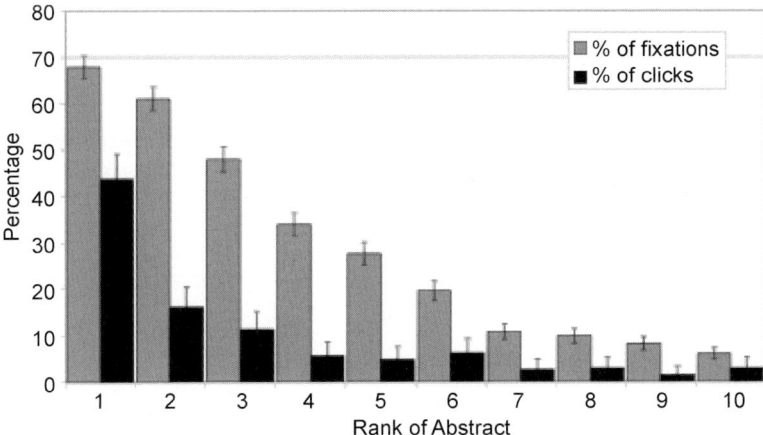

Fig. 16.6. Comparison of the surrogates a user fixates with what a user clicks

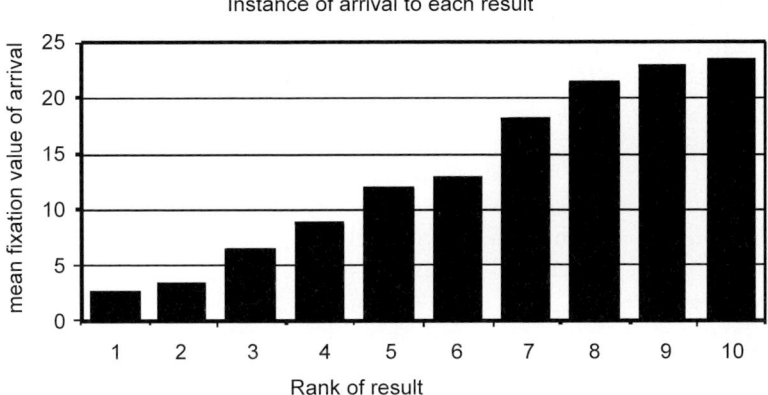

Fig. 16.7. Instance of arrival to individual surrogates, depicting a linear viewing order

viewed the surrogate, i.e., the number of fixations made before the searcher saw each rank. Figure 16.7 indicates that individuals viewed the first and second-ranked results early on, within the second or third fixation. There is a large gap before viewing the third-ranked surrogate. Furthermore, the page break also manifests itself in this graph, as the instance of arrival to results seven through ten is much higher than the other six, likely due to the fact that they are displayed below the fold, and few searchers scroll below to read the surrogates.

While very useful as a descriptive measure, this analysis does not account for regressions, where a user returns to look at a result a second or third time. It also does not show how users might make pairwise comparisons when selecting a result. There is some existing eye monitoring research in the field of consumer marketing which indicates that pairwise comparisons may also be common in selecting search results. Russo and LeClerc [544] outlined several processes that individuals employ before making decisions among various consumer products, the most common strategy being comparisons of adjacent products. It is probable that this same type of behavior also manifests itself during the process of viewing search engine results, where users may feel the need to justify their choice by quickly checking what the next or previous surrogate seems to offer.

Lorigo et al. [379] extrapolated upon this descriptive approach to viewing order and analyzed the overall sequence of viewing on a search results page. Some previous research was done using scanpath analysis to determine average viewing paths on web stimuli [244, 307]. However, the work done by Lorigo et al. [379] was the first to use scanpath analysis to more completely understand viewing on a search result page. They classified viewing behavior on a SERP into one of the following three groups: nonlinear scanning, linear scanning, and strictly linear scanning [379]. Linear scanning may contain regressions to previously visited surrogates, but adheres to the rule that a surrogate of rank n is not viewed until all surrogates of a smaller rank have also been seen. Strictly linear scanning does not include regressions to earlier surrogates. Also, a scanpath preceding a selection or click is said to be complete if the path contains all surrogates of rank less than or equal to the rank of the selected (clicked) result. Using these scanpaths, we can better characterize how users are likely to make decisions about what results to select. It is worthwhile to note that the analyses discussed here focus on text searches, where results are presented linearly. To better understand the growing mix of content on the web, such as video and image search, for example, where a grid layout is typically used, additional eye monitoring research would be invaluable.

Using the above classification, Lorigo et al. [379] discovered that participants viewed query surrogates in the strict order of their ranking, or strictly linearly, in only about one fifth of the cases, with roughly two thirds of the scanpaths being nonlinear. Hence, even though scanpath sequences are relatively short (visiting 3 surrogates on average), participants were generally not viewing the surrogates in the order intended by the search rank order. Jumps and skips were also short (typically skipping only one surrogate in its path) but prevalent. Visual highlights and cues in the surrogates, such as bolding should not be underestimated as means to grab the attention of the viewer.

Interactions with a SERP occur rapidly. Decisions as to what result to click on or how to refine a query occur in a matter of seconds. While a fixation by fixation analysis may seem tedious, these fixations tell a story about a process as ubiquitous as online search. The example scanpath in Fig. 16.8, mapped to rank values, is: $1 \rightarrow 2 \rightarrow 1 \rightarrow 1 \rightarrow 2 \rightarrow 3 \rightarrow 3 \rightarrow 2 \rightarrow 2 \rightarrow 2$, with length of 10 (compared to the reported average length of 16 in Lorigo

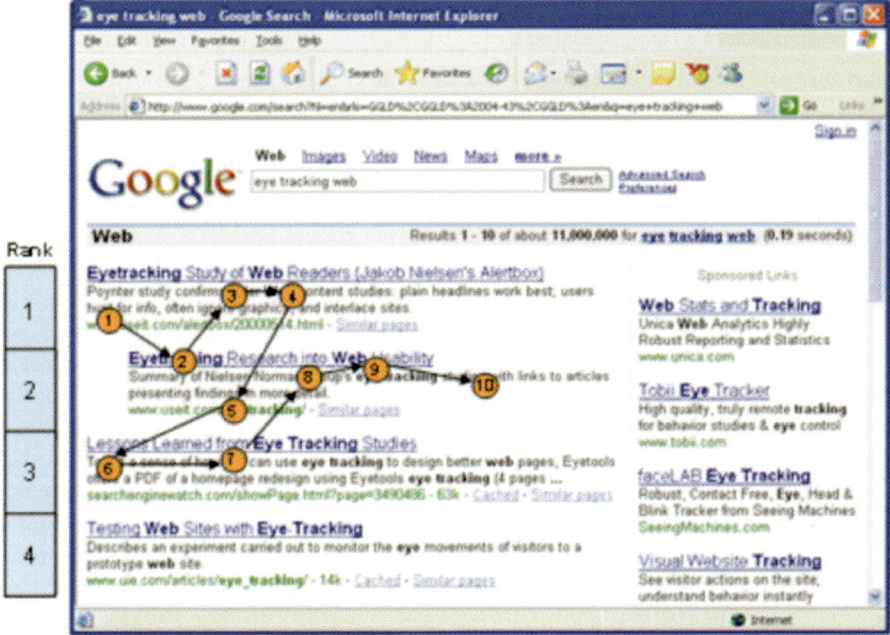

Fig. 16.8. Representation of sample data used for scanpath analysis

et al. [379]). If we ignore repeat fixations within a surrogate, we obtain a compressed sequence of length 6: $1 \rightarrow 2 \rightarrow 1 \rightarrow 2 \rightarrow 3 \rightarrow 2$. This shows the process of evaluation, with a reported average length of 6 in that same study. If we further ignore repeat surrogates, we obtain the minimal sequence which is the order in which new information was observed: $1 \rightarrow 2 \rightarrow 3$. Looking at scanpaths in these ways helps to gain a better understanding of the preferences and efficiencies of the participants.

16.3.6 Result Position

The presentation of the abstracts on the search results page is very likely to influence the manner in which users navigate through and select an online document. In his article, The Intelligent Use of Space, Kirsh [323] outlines several key features about managing and understanding space that may be relevant to the context of document selection in an online information retrieval system. Kirsh points out that one of the goals in organizing both physical and information spaces is to structure the space so that a user's option set (the number of viable alternatives available to the user) is reduced and more effectively managed. A reduction in the user's option set is typically accomplished in an information retrieval interface by rank-ordering the retrieved results. By having the system highlight the opportunistic actions that a user should take, the amount of effort that the searcher needs to expend when making a decision is reduced. Because most searchers understand that the top-ranked result has been rated most highly by the information retrieval system, one of the steps to decision-making organizing the information is thus eliminated. All of the studies that have used eye monitoring to assess the viewing of search results generally have been able to support the assumption that users rely heavily, almost exclusively, on the ranking offered by the search engine. Joachims took this one step further and switched the order of search results to see how significantly the presentation of results impacts a searcher's viewing behavior [299]. Their study generated three conditions: a normal condition, whereby the search results were left intact as retrieved from the search engine, a reversed condition, where the tenth ranked result was placed in the first position, the ninth in the second, etc, and finally a swapped condition in which only the first and second result were switched. This last condition was included based on findings that the first two results are given nearly equal attention, yet the second result lags disproportionately behind in terms of clickthrough. A comparison of viewing and click behavior by rank is depicted in Fig. 16.9. Interestingly, through eye monitoring analysis, they found that user search behavior was affected by the quality of the results, especially in the completely reversed condition. The swapped condition showed that on average, users were slightly more critical of the results being presented to them, but still clicked on the first result with a greater frequency. In the reversed condition subjects scanned significantly more surrogates than in the normal condition, and clicked on a lower-ranked result, on average. The average rank of a clicked document in the normal condition is 2.66 compared with 4.03 in the reversed condition. The study participants did not suspect any manipulation when asked after the session had concluded. The researchers concluded that users have substantial trust in the search engine's ability to estimate the relevance of a page, which influences their clicking behavior [299].

16.3.7 Task Types and Other Influences

Just as the presentation of the surrogates on the SERP influences navigation behavior, so too can the underlying search task, and also characteristics of each viewer. In his 2002 study of web search, Broder [69] describes three classes of search: navigational, informational, and transactional. In navigational search, the goal is to find a particular web page or URL. In informational search, the goal is to find a particular piece of information that may be on one or multiple web pages. In transactional searches, users are motivated by a desire to conduct a transaction such as making a purchase or trading a stock. Informational search is believed to be the most common type of search performed today [69, 537]. Eye Monitoring in online search has revealed behavioral differences with respect to task type and user characteristics. Pupil diameter, an indicator of cognitive exertion, was observed

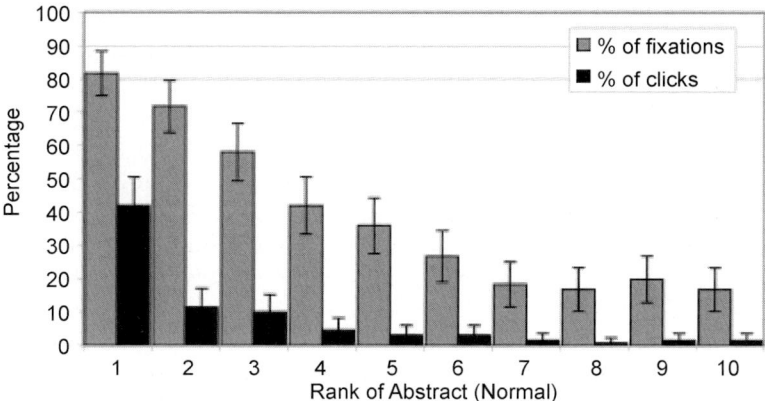

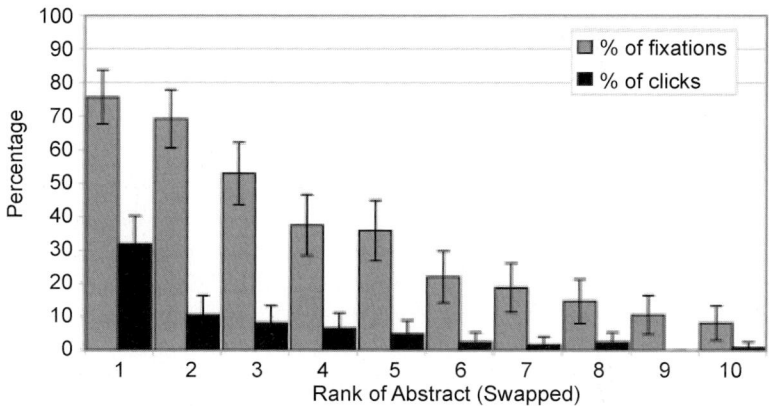

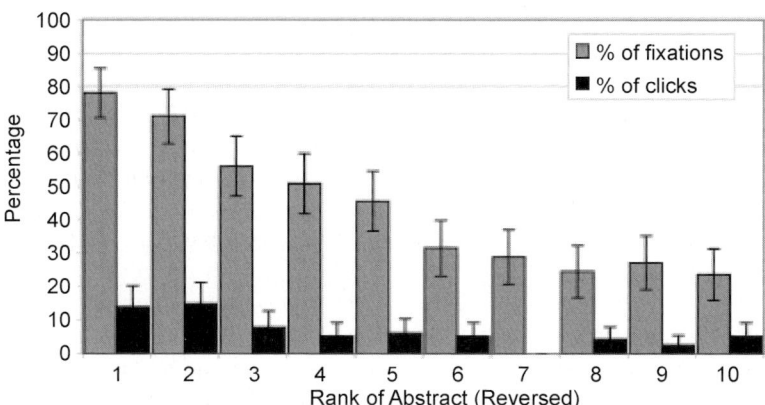

Fig. 16.9. Comparison of what the surrogates a user views and clicks, in the normal, reverse, and swapped condition

to be higher for informational tasks. This is likely because users may need to read more critically when interpreting results for an informational need. This offers behavioral evidence that the cognitive effort involved in informational tasks is greater than that of navigational tasks [205]. Additionally, Lorigo et al. [379] observed informational search tasks took longer than navigational tasks on average, which is not surprising because informational needs are often more complex. However, the time spent on SERPs alone was significantly greater for navigational tasks. This is likely because the answer to a navigational search task is more often found in a surrogate (via the title or URL, indicating the site identity) than for an informational search. The design choice of how much information is placed in a surrogate can have a large impact on user behavior. We have already noted the difference that age can make on the number of search results viewed. Additionally, there seems to be gender differences in the gaze patterns of the participants, in that males were more linear in their scanpaths than females, and were more likely to view the 7–10th ranked surrogates [379]. These initial findings indicate that the search within the search results may be driven at least in part by user-specific preferences.

16.3.8 Attention to Results Based on Rank

The previous findings have indicated how much attention is given to surrogates on the result page based on their rank. These data is based on all task instances, meaning that if a user did not scroll, their lack of viewing the last half of the page contributes to the lower overall fixation time in those bottom results. However, Fig. 16.10 offers a different interpretation by only accounting for the instances in which a particular result was looked at. Therefore, if a user did in fact view a surrogate, how much time did she spend in that given surrogate, relative to the others that she viewed? Figure 16.10 depicts the amount of time-based attention given to each of the results. There is a dip within the middle-ranked results, indicating that middle results are viewed less exhaustively than the results on either periphery, especially the first two results.

In addition to measuring fixation data, knowing how carefully users actually attend to the information in each of the surrogates, in terms of pupil dilation, also tells us whether some results are viewed more attentively, or with more interest, than others. Thus, pupil dilation, as well as total time spent in each surrogate, can be used as measures of interest and cognitive processing. Figure 16.11 below depicts the mean pupil dilation on each of the surrogates viewed. This graph follows a trend similar to the one previously described, offering more evidence that middle results are processed less critically. Pupil dilations for the middle ranked surrogates are smaller than the first and last results on SERP, especially the top two.

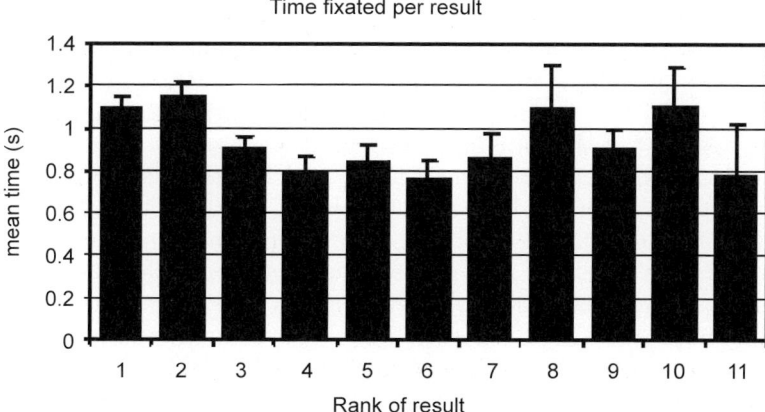

Fig. 16.10. Amount of time spent in each surrogate. Note that the middle results receive lower overall fixation time than the results on the periphery

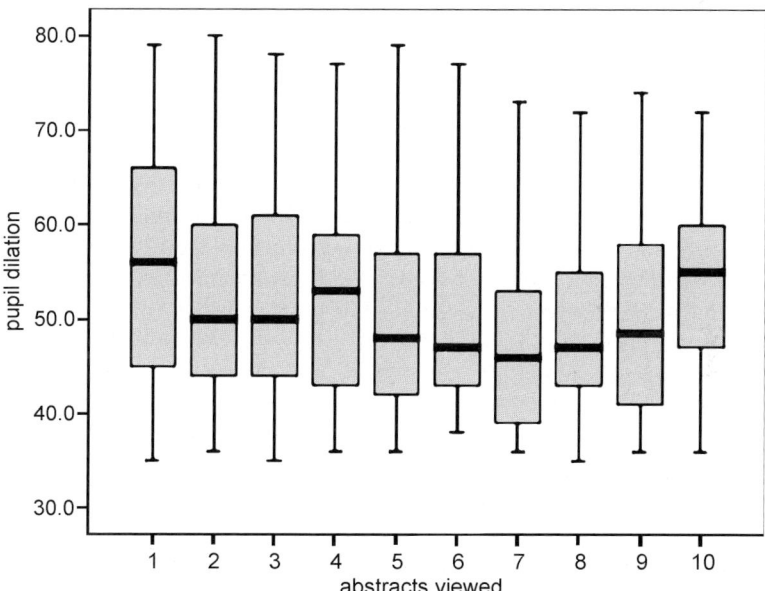

Fig. 16.11. Levels of pupil dilation when viewing surrogates on the SERP. Note that pupil dilation drops after viewing the first ranked surrogate

16.3.9 Most Viewed Information in the Surrogate

The majority of search engines format their results with a title, snippet (descriptor of text from the referring web page), and URL. In addition to investigating which of the presented surrogates were viewed, it could also be useful to know what specific content in the surrogates is relied on most to make a relevance judgment. To assess this, attention to each component of the surrogate was measured by the number of fixations in each region [205]. Overall, the snippet received the most attention (43% of total fixations), with the title and URL following close behind (30% and 21% of total fixations, respectively). 5% of the total fixations fell on other aspects of the surrogate, including such things as the cached link. However, it should be noted that the snippet is proportionately larger than both the URL and title, and is therefore likely to capture more fixations. Furthermore, because the information contained in the snippet is in the form of sentences and full-text content, it is presumably more difficult to cognitively parse than either the titles or URL. The titles and URL can be more readily perceived through an efficient visual scan. Finally, these findings should be interpreted in light of accuracy limitations inherent to the eyetracker, as previously discussed. Most eyetrackers are only accurate to within a degree of visual angle and thus we cannot be 100% certain of these percentages, but should only use them as a rough estimate for interpreting the impact of each textual component on viewing behavior.

16.4 Combining Eye Monitoring with Implicit Measures

Obtaining a low-cost estimate of how users behave in online contexts is desirable, especially for understanding the process of online search. Passive Eye Monitoring is an expensive analysis tool to use on a regular basis, so in some contexts it would be ideal for researchers to use eye monitoring on a periodic basis to enhance their understanding of other low cost, implicit methods. Query logs are an example of one such method, as they capture a great deal of data. In the context of online search, log data can provide useful information, such as the results a user clicks, how long users spent on a page, etc.

16.4.1 Eye Monitoring and Server Log Data

Some information retrieval researchers use server log data to assess the relevance of the documents that are returned to the user. For example, one of the fundamental assumptions is that if a user only clicks through to the third result, then that result must be more relevant than the two listed above it. While assumptions like this (that users look at surrogates in a linear order, users do not look at the documents below the selected one) are the foundation for the algorithms for this work, they had never been behaviorally affirmed. Eye Monitoring is one way to add value to these algorithms by specifically assessing the likelihood of these assumed behaviors. Thorsten Joachims was the first to use eye monitoring for this very purpose to more accurately and completely understand the underlying dimensions of server log data [300]. With the help of eye monitoring, he demonstrated that there are some problems assuming result clicks can be interpreted as absolute relevance judgments. Instead, they are highly dependent on the user's trust bias towards the ranking of a result ranking. Joachims then offered ways to extract relative feedback by looking at pairwise comparisons between result surrogates. The only prior work using eye monitoring to assess relevance in information retrieval was Salvogarvi, who looked at pupil dilation to determine result relevance, assuming that larger dilations could indicate greater perceptions of relevance. However, this work was strictly within the context of eye monitoring, as he did not correlate this method with larger scale data [545].

16.4.2 Eye Monitoring and Mouse Movement Data

In addition to combining eye movement data clickthrough data, exploring the correlation between continuous eye and mouse movements could prove to be very useful. Specifically, do the mouse and eye move together around the page? If there is a tight correlation, mouse movements could be a lower cost way of predicting what the eye is observing. Chen et al. [92] looked at eye and mouse movements on a set of general web pages. They divided each web page into logical regions, and measured the total time when the mouse and eye were both in a given region. They found that the eye and mouse were only in the same regions slightly above chance, but that if a user made a sudden mouse movement, there was a 84% chance that the user also looked to where the mouse moved. Rodden and Fu [528] explored this relationship of mouse to eye movements within the context of Google search result pages, and found that there is indeed some relationship between the two measures, but likely not enough to use mouse movements as a substitute for eye monitoring. They used scanpath analysis, similar to Lorigo et al. [379], but extended their analysis to compare mouse paths with the eye scanpath. They noted some interesting relationships between the mouse and eye, classifying three types of mouse and eye correlations. First, searchers sometimes use the mouse to mark an interesting result, second, users may keep the mouse still while reading, or lastly, move the mouse along as they read text.

16.5 Conclusion

Due to technological advances, eye monitoring is now a viable option for many user experience researchers and industry practitioners. Passive Eye Monitoring has been used to understand user behaviors in a number of online contexts, including news, homepages, and search results. While some challenges remain that limit ubiquitous use of eye monitoring (namely the high cost and lack of efficient analysis tools), significant contributions have already been made which would not have been possible without this tool. Specifically in online search, researchers now have more a more complete interpretation of what happens before a user selects one of the search results presented to them. Prior to eye monitoring, researchers could only rely on server log data to understand distinct user actions. With eye monitoring, researchers can now get a deeper cognitive and behavioral understanding of how individuals process the information presented to them online. This will add a valuable level of understanding, particularly with the growth in individuals relying on the Internet for information.

Studying Cognitive Processes in Computer Program Comprehension

Roman Bednarik and Justus Randolph

17.1 Introduction

Computer programming is a cognitively demanding task [127, 257]. Programmers have to constantly apply their knowledge and skills to acquire and maintain a mental representation of a program. Modern programming environments (sometimes called integrated development environments (IDE)) often present, in several adjacent windows, the information related to the actual program. These windows often contain a variety of representations of the program (e.g. the program text or a visualization of program variables), all of which need to be taken into consideration by the programmer. Investigating a programmer's visual attention to these different representations can lead to insights about how programmers acquire the skill of program comprehension, which is considered to be the crux of successful computer programming. Research questions related to the role of visual attention during program comprehension might focus, for instance, on whether visual attention patterns differentiate good and poor comprehenders, or what are the specific features of a program text that skilled programmers attend while debugging a flawed program.

Eye-movement tracking has been employed across a variety of domains to capture patterns of visual attention [146]. Among the many domains, studies of visual attention during reading is, probably, the domain that has benefited the most from eye monitoring technology (c.f. [510] for a review). In addition to studies of reading, other domains have benefited substantially from adopting eye monitoring methods to investigate visual attention, including studies of driving (e.g. [532, 637]) or studies of the usability of computer interfaces (e.g. [111, 199, 201]).

Despite the many benefits of eye monitoring, there are also several disadvantages and challenges to its application. Although the technical challenges to eye monitoring research are rapidly being overcome, there are several that still stand in the way. For example, Jacob and Karn [286] argue that the is

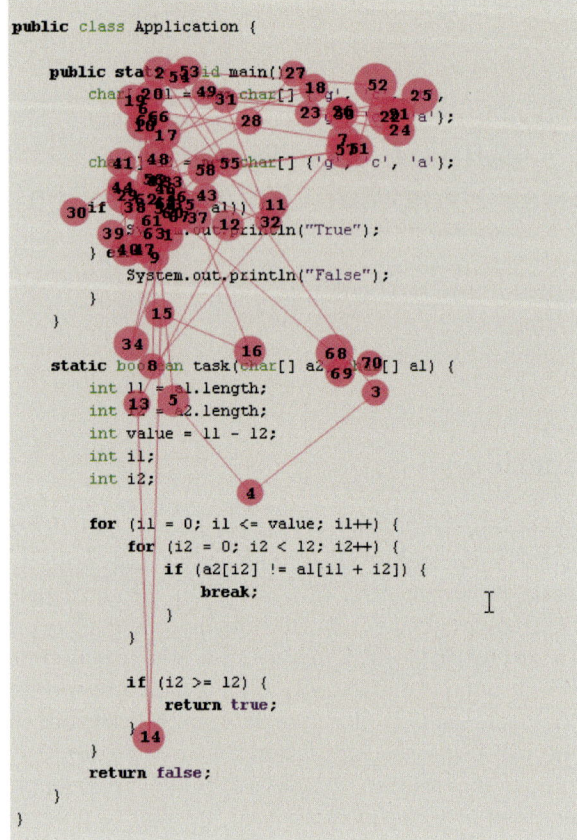

Fig. 17.1. Gaze-plot of one minute of eye-movement data during program comprehension

one of the principal challenges for eye monitoring studies. This indeed seems to be the case in the domain of studying, as we discuss in this chapter.

A typical course of eye-movements during a program comprehension session is illustrated in Fig. 17.1. The gaze-plot in Fig. 17.1 displays approximately one minute from the beginning of a Java program comprehension task; fixations, numbered consecutively, are displayed as circles and their diameter corresponds with the duration. In this particular case, the programmers were instructed to comprehend the program as best as they could with the purpose of writing a comprehension summary. From the figure it can be observed that the programmer first briefly overviewed the whole source code and then carefully investigated the first part of the program.

In many laboratory eye monitoring studies, participants are free to select which programming strategy and approach they engage in, to emulate what happens in real life programming tasks. The result is that if a certain programming activity is the phenomenon of interest, only the data of participants who willfully chose to do that certain programming procedure can be used for analysis. So for example if only 10% of participants tend to do a certain programming task, and 30 cases are needed to ensure the proper power for parametric tests, then 300 participants would be needed to get a sufficient amount of data. However, given the time and expense involved in eye monitoring studies, collecting data from this large of a number of cases is simply not feasible. The consequence, along with the fact that portions of data might be missing due to technical difficulties, is that eye monitoring data sets in program comprehension tend to be sparse.

The sparseness of the data sets leads to serious problems in applying conventional parametric statistical methods. In this paper we discuss two "back-to-basics" approaches that we have found to have been helpful in analyzing the sparse data sets generated from eye monitoring studies of program comprehension, especially those analyses dealing with the identification of trends. Those approaches are conducting visual analyses of data (in the single-case tradition) or conducting a binomial test. We admit that these methods are just a few of many suitable ways to analyze sparse data sets. The utility of the methods presented here is grounded in their parsimony and ease of conceptual understanding. They constitute what we believe to be a "minimally sufficient analysis" [7] of the types of sparse data sets common to eye monitoring studies of program comprehension. Although these approaches were applied to eye monitoring program comprehension data sets, there is no reason that they would not generalize to other domains where eye monitoring data sets tend to be sparse and have few cases. For example, studies of usability, as we also discuss later, are conducted using similar research settings and designs as studies of program comprehension. In both cases, users work under minimal restrictions, their tasks are as natural as possible, and the number of participants is similar.

This chapter is organized as follows. Before we explain the approaches to analysis that we have taken, we first give an overview of the methods employed in visual attention research in programming and then provide some examples. Next we present information about the visual analysis of eye monitoring data and about conducting a binomial test to identify trends.

17.2 Related Work

17.2.1 Capturing Visual Attention During Computer Programming

At the moment, there are three types of techniques that have been employed to estimate the current location of user's focus of visual attention. The first category includes systems based on tracking the direction of user's gaze (i.e. eye monitoring or tracking systems), the second category includes systems based on focus windows (where all of the screen is blurred except for one unrestricted window), and the third category includes systems that record verbal protocols [169] as users problem-solve. In this chapter, we focus on the analysis of data that result from the eye monitoring and focus window systems, and do not further consider the analysis of verbal protocols.

Most modern eye monitoring systems are based on video and image processing of eye movements; those systems capture the reflections from infrared emitters pointed at the eye. Typically, the vendors of eye monitoring systems also provide a software package that facilitates the recording of gaze location and the analysis of the resulting stream. An extensive treatment of the technical issues involved in eye monitoring can be found in [146].

The Focus Window Technique (FWT) is an alternative method for tracking visual attention. The technique was designed to reduce the technical problems with eye monitoring. In this method, the screen is blurred except for a certain section of the screen. The Restricted Focus Viewer (RFV) [292] is a system based on the FWT. In the RFV technique, a user controls the location of the only focused area with a computer mouse. The RFV produces a stream of data that is similar to that of the eye-tracker: the resulting file contains time-stamped data that contains the location of the focused spot, mouse clicks, key presses, and other events. The RFV or systems based on it have been applied in several studies (e.g. [535]); Bednarik and Tukiainen [36], however, showed that the focus window technique, when employed to track visual attention during debugging, interferes with the natural strategies of programmers. Jones and Mewhort [306] also showed that the FWT has to be properly calibrated to prevent the user from exploiting information that should have been made inaccessible through blurring.

17.2.2 Studying Visual Attention in Programming Behavior

Normally, computer programmers work with graphical, highly interactive development systems. These systems present program-related information in multiple windows that provide different representations of the program and its execution. While programming, programmers have to coordinate these windows to build and maintain a coherent model of the program. It is not clear, however, what the role of visual attention is in this process. Studies in the domain of the psychology of programming therefore attempt to gain insights into visual attention patterns during program maintenance tasks and during interaction with multi-representational development environments.

In this section, we describe how studies of programming behavior and visual attention are usually conducted, including information about the number of participants, the materials, eye monitoring apparatus used, and the research designs that are typically involved. We came about this information by reviewing a purposive sample of previous studies that reported an empirical investigation of the role of visual attention in some programming-related task. Although we did not conduct a systematic search of the literature, based on our familiarity with the field and because the field is a small one, we are confident that these studies, if not a comprehensive sample, are at least representative of the situations and methods of analysis being currently applied in visual attention studies of programming. One of the focuses of the review was on how the analyses of the data had been performed.

Table 17.1 summarizes 11 studies of programming that had incorporated a visual-attention-tracking technique. It presents the number of participants involved, the type of experimental design, the type of analysis, and duration of task(s) from which the eye monitoring data were aggregated into a measure and analyzed. If the maximal duration of a single session was reported instead of the actual mean duration, we used the maximal duration. The studies of Romero et al. [534, 535] employed the RFV, while the other studies used an eye monitoring device to record visual attention.

In terms of the number of experimental participants, in studies using an eye-tracker there were 13 participants on average; however, in studies using an RFV there were about 45 participants on average. The average number of participants in the eye monitoring studies reviewed here was close to the number that has been found in other reviews. For example, the average num-

Table 17.1. Summary of 11 empirical studies of programming that employed eye monitoring

Authors (year)	Subjects	Design	Analysis method	Duration of condition (min)
Crosby and Stelovsky (1990)	19	between-subject	difference in means, visual	–
Romero et al. (2002)	49	mixed	ANOVA	10.0
Romero et al. (2003)	42	within-subject	ANOVA	10.0
Bednarik and Tukiainen (2004)	10	within-subject	difference in means	10.0
Nevalainen and Sajaniemi (2004)	12	within-subject	difference in means	8.3
Bednarik et al. (2005, 2006a)	16	between-subject	ANOVA	9.8–17.6
Nevalainen and Sajaniemi (2005)	12	within-subject	ANOVA	26.3
Nevalainen and Sajaniemi (2006)	16	between-subject	ANOVA	12.5–19.2
Bednarik et al. (2006b)	16	between-subject	correlations	–
Uwano et al. (2006)	5	between-subject	visual	–

ber of participants in the review of 24 eye monitoring studies of usability reported by Jacob and Karn [286] was 14.6. In our sample, the average duration of a session from which the visual attention data were analyzed, if reported, was about 13.5 minutes.

As Table 17.1 shows, between-group or a within-group experimental designs were common, mixed designs were rare, and no study used a single-case design. We also found that parametric statistical methods were the norm. Very few studies had taken a visual approach to the analysis of the visual attention.

In summary, based on our sample, a typical eye monitoring study of behavioral aspects of programming involves about ten to sixteen participants. Participants are given a task, most frequently to comprehend or to debug a relatively short program. The interface of the stimulus environment is divided into several (usually large) nonoverlapping areas of interest, and eye monitoring data is recorded and aggregated with respect to these areas. Researchers then often conduct some type of parametric statistical procedure to compare the means of the eye monitoring measures between the areas of interest.

We believe that although the parametric approaches to the analysis of eye monitoring data are useful in certain cases, nonparametric procedures and visual analyses are important as well. For example, only two studies in our sample conducted a visual analysis of the resulting data. In the remaining studies, typically, the eye monitoring data is aggregated into a single measure across the duration of a session. For example, the measure of relative fixation counts on a certain area is computed as a proportion of the number of fixations anywhere on that area (e.g. on the code window) to the all fixations registered during the session. In programming tasks, however, the changes in visual attention can also be miniature, compared to the focus of the analyses of data from the larger areas. Crosby and Stelovsky [113], for example, showed differences in targeting visual attention on comments–therefore on the smaller areas within the program text. When eye monitoring measures are aggregated as described above, the miniature differences might disappear. An eye monitoring researcher can, of course, decide to increase the number of areas and design the matrix so that it still covers the screen but the areas are smaller (see for example Bednarik et al. [34]). In the case of increasing the number of areas to be analyzed, however, the resulting data become sparse, because even more of the areas (and cells with measures) become empty.

To deal with the challenges described above, we suggest two alternative solutions for the analysis of eye monitoring data: visually analyzing eye monitoring data at two levels of detail, and conducting a binomial test to investigate trends in sparse eye monitoring data.

17.3 Minimally Sufficient Analyses
of Sparse Eye Monitoring Data

As we mentioned before, one unfortunate characteristic of eye monitoring data sets is that they are often sparse (i.e., missing a large amount of data) or have few cases. The sparseness of the data often results from errors in eye monitoring measurements (such as loss of gaze due to the participant wearing eye-glasses or contact lenses, or head movements during recording) or from specialized types of research situations in which eye monitoring participants can choose how many measurements they will participate in. The low number of cases results from the intensity of eye monitoring measurements. For example, in studies of programming or in usability research the duration of a condition during which data is recorded is normally between ten minutes to a half an hour per condition. Because it takes approximately ten times more time to analyze and interpret the resulting data, it is no surprise that in most situations it is not practical to collect a very large number of measurements from a large number of participants.

Table 17.2 is a hypothetical example of an authentic eye monitoring data set that is sparse and has few cases. The data are hypothetical but are based on an authentic data set. We chose to use a hypothetical data set because it better illustrated the properties and methodology that we discuss here than the authentic data set. Table 17.2 shows the results of the ratio of fixation counts on two areas of interest from 12 participants over four possible measurements.

In this hypothetical situation, participants in each measurement were asked to do a programming task; each task was similar to the others. Partici-

Table 17.2. Hypothetical eye monitoring data set

| Participant | Fixation Ratio | | | |
	Measurement 1	Measurement 2	Measurement 3	Measurement 4
A	1.3	1.5	1.6	1.7
B	2.3	2.4	–	–
C	2.5	2.6	2.7	–
D	1.6	1.7	–	–
E	1.9	2.0	2.1	–
F	1.5	1.6	1.7	1.8
G	2.0	2.1	1.9	–
H	1.8	1.5	1.6	1.7
I	1.9	2.0	–	–
J	3.0	–	–	–
K	2.7	–	–	–
L	2.8	–	–	–

pants were allowed to repeat similar tasks up to four times so that they could master those programming tasks. Some participants, such as participants J, K, and L decided that they only needed to do the task one or two times. Other participants, like participants A, C, and E, decided that they needed to do three or four programming tasks to achieve mastery. The optional repetition of programming tasks was an inherent attribute of the research question: Do the fixation ratio times show an increase as optional programming tasks are repeated? The optional aspect in that research question is what causes the sparseness of the data; some participants will only complete one measurement, other participants will complete two or more measurements. The low number of measurements has to do with the fact that it took so much time and resources to complete a single measurement.

The question, then, is how does one analyze data sets such as the one presented in Table 17.2. Traditionally, a repeated-measures analysis (i.e., an all-within-design) would be used for this type of design. But when using that approach on this particular data set, because it is sparse, the power would be very low after excluding cases with missing data. In this case, the statistical power would have been only 0.12. Replacing the missing data would be questionable in this case because there are missing data in about 38% of the cells.

Instead of treating the data as repeated measures, one might argue that the data for each attempt could be treated as independent and means for each measure could be compared as if one were doing a one-way ANOVA analysis (i.e., an all between analysis) with four groups. The groups would be Measurement 1 ($n = 12$), Measurement 2 ($n = 9$), Measurement 3 ($n = 6$), and Measurement 4 ($n = 3$). However, this would seriously undermine the statistical assumptions involved in this kind of parametric analysis (especially, treating dependent measures as independent measures.) Even if one were still able to conclude that the assumption violation were acceptable, the statistical power would still be poor–in this case it would be 0.17.

Also, the choice between a repeated-measures analysis (all-within design) with three full cases (i.e., participants A, F, and G) and a one-way ANOVA type analysis with four groups and 30 total data points makes a striking difference in what the data show. In Fig. 17.2 the means across each measurement for the four full cases are taken, as would happen in a repeated-measures analysis without the replacement of data. Figure 17.2 shows that there is an upward trend over time. In Fig. 17.3, the means within each measurement of all the cases are taken, as would happen if a one way ANOVA type analysis were used. Contrary to Fig. 17.1, Fig. 17.3 shows that there is a downward trend.

Because of the low power and violation of statistical assumptions when using parametric types of analyses for the analysis of sparse data sets with few cases, we have found that the most reasonable solution is to get back to the basics and do minimally sufficient analyses. The minimally sufficient

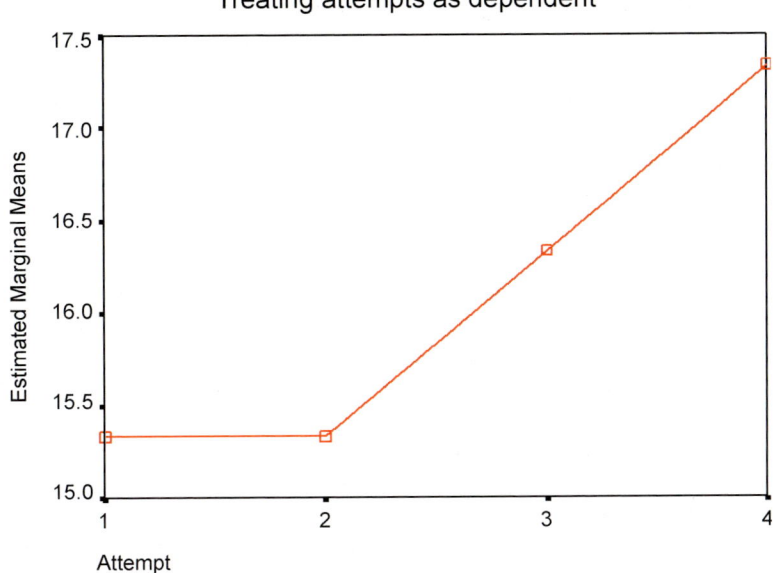

Fig. 17.2. Plot of measurement means of the four complete cases

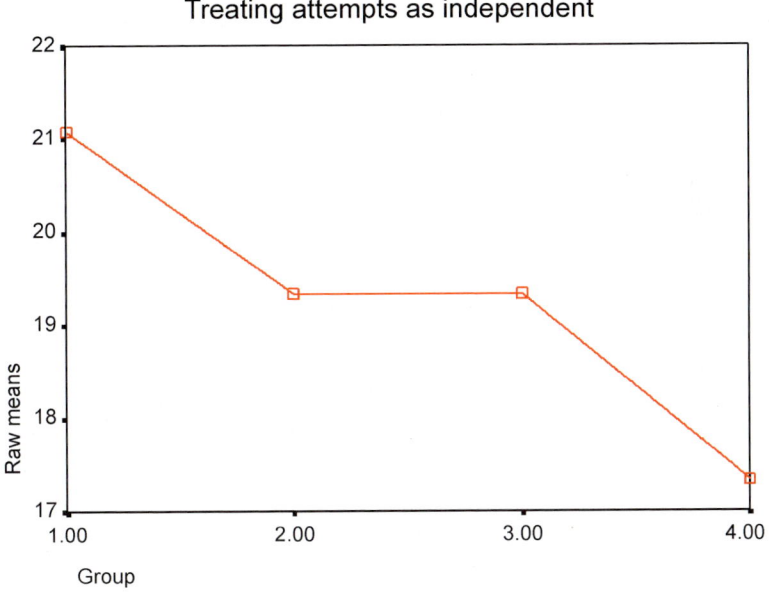

Fig. 17.3. Plot of measurement means of all cases

analytic solutions we have found for these types of data sets are visually inspecting the data and using a binomial test. (Using the Page test [477] is another parsimonious solution, but since it is much more complex than using the binomial distribution or visual analyses, we do not present it here.) How one would use those two approaches (i.e., visual analysis and the binomial test) in terms of the data set presented in Table 17.2 is presented below.

17.3.1 Visual Analysis of the Eye Monitoring Data

Simply visually inspecting a graph of the data can provide a wealth of information that would be overlooked by looking at the raw numbers in a data set. Figure 17.4 shows the data when they are arranged in a series and plotted by attempt and fixation ratio. First it appears that there is an upward trend – the fixation ratio increases each subsequent trial. Also, Fig. 17.4 shows that participants who only did one trial were those with the highest fixation ratios to begin with and that the participants who did the most optional programming tasks were the ones with the lowest fixation ratios. The visual analysis of single cases approach presented here could also be extended to include types of research studies other than trend studies. For example, one could

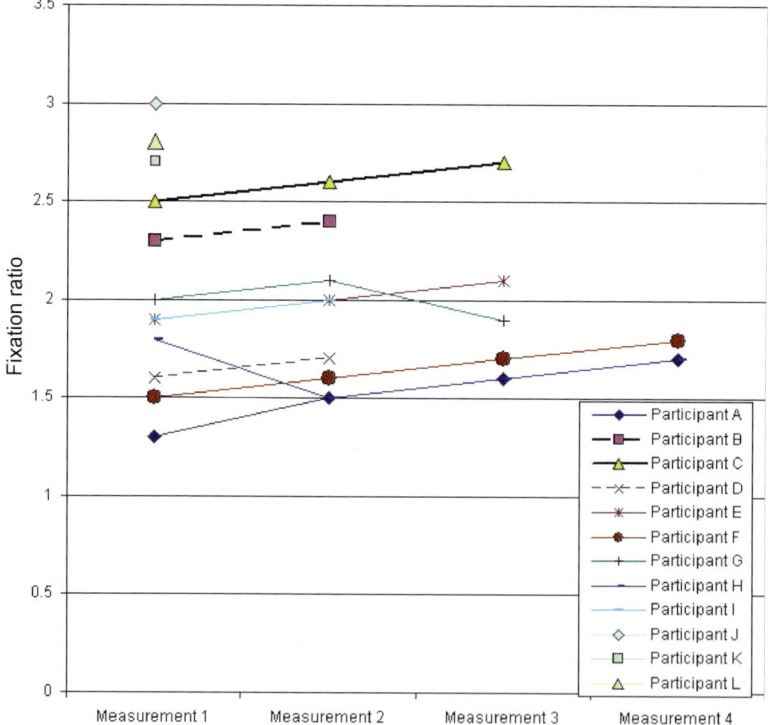

Fig. 17.4. Line graph of fixation ratios by measurement

use an ABA design where the A phase is a control condition and the B phase
is the treatment condition, or any of the other designs successfully used in
single-case research (see [22] for more information on the variety of single-case
designs).

While the previous approach operates with the eye monitoring data on
the measure level, there is, however, yet another, finer level of detail that
a researcher can use in order to analyze the eye monitoring data. Single
fixations can be plotted against a time line, to discover the subtle differences
and patterns in visual attention. Figure 17.5 shows about fifteen seconds
of fixation data from two participants during comprehension of a program,
where the vertical axis displays the Y-coordinate (the vertical coordinate of
the gaze point). Thus, when aligned side-by-side with the window containing
the source code, the graph represents the approximate lines of code and order
in which the participants visually attended to them. The horizontal dashed
line shown in Fig. 17.5 (with the Y-coordinate of 238) represents the long-
term average of both of the participants' visual attention on the Y-coordinate.
It is worth mentioning that parametric statistical analyses would use these
same averages to compare the participants. From the course of the visual
attention lines, however, we can observe that the two participants markedly
differed in their comprehension strategies and their actual visual attention
focus had only rarely been found near the average line.

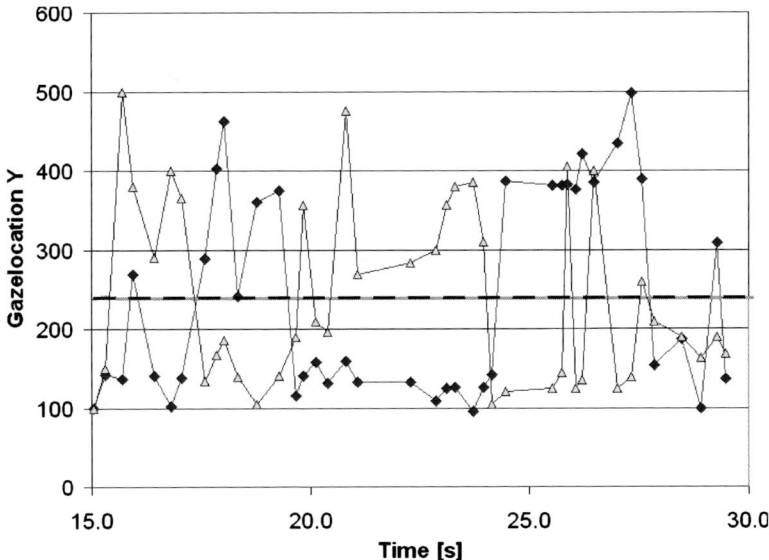

Fig. 17.5. A 15-second segment of the Y-coordinate of the visual attention focus
of two participants comprehending a program

17.3.2 Analyzing Trends of the Sparse Eye Monitoring Data

If the visual inspection of single-case data were not convincing enough, the binomial distribution can be used to easily determine the probability of getting a certain number of increases between measures given chance. If there were no trends in the data set, one would expect about the same number of increases and decreases to occur, if there were an increasing trend one would see a greater number of increases than decreases, and if there were a decreasing trend one would see a smaller number of increases than decreases. When the outcomes are only either a success or failure, the binomial distribution can be used to determine the probability that out of a given number of trials a certain number of successes (or failures) will occur. For example, the binomial distribution can be used to determine the probability that, given a fair coin, heads will come up 5 out of 5 times that the coin is tossed (the probability is a little bit over $3/100$). The binomial distribution is given in Eq. 17.1. In Eq. 17.1, p is the probability of success, N is the number trials, and x is the number of successful trials out of N trials.

$$P(X \geq x) = \sum_{i=x}^{N} P(X = i) = \sum_{i=x}^{N} \frac{N!}{i!(N-i)!}(p)^i(1-p)^{(N-i)} \qquad (17.1)$$

To use the binomial distribution to find the probability of getting the number of increases that were observed, one must convert the data set into binomial trials. Table 17.3 shows what the data set in Table 17.2 would look like if it were converted into binomial trials, where i indicates that there was an increase between earlier and later measurement and where d indicates that there was a decrease between earlier and later measurements. The data in Table 17.3 show that there were 30 trials ($N = 30$), 22 increases ($i = 22$), and we assume that the probability of success is .50 ($p = .50$). Using Eq. 17.1, the cumulative probability of getting 22 successes or more out of 30 trials where the probability of success is .50 is .008. (To find the cumulative probability of 22 or more successes out of thirty one has to add the probabilities of getting 22 of 30, 23 of 30, ..., 30 out 30.) Figure 17.6 shows the binomial distribution for getting 22 out of 30 successful trials when the probability of success is .50 and the standard deviation is 2.74. As shown in Fig. 17.6, the expected number of successful trials given chance is 15. The standard deviation is 2.74. The observed number of trials, 22, was much greater; it was 2.56 standard deviations greater than the expected number of successes. In short, the probability of getting 22 out of 30 increases by chance in this case was very small; therefore, one can conclude that there was an increasing trend in the data set.

There are, however, also variations to the presented approach of transforming measures to trials, each of those having a set of advantages and disadvantages. For instance, one approach to transforming the data set into trials is to use differences only between consecutive measures (e.g. 1–2, 2–3,

Table 17.3. Hypothetical data set converted to binary trials, i = increase, d = decrease

Case	1–2	1–3	1–4	2–3	2–4	3–4	Total increases	Possible increases
A	i	i	i	i	i	i	6	6
B	i	–	–	–	–	–	1	1
C	i	i	–	i	–	–	3	3
D	i	–	–	–	–	–	1	1
E	i	i	–	i	–	–	3	3
F	i	i	i	i	i	i	6	6
G	i	d	–	d	–	–	1	3
H	d	d	d	d	d	d	0	6
I	i	–	–	–	–	–	1	1
J	–	–	–	–	–	–	0	0
K	–	–	–	–	–	–	0	0
L	–	–	–	–	–	–	0	0
						Totals	22	30

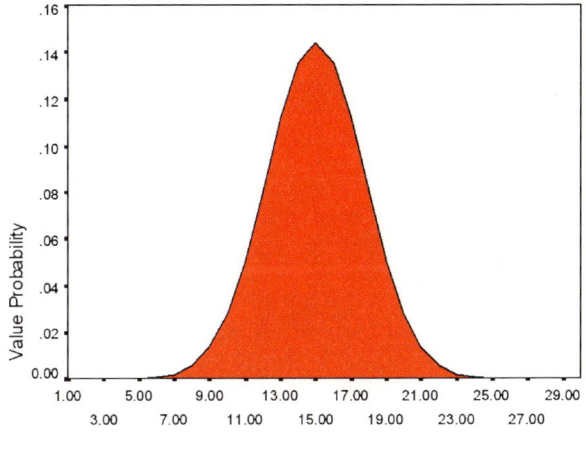

Number of increases

Fig. 17.6. Binomial distribution for 30 trials when probability of success is .50. The expected value is 15 and the standard deviation is 2.74. The cumulative probability of getting 22 or more increases out of thirty is 0.008

3–4). In that case the method can discover a prevailing number of increases or decreases, but not an overall trend in the measures across all possible combinations. Our method of transformation of measures into trials takes each possible pair of measures within a subject and transforms those into trials. That is, for example the pair 1–4 is also included as an increase if the fourth trial had been greater than the first trial, or as a decrease, otherwise.

As a result, while identifying the overall trends, our method leads to a greater number of trials than the consecutive measure transformation method. Therefore, the p-value using the all possible combinations method will also differ from the p-value using the consecutive measure method discussed above. Our method has the disadvantage of not having mathematically independent trials. For example, if there is an increase between Measures 1 and 2 and an increase between Measures 2 and 3, then, by definition, there will be an increase between Measures 1 and 3.

We admit that there are also many general disadvantages to using the binomial distribution to analyze these types of data. Namely, the conversion of continuous values into binary values loses a significant amount of information. Using the binomial approach for example, one cannot comment on how great the increases were, but only one on the probability that a certain amount of increases would have occurred given chance. We also admit that there are other, more-complicated nonparametric procedures that could be used in this case to identify trends (like the Page test [477]). However, we agree with the American Psychological Association [7] and the American Psychological Association's Task Force on Statistical Inference Testing [681] on the importance of conducting a minimally sufficient analysis. In the case of data sets that are sparse and have few cases, we propose that at least two of the candidates for a minimally sufficient analysis is through the visual inspection of data and using basic nonparametric tools, such as the binomial distribution.

17.4 Conclusion

While eye monitoring data have some properties that make them difficult to analyze using conventional techniques, they also have qualities that enable a researcher to easily create a chronological narrative from the data. Our review showed, however, that researchers employing eye monitoring in studies of programming tend to analyze eye monitoring data at the most aggregated levels of analysis and seem to ignore analyzing the data through visual inspection.

In summary, we have shown one method for statistically identifying trends in eye monitoring datasets that are sparse. In addition, we have also proposed to visually inspect eye monitoring data either by examining each participant's aggregated scores across measures or by visually inspecting a graph of an individual's data over time and plotted against given areas of the screen. The visual analysis methods presented here can reveal intricate and subtle patterns of individual visual attention behavior that, when woven together, can provide a more detailed narrative than when parametric procedures are used alone.

Eye Monitoring Studies on TV News and the Web Converge: A Tale of Two Screens

Sheree Josephson and Michael E. Holmes

18.1 Introduction

The merger of the "lay-back culture of the living room with the bustling activity of the lean-forward Net" [367] is underway. American television is starting to look a lot like the Web and the Web is looking more and more like television. This visual convergence will likely accelerate with the impending penetration of interactive TV (iTV) into the U.S. market and the growth of "Web 2.0" sites.

We have been interested in the eye-scan behavior of TV viewers and Internet users in the increasingly complex visual landscapes found on television and computer screens. Until recently, the TV screen was filled with mostly video, while the computer screen displayed mostly text and graphical images. However, today it is commonplace for news shows, especially those on cable channels, to use numerous on-screen enhancements like those traditionally associated with the Web (e.g., split-screen layouts, banner graphics and moving text). In addition, one of the most common uses of the Internet today is the downloading and viewing of short videos, and increasingly Web users are offered the chance to stream or download television programs and movies as well. Indeed, the era of Internet TV may soon be upon us as major service providers are exploring the potential of Internet Protocol Television (IPTV). The difference between media platforms ultimately will not be the channel (broadcast vs. traditional cable vs. IP) but the screen size. Text and video content delivery already extends from the small screens of mobile phones and portable devices through the medium screens of computer displays to the large screens of flat-screen TV and the super-large screens of public displays such as the Jumbotron. The good news is that a portion of what we have learned from studying television will apply to the Web, and some of what we have learned from studying Web pages will apply to television; hence, what we have learned from both media forms will apply to the emerging media vehicles of iTV and IPTV. The bad news is that as media merge and converge, we expect the challenges facing eye monitoring (or eye gaze tracking) researchers to increase exponentially.

This chapter will summarize two research reports – one on television news and another on Web portal pages. These research studies, grounded in visual perception and information-processing theories, have added to the understanding of how information acquisition takes place in today's "lean-back living room" and on today's "lean-forward Net." In each study the focus is on comparing viewing of different designs within a particular genre. In the first study, we compare visual attention to three levels of "enhancements" of TV news screen design. In the second study, we compare repeated viewings of three kinds of Web pages: a traditional Web portal, a retail product home page, and the home page of an online news site. We hope the theoretical approaches, methodology and results from both of these studies will serve to inform future research on the media forms emerging from the convergence of television and the Internet. The chapter will conclude with a discussion of the challenges facing eye monitoring researchers who are studying TV or the Web and those who will be engaged in studying the single screen of the emerging new media.

18.2 On-Screen TV News Enhancements

After the attacks of September 11th and during the war in Iraq, network television news operations were faced with the challenge of telling complex stories with countless details and fast-breaking developments. Networks and cable stations resorted to "on-screen enhancement" methods for concurrently presenting as much information as possible. Viewers often saw a split screen, with a news anchor or reporter presenting information on one side and a live scene shot on the other side. This was often accompanied by a bulleted summary of news developments, which at times complemented and other times supplemented what was being reported in the majority of the screen. A continuous crawl or ticker of headlines moved along the bottom of the screen with even more information. As a final touch, the TV news organizations inserted their corporate logos and/or story logos especially designed to promote the news event.

The authors, like most Americans, watched hours upon hours of this complex news coverage. After several marathon news-viewing sessions during which only snippets of information were obtained, there was a sense of exhaustion of having worked so long and hard at getting so little information. The authors wondered whether other news viewers were having the same experience, and if this style of news presentation was overwhelming – perhaps even detrimental – to the acquisition of information.

Multiple-channel communication such as described above has been of interest to communication theorists and instructional designers for decades.

Multiple-channel communication theories consider whether people can accommodate simultaneous audio and visual stimuli (pictorial and written) and, if so, the amount and types of information that can be processed. Given this body of research and our personal experiences watching news coverage of 9-11 and the war in Iraq, we wondered how viewers scanned TV crammed with on-screen enhancements and whether the complex news presentation impacted what viewers learned from the networks' news coverage.

18.2.1 Television and Dual-Processing Theory

In its simplest form, television viewing is a complex activity. Information is presented to viewers in multiple formats – in spoken and written words, and in moving and still pictures. The dual-processing theory of working memory can be used to make predictions about our ability to process these multiple channels of information simultaneously.

Mayer and Moreno [402] describe four assumptions of dual-processing theory:

> First, working memory includes an auditory working memory and a visual working memory, which are analogous to the phonological loop and visuospatial sketch pad, respectively in Baddeley's theory of working memory [16]. Second, each working memory store has a limited capacity, consistent with Sweller's [86, 595–597] cognitive load theory. Third, meaningful learning occurs when a learner retains relevant information in each memory store, organizes the information in each store into a coherent representation, and makes connections between corresponding representations in each store, analogous to the cognitive processes of selecting, organizing and integrating in Mayer's [401, 403] generative theory of multimedia learning. Fourth, connections can be made only if corresponding pictorial and verbal information is in working memory at the same time, corresponding to referential connections in Paivio's [99, 478] dual-coding theory.

Moore, Burton and Meyers point out that a fair amount of multiple-channel research has been conducted using television as a stimulus [418]. Some of the research (pre-dating the use of on-screen enhancements) found increased recall and retention of information in multiple channels when the information was assimilated and redundant [140, 181, 517]. Interference was found when cues were unrelated across audio and verbal channels, [559]. Other research [665] found multiple-channel presentations attracted more attention, but confirmed perceptual interferences occur across multiple channels when the content is not assimilated.

Practice has been found to help viewers deal with material presented simultaneously in different channels. Takeda found effective processing both of pictorial and verbal information was brought about by experience in viewing movies with subtitles [601]. Similarly, other researchers have concluded that when people watch television they can with practice effectively distribute attention between different channels of information [151, 152].

The results of these TV studies correspond with the four assumptions regarding dual-processing theory. First, television is a medium that uses pictures and sound to present information. Second, television is a medium that could easily overload a limited working memory. Third, when visual and auditory information are related, increased learning occurs. Fourth, it helps if related pictorial and verbal information is in working memory at the same time. The effect of practice in dealing with complex information displays like those on cable TV news programs that combine unrelated visual and verbal information has not yet been addressed by dual-coding theory.

18.2.2 Visual Attention and Dual-Processing Theory

We wondered whether news viewers are able to handle the bombardment of information presented via enhanced screen designs. We posed four research questions related to visual attention and recall:

RQ1:   Is distribution of visual attention to screen areas during TV news viewing associated with the presence of on-screen enhancements?

RQ2:   Are eye-path similarities associated with the presence of on-screen enhancements?

RQ3:   Is amount of recall of information in the audio content of TV news stories associated with the presence of on-screen enhancements?

RQ4:   Do viewers prefer TV news designs with on-screen enhancements?

To answer these questions, we conducted an eye monitoring and information recall study. Eye movements were measured with a corneal-reflection eye monitoring system. The eye monitoring system tracks fixations (brief pauses in eye movement associated with cognitive processing of foveal vision) and saccades (movements between fixations). From the eye-movement data, the time devoted to the different visual elements can be measured and the fixation paths can be noted and compared. Attention allocated to the audio track was measured with a recall test. Finally, preferences for on-screen enhancements and the perceived effort exerted to process on-screen information were assessed with Likert-scale questions.

These approaches align with those used by Weirwille and Eggemeier [671] and Sweller et al. [598] who cite three major categories of "mental effort measurement techniques". These include physiological measures, task- and performance-based indices, and subjective responses. Physiological techniques are based on the assumption that changes in cognitive functioning are reflected in physiological measures such as eye activity. Task- and performance-based techniques are based on learner performance of the task of interest. Subjective techniques are based on the assumption that people are able to be introspective on their cognitive processes and report the amount of mental effort expended. Typically, these techniques use rating scales to report the experienced effort of the capacity expenditure.

18.2.3 Method

18.2.3.1 Subjects

The subjects were 36 students recruited from a mid-sized university in the western United States. The average age of the participants was 24.3 years. The gender breakdown was 23 males and 13 females. On average, participants reported they watched 3.6 hours of TV news each week. Participants received extra-credit points in a college course and a small stipend for participating in the study.

18.2.3.2 Apparatus

The eye-movement data were collected using an ISCAN RK-426PC Pupil/ Corneal Reflection Tracking System (*ISCAN, Inc.*), which uses a corneal reflection system to measure the precise location of a person's eye fixations when looking at a visual display on a computer monitor. The eye monitoring equipment does not require attachments to the head or a bite bar or chin rest. It uses a real-time digital image processor to automatically track the center of the pupil and a low-level infrared reflection from the corneal surface. The system collects data at 60 Hz or about every 16.7 milliseconds, and at amanufacturer-reported resolution of plus/minus 0.3 degrees.

18.2.3.3 Fixation and Saccade Criteria

The minimum fixation duration was set at 100 milliseconds. Although there are no definite studies establishing 100 milliseconds as the minimum amount of time necessary for a pause to be considered a fixation, 100 milliseconds has become a widely accepted "rule of thumb" that numerous researchers believe produces reliable categorization of fixations from raw data. Researchers studying media images include Fischer et al. [182], who used this measurement in a study on print advertisements, and Baron [23], who used it in a study on television. Lohse and Johnson used 100 milliseconds as the minimum fixation duration in a marketing study. Researchers studying reading have concurred with this benchmark number [581]. They have argued that a duration of 100 milliseconds is a good criterion for distinguishing true fixations from corrective movements and eye drift.

In addition, the maximum shift that operationally defined a saccade was an area of 10 × 6 pixels, a measurement that was derived from slightly adjusting the manufacturer's suggested spatial setting because our study did not require such a small parameter.

18.2.3.4 Stimulus Materials

Three television news stories were recorded from public television for use in this study. The stories selected were all "breaking news" stories from countries other than the United States and included information that U.S. university students were not likely to have been exposed to by a domestic news source. None of the study participants were familiar with the content, although the stories were of significant international news events.

Three versions of each story were created using varying amounts of visual and verbal information. All versions contained a graphic indicating the source of the news report, which we edited into a generic source label ("Title" area, Fig. 18.1) and a revolving globe indicating an international story ("Globe" area, Fig. 18.1).

Version 1 ("Standard") consisted of the original visual and auditory information. The only written information shown on the screen was briefly-displayed reporter and news source names. Version 2 ("Standard with Crawler") of each story duplicated Version 1 but included of a textual crawler of unrelated information scrolling horizontally across the bottom of the screen. The content of the crawler was taken from a U.S. cable channel. None of the crawler content was time-dependent, widely known, or related to the main news story content. Version 3 (Standard with Headlines and Crawler) of each story included the news story, a crawler, and changing headlines, located above the crawler, highlighting six of the major points of information of the news story. The headline content was generated by two former journalists with experience in television news reporting, one with extensive experience writing news headlines. In dual-processing terms, the three versions represent (1) video with related audio; (2) video with related audio and unrelated textual content; and (3) video with related audio, unrelated textual content, and related textual content.

For purposes of eye monitoring, areas of interest (AOIs) were defined according to the most complex screen design (Fig. 18.1). Note that in versions 2 and 3, the on-screen enhancements obscured portions of the news video visible in Version 1.

The three stories (designated England, Belgium and Russia for their country of focus) were 2.3, 2.2, and 2.2 minutes in length. Crawler content was repeated twice during that time, while the six headlines also rotated two times. News stories longer than what are typically seen on U.S. television were used to gather a sufficient amount of eye monitoring data that might more accurately reflect how TV stories are naturally viewed. The news stories were conventional in format: a combination of studio scenes of news anchors, video of on-location events and reporters, brief interview excerpts, and maps identifying story locations.

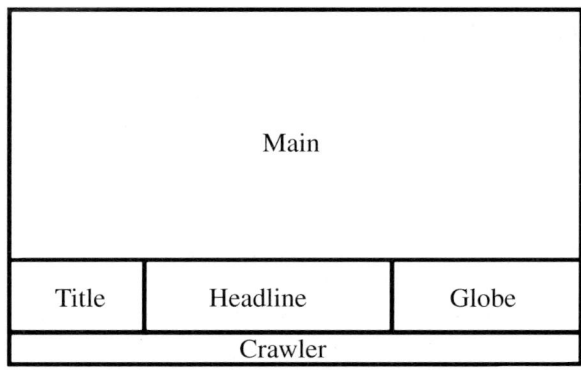

Fig. 18.1. TV news screen areas of interest

Data Collection

Participants entered the eye-movement laboratory separately. The eye-movement apparatus was calibrated to each participant before eye movements were recorded. Due to limitations of the eye monitoring equipment, participants viewed the news story on a computer screen at a viewing distance of approximately 18–24 inches, rather than on a television screen at the typical 8- to 10-foot viewing distance. Participants were told to imagine they were at home watching television news and were instructed to watch each news story as naturally as possible.

Eye movements were recorded while participants one by one viewed three versions of each of the three television news stories. Eye-tracking coordinates were recorded by the computer program. As noted above, for the purposes of this study a fixation was defined as a pause of at least 100 milliseconds within an area of 10 pixels by 6 pixels.

A Latin-square design was used to balance the number of participants and order of exposure to stories and screen designs in each cell of the 3×3 (story $\times$ screen) study design.

After the showing of each news story, participants completed an information recall test. They were asked to summarize all of the information they could remember from the entire screen. Then they were asked to answer questions based on what they saw and heard. Six questions corresponded with the major points of information used as headlines in Version 3. Five additional questions dealt with other information from the news story. Participants were instructed not to guess if they were not confident of the answer.

Following the showing of the third news story and third recall test, participants answered a short survey requesting information about gender, age and number of hours they watch television news in an average week. Next, they rated their preference for the three different news displays on five-point Likert scales. The final question asked: "Does the use of headlines and crawlers make it more difficult for you to pay attention to the video and sound regarding the main story?"

18.2.4 Results

18.2.4.1 Distribution of Visual Attention

The distribution of visual attention was characterized by the ratios of total time of fixations within the defined areas of interest to overall story length. We tested for differences in fixation time ratios using MANOVA as implemented in the GLM module of SPSS (SPSS 2005). The fixation time ratios are ipsative; given the finite length of a news story, increased fixation time in one AOI requires reduced time in one or more other areas. This would produce a problem of variance-covariance matrix of insufficient rank if all five dependent variables were included in the MANOVA. Therefore, only four of the five areas of interest were included. The content of the omitted area of interest ("Title", Fig. 18.1) is unchanged across the three levels of screen design, and the area receives little fixation time (typically less than one-half of one percent of total viewing).

Preliminary tests revealed the data failed Box's test for equality of covariance matrices ($M = 353.23$, $p < .01$) with dependent variables of fixation time ratios for the headline, globe, and crawler areas of interest failing Levene's test for equality of error variances. A log transformation was therefore performed on the fixation time ratio scores. The transformed data failed Box's test but showed improvement ($M = 128.05$, $p = 0.03$). Levene's test revealed no transformed variables violated the assumption of equality of error variance. The analysis proceeded with the four transformed dependent variables, though with unequal group sizes due to the loss, in the logarithmic transformation, of cases with zero fixations in a given AOI.

Results of a MANOVA considering both story and screen design in distribution of visual attention revealed a main effect for screen design only, $F(8, 110) = 9.79$, $p < .01$. There were no significant group differences based on story and no significant interactions between story and screen design. Univariate analyses indicated screen design influenced fixation time in the main AOI, $F(2, 57) = 5.92$, $p < .01$, headline AOI, $F(2, 57) = 13.47$, $p < .01$, and crawler AOI, $F(2, 57) = 19.33$, $p < .01$.

Simple contrasts to the standard screen revealed that viewers of the screen with crawler had smaller fixation time ratios in the main and headline AOIs and considerably greater fixation time ratios in the crawler area. Viewers of the screen with headlines and crawler spent less fixation time in the main AOI and more time in the crawler area. Simple contrasts of the screen with crawler and the screen with crawler and headlines indicated viewers of the latter had larger fixation time ratios in the headline AOI (Fig. 18.2).

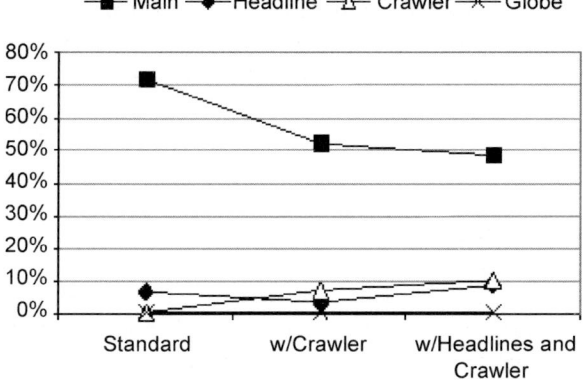

Fig. 18.2. Fixation time ratios (fixation time/story length) by screen design

18.2.4.2 Eye-Path Comparison

We compared viewer eye paths across the three TV news versions using a string-edit method, a technique that measures resemblance between sequences by Levenshtein distance, a metric based on the smallest possible cost of operations of insertions, deletions and substitutions required to "align" or transform one sequence into another [553]. While a few researchers [68, 547] have used string-edit methods to study eye-path sequences, to our knowledge this is the first study to apply this approach to television viewing. String-edit sequence comparison, in conjunction with scaling and clustering techniques suited for the resulting proximity data, provides a visual display of inter-sequence distances and identifies clusters of similar sequences.

Each area of interest was assigned a letter designation so that a sequence alphabet of fixations could be designated for each string of eye-scan data. S.A. Brandt and Stark [68] point out that stimulus materials used when a sequence alphabet is required must possess " a set of subfeatures whose positional encoding required careful review of the spatial layout by the subjects" (p. 34). The TV screen divided into five areas of interest accomplished this requirement. In this study exact coordinates are of less importance than knowledge regarding the area of interest. In addition, defining target regions avoids the complex question of exactly how much of the visual field is perceived in each fixation.

The next step was to define the eye path sequence for each participant's viewing of each news story by recording the sequence of fixations by the target area within which the fixation occurred (called "target tracing" by Salvucci and Anderson [547]. For example, a viewing beginning with a single fixation in area "A" followed by three fixations in area "C" would generate a sequence beginning "ACCC...".

Optimal matching analysis (OMA) was used to compare these coded sequences in order to generate a matrix of distance indexes for each possible pair of sequences for each stimulus. OMA is a generic string-edit tool for finding the Levenshtein distance between any two sequences composed of elements from the same sequence alphabet. Sequences similar in composition and sequence will, when compared, have smaller distances; the more different two sequences, the greater the Levenshtein distance. (See Figs. 18.3 and 18.4.) To adjust for the role of sequence length in shaping the total cost of alignment, the inter-sequence distance is determined by dividing the raw total alignment cost by the length of the longer sequence in the sequence pair. Given the range of insertion/deletion and substitution costs used in this study, and the normalization of the distance index by the length of the longer of the two sequences, the distance index ranges from 0 for identical sequences to 1 for maximally dissimilar sequences. Examples, illustrations and applications of OMA can be found in Kruskal [343], Sankoff and Kruskal [553], and Abbot and Forrest [3].

Alignments may use a combination of substitutions, insertions and deletions to produce the Levenshtein distance. S.A. Brandt and Stark [68] set

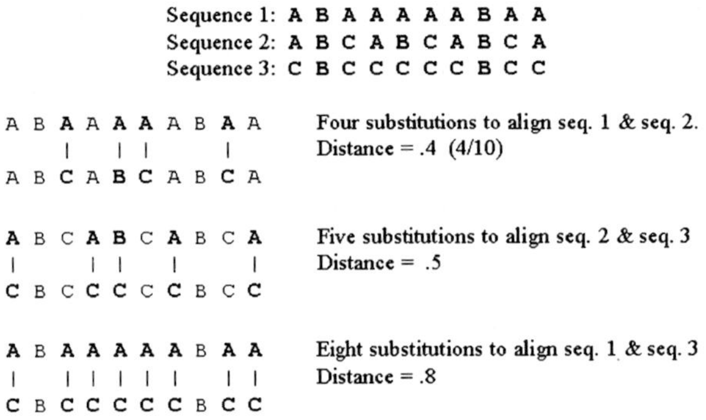

Sequence 1: **A B A A A A A B A A**
Sequence 2: **A B C A B C A B C A**
Sequence 3: **C B C C C C C B C C**

A B **A** A **A** A **A** B **A** A   Four substitutions to align seq. 1 & seq. 2.
    | .| | | Distance = .4 (4/10)
A B **C** A **B** C **A** B **C** A

A B **C** A B **C** A **B** C **A**   Five substitutions to align seq. 2 & seq. 3
| | | | | Distance = .5
C B C **C** C **C** C B C **C**

A B **A A A A A** B **A A**   Eight substitutions to align seq. 1 & seq. 3
| | | | | | | | Distance = .8
C B **C C C C C** B **C C**

Fig. 18.3. Sequence alignment or matching through insertions, deletions, and substitutions

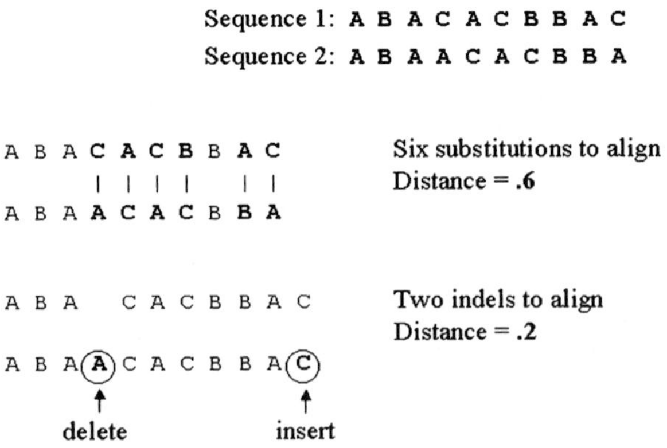

Sequence 1: **A B A C A C B B A C**

Sequence 2: **A B A A C A C B B A**

A B A **C** A C B B **A C**   Six substitutions to align
    | | | | | | Distance = **.6**
A B A **A** C A C B B **A**

A B A   C A C B B A C   Two indels to align
                                Distance = **.2**
A B A(A)C A C B B A(C)
     ↑   ↑
   delete   insert

Fig. 18.4. Levenshtein distance: the lowest cost alignment

equal substitution costs for all pairs of sequence elements. In this study, substitution values were based on the inverse of the physical adjacency of target areas; that is, the number of target regions in a direct path between the two points. The substitution cost between a fixation in a given target region and a fixation in a contiguous region was less than that between fixations in non-contiguous regions. The one exception to this scheme was the assignment of the same substitution costs to the headline and globe AOIs to avoid over-weighting of infrequent fixations in globe AOI.

SPSS software, version 13 (SPSS 2005) was used to perform non-metric multidimensional scaling (PROXSCAL) and hierarchical cluster analysis, Ward's method (CLUSTER) on the eye-path distance matrices for each of the news stories. Scaling arranges sequences in n-dimensional space such that the spatial arrangement approximates the distances between sequences; cluster analysis defines "neighborhoods" of similar cases within that space. Scree plots of stress for one- to five-dimension MDS solutions revealed, in each analysis, an elbow in the stress curve at two dimensions. The two-dimensional MDS solutions for eye-path comparisons for each story are displayed in Fig. 18.5. One dimension of the MDS solution (the x-axis) was strongly correlated with sequence length, i.e., the number of fixations in the sequence (Table 18.1). The dimension was also significantly correlated with fixations in the main AOI (which was itself strongly correlated to overall sequence length because it was the most frequently occurring fixation event). The other dimension (the y-axis) was significantly correlated with the number of fixations in the crawler AOI in each story, the number of fixations in the headline AOI (for two stories), and with fixations in main AOI (for the England story), though the correlation in this instance is much weaker than that for fixations in the crawler AOI.

Fixation sequence length is independent of screen design, $F(2, 105) = 0.741$, $p = 0.479$, so the primary evidence of screen design influence on the MDS solution is the positioning of cases on the y-axis. Standard-screen fixation sequences show less spread in the y-axis than sequences for the other two screen designs.. Standard-screen sequences have few fixations in the crawler and headline AOIs, consistent with our interpretation of the dimensions.

Dendograms and agglomeration schedules for the hierarchical clusterings of the fixation sequence distance matrices were examined to identify clustering stages which yielded a sharp increase in distance coefficients. Analysis of the England story results yielded three clusters at stage five of the 35-stage clustering. The Belgium sequences formed four clusters at stage 14 with one sequence (a short, simple fixation sequence) not clustered, and the Russia sequences formed four clusters at stage five. Cluster analysis results are mapped onto the MDS results in Fig. 18.5. Positioning of the clusters along the x and y dimensions suggests sequence length was a primary influence on cluster formation, with sequence composition a secondary influence.

Table 18.1. Correlations of MDS dimension coordinates and number of fixations in areas of interest

		Correlation to MDS dimensions (pearson's r)	
England	Total	.96**	−.14
	Main	.83**	.42**
	Title	.34*	.12
	Headline	.03	−.05
	Globe	.23	−.09
	Crawler	.13	−.91**
Belgium	Total	.93**	−.03
	Main	.80**	.29
	Title	.02	−.23
	Headline	.06	−.56**
	Globe	.22	.23
	Crawler	−.04	−.60
Russia	Total	−.71**	.06
	Main	.76**	.06
	Title	−.76	−.03
	Headline	.15	.37*
	Globe	−.16	.14
	Crawler	.17	−.36

* $p < .05$, ** $p < .01$

The sequential analysis thus far has reported sequence variations arising from differences in sequence length and composition; however, it does not address the influence of the order of fixations on the sequence comparison. One approach is to examine transition probabilities for the sequences; however, a Markovian analysis is beyond the scope of this chapter. Instead, we generated three additional distance matrices by comparing fixation sequences within each screen design. Each distance matrix was then examined to yield the "most central" sequence. The most central sequence has the lowest mean Levenshtein distance from other sequences and therefore may serve as a representative sequence for that group. Visual traces of the most central fixation sequences are displayed in Fig. 18.6.

The central sequence for viewings of the standard screen (Fig. 18.6, top image) displays the dominance of main AOI fixations. There are few transitions to the title and crawler AOIs and none to the logo AOI. Interestingly, this viewer devotes fixations to the title AOI; this is perhaps an order-of-exposure effect, as this is the participant's first story and the title graphic is unfamiliar.

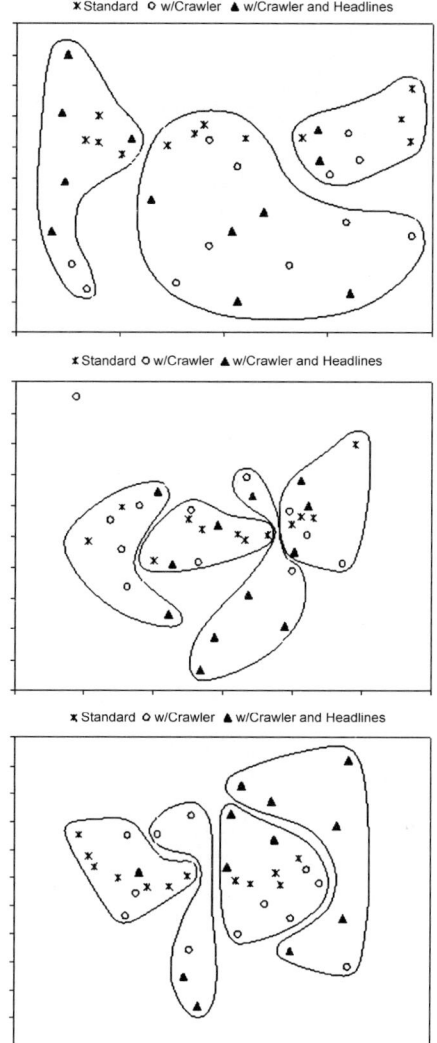

Fig. 18.5. MDS solutions in two dimensions for eye path comparisons with cluster analysis results superimposed. (*top*) England story (stress = 0.036); (*middle*) Belgium story (stress = 0.038); (*bottom*) Russia story (stress = 0.046)

The central sequence for viewings of the standard screen with crawler enhancement differs from the previous trace. First, the fixations are distributed across more of the main AOI. This may be a story effect: video in this story featured a sea shore with its horizon line just below the screen mid-point, and there were two instances of interview excerpts with the interviewees located left or right of screen center. Second, the trace moves more often to the

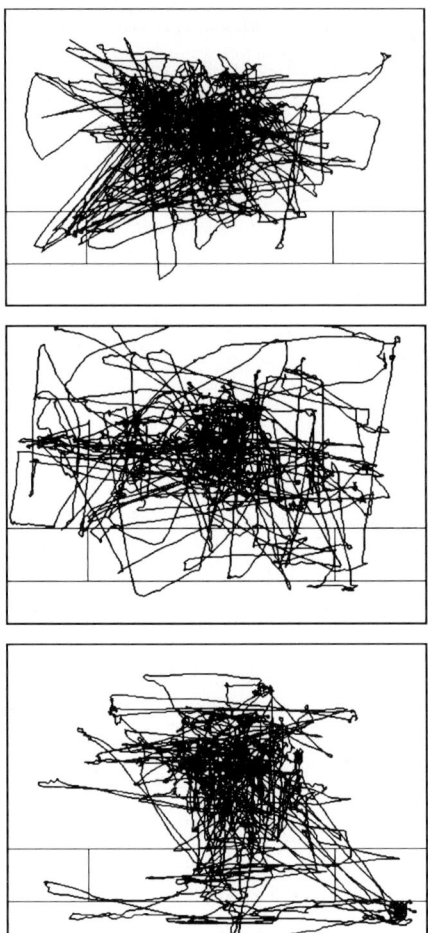

Fig. 18.6. Eye path trace of the most central fixation sequence for each screen design. (*top*) Standard screen (Belgium story; mean Levenshtein distance = 0.245, SD = 0.103); (*middle*) Standard screen with crawler enhancement (England story; mean Levenshtein distance = 0.309, SD = .091); (*bottom*) Standard screen with headline and crawler enhancements (Moscow story; mean Levenshtein distance = 0.308, SD = 0.080)

headline AOI but there are still relatively few fixations in the crawler AOI. This sequence does not reflect a case of the crawler capturing and keeping the visual attention of the viewer.

Visual attention is captured by the crawler in the central sequence for the screen with headline and crawler enhancements. Fixations cover a smaller portion of the main AOI. Horizontal eye movements from reading the story headlines and crawler text are evident. The density of the trace in the lower right corner results from reading words as they first appear in the crawler. Portions of the path traversing the lower corner appear to return to and from the main AOI or to scan to the left in the crawler AOI; transitions with the headline AOI appear to come from the central area of the crawler.

18.2.4.3 News Story Information Recall

News story information recall was measured in three ways: number of items correctly recalled, number of correct responses to questions about primary or "headline" information, and number of correct response to questions about other information content. Recall variables were positively correlated (Table 18.2).

Results of a MANOVA considering story and screen design in news story information recall revealed a main effect for story, $F(6, 196) = 7.47$, $p < .01$, and for screen design, $F(6, 196) = 4.23$, $p < .01$. There were no significant interactions between story and screen design. Univariate analyses indicated story topic influenced number of information items recalled, $F(2, 99) = 8.35$, $p < .01$, and other questions answered correctly, $F(2, 99) = 12.91$, $p < .01$. Screen design influenced number of headline questions answered correctly, $F(2, 99) = 5.46$, $p < .01$.

Simple contrasts for story topic revealed more information items were recalled and more non-headline questions answered correctly for the Belgium and Russia stories than for the England story. Simple contrasts for screen design revealed more headline questions correctly answered by viewers of the screen with headlines and crawlers than by viewers of the standard screen or by viewers of the screen with the crawler enhancement only. Conversely, fewer non-headline item questions were answered correctly by viewers of the screen with headlines and crawler (Fig. 18.7).

18.2.4.4 Screen Design Preferences

Participants evaluated all three screen designs favorably, with "liking" ratings well above midpoint on a five-point scale (Standard, $M = 3.6$, $SD = 1.0$; w/Crawler, $M = 3.2$, $SD = 1.0$; w/Headlines and Crawler, $M = 3.5$, $SD =$

Table 18.2. Correlations between recall measures

	Items recalled	Headline questions correct	Other questions correct
Items recalled			
Headline questions correct	0.477*		
Other questions correct	0.480*	0.321*	

* $p < 0.01$

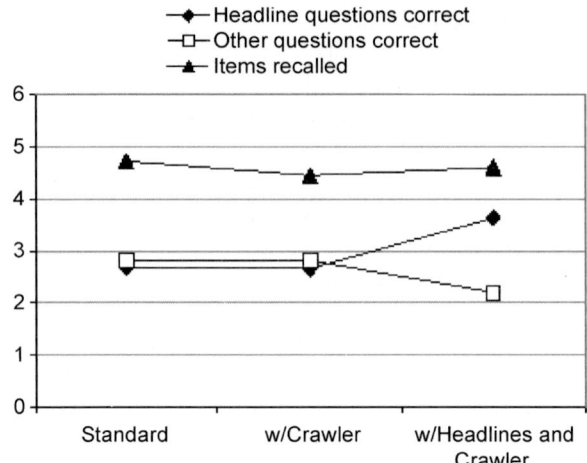

Fig. 18.7. Mean information recall measures by screen design

1.4). This finding may reflect the age demographic of study participants; younger viewers find on-screen enhancements more appealing than do older viewers [404]; note, however, that this study does not include an older viewer group for comparison so we are unable to provide an empirical test of this explanation).

Forty-two percent of participants rated the two designs with on-screen enhancements higher than the standard screen, while 31 percent of participants rated the standard screen higher than either enhanced design. Four participants (11 percent) rated the standard screen and the screen with headlines and crawler over the screen with crawler, while only three participants (8 percent) rated the "standard with crawler" screen design higher than both the standard screen and the screen with headlines and crawler.

18.2.5 Analysis and Conclusion

Research question 1 asked "Is distribution of visual attention to screen areas during TV news viewing influenced by the presence of on-screen enhancements?" Unsurprisingly, the presence of a crawler produced more fixation time at the bottom of the screen and the presence of a headline bar drew more visual attention to that area of the screen. In both cases the effect is at the expense of fixation time in the main screen area. Presence of a textual headline did not draw fixation time away from the crawler.

Research question 2 asked "Are eye-path similarities associated with the presence of on-screen enhancements?" The co-positioning of eye paths from different screen designs within clusters suggests design may influence path similarity by altering the distribution of fixations across the screen, but other factors such as sequence length have considerable effect on the Levenshtein distances.

Contrasts of the "most central" fixation sequences revealed eye paths including more of the lower part of the screen with the addition of a crawler or headlines and crawler. In the central sequence for the headlines and crawler design, saccades between the main area and the crawler tended toward the lower right corner of the screen where words first appear, whereas transitions between crawler to headline involved the crawler's center area.

Research question 3 asked "Is information recall for audio content of TV news stories influenced by the presence of on-screen visual enhancements?" Presence of a crawler did not diminish recall of main story content. Presence of a textual headline area enhanced recall of key information points of the main story; this suggests the reinforcing power of redundancy in the case of multiple channels of related information. However, diminished recall of non-headline content suggests an interference effect as well.

Research question 4 asked "Do viewers prefer TV news designs with on-screen enhancements?" Most participants believed on-screen enhancements increased the effort required to attend to the main story; however, they indicated overall liking for standard and enhanced screen designs. While the three designs were favorably evaluated, there are two preference camps: those who like the standard design more than enhanced designs, and those who like enhanced designs more than the standard design.

As television news presentation becomes increasingly more complex with the use of numerous on-screen enhancements such as headline bars and bottom-of-the-screen crawlers, it becomes increasingly important to attempt to understand whether these devices simply clutter the screen or add important content for the viewers. eye monitoring data showed that screen design did change the distribution of visual attention depending on the amount of on-screen enhancement. Subjects viewing a standard screen fixated more often in the main area of the screen compared to subjects viewing a screen with a crawler. Similarly, subjects viewing a screen with headlines and crawler spent less fixation time in the main area and more in the crawler AOI. A comparison of the screen with crawler and the screen with crawler and headline indicated viewers of the latter had larger fixation time ratios in the headline AOI.

In what appears to be the first study to apply string-edit methods to the study of TV news viewing, we found additional interesting differences in the gaze-path sequences for the three levels of on-screen enhancements. Standard-screen sequences have few fixations in the crawler and headline AOIs. In contrast, visual attention is captured frequently by the crawler in the version of the story with headline and crawler enhancements, indicating that the crawler may receive additional attention because the nearby headlines attract attention to the bottom of the screen. It is possible, as well, that the relatively static headline serves as a visual buffer reducing the saliency of movement in the video in the main portion of the screen; that is, once the crawler captures visual attention, the static buffer may reduce the re-capture of attention by the main screen area.

Dual-processing theories can be used to predict what happens to information recall when television viewers are presented with screens filled with numerous visual enhancements. According to the theory, viewers can process material being presented visually and aurally in separate working memory stores if the information is not so great as to create cognitive overload and when the information is related. Our results show that screen design impacted recall of content. Headlines aided recall of those pieces of information being summarized. However, subjects exposed to headline summaries were less likely to recall other story points, exhibiting an information interference effect.

It is important to note that screen structure in contemporary TV news design is typically stable. The rich mixture of video, graphic and textual elements and their respective positions on the screen are part of the visual identity branding of the news program. News content changes rapidly but the visual design of the news does not. This means loyal viewers see the same design repeatedly; the impact of such repeated viewing on scan paths has not been studied for television but has been explored for Web portals. Lessons learned from the web page research may prove valuable as designers face, for the first time, the challenge of designing for TV screens as portals to interactive content (imagine our "enhanced" news screen design with one more layer of complexity: a menu system, controlled by the TV remote, providing access to additional stories or related information and allowing immediate viewer feedback on news stories).

18.3 Repeated Scanning of Web Portal Pages

Since news sites emerged on the Web scene in 1994, followed by banner advertising in late 1994 and high-traffic portal pages in 1996, Internet users have been repeatedly exposed to certain visual displays of information on their computer screens. Users of online newspapers read multiple news stories with screen after screen of text displayed in similar visual patterns. Heavy users of a product or service are likely to call up the corporate home page for updated product information or to make a purchase online multiple times before the image is changed. And Web users access the Internet through the same search engine or portal page time after time seldom, if ever, changing their starting-page default.

These common practices of Internet users make the Web a natural place to test a somewhat controversial and often-discussed theory of visual perception, that of the existence of "scanpaths." Noton and Stark [459, 460] defined "scanpaths" as repetitive sequences of fixations and saccades. Fixations are when the eye is relatively immobile and indicate the area where attention is being allocated [511]. Saccades are the quick jumps of the eye from area to area, during which vision is essentially suppressed [699].

Noton and Stark's scanpath theory [460] predicts that a subject scans a new stimulus during the first exposure and stores the sequence of fixations in memory as a spatial model, so that a scanpath is established. When the subject is re-exposed to the stimulus, the first few eye movements tend to follow the same scanpath established during the initial viewing of the stimulus, which facilitates stimulus recognition.

Research also indicates that when a subject is presented with a blank screen and told to visualize a previously seen figure, the scanpath is similar to when he or she viewed the figure [581]. According to the scanpath theory, a spatial model is assumed to control the sequences of eye movements. Noton and Stark [459] asserted:

> [T]he internal representation of a pattern in memory is a network of features and attention shifts, with a habitually preferred path through the network, corresponding to the scanpath. During recognition, this network is matched with the pattern, directing the eye or internal attention from feature to feature of the pattern. (p. 940)

Noton and Stark argued that control of the eye by specific features in the visual stimulus is improbable because of the differences in scanpaths of different subjects for a given pattern. They also rejected the explanation that subjects are driven by stimulus-independent habits because of variation in scanpaths of a given subject for different stimulus patterns.

Ellis and Smith [165] elaborated on Noton and Stark's scanpath theory by suggesting that scanpaths can be generated by completely random, stratified random, or statistically dependent stochastic processes, but they did not test these conjectures. A completely random process assumes that each element of a visual has an equal probability of being fixated on during each fixation. A stratified random process assumes that the probabilities of visual elements being fixated reflect the attentional attractiveness of those elements, but they do not depend on previous fixations. The statistically dependent stochastic process specifies that the position of a fixation depends on previous fixations. In view of the perceptual processes that are assumed to underlie eye movements, Rayner [511] and Stark and Ellis [582] believed it is unlikely that saccades from one fixation point to another are generated by either completely random or stratified random processes and looked toward statistically dependent stochastic processes as explanation.

Early studies on eye movements made while subjects viewed scenes and pictures also have provided evidence that visual exploration or search is not random. Eye movements are related to the content of the scene [67, 77, 376, 699], although, as was noted, they are not controlled by it [459]. The pattern of fixations and saccades could be changed by altering the pictures or the task. Content that contains unique detail also dramatically influences the pattern of fixations and saccades, since such detail draws more attention than common or expected visual information. Viewers tend to fixate on unique regions of visual scenes sooner, more frequently, and for longer durations than any other area of the visual scene [8, 387].

Some studies focus on the role that peripheral vision plays in determining where a subject will look next. Parker [485] speculated that peripheral vision might be the major force driving the scanpath. However, eye-movement studies on ambiguous and fragmented figures showed that the same physical stimulus results in different scanpaths depending upon the changing perceptual representation of the viewer [166, 582]. Therefore, peripheral vision may not play a major role in generating the scanpath. S. A. Brandt and Stark [68] pointed out that since there was no actual diagram or picture in their visual imagery study that "[i]nput from foveal or peripheral vision cannot play a role in generating scanpath eye movements during imagery" (p. 32).

Although scanpath theory has not been studied extensively in recent years, several new studies provide support for scanpath theory. Pieters et al. [495] found that scanpaths remain constant across advertising repetitions and across experimentally induced and naturally occurring conditions, and like S.A. Brandt and Stark and others [68, 191, 710] also demonstrated firm evidence for scanpath sequences in both real and imagined stimuli.

18.3.1 The Internet and Scanpath Theory

In recent years researchers have used Markov models and string-edit methods to test the scanpath theory. For example, Pieters et al. [495] used Markov models to compare scanpaths in a study of repeated exposures to print advertisements. Stark and Ellis [582] also used Markov analysis to quantify the similarity of eye movements. S.A. Brandt and Stark [68] applied string-edit analysis to compare the viewing pattern of a diagram of an irregularly checkered grid displayed on a computer screen with the eye movements made while subjects imagined that particular grid. Using string-edit analysis, Zangemeister et al. [710] and Gbadamosi et al. [191] found evidence for scanpath sequences in their subjects' eye movements while similarly performing real viewing and visual imagery.

A Markov process is a stochastic model for the probabilities that the viewers' eyes will move from one visual element to another. The assumption is that scanpaths across visual elements can be described by a first-order Markov process; that is, each eye fixation only depends on the previous one. In addition to Markov dependence, there are two other constrained stochastic conditions that are possible: reversibility and stationarity. Reversibility means that saccades from element A to B occur as often as saccades from B to A [165], and stationarity predicts the scanpaths of viewers exposed repeatedly to the same visual remain constant across exposures.

Pieters et al. [495] concluded that scanpaths remain constant across repeated exposure to advertising stimuli and across experimentally induced and naturally occurring conditions. They also concluded that scanpaths obey a stationary, reversible, first-order Markov process. Using a different means of comparing scanpaths – the string-edit method – S.A. Brandt and Stark [68] also found evidence supporting scanpath theory. Specifically, they found that eye movements during imagery are not random but reflect the content of the visualized scene. They concluded, therefore, that an "internalized, cognitive perceptual model must be in control of these scanpaths" (p. 32).

Abbott and Hrycak [2] noted several advantages of string-edit methods for studying event sequences and outlined several limitations with Markovian sequence models. First and foremost, they argued, the sequence-generating process may have a longer history than the immediate past typically used in Markov analysis. Second, Markov models describe the stochastic processes that generate observed sequences and can be used to explore the goodness of fit of a predicted model, but don't address the questions of whether there is a typical event sequence for a given process. Abbott and Hrycak argued that the direct testing of the Markov model – in terms of actual resemblance between generated and observed sequences – requires a technique for assessing similarity between sequences, categorizing sequences, and identifying typical sequences. String-edit analysis affords all of these techniques.

We tested Noton and Stark's [459] scanpath theory on different kinds of images widely used on the World Wide Web: a news page, an advertising page and a portal page. We compared recorded scanpaths using a string-edit method, a technique that measures resemblance between sequences by means of a simple metric based on the insertions, deletions and substitutions required to transform one sequence into another [553]; this generates a distance index, or measure of dissimilarity.

Although several researchers such as S.A. Brandt and Stark [68] and Salvucci and Anderson [547] have used string-edit methods to study eye-path sequences, relatively few studies using this method have been reported, despite the fit between eye monitoring data and string-edit methodology. To our knowledge, this was the first study in which repeated exposures to Web page visual stimuli was examined using eye tracking for measurement and string-edit methods for analysis.

String-edit sequence comparison, in conjunction with scaling and clustering techniques suited for the resulting proximity data, provides a visual display of the inter-sequence distances and identifies clusters of similar sequences. If scanpaths are stable over repeated viewings and are not driven wholly by stimulus features (and are, therefore, variable across subjects), sequences for a given subject and stimulus should group together in neighborhoods in the scaling and should share cluster membership. Sequences for a given subject should not be in separate neighborhoods and clusters, although a given neighborhood or cluster may contain multiple subjects, reflecting the influence of stimulus features across subjects.

18.3.2 Method

18.3.2.1 Subjects

The subjects were eight students at a large western university (four males and four females). Their average age was 22.5 years. They were compensated for participating in the three-session study. All subjects were regular users of the Internet, reporting an average of almost nine hours a week of usage.

18.3.2.2 Apparatus

The eye-movement data were collected using an ISCAN RK-426PC Pupil/ Corneal Reflection Tracking System (ISCAN, Inc. 1998). See above for details.

18.3.2.3 Stimuli

The three Web pages used as stimuli were chosen for a number of reasons. Each page represents a distinct category of visual imagery on the Web. The portal page, used as a starting point for content on the Web, consists of a large number of hyperlinks and dialogue boxes for search functions and e-mail. The advertising page is highly visual and extremely colorful and is used to "build the brand" and sell the product. The news page displays mostly typography of various sizes for headlines, bylines, and body copy and is designed to convey information in an efficient manner.

The portal page and the on-line advertisement were completely contained in the first view, not forcing the viewer to scroll, thus simplifying the data analysis. On the news page, scrolling was required to view the entire news story, but the remainder of the Web page content remained stationary through the use of frames. Only the area containing the body of the story was scrollable.

These three Web pages easily facilitated superimposing simple grids over their images (see Fig. 18.8). This was necessary since string-edit analysis requires defining a sequence alphabet; in this case, a set of target regions in each stimulus. The Web pages selected were deemed to be relatively simple in layout, yet contained enough complexity that the sequential processing of the subfeatures could occur, thus producing the necessary sequential eye movements that define a scanpath. S.A. Brandt and Stark [68] emphasized that the stimulus materials used in this sort of analysis required "a set of subfeatures whose positional encoding required careful review of the spatial layout by the subjects" (p. 34).

The use of grids was also logical because of the use of Web pages as stimuli. In the news and advertising fields, for example, it is usually more useful to know what elements are looked at and how frequently they are viewed than to know the exact coordinates of each fixation. In addition, defining target

regions avoids the complex question of exactly how much of the visual field
is perceived in each fixation.

However, since the three stimuli shown in Fig. 18.8 differ considerably
in the size, variety, and configuration of their target regions, the sequence
comparisons can be performed only for scanpaths within each stimulus; cross-
stimulus sequence comparisons are not meaningful. The differences in stimuli
may also influence overall results in sequence similarity. A stimulus with
a large region that may capture the extended attention of the viewer, such
as the text region on the news page (Fig. 18.8), may result in lower sequence
dissimilarities, owing to less variety in scanpaths.

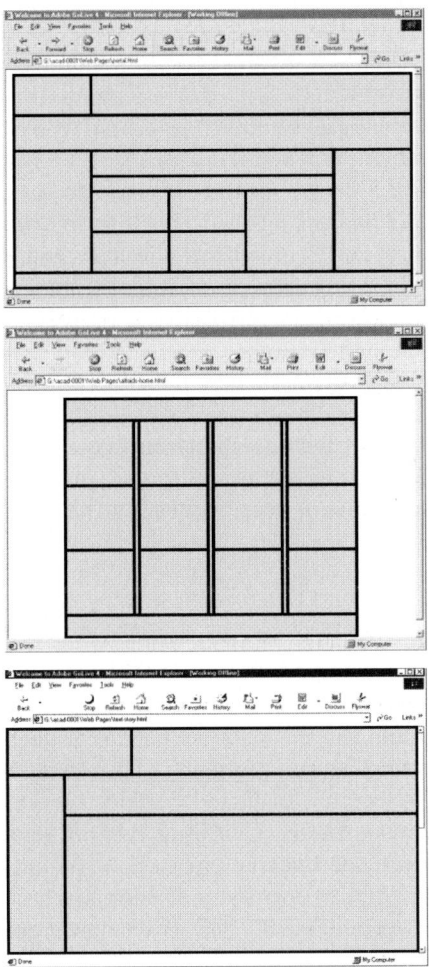

Fig. 18.8. Target region grids for portal, advertising, and news Web page stimuli.
(*top*) Portal page; (*middle*) Advertising page; (*bottom*) News page

18.3.2.4 Data Gathering

Each of the eight subjects reported to the eye monitoring lab three different days (a Tuesday, a Friday, and a Monday) separated by 48 hours. One week was selected as an appropriate length of time because basic research on scanpath theory finds stable effects, across exposure durations, of up to a week [581]. During each visit, the subjects viewed the same three Web pages while their eye movements were recorded. They viewed the Web pages in a different order of exposure on each visit. At the lab, a subject was seated in front of the computer monitor and was told that he or she would be looking at three Web pages for a brief period of time. Each subject was instructed to imagine that he or she had been surfing the Web and had encountered the page. The subject was told to view the page as he or she would in that situation. The subjects were not given specific tasks in viewing the Web pages, because we were interested in scanpaths independent of directed search behavior, similar to what might be found in casual Web surfing. Each subject was also told not to click on any links.

Eye movement data for a 15-second exposure to each page were recorded. Previous research [377] has established that eye fixations on essential information in a visual display occur within the first few seconds. S.A. Brandt and Stark [68] gave their subjects 20 seconds of viewing for familiarization and 10 seconds for imagining. In this study, a pretest was used to determine that 15 seconds was enough time for the subjects to examine the Web pages. Fifteen seconds also provided sufficient data to compare the eye paths with string-edit tests.

When the subjects returned on subsequent days, they were told to imagine they were surfing the Web and had chosen to revisit each site. They were instructed to view the pages as they would in a return visit to the site.

18.3.2.5 Procedure

Sequence comparison. The first step in comparing the eye-path sequences was to define a sequence alphabet for each Web site. This was accomplished by assigning each target area on each Web page an alphabetic code. The second step was to define the eye-path sequence for each subject's viewing of each Web page by recording the sequence of fixations by the defined target area within which the fixation occurred (called "target tracing" by Salvucci & Anderson [547]. For example, a viewing beginning with a single fixation in Area A followed by three fixations in Area C would generate a sequence beginning "ACCC...". Again, optimal matching analysis (OMA) was used to compare these coded sequences. See the television study reported above for more details on this analytic procedure.

In their application of the string-edit method, S.A. Brandt and Stark [68] set equal substitution costs for all pairs of sequence elements. In this study, substitution costs could have been set in at least three ways: (1) at a value equal for all substitutions, following S.A. Brandt and Stark; (2) by distinctions between content forms of the regions, so that costs for regions in related content categories would be lower (e.g., top-level site navigation menus vs. page content navigation menus), whereas costs between categories would be higher (e.g., navigation regions vs. page text regions); and (3) by a measure of distance such that regions "closer" to each other could be substituted at less cost than regions further apart. We believed that uniform substitution costs lacked requisite variety; that is, uniform substitution costs treat the difference between two fixations in adjacent target areas as equivalent to the difference between two fixations in widely separated areas. The second approach was unsuitable for testing a theory of physical scanpaths, since the distance indexes would emphasize differences in content viewed rather than differences in the actual paths through which it was viewed. We applied the third approach; substitution values were based on the inverse of the physical adjacency of target areas; that is, the number of target regions in a direct path between the two points. For example, a pair of contiguous target areas, on a page for which the longest direct path between the centers of two target areas traversed five regions, would be assigned a substitution cost of .20. This assumes that for the purposes of comparing scanpaths, the difference between a fixation in a given target region and a fixation in a contiguous region is less than that between fixations in noncontiguous regions.

Next, WinPhaser software [259] was used to generate a matrix of distance indexes for each possible pair of sequences for each stimulus. WinPhaser's OMA package uses a dynamic programming algorithm by *Andrew Abbott* of the University of Chicago. UCINET software was used to perform nonmetric multidimensional scaling and hierarchical cluster analysis on the distance matrices. Scaling arranges the sequences in n-dimensional space such that the spatial arrangement approximates the distances between sequences; cluster analysis helps to define neighborhoods of similar cases within that *n*-dimensional space.

18.3.2.6 Results

Twenty-four scanpath sequences were generated for the portal stimulus and for the advertising stimulus (three sequences for each of 8 subjects). There were 23 sequences generated for the news stimulus; 1 subject's second viewing of the stimulus was not recorded owing to equipment failure. Scanpath sequences for the portal stimulus ranged in length from 21 to 62 fixations ($M = 39.5$, $SD = 10.3$).

Scanpath lengths for the advertising stimulus ranged from 21 to 54 fixations ($M = 39.3$, $SD = 7.9$). The scanpaths for the news stimulus ranged from 22 to 70 fixations in length ($M = 49.8$, $SD = 13.5$). Sequence distances for the portal stimulus scanpaths ranged from .27 to .75 ($n = 276$, $M = .47$, $SD = .09$); the advertising stimulus scanpaths ranged from .33 to .74 ($n = 276$, $M = .50$, $SD = .08$). The news stimulus scanpaths, although showing lower overall distance indexes ($n = 253$, $M = .41$, $SD = .16$), produced a greater range of sequence distances, from .06 to .74. As was noted above, if the viewer reads the text block, the scanpath sequence settles into a series of fixations in the same target region. Pairwise comparisons of such sequences generate lower distance indexes, because few operations are required to align the sequences.

The distance indexes for the multiple viewings of each stimulus by each subject are displayed in Table 18.3. Two features should be noted: First, the results for the news stimulus reveal consistently low distance indexes for viewings by three subjects.

Distance indexes for repeated viewings of portal, advertising and news stimuli by each participant.

Their scanpaths are dominated by fixations in the text region of the stimulus, suggesting close reading. In contrast, other subjects devoted more initial fixations to regions surrounding the text or continued to move between the text region and other regions, rather than having their attention captured by the text.

Second, for each stimulus, the distance index for the comparison of the second and the third viewing tends to be the lowest of the three possible within-subjects comparison pairs (first to second viewing, second to third viewing, first to third viewing), suggesting an initial drift and subsequent stabilizing of the scanpath over time.

A 3×3 repeated measures analysis of variance (ANOVA) was performed on the distance indexes reported in Table 18.3. Mauchly's test of sphericity was not statistically significant for stimulus (portal, advertising, and news; $W(2) = .632$, $p = .317$) or for comparison pair (first vs. second viewing, second vs. third viewing, and first vs. third viewing); $W(2) = .494$, $p = .171$) but was statistically significant for the interaction of stimulus type and comparison pair ($W(9) = .011$, $p = .026$). Owing to this finding, the Greenhouse–Geisser correction was applied in the ANOVA. The main effect of stimulus was statistically significant ($F(2,12) = 11.676$, $\mathbf{e} = .731$, $p = .007$). The main effect of comparison pair was also significant ($F(2,12) = 5.80$, $\mathbf{e} = .664$, $p = .031$). The interaction of stimulus type and comparison pair was not statistically significant ($F(2,12) = 0.156$, $\mathbf{e} = .433$, $p = .829$).

Post hoc pairwise comparisons of stimuli, with Bonferroni correction for multiple comparisons, revealed a statistically significant difference between the distance index means for the advertising stimulus ($M = .484$, $SD = .046$), and the news stimulus ($M = .305$, $SD = .305$; $difference = .179$, $p = .012$). The advertising stimulus (Fig. 18.8) is a regular grid of independent images and offers, of the three stimuli, the least content-based impetus to top-down, left-right viewing, thereby placing fewer constraints on viewing differences among subjects. The news stimulus (Fig. 18.8), in contrast, contains a large textual region that may, as was noted, drive greater similarity between subjects in their viewing of the page.

Table 18.3. Distance indexes for repeated viewings of portal, advertising and news stimuli by each participant

Participant	Portal Stimulus Sequence Distance Index by Scanpath Pair			
	1st to 2nd	2nd to 3rd	1st to 3rd	Mean
1	0.32	0.43	0.40	0.38
2	0.37	0.37	0.42	0.39
3	0.46	0.34	0.51	0.44
4	0.44	0.33	0.42	0.40
5	0.46	0.58	0.41	0.48
6	0.44	0.27	0.46	0.39
7	0.46	0.38	0.38	0.41
8	0.49	0.41	0.51	0.47
Mean	0.43	0.39	0.44	

Participant	Advertising Stimulus Sequence Distance Index by Scanpath Pair			
	1st to 2nd	2nd to 3rd	1st to 3rd	Mean
1	0.57	0.44	0.60	0.54
2	0.43	0.39	0.48	0.43
3	0.48	0.44	0.47	0.47
4	0.57	0.57	0.59	0.57
5	0.63	0.34	0.66	0.54
6	0.43	0.47	0.49	0.46
7	0.46	0.37	0.47	0.43
8	0.44	0.46	0.52	0.47
Mean	0.50	0.44	0.53	

Participant	News Stimulus Sequence Distance Index by Scanpath Pair			
	1st to 2nd	2nd to 3rd	1st to 3rd	Mean
1[a]			0.16	
2	0.14	0.19	0.17	0.16
3	0.42	0.50	0.43	0.45
4	0.39	0.34	0.41	0.38
5	0.49	0.14	0.48	0.37
6	0.43	0.25	0.52	0.40
7	0.20	0.25	0.10	0.18
8	0.21	0.22	0.13	0.18
Mean	0.32	0.27	0.30	

[a] The second viewing session did not produce useable eye monitoring data for participant 1

Post hoc tests showed no statistically significant pairwise differences among the comparison pairs; however, if paths stabilize over time, distance indexes should follow a pattern of (distance of viewing 1 vs. viewing 3). (distance of viewing 1 vs. viewing 2). (distance of viewing 2 vs. viewing 3). This was supported by a within-subjects polynomial contrast that revealed a statistically significant linear trend for comparison pairs ($F(1,6) = 6.817$, $p = .040$).

Plots of the multidimensional scaling solution in two dimensions are displayed for the portal stimulus (Fig. 18.9a), the advertising page stimulus (Fig. 18.9b), and the news page stimulus (Fig. 18.9c). These are spatial representations of the structure of the distance indexes; each sequence is represented by a point in a multidimensional space. The points are arranged so that the distances between all pairs of points are an optimal fit to the dissimilarities of the sequence pairs. Similar sequences are represented by points that are close together, whereas more dissimilar sequences are represented by points that are farther apart.

The two-dimensional solution to multidimensional scaling was used for convenient display because we were more interested in recognizing neighborhoods than in defining dimensions. The figures also indicate the most central eye-path sequence (i.e., the sequence or sequences with the least mean distance from other sequences in the multidimensional scaling solution). In addition, in each set, the most similar sequence pairs are noted, as well as the two most dissimilar sequence pairs.

It is worthwhile to examine several of these cases, since they underscore the operation of the string-edit method used here and aid in the interpretation of the results. Figure 18.10 displays eye-path sequences for four viewings of the portal stimulus. According to our string-edit results, the paths for Subject 4 and Subject 5 are the most dissimilar in the set. Visual inspection suggests that this stems from differential attention to upper and lower areas of the display. Conversely, the lower two paths in Fig. 18.10 are more similar; they are from the same person (Subject 8) and represent "central" sequences in the set. They are characterized by more similarity than the previously noted paths and show attention distributed across the width of the middle tier of regions.

Visual examination of the spatial arrangement of the sequences reveals support for scanpath stability. In the case of the portal stimulus (Fig. 18.8), for Subjects 1, 3, 4, 6, and 7, we find the eye paths from the three separate viewings co-located in relatively small areas. In addition, two of the three eye-path sequences are neighbors for Subject 2. The advertising page (Fig. 18.8) stimulus reveals somewhat less stability of paths, since only Subjects 2, 6, and 8 appear to form small neighborhoods for all three sequences. The news page (Fig. 18.8), which invites top-down, left-right processing of its textual contents, displays tighter co-location of most cases. This is especially notable

a   Eye path sequences for the portal page stimulus

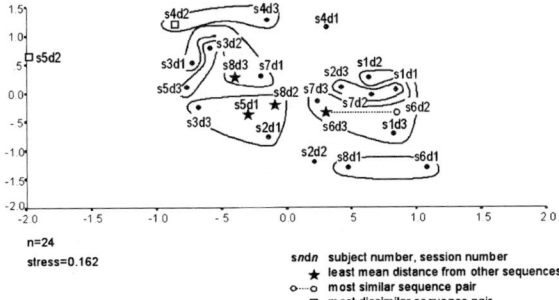

b   Eye path sequences for the advertising page stimulus

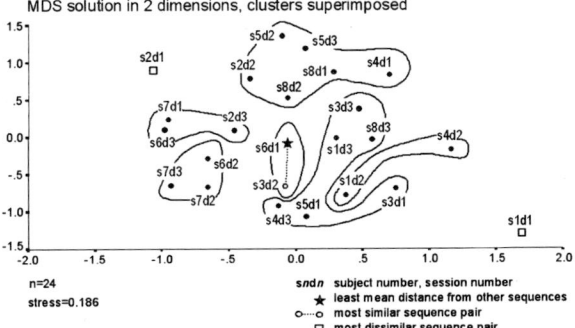

c   Eye path sequences for the news page stimulus

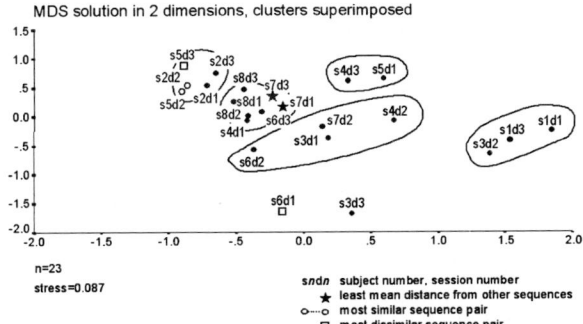

Fig. 18.9a–c. Multidimensional scaling solution in two dimensions with cluster analysis results superimposed

for Subjects 2, 7, and 8; however, Subjects 1, 4, and 5 all provide a pair of closely located sequences as well.

Cluster analysis results provide another approach to discerning neighborhoods of similar sequences and are superimposed on the multidimensional scaling results in Figs. 18.5, 18.6, and 18.7. The hierarchical clustering of the

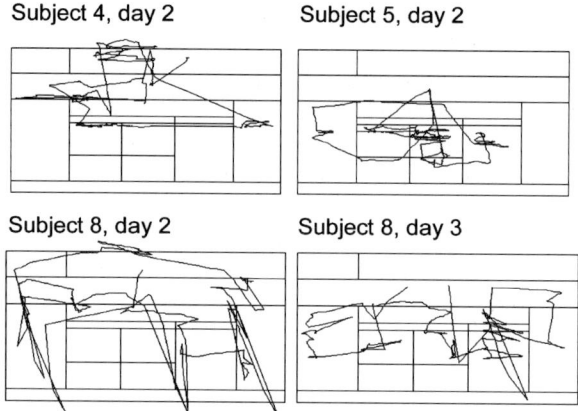

Fig. 18.10. Example eye path sequences for the portal stimulus

portal sequences yielded seven clusters at Step 13 of the 23-step clustering solution. Three sequences had not joined a cluster at this step. The advertising sequences formed six clusters at Step 14 of the 21-step solution with two sequences not clustered, and the news sequences formed four clusters at Step 16 of the 21-step solution with two sequences not clustered.

The clustering results suggest that a three-dimensional solution may have been superior, since the "worm-shaped" clusters in some of the diagrams may reflect the greater stress of forcing the items into the less-optimal fit of two-dimensional space (i.e., a cluster that would fit into a small three-dimensional space is distorted or stretched in a two-dimensional representation). The clustering reveals families of sequences across subjects; note, for example, the cluster composed of a pair of eye-path sequences from Subject 1 and a pair from Subject 6 in Fig. 18.5 (portal stimulus). Similar features are found in the other plots. Figure 18.6 (advertising stimulus) reveals a larger cluster to be composed of sequences from 4 subjects (Subjects 2, 4, 5, and 8). In Fig. 18.7 (news page stimulus), the largest cluster includes all three viewings for 2 subjects, and two viewings for 2 additional subjects; yet one of the other clusters is composed of single sequences from 4 different subjects. Such cross-subject scanpath similarity may bear witness to the interplay of design features and individual scanpath preferences.

The eye-path sequences for the news stimulus (Fig. 18.9) are more similar to each other than are the most similar sequences for the portal or advertising stimuli, perhaps because the text region tended to keep the fixations once it had captured them. Given this "pull," remaining within-subjects differences in the scanpath for the news stimulus are interesting, since they may suggest either (1) lack of stability, over time, of the person's scanpath (e.g., drift and subsequent stabilizing) or (2) a confounding memory effect wherein familiar textual material is scanned differently from new material (e.g., attention decrement).

In a similar fashion, the strong within-subjects resemblances between sequences for the portal and the advertising stimuli, which in design are less governed by left-right/top-down conventions, are notable, since they suggest either (1) scanpath stability or (2) particular page features tending to capture attention in the same general sequence. Note, however, that when different subjects show high within-subjects scanpath resemblance, but nevertheless with marked difference between subjects, we can conclude personal scanpath preferences do indeed have some explanatory power.

18.3.3 Conclusion

The presented results in this chapter are mixed. Some individuals show scanpaths that resemble each other over time. However, it is found many instances in which the most similar sequences were from different subjects, rather than from the same subject, suggesting strong stimulus influences. On the other hand, the clusters tend to include pairs of sequences from the same subject. The fact that clusters of sequences also typically contain paths from multiple subjects suggests that other forces may be important, such as features of the Web page or memory. These could be tested in future research with carefully manipulated page versions.

The linear trend to increased similarity across viewings suggests that paths appear to drift more from the first to the second viewing than from the second to the third viewing for a given subject. The following questions present themselves. Is there an increased tendency with the passing of time to ignore material on the top of Web pages? Do the paths become simpler on subsequent viewings? Do viewers dwell longer on selected regions of particular interest? Does task fatigue result in shifts in visual attention? Does a scanpath become increasingly habituated?

In summary, on the World Wide Web, with somewhat complex digital images, some viewers' eye movements appear to follow a "habitually preferred path" across the visual stimulus, as asserted in the scanpath theory of Noton and Stark [459]. Given the still-considerable variation in paths between and across subjects and the differences found across stimuli, much more research is needed to explore the influence of scanpaths, content, and form on sequences of eye movement.

18.4 Passive Eye Monitoring Challenges

Our eye monitoring research with TV screens and with Web pages presented methodological challenges specific to each platform; as the platforms converge these methodological issues will confront iTV and Web 2.0 researchers. Novel challenges will be presented by mobile device screens as well.

18.4.1 TV Studies

The most important limitation in our television study regarding the effects of on-screen enhancements was the ecological validity of the viewing setup. The viewing situation was more similar to a typical computer-use situation than to normal television viewing. We typically watch TV from comfortable chairs at a considerable distance from the screen (TV is the "8- to 10-foot" interface, while computer displays are the "2-foot interface"). Viewing video content at close range on a computer monitor may evoke scanning habits developed for computer screens rather than "normal" television viewing habits. We intend to replicate the study in a homelike viewing setting with the stimuli displayed on a television at a conventional distance from viewers. If results are similar, we have some assurance of the value of TV eye monitoring studies using desktop eye monitoring equipment. If results are markedly dissimilar, computer-based studies of TV content viewing will be called into question and more testing of differences between the computer and TV viewing experiences will be called for. Of course, the surge in availability of TV content on the Web may reduce this concern about ecological validity – watching TV content on computer screens isn't yet common, but it is likely to become so.

We separately analyzed the influence of screen design on distribution of fixations across areas of interest and on recall for story content. Direct linking of measures of visual attention to information recall requires a more sophisticated design that tracks the information content available during periods of visual attention to the main story. Our fixation analysis focused on the distribution of fixations across areas of interest. Future research should explore the alignment of eye paths with the timing of visual information in the main story and visual information in the crawler and headlines. Such alignments may lead to identification of direct influences of distribution of fixations on recall of visual content.

TV studies offer the eye monitoring researcher another source of information to consider: the auditory track of the story. On-screen headlines typically provide a visual display of main story points. This display and the audio story are mutually reinforcing; however, crawler content is a potential perceptual interference with audio content as well as with main story visual content. More advanced experimental designs will be necessary to explore the interactions of visual and auditory elements.

18.4.2 Web Studies

In our Web study regarding repeated visits of portal pages, scrolling was not required. This allowed us to simplify the data analysis by presenting the Web user with a limited, stationary visual field. However, we seldom encounter Web pages where scrolling is not characteristic. "Avoid making users scroll" is a design principle more often breached than honored.

Vertical scrolling through information in the Web environment probably does not come naturally to viewers who are more accustomed to pursuing moving information in their natural environment in a horizontal fashion. Our visual field is horizontal. Most of what we need to be aware of in our natural environment in order to survive can be seen on a horizontal plane. Most of the movement our eyes pursue is horizontal. For humans this pursuit of horizontal movement can be largely accomplished through eye movements, whereas the pursuit of vertical movement often requires head movements and larger-than-normal saccades.

In our study of scanpaths, the location and order of each fixation is important. The fixations made while a Web user is rapidly scrolling may simply be fixations of convenience to accommodate the scrolling behavior. The eye has to land somewhere at least five times each second until it locates a major informational signpost, causing the scrolling to cease and the processing of information to occur.

In addition, in our Web study viewers were not allowed to follow links. They were simply required to look at the page presented to them. Obviously, following links would allow them to move to different pages. This presents an enormous problem with a methodology reliant on designing a sequence alphabet based on grids used to form the visual structure of a particular page. Each page linked to would employ a different organizational structure, grid and therefore sequence alphabet, complicating and obscuring the data analysis used to compare scanpaths.

But scrolling and linking are not the only challenges presented in eye monitoring studies of the Web. The content on a Web page is not created equally. There is textual information of various kinds and sizes – headlines, body copy, navigation, links to related articles just to name a few. There is visual information of various kinds as well – graphics and photographs. If the visual grids contain verbal information more fixations may be required by the user to make sense of the information. Just a few fixations or even just one may be all that is necessary to process visual information. When you consider color, density of information, location of information on the screen and overall purpose of the Web site, the complexity of an eye monitoring study on the Web becomes even more apparent. In order to use the string-edit technique for sequence comparison, decisions have to be made regarding the basis for substitution costs in sequence alignments (e.g., should costs be based on content types or target adjacency?). A choice has to be made. There are trade-offs and no easy decisions. Future research should explore how the different alternatives impact analysis outcomes.

18.4.3 Compounded Challenges in Converged Media

As if the challenges in studying television and the Web using eye tracking are not enough, these challenges will be amplified in studies of emerging new media which possesses the characteristics of both television and the

Web. This single screen will display verbal information and visual information coupled with movement and sound. Split and layered screens of information will become the norm. The eye monitoring apparatus and analysis required to study this new media will be required to increase in accuracy and complexity accordingly.

Screen size, itself, raises interesting issues for eye monitoring studies. With the extra-large screens that may be used to display this merged media, managing visual angle may become a problem, whereas with small mobile screens calibration and accuracy of fixation become the issues. Commercially-available mobile eye monitoring systems allow the researcher to determine if the user is looking at the hand-held screen or the environment but point of gaze within the screen may be difficult to determine precisely.

In short, researchers employing eye monitoring methods are likely to look back at research on the viewing of print (e.g., newspapers, Yellow Pages[TM] and print advertisements) and static web pages as "the good old days when it was easy." The integration of dynamic video content with static and interactive design elements in interactive television and video-rich Web sites means our research designs and research tools must become more sophisticated if we are to capture and understand how viewers look at these media.

Acknowledgements

Chapter 2
Portions of this work were supported by NIMH grant R01 MH51435 to the University of Pittsburgh and Carnegie Mellon University. A previous version of this chapter was published in *IEEE Transactions on Pattern Analysis and Machine Intelligence* (©IEEE).

Chapter 3
The completion of this research was made possible thanks to the Natural Sciences and Engineering Research Council of Canada (NSERC), a NATEQ Nouveaux Chercheurs Grant, and a start-up grant from Concordia University.

Chapter 6
We thank the support from CAPES (Coordenação de Aperfeiçoamento de Pessoal de Nível Superior) and FAPESP (Fundação de Amparo à Pesquisa do Estado de São Paulo)

Chapter 7
This research was conducted at the Columbia Vision and Graphics Center in the Department of Computer Science at Columbia University. It was funded by an NSF ITR Award (IIS-00-85864). A previous version of this chapter has been published in Int'l Journal of Computer Vision [449] and IEEE Computer Vision and Pattern Recognition Conference [451].

Chapter 8
Riad I. Hammoud would like to thank his co-authors Paivi Majaranta, Dan Witzner Hansen and Volker Kruger for their collaboration on a joint Springer book chapter in 2005-2006 titled: "Human-Video interaction through eye-gaze: Methods, Interfaces and Perspectives". A part of this unpublished work has been edited thoroughly, and further extended and included in this Springer book chapter on "Gaze-Based Interaction".

Chapter 9
Dr Mike Jones wishes gratefully to acknowledge the Paul Instrument Fund of
the Royal Society (UK) and the Special Trustees of the United Bristol Health-
care NHS Trust for funding that supported development of this project. The
substantial part of this work was performed whilst Mike Jones was a Research
Fellow at Griffith University, Australia. Dr Stavri Nikolov wishes to thank
the University of Bristol, UK, where he worked for many years whilst early
development of the system was carried out.

Chapter 10
Our research on gaze guidance and gaze-based communication has received
funding from the European Commission within the projects GazeCom (contract no. IST-C-033816) and COGAIN (contract no. IST-2003-511598) of
the 6th Framework Programme. All views expressed herein are those of the
authors alone; the European Community is not liable for any use made of
the information. We thank SensoMotoric Instruments GmbH for their eye-
tracking support; data were obtained using their iView X Hi-Speed system.

   This work is based on an earlier work: Gaze-Contingent Temporal Filtering of Video, in Eye Tracking Research & Applications 2006, © ACM, 2006.
http://doi.acm.org/10.1145/1117309.1117353

Chapter 13
The work presented here is part of the SAfety VEhicle(s) using adaptive Interface Technology (SAVE-IT) program that was sponsored by the National
Highway Traffic Safety Administration (NHTSA) (Project Manager: Michael
Perel) and administered by the John A. Volpe National Transportation Systems Center (Project Manager: Mary D. Stearns). We are grateful for the
valuable comments offered by the graduate students in the Cognitive Systems Laboratory as well as for the assistance of Teresa Lopes at University
of Iowa Public Policy Center in preparing this manuscript.

Chapter 15
The authors would like to thank George Mylonas, Adam James and Louis
Atallah for their contribution to this chapter.

Chapter 16
Many thanks to Geri Gay, Helene Hembrooke, Thorsten Joachims, Fabio
Pellacini, Kerry Rodden, and Daniel Russell for their invaluable contributions
and discussions.

Biographies

In the following the list of contributors and their biographies:

M. Stella Atkins received the BSc degree in Chemistry from Nottingham University in 1966 and the Ph.D. degree in computer science from the University of British Columbia in 1985. She is a Professor in the School of Computing Science at Simon Fraser University (SFU), and director of the Medical Computing Lab at SFU. Her research interests include telehealth; medical image display and analysis involving medical image segmentation, image registration, denoising, medical image perception, and radiology workstation design; and the use of eyegaze trackers which provide a new imaging modality for seeing inside the brain.

Erhardt Barth is currently a scientist at the Institute for Neuro- and Bioinformatics, University of Lübeck, Germany. His main research interests are in the areas of human and computer vision. He obtained a Ph.D. in electrical and communications engineering from the Technical University of Munich, and has conducted research at the Universities of Melbourne and Munich, the Institute for Advanced Study in Berlin, the NASA Vision Science and Technology Group in California, and the Institute for Signal Processing in Lübeck. In May 2000 he received a Schloessmann Award.

Roman Bednarik is a researcher at the groups of Software Engineering and Educational Technologies at the Department of Computer Science and Statistics, University of Joensuu. He is conducting research related to eyetracking, human–computer interaction, and educational technologies.

Martin Böhme received his degree in computer science from the University of Lübeck, Germany in 2002. He spent six months at Kyoto University before returning to the University of Lübeck, where he is a research assistant at the Institute for Neuro- and Bioinformatics, working on gaze-contingent displays, eye-tracking, and 3D-Time-of-Flight-based tracking.

Nizar Bouguila received an engineering degree from the University of Tunis in 2000, the M.Sc. and Ph.D degrees from Sherbrooke University in 2002 and 2006, respectively, all in computer science. He is currently an Assistant Professor with the Concordia Institute for Information Systems Engineering (CIISE) at Concordia University, Montreal, Qc, Canada. His research interests include image processing, machine learning, 3D graphics, computer vision, and pattern recognition. He authored several papers in international journals and conferences. In 2006, Dr. Bouguila received both the NSERC and NATEQ postdoctoral fellowships. In 2007, he received the best Ph.D. thesis award in engineering and natural sciences from Sherbrooke University and was a runner-up for the prestigious NSERC doctoral prize.

Flavio Coutinho received his B.S. and M.S. degrees in Computer Science from University of Sao Paulo, Brazil, in 2001 and 2006. The focus of his M.S. dissertation was on free head motion eye gaze tracking and currently he is a Ph.D. student and a member of the Laboratory of Technologies for Interaction (LaTIn) at University of Sao Paulo, where he continues his research on gaze tracking.

Jeffrey Cohn is Professor of Psychology and Associate Professor of Psychiatry at the University of Pittsburgh and Adjunct Faculty at the Robotics Institute, Carnegie Mellon University. He earned his Ph.D. in Clinical Psychology from the University of Massachusetts in Amherst and completed his Clinical Internship at the University of Maryland Medical Center. For the past 20 years, he has conducted investigations in the theory and science of emotion, depression, and nonverbal communication. He has co-led interdisciplinary and inter-institutional efforts to develop advanced methods of automatic analysis of facial expression and prosody; and applied these tools to research in human emotion, social development, non-verbal communication, psychopathology, biomedicine, and biometrics. His research has been supported by grants from the National Institute of Mental Health, the Canadian Institutes of Health Research, the National Institute of Child Health and Human Development, the National Science Foundation, the Central Intelligence Agency, the Defense Advanced Research Projects Agency, and the Naval Research Laboratory.

Michael Dorr received his degree in computer science from the University of Lübeck, Germany in 2004. He is currently a research assistant at the Institute for Neuro- and Bioinformatics at the University of Lübeck, Germany. His research interests are the psychophysics of motion estimation and the modeling of eye movements and attention. In 2001, he was awarded the "Student Poster Prize" at the Tübingen Perception Conference.

Matthew K. Feusner is a Programmer/Analyst in the Department of Ophthalmology at the University of California, San Francisco. He earned his Bachelor of Science and Master of Engineering degrees in Computer Science at Cornell University in 2004 and 2005. After working on eyetracking scanpath analysis and visualization at the Human–Computer Interaction Group at Cornell, he is presently involved in eyetracking analysis for research in strabismus, amblyopia, and neuro-ophthalmology.

Laura A. Granka is a User Experience Researcher at Google, Inc., where she specializes in search and eyetracking. She earned both her B.S. and M.S. at Cornell University, where she did eyetracking research with the Human–Computer Interaction Group. One of her primary research interests is to better understand user behavior during online information retrieval.

Kenji Hara was born in Kanagawa, Japan, on October 15, 1980. He received B.E. and M.E. degrees in Information Science in 2004 and 2006, respectively, from Tokyo University. He joined the NTT Cyber Solutions Laboratories in 2006. He is now an employee of the NTT Cyber Solutions Laboratories, Network Appliance and Services Project, where he has been engaged in the research of communication appliances and their interface design.

Dr. Riad I. Hammoud is a research scientist, author, accomplished entrepreneur, futurist and advisor. He is currently a *senior research scientist* at the World Headquarters of Electronics and Safety Division of Delphi Corporation, working on safety and security systems for automotive. Since February 2001, he holds a Ph.D. degree in "Computer Vision and Robotics" from INRIA Rhone-Alpes, and a MS degree in "Control of Systems" from Université de Technologie de Compiègne, France. After his Ph.D., he launched a startup on the campus of Grenoble on "Interactive Video Tech". Around mid 2001, he moved to US and joined Rensselaer Polytechnic Institute (Troy, NY, USA) and Indiana University (Bloomington, IN, USA) as visiting and postdoctoral fellow. His research interests include automatic target classification in and beyond the visible spectrum, object tracking, biometrics, and real-time optimization techniques for safety, security and entertainment applications. His research is performed under confidential terms and has been funded by Alcatel Alsthom Research, INRIA, Honda, US Air Force, Indiana University, and Delphi Electronics & Safety. He published over fourty referred publications in journals, conferences, books and workshops, on various topics including object and image classification, video analysis, eye and pedestrian tracking, stereo vision, statistical modeling, face biometrics, surveillance and driver monitoring systems. He holds two United States patents, and over twenty patents pending, defense publications and trade secrets. Dr. Riad Hammoud authored several Springer books: "Interactive Video: algorithms and technologies", "Face Biometrics for Personal Identification: Multi-Sensory Multi-Modal Systems", "Passive Eye Monitoring", and "Applied Perception in Thermal-Infrared Imagery" (to appear in 2008). Dr. Riad Hammoud has been organizing and chairing several IEEE, SPIE and ACM International workshops and conference sessions (OTCBVS, ETRA, ATR, IVAN, IVRCIA). He is been serving on the reviewing committee of several journals in computer vision (PAMI, IJCV, CVIU), as well as national and international conferences (IEEE Intelligent Vehicles 2006, SPIE Defense and Security Symposiums, ACM ETRA Symposiums, IEEE Advanced Video and Signal Surveillance).

He was appointed in 2004 as *guest editor* of a special issue of Springer *International Journal of Computer Vision* (IJCV), and in 2005 as *guest editor* of a special issue of Elseiver *Computer Vision and Image Understanding Journal* (CVIU). He participated to the *Society Automotive Engineers* (SAE) conference, as a *panelist*, on November 1st, 2005. Recently, he gave an invited talk at the Houston University, Texas, on "Video technologies on the move" (February 2006), and he participated to the defense jury of a Ph.D. student at the University of Sherbrooke, Canada (March 2006). Dr. Riad Hammoud was nominated by US government as an *outstanding researcher/professor*, in May 2005. He received numerous awards from Delphi Corporation including the "best technical publicity/paper award" in 2006. He is the architect of the core algorithms of two vision-based safety products of Delphi Electronics & Safety: Driver Fatigue Monitoring and Driver Distraction Detection. Most of his work for Delphi E&S on vision products has not been authorized for outside disclosure.

Bogdan Hoanca is an Assistant Professor of Management Information Systems at the University of Alaska Anchorage (UAA). Before joining UAA, he co-founded, started up and sold a company that builds components for fiber optic communications. He also helped start and consulted with a number of other startup companies in optical fiber communications. Bogdan received a Ph.D. in Electrical Engineering from the University of Southern California in 1999. His current research interests revolve around technology, in particular e-learning and societal implications of technology, as well as privacy and security.

Hirohito Inagaki was born in Tokyo, Japan, on October 3, 1961. He received B.E. and M.E. degrees in Electrical Engineering in 1984 and 1986, respectively, from Keio University. Since Joining the NTT electrical Communication Laboratories in 1986, he has been engaged in the research and development of home and business terminal applications. From 2002 through 2004, he was on transfer to NTT West Corporation, Service Development Department as a senior manager of Information terminal development. He is now a senior manager of the NTT Cyber Solutions Laboratories, Network Appliance and Service Project, where he has been engaged in the research and development of information appliances architecture and its interface design.

Qiang Ji received his Ph.D. degree in electrical engineering from the University of Washington in 1998. He is currently an associate Professor with the Department of Electrical, Computer, and Systems engineering at Rensselaer Polytechnic Institute (RPI). Prior to joining RPI in 2001, he was an assistant professor with Dept. of Computer Science, University of Nevada at Reno. He also held research and visiting positions with Carnegie Mellon University, Western Research Company, and the US Air Force Research Laboratory. Dr. Ji's research interests are in computer vision, probabilistic reasoning with Bayesian Networks for decision making and information fusion, human computer interaction, pattern recognition, and robotics. He has published over 100 papers in peer-reviewed journals and conferences. His research has been funded by local and federal government agencies including NSF, NIH, AFOSR, ONR, DARPA, and ARO and by private companies including Boeing and Honda. He serves as an associate editor for IEEE Transactions on Systems, Man, and Cybernetics Part B, Pattern Recognition Letters journal, and Pattern Analysis and Applications journal. He is also on the editorial board of Image and Vision Computing journal. Dr. Ji is a senior member of the IEEE.

Dr. Mike Jones is a Senior Software Engineer with Google, Australia. Prior to that he was a Research Fellow within the School of Information Technology, Griffith University, Gold Coast, Australia. From 1996 until 2002 he was a Clinical Scientist with the Department of Medical Physics and Bioengineering within the United Bristol Healthcare NHS Trust, UK. Dr Jones has been active in the areas of 3-D visualisation and gaze-contingent displays for many years, focusing mainly on medical applications. Since 1998 he has been developing direct volume rendering (DVR) algorithms and various software packages now exist, which, on suitable hardware, can provide real-time interactive visualisation. Two of his main research areas are 3-D gaze-tracking and 3-D gaze-contingent displays.

Sheree Josephson is Professor of Communication at Weber State University in Ogden, Utah. Most of Dr. Josephson's research has involved using eye tracking to study how people process visual information (primarily in media environments). Before becoming a professor, Dr. Josephson was a print journalist for 18 years.

Takeo Kanade is U. A. Helen Whitaker University Professor of Computer Science and Robotics at Carnegie Mellon University. He received his Doctoral degree in Electrical Engineering from Kyoto University, Japan, in 1974. After holding a junior faculty position at Department of Information Science, Kyoto University, he joined Carnegie Mellon University in 1980, where he was the Director of the Robotics Institute from 1992 to 2001. Dr. Kanade has worked in multiple areas of robotics: computer vision, multi-media, manipulators, autonomous mobile robots, and sensors. He has written more than 250 technical papers and reports in these areas, as well as more than 15 patents. He has been the principal investigator of more than a dozen major vision and robotics projects at Carnegie Mellon. Dr. Kanade has been elected to the National Academy of Engineering, and to American Academy of Arts and Sciences. He is a Fellow of the IEEE, a Fellow of the ACM, and a Founding Fellow of American Association of Artificial Intelligence (AAAI), and the former and founding editor of International Journal of Computer Vision. He has received several awards including the C&C Award, Joseph Engelberger Award, Allen Newell Research Excellence Award, JARA Award, Marr Prize Award, and FIT Funai Accomplishment Award. Dr. Kanade has served for government, industry, and university advisory or consultant committees, including Aeronautics and Space Engineering Board (ASEB) of National Research Council, NASA's Advanced Technology Advisory Committee, PITAC Panel for Transforming Healthcare Panel, Advisory Board of Canadian Institute for Advanced Research.

John D. Lee received the B.A. degree in psychology and the B.S. degree in mechanical engineering from Lehigh University, Bethlehem, PA, in 1987 and 1988, respectively, and the M.S. degree in industrial engineering and the Ph.D. degree in mechanical engineering from the University of Illinois at Urbana-Champaign, in 1989 and 1992, respectively. He is currently a Professor in the Department of Mechanical and Industrial Engineering and the Director of the Cognitive Systems Laboratory at the University of Iowa, Iowa City, IA. His research enhances the safety and acceptance of complex human¨Cmachine systems by considering how technology mediates attention. Specific research interests include technology-mediated attention, trust in technology, supervisory control, and collision warning systems and driver distraction.

Lori A. Lorigo is a Knowledge Manager at the Tuck School of Business at Dartmouth and visiting researcher at Cornell University, in the Human Computer Interaction Group. She obtained her Ph.D. in Information Science from Cornell in 2006. Her research interests include information search and discovery, human–computer interaction, knowledge management, and information evolution and visualization. Her work with eyetracking has focused on online search engines, scanpath analysis, and data visualization.

Yulan Liang is a Ph.D. student in the Department of Mechanical and Industrial Engineering at the University of Iowa, Iowa City, IA. Before that, she received the B.E. and M.S. degrees from Tsinghua University, China, in 2000 and 2002, respectively. Her current research interests include user modeling and performance prediction, driver distraction, and data mining in human behavior.

Thomas Martinetz studied physics and mathematics in Munich and Cologne. From 1988 to 1991 he was with the Theoretical Biophysics Group at the Beckman Institute of the University of Illinois at Urbana-Champaign, focusing on research on self-organising neural networks. In 1991 he joined the Neuroinformatics Research Center of Siemens AG in Munich. In 1996 he became head of the Complex Systems Group at the Institute for Neuroinformatics of the University of Bochum. Since 1999 he is director of the Institute for Neuro- and Bioinformatics at the University of Lübeck. The main lines of research at his institute are in biological information processing, in particular in neural vision, pattern recognition, and learning.

Kenrick Mock received his Ph.D. in Computer Science from the University of California, Davis, in 1996. He currently holds the position of Associate Professor of Computer Science at the University of Alaska Anchorage. His research centers on complex systems, information management, artificial intelligence, computer security, and technological innovations in education. Dr.

Mock has previously held positions as a research scientist at Intel Corporation and as CTO of an Internet startup company, Unconventional Wisdom.

Carlos Hitoshi Morimoto received his B.S. and M.S. degrees in electric engineering from the University of Sao Paulo, Brazil, in 1988 and 1991 respectively, and the Ph.D. degree in computer science from the University of Maryland at College Park in 1997. He then joined the IBM Almaden Research Center, San Jose, CA, where he was involved in research projects related to eye gaze tracking technology, such as remote eye gaze tracking and eye gaze enhanced computer interfaces. Since 1999 he has been an Assistant Professor at the Department of Computer Science at the University of São Paulo, Brazil, where he started the Laboratory of Technologies for Interaction (LaTIn), which main focus has been on topics related to computer vision applied to human interaction.

Tsuyoshi Moriyama is Associate Professor in the Department of Media and Image Technology at Tokyo Polytechnic University. He received his Doctoral degree in Electrical Engineering from Keio University, Japan, in 1999. After JSPS research fellow in Institute of Industrial Science at University of Tokyo, Japan, Postdoctoral Fellow in the Robotics Institute at Carnegie Mellon University, U.S., and he joined Tokyo Polytechnic University in 2007. Dr. Moriyama has worked in many multidisciplinary projects, including analysis/synthesis of emotion in speech, automated analysis of facial expressions in images, automated summarization of movie films, software engineering, and design and development of multimedia contents. In addition to activities on education and research, he has also dedicated himself to musical activities as a tenor, including performances with the Wagner Society Male Choir of Japan 1990–2000 and the Pittsburgh Camerata 2001–2003. Dr. Moriyama is a Member of the IEEE, and a Member of IEICE of Japan. He received IEICE Young Investigators Award 1998.

Jeffrey B. Mulligan received the A.B. degree in physics from Harvard College in 1980 and the Ph.D. degree in psychology from the University of California at San Diego in 1986. He is currently a computer engineer at the NASA Ames Research Center, where his primary research interests are human and machine vision, and the inference of cognitive activity from behavioral observations such as eye movements. He is a member of the Optical Society of America, Vision Sciences Society, and the Society for Information Display.

Shree K. Nayar received his Ph.D. degree in Electrical and Computer Engineering from the Robotics Institute at Carnegie Mellon University in 1990. He is currently the T. C. Chang Professor of Computer Science at Columbia University. He co-directs the Columbia Vision and Graphics Center. He also heads the Columbia Computer Vision Laboratory (CAVE), which is dedicated to the development of advanced computer vision systems. His research is focused on three areas; the creation of novel cameras, the design of physics based models for vision, and the development of algorithms for scene understanding. His work is motivated by applications in the fields of digital imaging, computer graphics, and robotics. He has received best paper awards at ICCV 1990, ICPR 1994, CVPR 1994, ICCV 1995, CVPR 2000 and CVPR 2004. He is the recipient of the David Marr Prize (1990 and 1995), the David and Lucile Packard Fellowship (1992), the National Young Investigator Award (1993), the NTT Distinguished Scientific Achievement Award (1994), the Keck Foundation Award for Excellence in Teaching (1995) and the Columbia Great Teacher Award (2006). He has published over 100 scientific papers and has been awarded several patents for inventions related to vision and robotics.

Marios Nicolaou received a BMedSci in Microbiology from the University of Nottingham in 1995 and Bachelors of Medicine and Bachelors of Surgery from the same university in 1998. He has completed basic surgical training in London and became a Member of the Royal College of Surgeons (MRCS) in 2001 and is currenlty undergoing specialist training in Plastic and reconstructive surgery. He has recently completed his Ph.D. in the field of eye tracking research in Surgery from Imperial College, London. His research interests include eye tracking in minimally invasive and plastic surgery, surgical skills assessment, surgical technology and informatics.

Dr. Stavri Nikolov is a Founder of Attentive Displays Ltd, its Managing Director and Head of Research. Prior to that for ten years Dr Nikolov was a Research Fellow/Senior Research Fellow in Image Processing at the University of Bristol, UK. His research interests over the years have spanned several areas including image analysis, image fusion, computer graphics, new methods for data visualisation and navigation, the use of gaze-tracking and VR in 2-D and 3-D image analysis and understanding, and the construction of attentive and interactive information displays. In the last 15 years he has led and participated in many international and national research projects in the UK, Portugal, Austria, and Bulgaria, in image processing, analytical data processing, medical imaging, image fusion, and gaze-contingent data visualisation and analysis. He has published more than 50 refereed or invited papers, including four invited book chapters, and also numerous technical reports in these areas. He has also given many invited lectures around the world on image processing, image fusion and information visualisation. Dr Nikolov has undertaken consultancies for a number of major European companies and has taken part in the development of various large software systems. He is the creator and co-ordinator of The Online Resource for Research in Image Fusion (www.imagefusion.org) and The Online Archive of Scanpath Data (www.scanpaths.org), together with Dr Jeff Pelz from RIT, USA. Dr Nikolov is a member of the British Machine Vision Association, the Applied Vision Association, the International Society of Information Fusion, ACM SIGGRAPH and IEEE.

Ko Nishino is Assistant Professor in the Department of Computer Science at Drexel University. He received his BE and ME degrees in Information and Communication Engineering and Ph.D. degree in Information Science from The University of Tokyo in 1997, 1999, and 2002, respectively. Prior to joining

Drexel University in 2005, he worked as a Postdoctoral Research Scientist in the Department of Computer Science at Columbia University. His research interests span computer vision and computer graphics. The main focus of his research is on photometric and geometric modeling of real-world objects and scenes. He has published a number of papers on related topics, including physics-based vision, image-based modeling and rendering, and recognition. He is a member of the IEEE and ACM.

Takehiko Ohno was born in Kanagawa, Japan, on October 1969. He received B.E and M.E. degrees from Tokyo Institute of Technology, Japan, in 1992 and 1994, respectively. In 1994, he joined NTT Basic Research Laboratories, and he has been engaged in cognitive psychology and human–computer interaction. His research interests include gaze-based interaction, gaze tracking technologies, computer-mediated communication, information appliances, and cognitive modeling of human–computer interaction. From 2004 through 2006, he was a research engineer of the NTT Cyber Solutions Laboratories. Presently, he is a manager of Information and Communication Technology Office at Technology Planning Department in NTT Corporation. He is a member of ACM, Information Processing Society of Japan (IPSJ) and Japanese Cognitive Science Society (JCSS).

Justus Randolph's background is in education research and evaluation methods. At the time of writing he was a planning officer at the University of Joensuu and is currently a researcher at Häme Polytechnic's eLearning Centre.

Yan Tong is currently pursing her Ph.D. at Rensselaer Polytechnic Institute, Troy, NY. She received the B.S. degree from Zhejiang University, P. R. China in 1997, and M.S. degree from University of Nevada, Reno in 2004. Current research interest focuses on computer vision, pattern recognition, and human computer interaction.

Professor Guang-Zhong Yang received Ph.D. in Computer Science from Imperial College London in 1991. He is Director and Founder of the Royal Society/Wolfson Medical Image Computing Laboratory at Imperial, co-founder of the Wolfson Surgical Technology Laboratory, and Director of Medical Imaging and Robotics, Institute of Biomedical Engineering, Imperial College London. His main research interests include medical imaging, sensing and robotics. He received several major international awards including the I.I. Rabi Award from the International Society for Magnetic Resonance in Medicine and the Research Merit Award from the Royal Society.

Jing Xiao received the BS degree in electrical engineering from the University of Science and Technology of China in 1996, the MS degree in computer science from the Institute of Automation, Chinese Academy of Science in 1999, and the Ph.D. degree in robotics from the Robotics Institute, Carnegie Mellon University in 2005. Since 2005 he has been a senior member of technical staff (research scientist) in Epson Research and Development,

Inc. His research interests include computer vision, pattern recognition, machine learning, human computer interface, computer animation, and related areas of research and application. He has authored or coauthored more than 50 publications and invention disclosures in these areas.

Dr. Harry Zhang is a Principal Staff Human Factors Engineer at Motorola Research Labs in Tempe, Arizona. At Motorola, he has been doing research in context-aware applications and performing usability studies on Bluetooth devices, car kits, cell phones, and navigation systems. Prior to Motorola, he was a Senior Human Factors Engineer at Delphi Electronics and Safety Systems in Kokomo, Indiana, leading human factors research and development in the areas of driver distraction, impairment (including driver drowsiness), and workload management system. Before joining Delphi, he was an assistant professor at Indiana University and carried out research in the areas of attention, human performance, vision, and computer modeling. He received his Ph.D. in Experimental Psychology from the University of Michigan in 1994.

Djemel Ziou received the B.Eng. degree in Computer Science from the University of Annaba (Algeria) in 1984, and Ph.D. degree in Computer Science from the Institut National Polytechnique de Lorraine (INPL), France in 1991. From 1987 to 1993 he served as lecturer in several universities in France. During the same period, he was a researcher in the Centre de Recherche en Informatique de Nancy (CRIN) and the Institut National de Recherche en Informatique et Automatique (INRIA) in France. Presently, he is full Professor at the department of computer science, Sherbrooke University, QC, Canada. He is holder of the NSERC/Bell Canada Research Chair in personal imaging. He has served on numerous conference committees as member or chair. He heads the laboratory MOIVRE and the consortium CoRIMedia which he founded. His research interests include image processing, information retrieval, computer vision and pattern recognition.

References

1. Vision systems improve road safety. Laser Focus World, 2004. http://lfw.pennnet.com/Articles.
2. A. Abbott and A. Hrycak. Measuring resemblance in sequence data. *American Journal of Sociology*, 96(1):144–185, 1990.
3. A. Abbott and J. J. Forrest. Optimal matching sequences for historical sequences. *Journal of Interdisciplinary History*, 16(3):471–494, 1986.
4. A. M. Al-Qayedi and A. F. Clark. Constant-rate eye tracking and animation for model-based-coded video. In *Proc. ICASSP'2000*, 2000.
5. H. Alm and L. Nilsson. The effects of a mobile telephone task on driver behaviour in a car following situation. *Accident Analysis and Prevention*, 27(5):707–715, 1995.
6. S. Amari and S. Wu. Improving support vector machine classifiers by modifying kernel functions. *Neural Networks*, 12:783–789, 1999.
7. American Psychological Association. Publication manual of the American Psychological Association (5th Ed.), 2002.
8. J. R. Antes. The time course of picture viewing. *Journal of Experimental Psychology*, 103:62–70, 1974.
9. Applied Science Laboratories (ASL). Applied science laboratories – eye tracking expertise. http://www.a-s-l.com/.
10. O. Arandjelovic, R. I. Hammoud, and R. Cipolla. Multi-sensory face biometric fusion (for personal identification). In *IEEE Computer Vision and Pattern Recognition (CVPR) Workshop on Object Tracking and Classification Beyond the Visible Spectrum (OTCBVS'06)*, page 52, 2006.
11. R. Argue, M. Boardman, J. Doyle, and G. Hickey. Building a low-cost device to track eye movement. Technical report, Faculty of Computer Science, Dalhousie University, December 2004.
12. M. S. Atkins, A. E. Kirkpatrick, A. C. Knight, and B. Forster. Evaluating user interfaces for stack mode viewing. In *Proceedings of SPIE Medical Imaging 2007*, volume 6515, 2007.
13. M. S. Atkins, A. Moise, and R. Rohling. An application of eyegaze tracking for designing radiologists' workstations: insights for comparative visual search tasks. *ACM Transactions on Applied Perception*, 3(2):136–151, 2006.

14. M. S. Atkins, A. Moise, and R. Rohling. Understanding search and errors in comparative visual search: Insights from eye-gaze studies. Technical Report TR-2007, Simon Fraser University School of Computing Science, Jan 2007.

15. A. Aula, P. Majaranta, and K-K. Raiha. Eye-tracking reveals the personal styles for search result evaluation. In M.F. Constabile & F. Paterno, editor, *Proceedings of the IFIP TC 13 International Conference on Human–Computer Interaction (INTERACT)*, pages 1058–1061. Springer-Verlag, 2005.

16. A. D. Baddeley. *Working Memory*. Oxford University Press, 1986.

17. T. Y. Baker. Ray tracing through non-spherical surfaces. *Proc. of The Royal Society of London*, 55:361–364, 1943.

18. S. Baker and S. K. Nayar. A Theory of Single-Viewpoint Catadioptric Image Formation. *IJCV*, 35(2):1–22, 1999.

19. P. Baldi and S. Brunak. *Bioinformatics: The machine learning approach, 2nd edition*. MIT Press, 2001.

20. D. Ballard, M. Hayhoe, and J. Pelz. Memory representations in natural tasks. *Cog. Neurosci.*, 7:66–80, 1995.

21. H. B. Barlow. Eye movements during fixation. *J. Physiol.*, 116:290–306, 1952.

22. D. H. Barlow and M. Hersen. *Single Case Experimental Designs (2nd Ed.)*. Pergamon, New York, USA, 1992.

23. L. Baron. Interaction between television and child-related characteristics as demonstrated by eye movement research. *Education, Communication and Technology: A Journal of Theory, Research and Development*, 28(4):267–281, 1980.

24. T. Baron, M. D. Levine, and Y. Yeshurun. Exploring with a foveated robot eye system. In *Proceedings of the 12th IAPR International Conference on Pattern Recognition*, volume 3, pages 377–380, 1994.

25. J. R. Barrett and F. Trainor. Eye movements can control localized image enhancement and analysis. *Radiographics*, 17:525–530, 1997.

26. E. Barth, M. Dorr, M. Böhme, K. Gegenfurtner, and T. Martinetz. Guiding the mind's eye: improving communication and vision by external control of the scnapath. In *Proceedings of the SPIE. The International Society for Optical Engineering*, volume 6057, pages 116–123, 2006.

27. J. C. Bass and C. Chiles. Visual skill: correlation with detection of solitary pulmonary nodules. *Invest Radiol*, 25:994–998, 1990.

28. R. Bates and O. Spakov. D2.3 implementation of cogain gaze tracking standards. Deliverable IST-2003-511598, Communication by Gaze Interaction (COGAIN), 2006. http://www.cogain.org/.

29. P. Baudisch, D. DeCarlo, A. T. Duchowski, and W. S. Geisler. Focusing on the essential: considering attention in display design. *Commun. ACM*, 46(3):60–66, 2003.

30. P. Baudisch, N. Good, V. Bellotti, and P. Schraedley. Keeping things in context: A comparative evaluation of focus plus context screens, overviews and zooming. In *Proceedings of CHI '02. ACM*, pages 259–266, 2002.

31. R. A. Baxter and J. J. Oliver. Finding overlapping components with mml. *Statistics and Computing*, 10(1):5–16, 2000.

32. R. Bednarik, T. Kinnunen, A. Mihaila, and P. Fnti. Eye-movements as a biometric. *In Proceedings of the 14th Scandinavian Conference on Image Analysis, SCIA 2005*, June 2005.

33. R. Bednarik, N. Myller, E. Sutinen, and M. Tukiainen. Effects of experience on gaze behaviour during program animation. In *Proceedings of the 17th Annual Psychology of Programming Interest Group Workshop (PPIG'05)*, pages 49–61, Brighton, UK, 2005.
34. R. Bednarik, N. Myller, E. Sutinen, and M. Tukiainen. Analyzing Individual Differences in Program Comprehension with Rich Data. *Technology, Instruction, Cognition and Learning*, 3(3-4):205–232, 2006.
35. R. Bednarik, N. Myller, E. Sutinen, and M. Tukiainen. Program visualization: Comparing eye-tracking patterns with comprehension summaries and performance. In *Proceedings of the 18th Annual Psychology of Programming Interest Group Workshop (PPIG'06)*, pages 68–82, Brighton, UK, 2006.
36. R. Bednarik and M. Tukiainen. Validating the Restricted Focus Viewer: A Study Using Eye-Movement Tracking. *Behavior Research Methods*, to appear.
37. R. Bednarik and M. Tukiainen. Visual attention and representation switching in java program debugging: A study using eye movement tracking. In *Proceedings of the 16th Annual Psychology of Programming Interest Group Workshop (PPIG'04)*, pages 159–169, Carlow, Ireland, 2004.
38. S. Belongie, J. Malik, and J. Puzicha. Shape Matching and Object Recognition Using Shape Descriptors. *IEEE Transactions on Pattern Analysis and Machine Intelligence*, 24, 2002.
39. H. Bensmail, G. Celeux, A. Raftery, and C. P. Robert. Inference in Model-Based Cluster Analysis. *Statistics and Computing*, 7:1–10, 1997.
40. J. Benway and D. Lane. Banner blindness: web searchers often miss "obvious" links. internetworking. newsletter 1.3, 1998.
41. K. S. Berbaum, E. A. Brandser, E. A. Franken, D. D. Dorfman, R. T. Cadwell, and E. A. Krupinski. Gaze dwell time on acute trauma injuries missed because of satisfaction of search. *Academic Radiology*, 8(4):304–314, 2001.
42. K. S. Berbaum, E. A. Franken, D. D. Dorfman, E. M. Miller, R. T. Caldwell, D. M. Kuehn, and M. L. Berbaum. Role of faulty decision making in the satisfaction of search effect in chest radiography. *Academic Radiology*, 5(1):9–19, 1998.
43. L. M. Bergasa, J. Nuevo, M. A. Sotelo, R. Barea, and M. E. Lopez. Real-time system for monitoring driver vigilance. *IEEE Transactions on Intelligent Transportation Systems*, 7(1):63–77, 2006.
44. N. T. Berlinger. Robotic surgery – squeezing into tight places. *New England Journal of Medicine*, 354;20:2099–2101, 2006.
45. D. Beymer and M. Flickner. Eye gaze tracking using an active stereo head. In *IEEE Conference on Computer Vision and Pattern Recognition*, volume II, pages 451–458, Madison, WI, 2003.
46. T. N. Bhaskar, F. T. Keat, S. Ranganath, and Y. V. Venkatesh. Blink detection and eye tracking for eye localization. In *TENCON 2003. Conference on Convergent Technologies for Asia-Pacific Region*, volume 2, pages 821–824, 15–17 Oct 2003.
47. C. M. Bishop. *Neural Networks for Pattern Recognition*. Oxford University Press, 1995.
48. C. M. Bishop. *Pattern Recognition and Machine Learning*. Springer, 2006.
49. M. J. Black and P. Anandan. The robust estimation of multiple smooth motions: parametric and piecewise-smooth flow fields. *Computer Vision and Image Understanding*, pages 75–104, Jan 1996.

50. V. Blanz and T. Vetter. A morphable model for the synthesis of 3d faces. *Siggraph 1999, Computer Graphics Proceedings*, pages 187–194, 1999.

51. G. Blonder. Graphical passwords. Technical report, United States patent 5559961, 1996.

52. M. Böhme, M. Dorr, C. Krause, T. Martinetz, and E. Barth. Eye movement predictions on natural videos. *Neurocomputing*, 69(16–18):1996–2004, 2006.

53. M. Bolduc and M. D. Levine. A real-time foveated sensor with overlapping receptive fields. *Real-Time Imaging*, 3(3):195–212, 1997.

54. W. M. Bolstad. *Introduction to Bayesian Statistics*. John Wiley and Sons, 2004.

55. R. A. Bolt. Eyes at the Interface. In *ACM CHI*, pages 360–362, 1982.

56. R. A. Bolt. Gaze-orchestrated dynamic windows. In *SIGGRAPH '81: Proceedings of the 8th annual conference on Computer graphics and interactive techniques*, pages 109–119, New York, NY, USA, 1981. ACM Press.

57. N. Bouguila and D. Ziou. A Hybrid SEM Algorithm for High-Dimensional Unsupervised Learning Using a Finite Generalized Dirichlet Mixture. *IEEE Transactions on Image Processing*, 15(9):2657–2668, 2006.

58. N. Bouguila and D. Ziou. A Powerful Finite Mixture Model Based on the Generalized Dirichlet Distribution: Unsupervised Learning and Applications. In *Proc. IEEE/IAPR International Conference on Pattern Recognition (ICPR)*, pages 280–283, 2004.

59. N. Bouguila and D. Ziou. High-Dimensional Unsupervised Selection and Estimation of a Finite Generalized Dirichlet Mixture Model Based on Minimum Message Length. *IEEE Transactions on Pattern Analysis and Machine Intelligence*, 2007. To appear.

60. N. Bouguila and D. Ziou. MML-Based Approach for High-Dimensional Learning using the Generalized Dirichlet Mixture. In *Proc. of the 2005 IEEE Computer Society Conference on Computer Vision and Pattern Recognition (CVPR) – Workshops – Volume 03*, page 53, 2005.

61. N. Bouguila and D. Ziou. Using Unsupervised Learning of a Finite Dirichlet Mixture Model to Improve Pattern Recognition Applications. *Pattern Recognition Letters*, 26(12):1916–1925, 2005.

62. N. Bouguila, D. Ziou, and E. Monga. Practical Bayesian Estimation of a Finite Beta Mixture Through Gibbs Sampling and its Applications. *Statistics and Computing*, 16(2):215–225, 2006.

63. N. Bouguila, D. Ziou, and J. Vaillancourt. Novel Mixtures Based on the Dirichlet Distribution: Application to Data and Image Classification. In Petra Perner and Azriel Rosenfeld, editors, *Machine Learning and Data Mining in Pattern Recognition (MLDM)*, pages 172–181, Leipzig, Germany, 2003. Springer, LNAI2734.

64. N. Bouguila, D. Ziou, and J. Vaillancourt. Unsupervised Learning of a Finite Mixture Model Based on the Dirichlet Distribution and its Application. *IEEE Transactions on Image Processing*, 13(11):1533–1543, 2004.

65. F. Bourel, C. Chibelushi, and A. Low. Robust facial feature tracking. *Proc. of 11th British Machine Vision Conference*, 1:232–241, 2000.

66. G. C. Brainard, J. P. Hanifin, J. M. Greeson, B. Byrne, G. Glickman, E. Gerner, and M. D. Rollag. Action spectrum for melatonn regulation in humans: Evidence for a novel circadian photoreceptor. *Journal of Neuroscience*, 21:6405–6412, 2001.

67. H. F. Brandt. Ocular patterns and their psychological implications. *American Journal of Psychology*, 53(2):260–268, 1940.
68. S. A. Brandt and L. W. Stark. Spontaneous eye movements during visual imagery reflect the content of the visual scene. *Journal of Cognitive Neuroscience*, 9:27–38, 1997.
69. A. Broder. A taxonomy of web search. *SIGIR Forum*, 36(2):3–10, 2002.
70. X. Brolly, J. B. Mulligan, and C. Stratelos. Model-based head pose estimation for air-traffic controllers. In *Proc. IEEE Intl. Conf. on Image Processing.*, volume 3, pages 1268–1269, 1993.
71. S. Brostoff and M. A. Sasse. Ten strikes and you're out: Increasing the number of login attempts can improve password usability. *In Proc. of the CHI 2003 Workshop on Human–Computer Interaction and Security Systems*, April 2003.
72. S. Bruckner, S. Grimm, A. Kanitsar, and M. E. Gröller. Illustrative context-preserving exploration of volume data. *IEEE Trans. Vis. and Comp. Graph.*, 12(6):1559–1569, 2006.
73. U. Bueker, R. Schmidt, and S. Wiesner. Camera-based driver monitoring for increased safety and convenience. In *SAE Int'l Conf., IVI Technology and Intelligent Transportation Systems (SP-2099)*, 2007 World Congress Detroit, Michigan, April 16–19 2007.
74. A. E. Burgess, X. Li, and C. K. Abbey. Nodule detection in two-component noise: toward partient structure. In *Proc SPIE Medical Imaging 1997*, volume 3036, pages 2–13, 1997.
75. D. G. Burkhard and D. L. Shealy. Flux Density for Ray Propagation in Geometrical Optics. *JOSA*, 63(3):299–304, Mar. 1973.
76. P. J. Burt and E. H. Adelson. The Laplacian pyramid as a compact image code. *IEEE Transactions on Communications*, 31(4):532–540, 1983.
77. G. T. Buswell. *How People Look at Pictures*. University of Chicago Press, Chicago, USA, 1935.
78. H. Byun and S. W. Lee. Applications of support vector machines for pattern recognition: A survey. In *Proceedings of Pattern Recognition with Support Vector Machines: First International Workshop, SVM 2002, Niagara Falls, Canada*, 2002.
79. B. N. Campbell, J. D. Smith, and W. G. Najm. Examination of crash contributing factors using national crash databases. Technical report, The Volpe National Transportation Systems Center, April 2002.
80. J. Canny. A Computational Approach to Edge Detection. *IEEE Transactions on Pattern Analysis and Machine Intelligence*, 8:679–698, 1986.
81. O. Capp, C. P. Robert, and T. Ryden. Reversible Jump MCMC Converging to Birth-and-Death MCMC and More General Continuous Time Samplers. *Journal of the Royal Statistical Society, B*, 65:679–700, 2002.
82. D. P. Carmody, H. L. Kundel, and L. C. Toto. Comparison scans while reading chest images: taught, but not practiced. *Invest Radiol*, 19:462–466, 1984.
83. G. Celeux and J. Diebolt. A Stochastic Approximation Type EM Algorithm for the Mixture Problem. *Stochastics and Stochastics Reports*, 41:119–134, 1992.
84. G. Celeux and J. Diebolt. The SEM Algorithm: a Probabilistic Teacher Algorithm Derived from the EM Algorithm for the Mixture Problem. *Computational Statistics Quarterly*, 2(1):73–82, 1985.
85. B. J. Challacombe, M. S. Khan, D. Murphy, and P. Dasgupta. The history of robotics in urology. *World J Urol*, 2006.

86. P. Chandler and J. Sweller. The split-attention effect as a factor in the design of instruction. *British Journal of Educational Psychology*, 62:233–246, 1992.

87. C.-C. Chang and C.-J. Lin. Libsvm: a library for support vector machines. http://www.csie.ntu.edu.tw/c̃jlin/libsvm.

88. E.-C. Chang, S. Mallat, and C. Yap. Wavelet foveation. *Applied and Computational Harmonic Analysis*, 9(3):312–335, 2000.

89. C. H. Chen and P. S. P. Wang. *Handbook of Pattern Recognition and Computer Vision*. World Scientific, Singapore, 3rd edition edition, 2005.

90. H. Chen, Y. Q. Yu, H. Y. Shum, S. C. Zhu, and N. N. Zheng. Example-based facial sketch generation with non-parametric sampling. In *Proc. IEEE International Conference on Computer Vision '01*, volume 2, pages 433–438, 2001.

91. J.-Y. Chen, C. A. Bouman, and J. C. Dalton. Hierarchical browsing and search of large image databases. *IEEE Transactions on Image Processing*, 9(3), 2000.

92. M. C. Chen, J. R. Anderson, and M. H. Sohn. What can a mouse cursor tell us more? correlation of eye/mouse movements on web browsing. In *CHI '01: Proceedings of the SIGCHI conference on Human factors in computing systems*, pages 281–282, New York, NY, USA, 2001. ACM Press.

93. X. Chen, P. J. Flynn, and K. W. Bowyer. IR and Visible Light Face Recognition. *Computer Vision and Image Understanding (CVIU)*, 99(3):332–358, 2005.

94. X. Chen, P. J. Flynn, and K. W. Bowyer. PCA-Based Face Recognition in Infrared Imagery: Baseline and Comparative Studies. In *Proc. IEEE International Workshop on Analysis and Modeling of Faces and Gestures (FGR)*, pages 127–134, 2003.

95. D. M. Chickering and D. Heckerman. Efficient approximations for the marginal likelihood of bayesian networks with hidden variables. *Machine Learning*, 29:181–212, 1997.

96. S. H. Choi, K. S. Park, M. W. Sung, and K. H. Kim. Dynamic and quantitative evaluation of eyelid motion using image analysis. In *Medical and Biological Engineering and Computing*, volume 41:2, pages 146–150, 2003.

97. G. Chow and X. B. Li. Towards a system for automatic facial feature detection. *PR*, 26(12):1739–1755, 1993.

98. C. M. Christoudias and T. Darrell. On modelling nonlinear shape-and-texture appearance manifolds. *Proc. of CVPR05*, 2:1067–1074, 2005.

99. J. M. Clark and A. Paivio. Dual coding theory and education. *Educational Psychology Review*, 3(3):149–210, 1991.

100. M. Cohen and K. Brodie. Focus and context for volume visualization. In *Theory and Practice of Computer Graphics, 2004 Proceedings*, pages 32–39, 2004.

101. J. Conway and N. Sloane. *Sphere Packings, Lattice, and Groups*. New York: Springer Verlag, 1993.

102. T. F. Cootes, G. J. Edwards, and C. J. Taylor. Active appearance models. *IEEE Trans. on PAMI*, 23(6):681–685, 2001.

103. T. F. Cootes and C. J. Taylor. A mixture model for representing shape variation. *Image and Vision Computing*, 17(8):567–573, 1999.

104. T. F. Cootes and C. J. Taylor. Constrained active appearance models. In *International Conference on Computer Vision 2001*, volume I, pages 748–754, Vancouver, Canada, July 2001.

105. T. F. Cootes, C. J. Taylor, D. H. Cooper, and J. Graham. Active shape models – their training and application. *Computer Vision and Image Understanding*, 61(1):38–59, 1995.

106. T. F. Cootes, G. V. Wheeler, K. N. Walker, and C. J. Taylor. View-based active appearance models. *Image and Vision Computing*, 20(9):657–664, 2002.

107. S. Cornbleet. *Microwave and Optical Ray Geometry*. John Wiley and Sons, 1984.

108. F. W. Cornelissen, K. J. Bruin, and A. C. Kooijman. The influence of artificial scotomas on eye movements during visual search. *Optometry and Vision Science*, 82(1):27–35, 2005.

109. T. N. Cornsweet and H. D. Crane. Accurate two-dimensional eye tracker using first and fourth Purkinje images. *J. Opt. Soc. Am.*, 63(8):921–928, 1973.

110. Court Technology Laboratory. Retinal scanning. Retrieved on September 12, 2006 at http://ctl.ncsc.dni.us/biomet%20web/BMRetinal.html.

111. L. Cowen, L. J. Ball, and J. Delin. An eye-movement analysis of web-page usability. In X. Faulkner, J. Finlay, and F. Détienne, editors, *People and Computers XVI: Memorable yet Invisible: Proceedings of HCI 2002*. Springer-Verlag Ltd, 2002.

112. N. Cristianini and J. S. Taylor. *An introduction to Support Vector Machines and other kernel-based learning methods*. Cambridge University Press, Cambridge, 2000.

113. M. E. Crosby and J. Stelovsky. How Do We Read Algorithms? A Case Study. *IEEE Computer*, 23(1):24–35, 1990.

114. J. Daugman. Combining multiple biometrics, 2006. Retrieved on September 12, 2006 at http://www.cl.cam.ac.uk/ jgd1000/combine/combine.html.

115. J. Daugman. Complete discrete 2-d gabor transforms by neural networks for image analysis and compression. *IEEE Trans. on ASSP*, 36(7):1169–1179, 1988.

116. J. Daugman. High confidence visual recognition of persons by a test of statistical independence. *IEEE Transactions on Pattern Analysis and Machine Intelligence*, 15(11):1148–1161, 1993.

117. J. Daugman. How iris recognition works, 2006. Retrieved on September 12, 2006 at http://www.cl.cam.ac.uk/ jgd1000/irisrecog.pdf.

118. D. Davis, F. Monrose, and M. Reiter. On user choice in graphical password schemes. *In Proceedings of the 13th USENIX Security Symposium*, August 2004.

119. H. Davson. *Physiology of the Eye*. Macmillan, 5th edition, 1990.

120. P. Debevec. Rendering synthetic objects into real scenes: Bridging traditional and image-based graphics with global illumination and high dynamic range photography. In *ACM SIGGRAPH 98*, pages 189–198, 1998.

121. P. Debevec, T. Hawkins, C. Tchou, H.-P. Duiker, and W. Sarokin. Acquiring the Reflectance Field of a Human Face. In *ACM SIGGRAPH 00*, pages 145–156, 2000.

122. E. B. Delabarre. A method of recording eye-movements. *American Journal of Psychology*, 9(4):572–574, 1898.

123. L. Dempere-Marco, X. P. Hu, S. MacDonald, S. Ellis, D. Hansell, and G. Z. Yang. The use of visual search for knowledge gathering in image decision support. *IEEE Transactions on Medical Imaging*, 21(7):741–754, 2002.

124. A. P. Dempster, N. M. Laird, and D. B. Rubin. Maximum Likelihood from Incomplete Data via the EM Algorithm. *Journal of the Royal Statistical Society, B*, 39:1–38, 1977.

125. J. Y. Deng and F. P. Lai. Region-based template deformation and masking for eye-feature extraction and description. *PR*, 30(3):403–419, 1997.

126. Department of Defense. Department of defense interface standard: Comon warfighting symbology: Mil-std-2525b. Technical Report DISA/JIEO/CFS, Department of Defense, Reston, VA, 1999.

127. F. Détienne. *Software Design: Cognitive Aspects*. Springer, November 2001.

128. C. Dickie, R. Vertegaal, J. S. Shell, C. Sohn, D. Cheng, and O. Aoudeh. Eye contact sensing glasses for attention-sensitive wearable video blogging. In *CHI '04: Extended abstracts on Human factors in computing systems*, pages 769–770, New York, NY, USA, 2004. ACM Press.

129. J. Diebolt and C. P. Robert. Estimation of Finite Mixture Distributions Through Bayesian Sampling. *Journal of the Royal Statistical Society, B*, 56(2):363–375, 1994.

130. F. Diedrich, E. Entin, S. Hutchins, S. Hocevar, B. Rubineau, and J. MacMillan. When do organizations need to change (part i)? coping with incongruence. In *Proceedings of the Command and Control Research and Technology Symposium*, pages 472–479, Washington, DC, 2003. National Defense University.

131. D. F. Dinges, M. M. Mallis, G. Maislin, and J. W. Powell. Evaluation of techniques for ocular measurement as an index of fatigue and the basis for alertness management. Technical report, National Highway Traffic Safety Administration, 1998.

132. R. Dodge and T. S. Cline. The angle velocity of eye movements. *Psychological Review*, 8(2):145–157, 1901.

133. B. Donmez, L. N. Boyle, J. D. Lee, and G. Scott. Assessing differences in young drivers' engagement in distractions. In *the 85th Transportation Research Board Annual Meeting, Washington, D.C.*, 2006.

134. F. Dornaika and F. Davoine. Simultaneous Facial Action Tracking and Expression Recognition Using a Particle Filter. In *Proc. IEEE International Conference on Computer Vision (ICCV)*, pages 1733–1738, 2005.

135. M. Dorr, M. Böhme, T. Martinetz, and E. Barth. Visibility of temporal blur on a gaze-contingent display. In *APGV 2005 ACM SIGGRAPH Symposium on Applied Perception in Graphics and Visualization*, pages 33–36, 2005.

136. M. Dorr, M. Böhme, T. Martinetz, K. R. Gegenfurtner, and E. Barth. Eye movements on a display with gaze-contingent temporal resolution. *Perception ECVP 2005 Supplement*, 34:50, 2005.

137. M. Dorr, M. Böhme, T. Martinetz, K. R. Gegenfurtner, and E. Barth. Visibility of spatial and temporal blur in dynamic natural scenes. *Perception ECVP 2006 Supplement*, 35:110, 2006.

138. M. Dorr, T. Martinetz, K. Gegenfurtner, and E. Barth. Guidance of eye movements on a gaze-contingent display. In Uwe J. Ilg, Heinrich H. Bülthoff, and Hanspeter A. Mallot, editors, *Dynamic Perception Workshop of the GI Section "Computer Vision"*, pages 89–94, 2004.

139. R. A. Drebin, L. Carpenter, and P. Hanrahan. Volume rendering. *Comp. Graph.*, 22:65–74, 1988.

140. D. G. Drew and T. Grimes. Audio-visual redundancy and tv news recall. *Communication Review*, 14(4):452–461, 1987.

141. X. Dreze and F. Hussherr. Internet advertising: Is anybody watching? *Journal of Interactive Marketing*, 17(4):8–23, Autumn 2003.

142. J. E. Driskell and B. Mullen. The efficacy of naps as a fatigue countermeasures: a meta-analytic integration. *Human Factors*, 47:360–377, 2005.

143. H. Drucker, B. Shahrary, and D. Gibbon. Support vector machines: relevance feedback and information retrieval. *Information Processing and Management*, 38(3):305–451, 2002.

144. I. L. Dryden and K. V. Mardia. *Statistical Shape Analysis*. John Wiley, Chichester, 1998.

145. A. T. Duchowski. A breadth-first survey of eye tracking applications. *Behavior Research Methods, Instruments, & Computers*, 34(4):455–470, 2002.

146. A. T. Duchowski. *Eye Tracking Methodology: Theory & Practice*. Springer-Verlag, Inc., London, UK, 2003.

147. A. T. Duchowski. Hardware-accelerated real-time simulation of arbitrary visual fields (poster). In *Eye Tracking Research & Applications (ETRA)*, page 59, 2004.

148. A. T. Duchowski, N. Cournia, and H. Murphy. Gaze-contingent displays: A Review. *CyberPsychology and Behaviour*, 7(6):621–634, 2004.

149. R. Duda and P. Hart. Use of the hough transform to detect lines and curves in pictures. *Communication of the Association of Computer Machinery*, 15(1):11–15, 1972.

150. R. O. Duda, P. E. Hart, and D. G. Stork. *Pattern Classification*. Wiley, New York, 2001.

151. G. d'Ydewalle and I. Gielen. *Image and Text*, chapter Attention Allocation with Overlapping Sound, pages 415–427. Springer Verlag, New York, 1992. Rayner, K. (Ed.).

152. G. d'Ydewalle, J. Rensbergen, and J. Pollet. *Eye Movements: From Physiology to Cognition*, chapter Reading a Message When the Same Message is Available Auditorily in Another Language: The Case of Subtitling, pages 313–321. Elsevier Science Publishers, Amsterdam, 1987. O'Regan, J. K., Lev-Schoen, A. and Pollet, J. (Eds.).

153. M. Eaddy, G. Blasko, J. Babcock, and S. Feiner. My own private kiosk: Privacy-preserving public displays. In *ISWC '04: Proceedings of the Eighth International Symposium on Wearable Computers*, pages 132–135, Washington, DC, USA, 2004. IEEE Computer Society.

154. Y. Ebisawa. Improved video-based eye-gaze detection method. *IEEE Transaction on Instrumentation and Measurement*, 47(4):948–955, 1998.

155. Y. Ebisawa and S. Satoh. Effectiveness of pupil area detection technique using two light sources and the image difference method. In A. Szeto and R. Rangayan, editors, *Proc. 15th Annual Int. Conf. IEEE Eng. Med. Biol. Soc.*, pages 1268–1269, 1993.

156. D. W. Eby and L. P. Kostyniuk. Safety vehicles using adaptive interface technology (task 1): Distracted-driving scenarios: A synthesis of literature, 2001 crashworthiness data system (cds) data, and expert feedback. Technical report, University of Michigan Transportation Research Institute, 2004.

157. N. Edenborough, R. I. Hammoud, A. Harbach, A. Ingold, B. Kisacanin, P. Malawey, T. Newman, G. Scharenbroch, S. Skiver, M. Smith, A. Wilhelm, G. Witt, E. Yoder, and H. Zhang. Driver state monitor from delphi. In *IEEE Computer Vision and Pattern Recognition (CVPR)*, pages 1206–1207, 2005.

158. N. Edenborough, R. I. Hammoud, A. Harbach, and the others. Drowsy driver monitor from delphi. In *Proc. IEEE Conference on Computer Vision and Pattern Recognition (CVPR), Demo Session*, Washington, DC, USA, June 27–July 2 2004.

159. P. Ekman. Facial Expression of Emotion. *American Psychologist*, 48:384–392, 1993.

160. P. Ekman and W. Friesen. *Facial Action Coding System*. Consulting Psychologists Press, Palo Alto, CA, 1978.

161. P. Ekman, J. Hagar, C. H. Methvin, and W. Irwin. *Ekman-Hagar Facial Action Exemplars*. unpublished data, Human Interaction Laboratory, Univ. of California, San Francisco. Publication year unknown.

162. P. Ekman and E. L. Rosenberg, editors. *What the Face Reveals*. Oxford University Press, New York, 1997.

163. M. El-Helw, M. Nicolaou, A. Chung, G. Z. Yang, and M. S. Atkins. A gaze-based study for investigating the perception of visual realism in simulated scenes. *ACM TAP*, In Press.

164. S. M. Ellis, X. Hu, L. Dempere-Marco, G. Z. Yang, A. U. Wells, and D. M. Hansell. Thin-section CT of the lungs: eye-tracking analysis of the visual approach to reading tiled and stacked display formats. *European Journal of Radiology*, 59:257–264, 2006.

165. S. R. Ellis and J. D. Smith. *Eye Movements and Human Information Processing*, chapter Patterns of Statistical Dependency in Visual Scanning, pages 221–238. Elsevier Science Publishers, Amsterdam, 1985. Groner, R. (Ed.).

166. S. R. Ellis and L. W. Stark. Eye movements during the viewing of necker cubes. *Perception*, 7(5):575–581, 1978.

167. Enquiro. Enquiro eye tracking report i: Google. http://www.enquiro.com/ research.asp, July 2005.

168. E. Entin, F. Diedrich, D. Kleinman, W. Kemple, S. Hocevar, B. Rubineau, and J. MacMillan. When do organizations need to change (part ii)? incongruence in action. In *Proceedings of the Command and Control Research and Technology Symposium*, pages 480–486, Washington, DC, 2003. National Defense University.

169. K. A. Ericsson and H. A. Simon. *Protocol analysis: Verbal reports as data*. MIT Press, Cambridge, MA, 1984.

170. M. Eriksson and N. P. Papanikotopoulos. Eye-Tracking for Detection of Driver Fatigue. In *Proc. IEEE Conference on Intelligent Transportation System (ITSC)*, pages 314–319, 1997.

171. C. W. Erwin. Studies of drowsiness (final report). Technical report, Durham, NC: The National Driving Center, Durham, NC, 1976.

172. K. Essig, M. Pomplun, and H. Ritter. A neural network for 3d gaze recording with binocular eye trackers. *The International Journal of Parallel, Emergent and Distributed Systems*, 21(2):79–95, 2006.

173. B. S. Everitt and D. J. Hand. *Finite Mixture Distributions*. Chapman and Hall, London, UK., 1981.

174. R. J. Fairbanks, S. E. Fahey, and W. W. Wierwille. Research on vehicle-based driver status/performance monitoring: Seventh semi-annual research report. Technical report, National Highway Traffic Safety Administration, 1995.

175. S. Fan, C. X. Mao, J. D. Zhang, and L. N. Chen. Forecasting electricity demand by hybrid machine learning model. In *Proceedings of Neural Information Processing, Pt 2*, pages 952–963, 2006.

176. D. Fell. Safety update: problem definition and countermeasure summary: Fatigue. Technical report, New South Wales Road Safey Bureau, Australia, 1994.

177. G. C. Feng and P. C. Yuen. Multi-cues eye detection on gray intensity image. *Pattern Recognition*, 34(5):1033–1046, 2001.

178. L. Ferman, H. Collewijn, T. C. Jansen, and A. V. van den Berg. Human gaze stability in the horizontal, vertical and torsional direction using voluntary head movements evaluated with a three-dimensional scleral induction coil technique. *Vision Res.*, 27:811–828, 1987.

179. Eye-activity measures of fatigue and napping as a fatigue countermeasure. Life Sciences & Biotechnology Update, Oct 1999.
http://findarticles.com/p/articles/mi_m0DFY/is_10_99/ai_55888446

180. M. A. T. Figueiredo and A. K. Jain. Unsupervised Learning of Finite Mixture Models. *IEEE Transactions on Pattern Analysis and Machine Intelligence*, 24(3):4–37, 2002.

181. O. Findahl. The effect of visual illustrations upon perception and retention of news programmes. ERIC Document Reproduction Service No. ED 054 631, 1981.

182. P. M. Fischer, J. W. Richards, E. J. Berman, and D. M. Krugman. Recall and Eye Tracking Study of Adolescents Viewing Tobacco Advertisements. *Journal of American Medical Association*, 261:90–94, 1989.

183. M. A. Fischler and R. C. Bolles. Random sample consensus: A paradigm for model fitting with applications to image analysis and automated cartography. *Communications of the ACM*, 24(6):381–395, 1981.

184. D. J. Fleet and A. D. Jepson. Computation of component image velocity from local phase information. *Int'l Journal of Computer Vision*, 5(1):77–104, 1990.

185. L. Flom and A. Safir. Iris Recognition System. US patent 4,641,349, 1987.

186. D. Fono and R. Vertegaal. Eyewindows: evaluation of eye-controlled zooming windows for focus selection. In *CHI '05: Proceedings of the SIGCHI conference on Human factors in computing systems*, pages 151–160, New York, NY, USA, 2005. ACM Press.

187. K. Fukunaga. *Introduction to Statistical Pattern Recognition*. Academic Press, New York, 2nd edition edition, 1990.

188. A. G. Gallagher, C. D. Smith, S. P. Bowers, N. E. Seymour, A. Pearson, S. McNatt, D. Hananel, and R. M. Satava. Psychomotor skills assessment in practicing surgeons experienced in performing advanced laparoscopic procedures. *J Am Coll Surg*, 197:479–488, 2003.

189. V. Gallese, L. Craighero, L. Fadiga, and L. Fogassi. Perception through action. *Psyche*, 5(21), 1999.

190. D. Gamerman. *Markov Chain Monte Carlo*. Chapman and Hall, London, 1997.

191. J. Gbadamosi, U. Oechsner, and W. H. Zangmeister. Quantitative analysis of gaze movements during visual imagery in hemianopic patients and control subjects. *J. Neurol. Rehabil.*, 3:165–172, 1997.

192. W. S. Geisler and J. S. Perry. A real-time foveated multi-resolution system for low-bandwidth video communication. *Proc: SPIE* 3299:294–305, 1998. Human Vision and Electronic Imaging III, Rogowitz, B. E. and Pappas, T. N. (Eds.)

193. W. S. Geisler and J. S. Perry. Real-time simulation of arbitrary visual fields. In *Eye Tracking Research & Applications (ETRA) Symposium. ACM*, pages 83–153, New Orleans, 2002.

194. W. S. Geisler, J. S. Perry, and J. Najemnik. Visual search: The role of peripheral information measured using gaze-contingent displays. *Journal of Vision*, 6:858–873, 2006.

195. J. Gemmell, K. Toyama, L. C. Zitnick, T. Kang, and S. Seitz. Gaze awareness for video-conferencing: A software approach. *IEEE Multimedia*, 7(4):26–35, 2000.

196. T. M. Gersch, E. Kowler, and B. Dosher. Dynamic allocation of visual attention during the execution of sequences of saccades. *Vision Research*, 44(12):1469–83, 2004.

197. G. A. Gescheider. *Psychophysics: the fundamentals*. Lawrence Erlbaum Associates, 3rd edition, 1997.

198. W. R. Gilks, S. Richardson, and D. G. Spiegelhater. *Markov Chain Monte Carlo In Practice*. Chapman and Hall, London, 1996.

199. J. Goldberg and X. P. Kotval. Computer Interface Evaluation Using Eye Movements: Methods and Constructs. *International Journal of Industrial Ergonomics*, 24:631–645, 1999.

200. J. Goldberg, M. Stimson, M. Lewenstein, N. Scott, and A. Wichansky. Eye tracking in web search tasks: design implications. In *ETRA '02: Proceedings of the 2002 symposium on Eye tracking research & applications*, pages 51–58, New York, NY, USA, 2002. ACM Press.

201. J. Goldberg and A. Wichansky. Eye Tracking in Usability Evaluation: A Practitioner's Guide. In J. Hyn, R. Radach, and H. Deubel, editors, *The Mind's Eye: Cognitive and Applied Aspects of Eye Movement Research*, pages 493–516. Elsevier Science, 2003.

202. S. K. Goldenstein, C. Vogler, and D. Metaxas. Statistical cue integration in dag deformable models. *IEEE Trans. on PAMI*, 25(7):801–813, 2003.

203. E. Gose, R. Johnsonbaugh, and S. Jost. *Pattern Recognition and Image Analysis*. Pretice Hall Inc., 1996.

204. R. Grace and S. Steward. Drowsy driver monitor and warning system. In *Proceedings of the First International Drivign Symposium on Human Factors in Driver Assessment, Training and Vehicle Design*, pages 64–69, 2001.

205. L. Granka. Eye-r: Eye-tracking analysis of user behavior in online search. Master's thesis, Cornell University, 2004.

206. L. Granka, T. Joachims, and G. Gay. Eye-tracking analysis of user behavior in www search. In *SIGIR '04: Proceedings of the 27th annual international ACM SIGIR conference on Research and development in information retrieval*, pages 478–479, New York, NY, USA, 2004. ACM Press.

207. L. Granka and K. Rodden. Incorporating eyetracking into user studies at google. Position paper presented at CHI Workshop, ACM, 2006.

208. K. Grauman, M. Betke, J. Gips, and G. Bradski. Communication via eye blinks – detection and duration analysis in real time. In *Proc. IEEE Computer Vision and Pattern Recognition (CVPR)*, 2001.

209. K. Grauman and T. Darrell. Fast contour matching using approximate earth movers's distance. *Proc. of CVPR04*, 1:220–227, 2004.

210. H. Gu, Q. Ji, and Z. Zhu. Active Facial Tracking for Fatigue Detection. In *Proc. of sixth IEEE Workshop on Applications of Computer Vision*, pages 137–142, 2002.

211. M. Guhe, W. Liao, Z. Zhu, Q. Ji, W. D. Gray, and M. J. Schoelles. Non-intrusive measurement of workload in real-time. In *Proceedings of Human*

Factors and Ergonomics Society 49th Annual Meeting, Orlando, FL, pages 1157–1161, 2005.

212. S. Hacisalihzade, L. Stark, and J. Allen. Visual perception and sequences of eye movement fixations: A stochastic modeling approach. *IEEE Transactions on Systems, Man, and Cybernetics*, 22(3):474–481, 1992.

213. M. Hadwiger, C. Berger, and H. Hauser. High-quality two-level volume rendering of segmented data sets on consumer graphics hardware. In *Visualisation 2000, IEEE VIS 2003*, pages 301–308, 2003.

214. M. A. Halstead, B. A. Barsky, S. A. Klein, and R. B. Mandell. Reconstructing Curved Surfaces from Specular Reflection Patterns Using Spline Surface Fitting of Normals. In *ACM SIGGRAPH 96*, pages 335–342, 1996.

215. R. I. Hammoud. A robust eye position tracker based on invariant local features, eye motion and infrared-eye responses. In *SPIE Defense and Security Symposium, Automatic Target Recognition Conference, Proceedings of SPIE* 5807:35–43, 2005.

216. R. I. Hammoud. *Building and Browsing of Interactive Videos*. PhD thesis, INRIA, Feb 27 2001. http://www.inrialpes.fr/movi/publi/Publications/2001/Ham01/index.html.

217. R. I. Hammoud. Guest editorial: Object tracking and classification beyond the visible spectrum. *(Springer) International Journal of Computer Vision (IJCV)*, 71(2):123–124, 2007.

218. R. I. Hammoud. *Interactive Video: Algorithms and Technologies*. Signals and Communication Technology (Hardcover). Springer-Verlag, New York Inc., 1 edition, June 2006. ISBN: 978-3-540-33214-5, English XVI, 250 p., 109 illus.

219. R. I. Hammoud, B. R. Abidi, and M. A. Abidi. *Face Biometrics for Personal Identification: Multi-Sensory Multi-Modal Systems*. Signals and Communication Technology (Hardcover). Springer-Verlag, New York Inc., 1 edition, Feb 2007. SBN: 978-3-540-49344-0,English, 2007, XVI, 276 p., 118 illus., 76 in colour.

220. R. I. Hammoud and J. W. Davis. Guest editorial: Advances in vision algorithms and systems beyond the visible spectrum. *Journal of Computer Vision and Image Understanding (CVIU)*, 106(2 and 3):145–147, 2007.

221. R. I. Hammoud and D. W. Hansen. *Physics of the Automatic Target Recognition*, volume III of *Advanced Sciences and Technologies for Security Applications (Optics & Lazers)*, chapter Biophysics of the eye in computer vision: methods and advanced technologies, pages 143–175. Springer, Sadjadi, F. and Javidi, B. (Eds.), 2007. ISBN: 978-0-387-36742-2.

222. R. I. Hammoud and P. Malaway. System and method for determining eye closure state, August 2007. United States Patent.

223. R. I. Hammoud and R. Mohr. Gaussian mixture densities for indexing of localized objects in a video sequence. Technical report, INRIA, France, March 2000. http://www.inria.fr/RRRT/RR-3905.html.

224. R. I. Hammoud and R. Mohr. Interactive tools for constructing and browsing structures for movie films. In *ACM Multimedia*, pages 497–498, Los Angeles, California, USA, Oct 30–Nov 3 2000. (demo session).

225. R. I. Hammoud and R. Mohr. Mixture densities for video objects recognition. In *International Conference on Pattern Recognition*, volume 2, pages 71–75, Barcelona, Spain, 3–8 September 2000.

226. R. I. Hammoud, A. Wilhelm, P. Malawey, and G. J. Witt. Efficient real-time algorithms for eye state and head pose tracking in advanced driver support systems. In *IEEE Computer Vision and Pattern Recognition (CVPR)*, page 1181, San Diego, CA, USA, June 2005.

227. G. B. Hanna and A. Cuschieri. Influence of the optical axis-to-target view angle on endoscopic task performance. *Surg Endosc*, 13:371–375, 1999.

228. G. B. Hanna and A. Cuschieri. Influence of two-dimensional and three-dimensional imaging on endoscopic bowel suturing. *World J Surg*, 24:444–8; discuss, 2000.

229. D. W. Hansen and R. I. Hammoud. An improved likelihood model for eye tracking. *Journal of Computer Vision and Image Understanding (CVIU)*, 106(2 and 3):220–230, 2007.

230. D. W. Hansen and R. I. Hammoud. Boosting particle filter-based eye tracker performance through adapted likelihood function to reflexions and light changes. In *Proc. IEEE Conference on Advanced Video and Signal Based Surveillance (AVSS'05)*, pages 111–116, 2005.

231. D. W. Hansen, J. P. Hansen, M. Nielsen, A. S. Johansen, and M. B. Stegmann. Eye Typing using Markov and Active Appearance Models. In *Proc. IEEE Workshop on Applications on Computer Vision (WACV)*, pages 132–136, 2003.

232. J. P. Hansen, A. W. Andersen, and P. Roed. Eye-gaze control of multimedia systems. In Y. Anzai, K. Ogawa, and H. Mori, editors, *HCII'95: Proceedings of HCI International '95*, pages 37–42, Amsterdam, 1995. Elsevier.

233. J. P. Hansen, D. W. Hansen, A. S. Johansen, and J. Elves. Mainstreaming gaze interaction towards a mass market for the benefit of all. In *Proceedings of the 11th International Conference on Human–Computer Interaction (HCII 2005)*. IOS Press, 2005.

234. J. P. Hansen, K. Itoh, A. S. Johansen, K. Torning, and A. Hirotaka. Gaze typing compared with input by head and hand. In *Eye Tracking Research & Applications Symposium 2004*, pages 131–138. ACM, 2004.

235. R. M. Haralick, K. Shanmugan, and I. Dinstein. Texture Features for Image Classification. *IEEE Transactions on Systems, Man and Cybernetics*, 8:610–621, 1973.

236. A. Haro, M. Flickner, and I. Essa. Detecting and Tracking Eyes by Using Their Physiological Properties, Dynamics, and Appearance. In *Proc. IEEE Conference on Computer Vision and Pattern Recognition (CVPR)*, pages 163–168, 2000.

237. R. Hartley and A. Zisserman. *Multiple View Geometry in Computer Vision*. Cambridge University Press, 2000.

238. H. Hauser. *Generalizing Focus+Context Visualization*, pages 305–327. Springer, Berlin, Heidelberg, 2006.

239. H. Hauser, G. I. Bischi, and M. E. Gröller. Two-level volume rendering. *IEEE Trans. Vis. and Comp. Graph.*, 7(3):242–252, 2001.

240. R. E. Haven, D. J. Anvar, J. E. Fouquet, and J. S. Wenstrand. Apparatus and method for detecting pupils. *US Patent 20040170304*, 2004.

241. Health-Care, 2004. http://www.lungcancer.org/health_care/focus_on_lc/screening/screening.htm.

242. T. Heap and D. Hogg. Wormholes in shape space: Tracking through discontinuous changes in shape. *Proc. of ICCV98*, pages 344–349, 1998.

243. H. von Helmholtz. *Physiologic Optics*, volume 1 and 2. Voss, Hamburg, Germany, third edition, 1909.

244. H. Hembrooke, M. Feusner, and G. Gay. Averaging scan patterns and what they can tell us. In *ETRA '06: Proceedings of the 2006 symposium on Eye tracking research & applications San Diego, California*, 41–41, 2006. ISBN: 1-59593-305-0.

245. H. Hembrooke, L. Granka, G. Gay, and E. Liddy. The effects of expertise and feedback on search term selection and subsequent learning: Research articles. *J. Am. Soc. Inf. Sci. Technol.*, 56(8):861–871, 2005.

246. W. R. Hendee. The perception of images: I wouldn't have seen it if i hadn't believed it. In *Proc SPIE Medical Imaging 2002*, volume 4686, 2002.

247. C. I. Henschke, D. I. McCauley, and D. F. Yankelevitz. Early lung cancer action project: overall design and findings from baseline screenings. *Lancet*, 354(99-105), 1999.

248. R. Herpers, M. Michaelis, K. H. Lichtenauer, and G. Sommer. Edge and key-point detection in facial regions. In *Proc. IEEE Face and Gesture '96*, pages 212–217, 1996.

249. J. P. Hespanha P. N. Belhumeour and D. J. Kriegman. Eigenfaces vs. fisher-erfaces: Recognition using class specific linear projection. *IEEE Transactions on Pattern Analysis and Machine Intelligence*, 19(7):711–720, 1997.

250. E. Hess and J. Polt. Pupil size as related to interest value of visual stimuli. *Science*, 132(3423):349–350, 1960.

251. E. Hess and J. Polt. Pupil size in relation to mental activity during simple problem-solving. *Science*, 1964.

252. D. L. G. Hill, P. G. Batchelor, M. Holden, and D. J. Hawkes. Medical image registration. *Physics in Medicine and Biology*, 46:R1–R45, 2001.

253. A. P. Hillstrom. Repetition effects in visual search. *Perception Psychophysics*, 62:800–817, 2000.

254. C. Ho and C. Spence. Olfactory facilitation of dual-task performance. *Neuroscience Letters*, 389:35–40, 2005.

255. B. Hoanca and K. Mock. Screen oriented technique for reducing the incidence of shoulder surfing. *In Proceedings of Security and Management (SAM05)*, June 2005.

256. B. Hoanca and K. Mock. Secure graphical password entry system for access from high-traffic public areas. *In Proceedings of the Eye Tracking and Research Applications (ETRA'06)*, March 2006.

257. J. M. Hoc, T. R. G. Green, R. Samurcay, and D. J. Gilmore. *The Psychology of Programming*. Academic Press, 1990.

258. D. Holman, R. Vertegaal, C. Sohn, and D. Cheng. Attentive display: paintings as attentive user interfaces. In *Extended Abstract of ACM Conference on Human Factors in Computing Systems (CHI)*, pages 1127–1130. ACM Press, 2004.

259. M. E. Holmes. Optimal matching analysis of negotiation phase sequences in simulated and authentic hostage negotiations. *Communication Reports*, 10(1):1–8, 1997.

260. M. E Holmes. Winphaser user's manual (version 1.0c), 1996.

261. N. Hopper and M. Blum. A secure human–computer authentication scheme. Technical report, Carnegie Mellon University, 2000. Retrieved on February 14, 2005 at http://www.andrew.cmu.edu/user/abender/humanaut/links.html.

262. J. A. Horne and L. A. Reyner. Sleep-related vehicle accidents. *British Medical Journal*, 310:565–567, 1995.

263. A. Hornof, A. Cavender, and R. Hoselton. Eyedraw: a system for drawing pictures with eye movements. In *ASSETS'04: Proceedings of the ACM SIGACCESS conference on Computers and accessibility*, pages 86–93, New York, NY, USA, 2004. ACM Press.

264. A. Hornof and L. Sato. Eyemusic: Making music with the eyes. In Yoichi Nagashima and Michael J. Lyons, editors, *NIME '04: Proceedings of the 2004 Conference on New Interfaces for Musical Expression*, pages 185–188. Shizuoka University of Art and Culture, 2004.

265. W. J. Horrey and C. D. Wickens. Examining the impact of cell phone conversations on driving using meta-analytic techniques. *Human Factors*, 48(1):196–205, 2006.

266. X. W. Hou, S. Z. Li, H. J. Zhang, and Q. S. Cheng. Direct appearance models. *Proc. of CVPR01*, 1:828–833, 2001.

267. I. Hsieh-Yee. Research on web search behavior. *Library and Information Science Research*, 23:167–185, 2001.

268. C.-W. Hsu, C.-C. Chang, and C.-J. Lin. A practical guide to support vector classification. http://www.csie.ntu.edu.tw/c̆jlin/libsvm.

269. C. H. Hu, H. L. Kundel, C. F. Nodine, E. A. Krupinski, and L. C. Toto. Searching for bone fractures: a comparison with pulmonary nodule search. *Academic Radiology*, 1:25–32, 1994.

270. X. P. Hu, L. Dempere-Marco, and G. Z. Yang. Hot spot detection based on feature space representation of visual search. *IEEE Transactions on Medical Imaging*, 22(9):1152–1162, 2003.

271. G. K. Humphrey and S. J. Lupker. Codes and operations in picture matching. *Psychological Research*, 55:237–247, 1993.

272. T. E. Hutchinson, K. P. White, W. N. Martin, K. C. Reichert, and L. A. Frey. Human–computer interaction using eye-gaze input. *Systems, Man and Cybernetics, IEEE Transactions on*, 19(6):1527–1534, 1989.

273. A. Hyrskykari. Detection of comprehension difficulties in reading on the basis of word frequencies and eye movement data. *In Proc. 12th European Conference on Eye Movements (ECEM 12)*, 2003. PB13.

274. A. Hyrskykari, P. Majaranta, A. Aaltonen, and K.-J. Räihä. Design issues of idict: a gaze-assisted translation aid. In *ETRA '00: Proceedings of the symposium on Eye tracking research & applications*, pages 9–14, New York, NY, USA, 2000. ACM Press.

275. A. Hyrskykari, P. Majaranta, and K.-J. Räihä. From gaze control to attentive interfaces. In *Proceedings of the 11th International Conference on Human–Computer Interaction (HCII 2005)*. IOS Press, 2005.

276. A. Hyrskykari, P. Majaranta, and K.-J. Räihä. Proactive response to eye movements. In Matthias Rauterberg, Marino Menozzi, and Janet Wesson, editors, *INTERACT '03: IFIP TC13 International Conference on Human–Computer Interaction*, pages 129–136, Amsterdam, 2003. IOS Press.

277. K. Ikeuchi and T. Suehiro. Toward an Assembly Plan from Observation, Part 1: Task Recognition with Polyhedral Objects. *IEEE Trans. Robotics and Automation*, 10(3):368–385, 1994.

278. I. Inamdar and M. Pomplun. Comparative search reveals the tradoff between eyemovements and working memory use in visual tasks. In *Proceedings of the 25th Annual Meeting of the Cognitive Science Society*, pages 599–604, 2003.

279. Itap. Information technology for active perception website, 2002. http://www.inb.uni-luebeck.de/Itap/.

280. L. Itti. Automatic foveation for video compression using a neurobiological model of visual attention. *IEEE Transactions on Image Processing*, 13(10):1304–1318, 2004.

281. J. Kniss, S. Premoze, M. Ikits, A. Lefohn, C. Hansen, and E. Praun. Gaussian transfer functions for multi-field volume visusalisation. In *Proceedings IEEE Visualization 2003*, pages 497–504, 2003.

282. R. J. K. Jacob. *Eye Movement-Based Human Computer Interaction Techniques: Toward Non Command Interfaces*, volume 4, page 151 190. Ablex Publishing, 1993. Hartson, H. R. and Hix D. (Eds.).

283. R. J. K. Jacob. *Eye tracking in advanced interface design*, chapter Advanced Interface Design and Virtual Environments, pages 258–288. Oxford University Press, Oxford, UK, 1995.

284. R. J. K. Jacob. The use of eye movements in human–computer interaction techniques: what you look at is what you get. *ACM Transactions on Information Systems*, 9(2):152–169, 1991.

285. R. J. K. Jacob. What you look at is what you get: Eye movement-based interaction techniques. In *Proc. ACM Conference on Human Factors in Computing Systems (CHI)*, pages 11–18. ACM Press, 1990.

286. R. J. K. Jacob and K. S. Karn. Eye tracking in human–computer interaction and usability research: Ready to deliver the promises. In Jukka Hyn, Ralph Radach, and Deubeln Heiner, editors, *The mind's eye: Cognitive and applied aspects of eye movement research*, pages 573–605, Amsterdam, 2003. Elsevier Science. http://www.cs.ucl.ac.uk/staff/j.mccarthy/pdf/library/eyetrack/eye_tracking_in_HCI.pdf.

287. R. J. K. Jacob and K. S. Karn. *The Mind's: Cognitive and Applied Aspects of Eye Movement Research*, chapter Eye tracking in human–computer interaction and usability research: Ready to deliver the promises, pages 573–605. Elsevier Science, 2003. Hyona, J., Radach, R. and Deubel, H. (Eds.).

288. B. Jähne and H. Haußecker, editors. *Computer Vision and Applications*. Academic Press, 2000.

289. A. K. Jain and A. Vailaya. Image retrieval using color and shape. *Pattern Recognition*, 29(8):1233–1244, 1996.

290. A. James. Saccadic eye movements during laparoscopic surgery: a preliminary study. *ECEM*, 2003.

291. T. von Jan, T. Karnahl, K. Seifert, J. Hilgenstock, and R. Zobel. Don't sleep and drive: Vw's fatigue detection technology. In *19th International Technical Conference on the Enhanced Safety of Vehicles (CD).*, Washington, DC, 2005.

292. A. R. Jansen, A. F. Blackwell, and K. Marriott. A tool for tracking visual attention: The Restricted Focus Viewer. *Behavior Research Methods, Instruments, and Computers*, 35(1):57–69, 2003.

293. W. Jansen. *The Internet Society: Advances in Learning, Commerce and Security*, volume 30, chapter Authenticating Mobile Device Users through Image Selection, page 10pp. WIT Press, 2004. Morgan K. and Spector, M. J. (Eds.).

294. J. Jerald and M. Daily. Eye gaze correction for videoconferencing. In *ETRA '02: Proceedings of the symposium on Eye tracking research & applications*, pages 77–81, New York, NY, USA, 2002. ACM Press.

295. I. Jermyn, A. Mayer, F. Monrose, M. Reiter, and A. Rubin. The design and analysis of graphical passwords. *In Proceedings of the 8th USENIX Security Symposium*, 1999.

296. Q. Ji, Z. Zhu, and P. Lan. Real-time nonintrusive monitoring and prediction of driver fatigue. *IEEE Transactions on Vehicle Technology*, 53(4):1052–1068, 2004.

297. F. Jiao, S. Z. Li, H. Y. Shum, and D. Schuurmans. Face alignment using statistical models and wavelet features. *Proc. of CVPR03*, 1:321–327, 2003.

298. T. Joachims. Optimizing search engines using clickthrough data. In *Proceedings of the ACM Conference on Knowledge Discovery and Data Mining*, pages 132–142. ACM, 2002.

299. T. Joachims, L. Granka, B. Pan, H. Hembrooke, and G. Gay. Accurately interpreting clickthrough data as implicit feedback. In *SIGIR '05: Proceedings of the 28th annual international ACM SIGIR conference on Research and development in information retrieval*, pages 154–161, New York, NY, USA, 2005. ACM Press.

300. T. Joachims, L. Granka, B. Pan, H. Hembrooke, F. Radlinski, and G. Gay. Evaluating the accuracy of implicit feedback from clicks and query reformulations in web search. *Transactions on Information Systems*, 25, 2007.

301. M. St. John, S. P. Marshall, S. R. Knust, and K. Binning. Attention management on a geographical display: Minimizing distraction by decluttering. Technical Report ETI:0301, EyeTracking, Inc., San Diego, CA, 2003.

302. M. G. Jones and S. G. Nikolov. Dual-modality two-stage direct volume rendering. In *Proceedings of the 6th International Conference on Information Fusion (Fusion 2003), Cairns, Queensland, Australia, July 8–11, 2003, International Society of Information Fusion (ISIF)*, pages 845–851, 2003.

303. M. G. Jones and S. G. Nikolov. Region-enhanced volume visualization and navigation. In *Medical Imaging 2000 (Image Display and Visualization) conference, San Diego, California, USA, 12–18 February, 2000, Proceedings of SPIE*, volume 3976, pages 454–465, 2000.

304. M. G. Jones and S. G. Nikolov. Volume visualisation via region-enhancement around an observer's fixation point. In *MEDSIP 2000 (International Conference on Advances in Medical Signal and Information Processing), Bristol, UK, 4–6 September*, volume 476, pages 305–312, 2000.

305. M. J. Jones and T. Poggio. Multidimensional morphable models: A framework for representing and matching object classes. *Int'l Journal of Computer Vision*, 29:107–131, 1998.

306. M. N. Jones and D. J. K. Mewhort. Tracking attention with the focus-window technique; the information filter must be calibrated. *Behavior Research Methods, Instruments, and Computers*, 36(2):270–276, 2004.

307. S. Josephson and M. Holmes. Visual attention to repeated internet images: testing the scanpath theory on the world wide web. In *ETRA '02: Proceedings of the 2002 symposium on Eye tracking research & applications*, pages 43–49, New York, NY, USA, 2002. ACM Press.

308. M. A. Just and P. A. Carpenter. A theory of reading: From eye fixations to comprehension. *Psychological Review*, 87:329–354, 1980.

309. M. Kampmann and L. Zhang. Estimation of eye, eyebrow and nose features in videophone sequences. In *In International Workshop on Very Low Bitrate Video Coding (VLBV 98)*, pages 101–104, Urbana, USA, 1998.

310. T. Kanade, J. F. Cohn, and Y. Tian. Comprehensive database for facial expression analysis. In *Proc. IEEE Face and Gesture '00*, pages 46–53, 2000.

311. S. B. Kang and K. Ikeuchi. Toward Automatic Robot Instruction from Perception – Mapping Human Grasps to Manipulator Grasps. *IEEE Trans. on Robotics and Automation*, 13(1), 1997.

312. A. Kasprowski and J. Ober. *BioAW*, volume LCNS 3087, chapter Eye Movements in Biometrics, pages 248–258. Springer-Verlag, 2004. Maltoni, D. and Jain, A. (Eds.).

313. R. E. Kass and A. E. Raftery. Bayes Factors. *Journal of the American Statistical Association*, 90:773–795, 1995.

314. M. Kass, A. Witkin, and D. Terzopoulos. Snakes: Active contour models. *Int'l Journal of Computer Vision*, 1(4):321–331, 1988.

315. P. L. Kaufman and A. Alm, editors. *Adler's Physiology of the Eye: Clinical Application*. Mosby, 10th edition, 2003.

316. A. Kaufman, D. Cohen, and R. Yagel. Volume graphics. *Computer*, 26:51–64, 1993.

317. A. Kendon. Some functions of gaze-direction in social interaction. *Acta Psychologica*, 26:22–63, 1967.

318. J. Kent. Malaysia car thieves steal finger. *BBC News*, 2006. Retrieved on September 12, 2006 at http://news.bbc.co.uk/2/hi/asia-pacific/4396831.stm.

319. D. Kieras and S. P. Marshall. Visual availability and fixation memory in modeling visual search using the epic architecture. In *Proceedings of the 28th Annual Conference of the Cognitive Science Society*, pages 423–428, Mahwah, NJ, 2006. Lawrence Earlbaum Associates.

320. D. Kieras and D. E. Meyer. An overview of the epic architecture for cognition and performance with application to human–computer interaction. *Human–Computer Interaction*, 12:391–438, 1997.

321. Y. Kim and A. Varshney. Saliency-guided enhancement for volume visualization. *IEEE Trans. Vis. and Comp. Graph.*, 12(5):925–932, 2006.

322. I. King and L. Xu. Localized principal component analysis learning for face feature extraction. In *Proc. of the Workshop on 3D Computer Vision*, pages 124–128, Shatin, Hong Kong, 1997.

323. D. Kirsh. The intelligent use of space. *Artif. Intell.*, 73(1–2):31–68, 1995.

324. S. G. Klauer, T. A. Dingus, V. L. Neele, J. D. Sudweeks, and D. J. Rasey. The impact of driver inattention on near-crash/crash risk: An analysis using the 100-car naturalistic driving study data. Technical report, Virginia Tech Transportation Institute, 2006.

325. S. G. Klauer, V. L. Neale, T. A. Dingus, D. Ramsey, and J. Sudweeks. Driver inattention: A contributing factor to crashes and near-crashes. In *Proceedings of Human Factors and Ergonomics Society 49th Annual Meeting, Orlando, FL*, pages 1922–1926, 2005.

326. D. Kleinman, P. Young, and G. Higgins. The ddd-iii: A tool for empirical research in adaptive organizations. In *Proceedings of the Command and Control Research and Technology Symposium*, pages 273–300, Monterey, CA, 1996. Naval Posgraduate School.

327. G. D. Kleiter. Bayesian Diagnosis in Expert Systems. *Artificial Intelligence*, 54(1–2):1–32, 1992.

328. P. R. Knipling and S. S. Wang. Revised estimates of the us drowsy driver crash problem size based on general estimates system case reviews. In *39th An-*

nual Proceedings, Association for the Advancement of Automotive Medicinea, Chicago, 1995.

329. J. Kniss, G. Kindlmann, and C. Hansen. Interactive volume rendering using multi-dimensional transfer functions and direct manipulation widgets. *IEEE Computer Graphics and Applications*, 21(4):52–61, 2001.

330. J. Kniss, G. Kindlmann, and C. Hansen. Multi-dimensional transfer functions for interactive volume rendering. *IEEE Trans. Vis. and Comp. Graph.*, 8(3):270–285, 2002.

331. E. Kocak, J. Ober, N. Berme, and W. S. Melvin. Eye motion parameters correlate with level of experience in video-assisted surgery: objective testing of three tasks. *J Laparoendosc Adv Surg Tech A*, 15:575–580, 2005.

332. S. M. Kolakowski and J. B. Pelz. Eye tracking in human–computer interaction and usability research: Ready to deliver the promises (section commentary). In J. Hyona, R. Radach, and H. Deubel, editors, *The Mind's Eye: Cognitive and Applied Aspects of Eye Movement Research*, pages 573–605. Elsevier Science, 2003.

333. P. T. Kortum and W. S. Geisler. Implementation of a foveated image-coding system for bandwidth reduction of video images. In *Proceedings of the SPIE. The International Society for Optical Engineering*, volume 2657, pages 350–360, 1996.

334. U. Köthe. Accurate and efficient approximation of the continuous Gaussian scale-space. In Carl E Rasmussen, Heinrich H Bülthoff, Martin Giese, and Bernhard Schölkopf, editors, *26th DAGM-Symposium*, volume 3175 of *LNCS*, pages 350–358, Heidelberg, 2004. Springer.

335. K. Kozak, J. Pohl, W. Birk, J. Greenberg, B. Artz, M. Blommer, L. Cathey, and R. Curry. Evaluatoin of lane departure warnings for drowsy drivers. In *Proceedings of the human factors society 50th annual meeting*, pages 2400–2404, San Francisco, CA, 2006.

336. J. Krissler, L. Thalheim, and P.-M. Ziegler. Body check: Biometrics defeated. *c't Magazine*, May 2002. Retrieved October 23, 2006 from http://www.heise.de/ct/english/02/11/114/.

337. E. A. Krupinski. Technology and perception in the 21st century reading room. *Journal of the American College of Radiology*, 3:433–440, 2006.

338. E. A. Krupinski. The importance of perception research in medical imaging. *Radiation Medicine*, 18(6):329–334, 2000.

339. E. A. Krupinski, H. L. Kundel, P. F. Judy, and C. F. Nodine. The medical image perception society – key issues for image perception research. *Radiology*, 209:611–612, 1998.

340. E. A. Krupinski and C. F. Nodine. *Gaze duration predicts locations of missed lesions in mammography*, chapter 3, pages 399–403. Elsevier, 1994. Gale, A. G., Astley, S. M. and Dance, D. R. (Eds.).

341. E. A. Krupinski and H. Roehrig. Pulmonary nodule detection and visual search: P45 and P104 monochrome vs. color monitor display. *Academic Radiology*, 9(6):638–645, 2002.

342. E. A. Krupinski, H. Roehrig, and T. Furukawa. Influence of monitor display luminance on observers performance and visual search. *Academic Radiology*, 6:411–418, 1999.

343. J. B. Kruskal. *Time Warps, String Edits, and Macromolecules: The Theory and Practice of Sequence Comparison*, chapter An Overview of Sequence Com-

parison, pages 1–44. Addison Wesley, 1983. Sankoff, D. and Kruskal, J. B. (Eds.).

344. T. Kumagai and M. Akamatsu. Prediction of human driving behavior using dynamic bayesian networks. *IEICE Transactions on Information and Systems*, E89-D(2):857–860, 2006.

345. H. L. Kundel. Reader error, object recognition, and visual search. In *Proc SPIE Medical Imaging 2004*, volume 5372, pages 1–11, 2004.

346. H. L. Kundel. Visual sampling and estimates of the location of information on chest films. *Invest Radiol*, 9(2):87–93, 1974.

347. H. L. Kundel and C. F. Nodine. Modeling visual search during mammogram viewing. In *Proc SPIE Medical Imaging 2004*, volume 5372, pages 110–115, 2004.

348. H. L. Kundel, C. F. Nodine, and D. Carmody. Visual scanning, pattern recognition and decision-making in pulmonary nodule detection. *Invest Radiol*, 13:175–181, 1978.

349. H. L. Kundel, C. F. Nodine, and E. A. Krupinski. Search for lung nodules: Visual dwell indicates locations of false-positive and false-negative decisions. *Invest Radiol*, 24:472–478, 1989.

350. H. L. Kundel, C. F. Nodine, and E. A. Krupinski. Computer-displayed eye-position as a visual aid to pulmonary nodule interpretation. *Invest Radiol*, 25(8):890–896, 1990.

351. H. L. Kundel, G. Revesz, and H. M. Stauffer. Evaluation of a television image processing system. *Invest Radiol*, 3(1):44–50, 1968.

352. Y. Kuniyoshi, N. Kita, K. Sugimoto, S. Nakamura, and T. Suehiro. A foveated wide angle lens for active vision. In *Proceedings of the IEEE International Conference on Robotics and Automation*, volume 3, pages 2982–2988, 1995.

353. K.-M. Lam and H. Yan. Locating and extracting the covered eye in human face images. *Pattern Recognition*, 29(5):771–779, 1996.

354. D. Lamble, T. Kauranen, M. Laakso, and H. Summala. Cognitive load and detection thresholds in car following situations: Safety implications for using mobile (cellular) telephones while driving. *Accident Analysis and Prevention*, 31:617–623, 1999.

355. C. Lankford. Gazetracker: software designed to facilitate eye movement analysis. In *ETRA '00: Proceedings of the 2000 symposium on Eye tracking research & applications*, pages 51–55, New York, NY, USA, 2000. ACM Press.

356. L. J. Latecki, R. Lakamper, and U. Eckhardt. Shape Descriptors for Non-Rigid Shapes with a Single Closed Contour. In *Proc. IEEE Conference on Computer Vision and Pattern Recognition (CVPR)*, pages 424–429, 2000.

357. B. Law, M. S. Atkins, A. E. Kirkpatrick, A. Lomax, and C. L. MacKenzie. Eye gaze patterns differentiate skill in a virtual laparoscopic training environment. In *Proceedings of Eye Tracking Research and Applications*, pages 41–47. ETRA, 2004.

358. LC Technologies. History of eyetracking technology. http://www.eyegaze.com/3Solutions/HistoryofET.htm. LC Technologies INC., 2004.

359. J. D. Lee, B. Caven, S. Haake, and T. L. Brown. Speech-based interaction with in-vehicle computers: The effect of speech-based e-mail on drivers' attention to the roadway. *Human Factors*, 43(4):631–640, 2001.

360. S. P. Lee, J. B. Badler, and N. I. Badler. Eyes alive. In *International Conference on Computer Graphics and Interactive Techniques*, pages 637–644, 2002.

361. R. J. Leigh and D. S. Zee. *The Neurology of Eye Movements*. F. A. Davis Co, Philadelphia, 1991.

362. P. LeRay. The bnt structure learning package v.1.4.

363. V. Levenshtein. Binary codes capable of correcting deletions, insertions, and reversals. *Soviet Physics Doklady*, 10(8):707–710, 1965–1966.

364. J. L. Levine. An eye-controlled computer. Technical Report RC-8857, IBM Thomas J. Watson Research Center, Yorktown Heights, N.Y, 1982.

365. M. Levoy. Display of surfaces from volume data. *IEEE Comp. Graph. and App.*, 8:29–37, 1988.

366. M. Levoy and R. Whitaker. Gaze-directed volume rendering. *Comp. Graph.*, 24:217–223, 1990.

367. S. Levy. Television reloaded. *Newsweek*, page 49, May 2005.

368. S. M. Lewis and A. E. Raftery. Estimating Bayes Factors via Posterior Simulation With the Laplace-Metropolis Estimator. *Journal of the American Statistical Association*, 92:648–655, 1997.

369. D. Li, J. Babcock, and D. Parkhurst. openeyes: a low-cost head-mounted eye-tracking solution. *In Proceedings of the Eye Tracking and Research Applications (ETRA'06)*, March 2006.

370. D. Li, D. Winfield, and D. J. Parkhurst. Starburst: A hybrid algorithm for video-based eye tracking combining feature-based and model-based approaches. In *Proceedings of the 2005 IEEE Computer Society Conference on Computer Vision and Pattern Recognition (CVPR'05) – Workshops*, page 79, 2005.

371. X. Li and Q. Ji. Active affective state detection and user assistance with dynamic bayesian networks. *IEEE Transactions on Systems, Man, and Cybernetics–Part A: Systems and Humans*, 35(1):93–105, 2005.

372. Y. Li, S. Gong, and H. Liddell. Modelling faces dynamically across views and over time. *Proc. of ICCV01*, 1:554–559, 2001.

373. Y. Liang, J. D. Lee, and M. L. Reyes. Non-intrusive detection of driver cognitive distraction in real-time using bayesian networks. In *Proceedings of TRB 86th Annual Meeting, Washington, D.C.*, pages 07–3101, 2007.

374. Y. Liang, M. L. Reyes, and J. D. Lee. Real-time detection of driver cognitive distraction using support vector machines. *IEEE Transactions on Intelligent Transportation Systems*, 2007. in press.

375. J. J. Lien, T. Kanade, J. F. Cohn, and C. Li. Detection, tracking, and classification of subtle changes in facial expression. *Journal of Robotics and Autonomous Systems*, 31:131–146, 2000.

376. E. Llewellyn-Thomas. Movements of the eye. *Scientific American*, 219:88–95, 1968.

377. G. R. Loftus. *Eye Movements and Psychological Processes*, chapter A Framework for a Theory of Picture Recognition. John Wiley and Sons, New York, 1976. Monty, R. A. and Senders, J. W. (Eds.).

378. W. E. Lorensen and H. E. Cline. Marching cubes: A high resolution 3D surface construction algorithm. *Comp. Graph.*, 21:163–169, 1987.

379. L. Lorigo, B. Pan, H. Hembrooke, T. Joachims, L. Granka, and G. Gay. The influence of task and gender on search and evaluation behavior using google. *Inf. Process. Manage.*, 42(4):1123–1131, 2006.

380. L. C. Loschky and G. W. McConkie. How late can you update? detecting blur and transients in gaze-contingent multi-resolution displays. In *Proceedings of*

the Human Factors and Ergonomics Society 49th Annual Meeting 2005, pages 1527–1530, 2005.

381. L. C. Loschky and G. W. McConkie. User performance with gaze contingent multiresolutional displays. In *Proceedings of the 2000 symposium on Eye tracking research & applications*, Palm Beach Gardens, Florida, United States, 97–103, 2000.
ISBN: 1-58113-280-8

382. L. C. Loschky, G. W. McConkie, J. Yang, and M. E. Miller. The limits of visual resolution in natural scene viewing. *Visual Cognition*, 12(6):1057–1092, 2005.

383. A. Lu, R. Maciejewski, and D. S. Ebert. Volume composition using eye tracking data. In *IEEE-VGTC Symposium on Visualization*, pages 115–122, 2006.

384. B. D. Lucas and T. Kanade. An iterative image registration technique with an application to stereo vision. In *Proc. Int. Joint Conf. Artificial Intelligence*, pages 674–679, 1981.

385. D. Luebke, M. Reddy, J. Cohen, A. Varshney, B. Watson, and R. Huebner. *Level of detail for 3D graphics*. Morgan-Kaufman, San Francisco, 2002.

386. M. J. Mack. Minimally invasive and robotic surgery. *JAMA*, 285:568–572, 2001.

387. N. H. Mackworth and A. J. Morandi. The gaze selects informative details within pictures. *Perceptions and Psychophysics*, 2:547–552, 1967.

388. P. P. Maglio and C. S. Campbell. Attentive agents. *Commun. ACM*, 46(3):47–51, 2003.

389. P. Majaranta and K.-J. Räihä. Twenty years of eye typing: systems and design issues. In *ETRA '02: Proceedings of the symposium on Eye tracking research & applications*, pages 15–22, New York, NY, USA, 2002. ACM Press.

390. S. G. Mallat. *A Wavelet Tour of Signal Processing*. Academic Press, second edition, 1999.

391. D. Manning and S. C. Ethell. Lesion conspicuity and afroc performance in pulmonary nodule detection. In *Proc SPIE Medical Imaging 2002*, volume 4686, pages 300–311, 2002.

392. J. M. Marin, K. Mengersen, and C. P. Robert. Bayesian modeling and inference on mixtures of distributions. In D. Dey and C. R. Rao, editors, *Handbook of Statistics 25*. Elsevier-Sciences, 2004.

393. S. R. Marschner and D. P. Greenberg. Inverse Lighting for Photography. In *IS&T/SID Color Imaging Conference*, pages 262–265, 1997.

394. S. P. Marshall. *Decision Making in Complex Environments*, chapter Measures of attention and cognitive effort in tactical decision making, pages 321–332. Ashgate Publishing, 2007. Cook, M., Noyes, J. and Masakawski, V. (Eds.).

395. S. P. Marshall. Identifying cognitive state from eye metrics. *Aviation, Space, and Environmental Medicine*, in press.

396. S. P. Marshall, J. G. Morrison, L. E. Allred, S. Gillikin, and J. A. McAllister. Eye tracking in tactical decision-making environments: Implementation and analysis. In *Proceedings of the Command and Control Research and Technology Symposium*, pages 347–354, Washington, DC, 1997. National Defense University.

397. S. P. Marshall, D. M. Wilson, S. R. Knust, M. Smith, and R. J. Garcia. Sharing decision-making knowledge in tactical situations. In *Proceedings of the Command and Control Research and Technology Symposium*, pages 693–699, Monterey, CA, 1998. Naval Postgraduate School.

398. A. M. Martinez and A. C. Kak. Pca versus lda. *IEEE Transactions on Pattern Analysis and Machine Intelligence*, 23(2):228–233, 2001.

399. Y. Matsumoto and A. Zelinsky. An algorithm for real-time stereo vision implementation of head pose and gaze direction measurement. In *Proc. Fourth IEEE International Conference on Automatic Face and Gesture Recognition*, pages 499–504, 2000.

400. N. Max. Optical models for direct volume rendering. *IEEE Trans. Vis. and Comp. Graph.*, 1(2):99–108, 1995.

401. R. E. Mayer. Multimedia learning: Are we asking the right questions. *Educational Psychologist*, 32(1):1–19, 1997.

402. R. E. Mayer and R. Moreno. A split-attention effect in multimedia learning: Evidence for dual processing systems in working memory. *Journal of Educational Psychology*, 90(2):312–320, 1998.

403. R. E. Mayer, K. Steinhoff, G. Bower, and R. Mars. A generative theory of textbook design: Using annotated illustrations to foster meaningful learning of science text. *Educational Technology Research and Development*, 43(1):31–43, 1995.

404. S. McClellan and K. Kerschbaumer. Tickers and bugs: Has tv gotten way too graphic? *Broadcasting and Cable*, 131(50):16–20, 2001.

405. G. McConkie and L. Loschky. Perception onset time during fixations in free viewing. *Behavior Research Methods, Instruments, & Computers*, 34(4):481–490, 2002.

406. G. W. McConkie and K. Rayner. The span of the effective stimulus during a fixation in reading. *Perception & Psychophysics*, 17:578–586, 1975.

407. C. McCulloch and S. Searle. *Generalized, Linear, and Mixed Models*. John Wiley and Sons, New York, 2001.

408. S. J. McKenna, S. Gong, R. P. Würtz, J. Tanner, and D. Banin. Tracking facial feature points with gabor wavelets and shape models. *Proc. of Int'l Conf. on Audio- and Video- based Biometric Person Authentication*, pages 35–42, 1997.

409. G. J. McLachlan and T. Krishnan. *The EM Algorithm and Extensions*. New York: Wiley, 1997.

410. G. J. McLachlan and D. Peel. *Finite Mixture Models*. New York: Wiley, 2000.

411. M. Meißner, U. Hoffman, and W. Straßer. Enabling classification and shading for 3D texture mapping based volume rendering using OpenGL and extensions. In *Proceedings of IEEE Visualization '99*, 1999.

412. C. Mello-Thoms, C. F. Nodine, and H. L. Kundel. What attracts the eye to the location of missed and reported breast cancers? In *Proceedings of Eye Tracking Research and Applications*, pages 111–117. ETRA, 2002.

413. X. Meng and D. van Dyk. The EM Algorithm – An Old Folk Song Sung to a Fast New Tune. *Journal of the Royal Statistical Society, B*, 59(3):511–567, 1997.

414. Microsoft Corp. Microsoft wireless intellimouse explorer with fingerprint reader, Feb 2005. Retrieved February 4, 2005 at http://www.microsoft.com/hardware/mouseandkeyboard/productdetails.aspx?pid=035.

415. K. Mikolajczyk and C. Schmid. A Performance Evaluation of Local Descriptors. *IEEE Transactions on Pattern Analysis and Machine Intelligence*, 27:1615–1630, 2005.

416. MIPS, 2005. http://www.mips.ws.

417. A. Moise, M. S. Atkins, and R. Rohling. Evaluating different radiology work-station interaction techniques with radiologists and laypersons. *Journal of Digital Imaging*, 18(2):116–130, 2005.

418. D. Moore, J. Burton, and R. Meyers. *The Theoretical and Research Foundations of Multimedia*, chapter Multiple-channel Communication. Simon and Schuster Macmillan, 1996. Jonassen, D. H. (Ed.).

419. C. H. Morimoto, A. Amir, and M. Flickner. Detecting eye position and gaze from a single camera and 2 light sources. In *International Conference on Pattern Recognition*, Quebec, Canada, August 2002.

420. C. Morimoto and M. Flickner. Real-time multiple face detection using active illumination. In *Proc. Fourth IEEE International Conference on Automatic Face and Gesture Recognition*, pages 8–13, 2000.

421. C. H. Morimoto, D. Koons, A. Amir, and M. Flickner. Frame-rate pupil detector and gaze tracker. In *IEEE ICCV'99 Frame-Rate Computer Vision Applications Workshop*, 1999.

422. C. H. Morimoto, D. Koons, A. Amir, M. Flickner, and S. Zhai. Keeping an eye for hci. In *Proc. XII Brazilian Symp. Comp. Graphics and Image Proc.*, pages 171–176, 1999.

423. C. H. Morimoto and M. R. M. Mimica. Eye gaze tracking techniques for interactive applications. *Computer Vision and Image Understanding*, 98(1):4–24, 2005.

424. J. G. Morrison, S. P. Marshall, R. T. Kelly, and R. A. Moore. Eye tracking in tactical decision-making environments: Implications for decision support. In *Proceedings of the Command and Control Research and Technology Symposium*, pages 355–364, Washington, DC, 1997. National Defense University.

425. L. Mroz and H. Hauser. RTVR: a flexible java library for interactive volume rendering. In *Proceedings of the conference on Visualization '01*, pages 279–286, 2001.

426. J. B. Mulligan. Image processing for improved eye tracking accuracy. *Behav. Res. Methods, Instr. Comp.*, 29:54–65, 1997.

427. M. P. Murphy. The bayes net toolbox.

428. G. P. Mylonas, A. Darzi, and G. Yang. Gaze contingent depth recovery and motion stabilisation for minimally invasive robotic surgery. *Lecture Notes in Computer Science*, 3150:311–319, 2004. Yang, G.-Z. and Jiang, T. (Eds.).

429. G. P. Mylonas, A. Darzi, and G. Z. Yang. Gaze-contingent control for minimally invasive robotic surgery. *Comput Aided Surg*, 11:256–266, 2006.

430. T. Nakano, K. Sugiyama, M. Mizuno, and S. Yamamoto. Blink measurement by image processing and application to warning of driver's drowsiness in automobiles. In *IEEE Intelligent Vehicles*, 1998.

431. O. Nakayama, T. Futami, T. Nakamura, and E. R. Boer. Development of a steering entropy method for evaluating driver workload. In *SAE Technical Paper Series: #1999-01-0892*, 1999. Presented at the International Congress and Exposition, Detroit, Michigan.

432. National Research Council. *Caffeine for the sustainment of mental task performance: Formulations for military operations*. National Academy Press, Washington, DC, 2001.

433. S. K. Nayar. Sphereo: Recovering depth using a single camera and two specular spheres. In *SPIE: Optics, Illumination and Image Sensing for Machine Vision II*, 1988.

434. S. Needleman and C. Wunsch A general method applicable to the search for similarities in the amino acid sequence of two proteins. *J Mol Biol*, 48(3):443–53, Mar 1970.

435. S. A. Nene and S. K. Nayar. Stereo Using Mirrors. In *IEEE ICCV 98*, pages 1087–1094, Jan 1998.

436. S. Nevalainen and J. Sajaniemi. An experiment on short-term effects of animated versus static visualization of operations on program perception. In *ICER '06: Proceedings of the 2006 international workshop on computing education research*, pages 7–16, New York, NY, USA, 2006. ACM Press.

437. S. Nevalainen and J. Sajaniemi. Comparison of three eye tracking devices in psychology of programming research. In *Proceedings of the 16th Annual Psychology of Programming Interest Group Workshop (PPIG'04)*, pages 151–158, Carlow, Ireland, 2004.

438. S. Nevalainen and J. Sajaniemi. Short-Term Effects of Graphical versus Textual Visualisation of Variables on Program Perception. In *Proceedings of the 17th Annual Psychology of Programming Interest Group Workshop (PPIG'05)*, pages 77–91, Brighton, UK, 2005.

439. R. Newman, Y. Matsumoto, S. Rougeaux, and A. Zelinsky. Real time stereo tracking for head pose and gaze estimation. In *Proceedings Int. Conf. on Automatic Face and Gesture Recognition*, Grenoble, France, March 2000.

440. M. Nicolaou. The assessment of visual behaviour and depth perception in surgery, 2006.

441. M. Nicolaou, L. Atallah, A. James, J. Leong, A. Darzi, and G. Z. Yang. The effect of depth perception on visual-motor compensation in minimal invasive surgery. *MIAR*, 2006.

442. M. Nicolaou, A. James, A. Darzi, and G. Z. Yang. *A Study of Saccade Transition for Attention Segregation and Task Strategy in Laparoscopic Surgery, MICCAI 2004* In Medical Image Computing and Computer-Assisted Intervention – MICCAI(2) 2004, 7th International Conference Saint-Malo, France, September 26–29, 2004, Proceedings, Part II, *Lecture Notes in Computer Science*, 3217:97–104, 2004. ISBN 3-540-22977-9.

443. J. Nielsen. F-shaped pattern for reading web content. alertbox, april 17, 2006, 2006.

444. J. Nielsen. Noncommand user interfaces. *Commun. ACM*, 36(4):83–99, 1993.

445. S. G. Nikolov, M. G. Jones, D. Agrafiotis, D. R. Bull, and C. N. Canagarajah. Focus+context visualisation for fusion of volumetric medical images. In *Proceedings of the 4th International Conference on Information Fusion (Fusion 2001), Montreal, QC, Canada, August 7–10, 2001, International Society of Information Fusion (ISIF)*, volume I, pages WeC3–3–10, 2001.

446. S. G. Nikolov, M. G. Jones, I. D. Gilchrist, D. R. Bull, and C. N. Canagarajah. *Gaze-Contingent Multi-Modality Displays for Visual Information Fusion: Systems and Applications*, chapter 15, pages 431–471. CRC/Taylor and Francis, Boca Raton, FL, 2006. Blum, R. S. and Liu, Z. (Eds.).

447. S. G. Nikolov, T. D. Newman, D. R. Bull, C. N. Canagarajah, M. G. Jones, and I. D. Gilchrist. Gaze-contingent display using texture mapping and OpenGL: system and applications. In *Eye Tracking Research & Applications (ETRA)*, pages 11–18, 2004.

448. K. Nishino, P. N. Belhumeur, and S. K. Nayar. Using Eye Reflections for Face Recognition Under Varying Illumination. In *IEEE International Conference on Computer Vision*, volume I, pages 519–526, 2005.

449. K. Nishino and S. K. Nayar. Corneal Imaging System: Environment from Eyes. *International Journal on Computer Vision*, 70(1):23–40, 2006.

450. K. Nishino and S. K. Nayar. Eyes for Relighting. *ACM Trans. on Graphics (Proceedings of SIGGRAPH 2004)*, 23(3):704–711, 2004.

451. K. Nishino and S. K. Nayar. The World in Eyes. In *IEEE Conference on Computer Vision and Pattern Recognition*, volume I, pages 444–451, 2004.

452. C. F. Nodine and E. A. Krupinski. Perceptual skill, radiology expertise, and visual test performance with NINA and WALDO. *Academic Radiology*, 5(9):603–612, 1998.

453. C. F. Nodine and H. L. Kundel. A visual dwell algorithm can aid search and recognition of missed lung nodules in chest radiographs. In *Visual Search 2*, pages 399–406, 1990.

454. C. F. Nodine and H. L. Kundel. Using eye movements to study visual search and to improve tumour detection. *Radiographics*, 7:1241–1250, 1987.

455. C. F. Nodine, H. L. Kundel, S. C. Lauver, and L. C. Toto. The nature of expertise in searching mammograms for breast masses. In *Proc SPIE Medical Imaging 1996*, volume 2712, pages 89–94, 1996.

456. C. F. Nodine, H. L. Kundel, C. Mello-Thoms, and S. P. Weinstein. Role of computer-assisted visual search in mammographic interpretation. In *Proc SPIE Medical Imaging 2001*, volume 4324, pages 52–55, 2001.

457. C. F. Nodine, H. L. Kundel, L. C. Toto, and E. A. Krupinski. Recording and analyzing eye-position data using a microcomputer workstation. *Behavior Research Methods, Instruments, and Computers*, 24:475–485, 1992.

458. C. F. Nodine, C. Mello-Thoms, S. P. Weinstein, and H. L. Kundel. Time course of perception and decision making during mammographic interpretation. *American Journal of Roentology*, 179(4):917–922, 2002.

459. D. Noton and L. W. Stark. Scanpaths in eye movements during pattern perception. *Science*, 171(3968):308–311, 1971.

460. D. Noton and L. W. Stark. Scanpaths in saccadic eye movements while viewing and recognizing patterns. *Vision Research*, 11(9):929–942, 1971.

461. T. Ohno. Eyeprint: Support of document browsing with eye gaze trace. In *Proc. ACM International Conference on Multimodal Interfaces (ICMI)*, pages 16–23. ACM Press, 2004.

462. T. Ohno. Eyeprint: Using passive eye trace from reading to enhance document access and comprehension. *International Journal of Human–Computer Interaction*, 2007 (to appear).

463. T. Ohno. Features of eye gaze interface for selection tasks. In *Proc. Third Asia Pasific Computer Human Interaction (APCHI)*, pages 176–181. IEEE Computer Society, 1998.

464. T. Ohno. One-point calibration gaze tracking method. In *Proc. ACM Symposium on Eye Tracking Research & Applications (ETRA)*, page 34. ACM Press, 2006.

465. T. Ohno. Weak gaze awareness in video-mediated communication,. In *Extended Abstract of ACM Conference on Human Factors in Computing Systems (CHI)*, pages 1709–1702. ACM Press, 2005.

466. T. Ohno and N. Mukawa. A free-head, simple calibration, gaze tracking system that enables gaze-based interaction. In *Proc. ACM Symposium on Eye Tracking Research & Applications (ETRA)*, pages 115–122. ACM Press, 2004.

467. T. Ohno and N. Mukawa. Gaze-based interaction for anyone, anytime. *Proceedings of the 9th International Conference on Human–Computer Interaction (HCII 2003)*, 4:1452–1456, 2003.

468. T. Ohno, N. Mukawa, and A. Yoshikawa. Freegaze: A gaze tracking system for everyday gaze interaction. In *Proc. ACM Symposium on Eye Tracking Research & Applications (ETRA)*, pages 125–132. ACM Press, 2002.

469. N. Oliver and E. Horvitz. *A comparison of HMMs and dynamic Bayesian networks for recognizing office activities*, chapter Lecture Notes in Computer Science, pages 199–209. Springer Berlin, Heidelberg, 2005.

470. N. Oliver, A. P. Pentland, and F. Berard. LAFTER: Lips and Face Real Time Tracker. In *Proc. IEEE Conference on Computer Vision and Pattern Recognition (CVPR)*, pages 123–129, 1997.

471. S. H. Or, W. S. Luk, K. H. Wong, and I. King. An efficient iterative pose estimation algorithm. *Image and Vision Computing*, 16(5):353–362, 1998.

472. K. F. Van Orden, T. P. Jung, and S. Makeig. Combined eye activity measures accurately estimate changes in sustained visual task performance. *Biological Psychology*, 52:221–240, 2000.

473. C. E. Osgood. *Method and Theory in Experimental Psychology*. Oxford University Press, 1956.

474. OSHA. Osha ergonomic solutions: Computer workstations etool – components – monitors. http://www.osha.gov/sltc/etools/computerworkstations/components_monitors.html.

475. C. O'Sullivan, J. Dingliana, and S. Howlett. *Eye-movements and Interactive Graphics*, In The Mind's Eyes: Cognitive and Applied Aspects of Eye Movement Research. Hyönä J., Radach, R. and Deubel, H. (Eds.) Elsevier Science, Oxford pages 555–571. 2003.

476. E. Osuna, R. Freund, and F. Girosi. Training Support Vector Machines: an Application to Face Detection. In *Proc. IEEE Conference on Computer Vision and Pattern Recognition (CVPR)*, pages 130–136, 1997.

477. E. B. Page. Ordered hypotheses for multiple treatments: A significance test for linear ranks. *Journal of the American Statistical Association*, 58:206–230, 1963.

478. A. Paivio. *Mental Representation: A Dual Coding Approach*. Oxford University Press, Oxford, 1986.

479. T. Pajdla, T. Svoboda, and V. Hlaváč. Epipolar geometry of central panoramic cameras. In *Panoramic Vision : Sensors, Theory, and Applications*. Springer Verlag, 2000.

480. S. E. Palmer. *Vision Science: Photons to Phenomenology*. MIT Press, 1999.

481. B. Pan, H. Hembrooke, G. Gay, L. Granka, M. Feusner, and J. Newman. The determinants of web page viewing behavior: an eye-tracking study. In *ETRA '04: Proceedings of the 2004 symposium on Eye tracking research & applications*, pages 147–154, New York, NY, USA, 2004. ACM Press.

482. M. Pantic and L. J. M. Rothkrantz. Expert system for automatic analysis of facial expression. *Image and Vision Computing*, 18(11):881–905, 2000.

483. Passmark Inc. How the passface system works, Feb 2005. Retrieved October 23, 2006 from http://www.realuser.com/cgi-bin/ru.exe/_/homepages/users/passface.htm.

484. K. S. Park and C. J. Lim. A simple vision-based head tracking method for eye-controlled human/computer interface. *International Journal of Human–Computer Studies*, 54:319–332, 2001.

485. R. E. Parker. Picture processing during recognition. *Journal of Experimental Psychology*, pages 284–292, 1978.

486. D. J. Parkhurst and E. Niebur. Variable-resolution displays: a theoretical, practical, and behavioral evaluation. *Human Factors*, 44(4):611–629, 2004.

487. E. Peli and G. A. Geri. Discrimination of wide-field images as a test of a peripheral-vision model. *Journal of the Optical Society of America A*, 18:294–301, 2001.

488. E. Peli, J. Yang, and R. B. Goldstein. Image invariance with changes in size: the role of peripheral contrast thresholds. *Journal of the Optical Society of America A*, pages 1762–1774, 1991.

489. F. Pellacini, L. Lorigo, and G. Gay. Visualizing paths in context. Technical Report #TR2006-580, Dept. of Computer Science, Dartmouth College, 2006.

490. D. G. Pelli, M. Palomares, and N. J. Majaj. Crowding is unlike ordinary masking: Distinguishing feature integration from detection. *Journal of Vision*, 4(12):1136–1169, 2004.

491. A. Perrig and D. Song. Hash visualization: A new technique to improve real-world security,. In *Proc. 1999 International Workshop on Cryptographic Techniques and E-Commerce (CryTEC '99)*, 1999.

492. J. S. Perry and W. S. Geisler. Gaze-contingent real-time simulation of arbitrary visual fields. In Bernice E Rogowitz and Thrasyvoulos N Pappas, editors, *Human Vision and Electronic Imaging: Proceedings of SPIE, San Jose, CA*, volume 4662, pages 57–69, 2002.

493. H. Pfister, B. Lorensen, C. Bajaj, G. Kindlmann, W. Schroeder, L. S. Avila, K. Martin, R. Machiraju, and J. Lee. The transfer function bake-off. *IEEE Comp. Graph. and App.*, 21(3):16–22, 2001.

494. P. J. Phillips, P. J. Flynn, T. Scruggs, K. W. Bowyer, J. Chang, K. Hoffman, J. Marques, J. Min, and W. Worek. Overview of the face recognition grand challenge. *Proc. of CVPR05*, 1:947–954, 2005.

495. R. Pieters, E. Rosbergen, and M. Wedel. Visual attention to repeated print advertising: A test of scanpath theory. *Journal of Marketing Research*, 36(4):424–438, 1999.

496. E. D. Pisano, C. Gatsonis, E. Hendrick, M. Yaffe, J. K. Baum, S. Acharyya, E. F. Conant, L. L. Fajardo, L. Bassett, C. D'Orsi, R. Jong, and M. Rebner. Diagnostic performance of digital versus film mammography for breast-cancer screening. *New England Journal of Medicine*, 353:1773–1783, 2005.

497. C. Poelman. The paraperspective and projective factorization method for recovering shape and motion. *Carnegie Mellon University Technical Report CMU-CS-95-173*, 1995.

498. M. Pomplun, E. Reingold, and J. Shen. Investigating the visual span in comparative search: the effects of task difficulty and divided attention. *Cognition*, 81:B57–B67, 2001.

499. M. Pomplun and H. Ritter. Three-level model of comparative visual search. In *Proceedings of the 21st Annual Meeting of the Cognitive Science Society*, pages 543–548, 1999.

500. M. Pomplun, L. Sichelschmidt, K. Wagner, T. Clermont, G. Rickheit, and H. Ritter. Comparative visual search: a difference the makes a difference. *Cognitive Science*, 25(1):3–36, 2001.

501. V. Ponsoda, D. Scott, and J. M. Findlay. A probability vector and transition matrix analysis of eye movements during visual search. *Acta Psychologica*, 88:167–185, 1995.

502. J. C. Popieul, P. Simon, R. Leroux, and J. C. Angué. Automatic processing of a car driver eye scanning movements on a simulated highway driving context. In *IEEE Intelligent Vehicles Symposium*, Dearborn, MI, 2000.

503. W. H. Press, S. A. Teukolsky, W. T. Vetterling, and B. P. Flannery. Numerical Recipes in C The Art of Scientific Computing. *2nd ed. Cambridge, UK: Cambridge University Press*, pages 408–412, 1992.

504. P. Qvarfordt. Eyes on multimodal interaction. 0345-7524 893, Department of Computer and Information Science, Linkoping University, Department of Computer and Information Science, Linkoping University, Sweden, 2004. Dissertation No. 893 (Ph.D. thesis).

505. P. Qvarfordt, D. Beymer, and S. Zhai. Realtourist – a study of augmenting human-human and human–computer dialogue with eye-gaze overlay. *Lecture Notes in Computer Science*, 3585:767–780, 2005.

506. P. Qvarfordt and S. Zhai. Conversing with the user based on eye-gaze patterns. In *CHI '05: Proceedings of the SIGCHI conference on Human factors in computing systems*, pages 221–230, New York, NY, USA, 2005. ACM Press.

507. T. Randen and J. H. Husoy. Filtering for Texture Classication: A Comparative Study. *IEEE Transactions on Pattern Analysis and Machine Intelligence*, 21(4):291–310, 1999.

508. P. S. Rau. Drowsy driver detection and warning system for commercial vehicle drivers: Field operational test design, data analysis, and progress. In *19th International Technical Conference on the Enhanced Safety of Vehicles (CD).*, Washington, DC, 2005.

509. I. Ravyse, H. Sahli, and J. Cornelis. Eye activity detection and recognition using morphological scale-space decomposition. In *Proc. IEEE International Conference on Pattern Recognition '00*, volume 1, pages 5080–5083, 2000.

510. K. Rayner. Eye movements in reading and information processing: 20 years of research. *Psychological Bulletin*, 124(3):372–422, 1998.

511. K. Rayner. *Eye Movement Research: Mechanisms, Processes and Applications*, chapter Eye Movements and Cognitive Processes in Reading, Visual Search and Scene Perception, pages 3–22. Elsevier Science Publishers, 1995. Findlay, J. M., Walker, R. and Kentridge, R. W. (Eds.).

512. K. Rayner. The perceptual span and peripheral cues in reading. *Cognitive Psychology*, 7:65–81, 1975.

513. M. A. Recarte and L. M. Nunes. Effects of verbal and spatial-imagery tasks on eye fixations while driving. *Journal of Experimental Psychology: Applied*, 6(1):31–43, 2000.

514. M. A. Recarte and L. M. Nunes. Mental workload while driving: Effects on visual search, discrimination, and decision making. *Journal of Experimental Psychology: Applied*, 9(2):119–137, 2003.

515. M. Reddy. *Perceptually Modulated Level of Detail for Virtual Environments*. Ph.D. thesis, University of Edinburgh, 1997.

516. D. A. Redelmeier and R. J. Tibshirani. Association between cellular-telephone calls and motor vehicle collisions. *The New England Journal of Medicine*, 336(7):453–458, 1997.

517. S. D. Reese. Improving audience learning from television news through between-channel redundancy. ERIC Document Reproduction Service No. Ed 229 777, 1983.

518. M. A. Regan, J. D. Lee, and K. D. Young. Driver distraction: Theory, effects and mitigation. (book in preparation).

519. E. M. Reingold and L. C. Loschky. Saliency of peripheral targets in gaze-contingent multi-resolutional displays. *Behavioural Research Methods, Instruments and Computers*, 34:491–499, 2002.

520. E. M. Reingold, L. C. Loschky, G. W. McConkie, and D. M. Stampe. Gaze-contingent multiresolutional displays: An integrative review. *Human Factors*, 45(2):307–328, 2003.

521. P. Rheingans and D. Ebert. Volume illustration: Nonphotorealistic rendering of volume models. *IEEE Trans. Vis. and Comp. Graph.*, 7(3):253–264, 2001.

522. D. C. Richardson and M. J. Spivey. Eye-tracking: Characteristics and methods (part 1); eye-tracking: Research areas and applications (part 2). Encyclopedia of Biomaterials and Biomedical Engineering, Wnek. G. and Bowlin, G. (Eds.), 2004.

523. S. Richardson and P. J. Green. On Bayesian Analysis of Mixtures with an Unknown Number of Components (With Discussion). *Journal of the Royal Statistical Society, B*, 59:731–792, 1997.

524. B. D. Ripley. *Pattern Recognition and Neural Networks*. Cambridge University Press, 1996.

525. J. Rissanen. Modeling by shortest data description. *Automatica*, 14:465–471, 1978.

526. C. P. Robert and G. Casella. *Monte Carlo Statistical Methods*. Springer-Verlag, 1999.

527. D. A. Robinson. A Method of Measuring Eye Movement Using a Scleral Search Coil in a Magnetic Field. *IEEE Trans. on Biomedical Engineering*, BME-10:137–145, 1963.

528. K. Rodden and X. Fu. Exploring how mouse movements relate to eye movements on web search results pages. In *Proceedings of the Conference on Research and Development in Information Retrieval (SIGIR)*, 2007.

529. K. Rodden and X. Fu. Eyetracking analysis of google web search. Internal research, Google Inc., 2006.

530. L. F. Rogers. Keep looking: satisfaction fo search. *American Journal of Roentgenology*, 175:287, 2000.

531. M. Rogers and J. Graham. Robust active shape model search. *Proc. of ECCV*, 4:517–530, 2002.

532. S. D. Rogers, E. E. Kadar, and A. Costall. Drivers' gaze patterns in braking from three different approaches to a crash barrier. *Ecological Psychology*, 17(1):39–53, 2005.

533. S. Romdhani, S. Gong, and A. Psarrou. Multi-view nonlinear active shape model using kernel pca. *Proc. of BMVC*, pages 483–492, Sep 1999.

534. P. Romero, B. du Boulay, R. Lutz, and R. Cox. The effects of graphical and textual visualisations in multi-representational debugging environments. In *HCC '03: Proceedings of the IEEE 2003 Symposia on Human Centric Computing Languages and Environments*, Washington, DC, USA, 2003. IEEE Computer Society.

535. P. Romero, R. Lutz, R. Cox, and B. du Boulay. Co-ordination of multiple external representations during Java program debugging. In *HCC '02: Proceedings of the IEEE 2002 Symposia on Human Centric Computing Languages and Environments (HCC'02)*, page 207, Washington, DC, USA, 2002. IEEE Computer Society.

536. A. Roorda, F. Romero-Borja, W. Donnelly III, H. Queener, T. Hebert, and M. Cambell. Adaptive optics scanning laser ophthalmoscopy. *Opt. Express*, 10:405–412, 2002.

537. D. Rose and D. Levinson. Understanding user goals in web search. In *WWW '04: Proceedings of the 13th international conference on World Wide Web*, pages 13–19, New York, NY, USA, 2004. ACM Press.

538. B. H. Rosenberg, D. Landsittel, and T. D. Averch. Can video games be used to predict or improve laparoscopic skills? *J Endourol*, 19:372–376, 2005.

539. A. Rosenfled. Image pattern recognition. *Proceedings of the IEEE*, 69(5):596–605, 1981.

540. J. Ross, M. C. Morrone, M. E. Goldberg, and D. C. Burr. Changes in visual perception at the time of saccades. *Trends in Neurosciences*, 24(2):113–121, 2001.

541. R. J. Rost. *OpenGL shading language*. Addison Wesley, Boston, Massachusetts, 1999.

542. L. M. Rowland, M. L. Thomas, D. R. Thorne, H. C. Sing, J. L. Krichmar, H. Q. Davis, S. M. Balwinski, R. D. Peters, E. Kloeppel-Wagner, D. P. Redmond, E. Alicandri, and G. Belenky. Oculomotor responses during partial and total sleep deprivation. *Aviat. Space Environ. Med.*, 76 (suppl.):C104–C113, 2005.

543. D. Royal. Volume i: Findings: National survey of distracted and drowsy driving attitudes and behavior: 2002. Technical report, The Gallup Organization, 2003.

544. E. J. Russo and F. LeClerc. An eye-fixation analysis of choice processes for consumer nondurables. *Journal of Consumer Research*, 21(2):274–290, 1994.

545. J. Salojarvi, I. Kojo, J. Simola, and S. Kaski. Can relevance be inferred from eye movements in information retrieval? In *Proceedings of the Workshop on Self-Organizing Maps (WSOM'03)*, pages 261–266, 2003.

546. D. Salvucci. Inferring intent in eye-based interfaces: tracing eye movements with process models. In *CHI '99: Proceedings of the SIGCHI conference on Human factors in computing systems*, pages 254–261, New York, NY, USA, 1999. ACM Press.

547. D. D. Salvucci and J. R. Anderson. Automated eye-movement protocol analysis. *Human–Computer Interaction*, 16:39–86, 2001.

548. D. D. Salvucci and J. R. Anderson. Intelligent gaze-added interfaces. In *CHI 2000: Proceedings of the Conference on Human Factors in Computing Systemns*, pages 273–280. ACM Press, 2000.

549. D. Salvucci and J. Goldberg. Identifying fixations and saccades in eye-tracking protocols. In *ETRA '00: Proceedings of the 2000 symposium on Eye tracking research & applications*, pages 71–78, New York, NY, USA, 2000. ACM Press.

550. D. D. Salvucci and K. L. Macuga. Predicting the effects of cellular-phone dialing on driving performance. *Congitive Systems Research*, 3:95–102, 2002.

551. K. Samuel, N. Balakrishman, and J. Norman. *Continous Multivariate Distributions*, volume 1. New York: Wiley-Interscience, 2000.

552. G. Sandini and G. Metta. *Retina-Like Sensors: Motivations, Technology and Applications*, pages 251–262. Springer, 2003. Barth, F. G., Humphrey, J. A. C. and Secomb, T. W. (Eds.).

553. D. Sankoff and J. B. Kruskal, editors. *Time Warps, String Edits, and Macromolecules: The Theory and Practice of Sequence Comparison*. Addison Wesley, Reading, MA, 1983.

554. T. F. Sanquist and M. C. McCallum. A review, evaluation and classification of fatigue countermeasures for transportation operators. In *Proceedings of the Human Factors and Ergonomics Society 48th Annual Meeting*, Santa Monica, CA, 2004.

555. A. Santella and D. DeCarlo. Robust clustering of eye movement recordings for quantification of visual interest. In *ETRA '04: Proceedings of the 2004 symposium on Eye tracking research & applications*, pages 27–34, New York, NY, USA, 2004. ACM Press.

556. J. Scheeres. Airport face scanner failed. *In Wired News*, May 2002. Retrieved February 4, 2005 at http://www.wired.com/news/privacy/0,1848,52563,00.html.

557. C. Schmid and R. Mohr. Local Grayvalue Invariants for Image Retrieval. *IEEE Transactions on Pattern Analysis and Machine Intelligence*, 19:530–535, 1997.

558. S. Sclaroff and J. Isidoro. Active blobs: region-based, deformable appearance models. *Computer Vision and Image Understanding*, 89(2-3):197–225, 2003.

559. W. J. Severein. Another look at cue summation. *Audio Visual Communication Review*, 15(4):233–245, 1967.

560. H. R. Sheikh, B. L. Evans, and A. C. Bovik. Real-time foveation techniques for low bit rate video coding. *Real-Time Imaging*, 9(1):27–40, 2003.

561. J. S. Shell, R. Vertegaal, D. Cheng, A. W. Skaburskis, C. Sohn, A. J. Stewart, O. Aoudeh, and C. Dickie. Ecsglasses and eyepliances: using attention to open sociable windows of interaction. In *ETRA'2004: Proceedings of the Eye tracking research & applications symposium*, pages 93–100, New York, NY, USA, 2004. ACM Press.

562. J. S. Shell, R. Vertegaal, and A. W. Skaburskis. Eyepliances: attention-seeking devices that respond to visual attention. In *CHI '03: Extended abstracts on Human factors in computing systems*, pages 770–771, New York, NY, USA, 2003. ACM Press.

563. J. Shi and C. Tomasi. Good features to track. *Proc. of CVPR94*, pages 593–600, 1994.

564. S. W. Shih and J. Liu. A novel approach to 3d gaze tracking using stereo cameras. *IEEE Transactions on systems, man, and cybernetics – PART B*, 34:234–245, 2004.

565. S. W. Shih, Y. Wu, and J. Liu. A calibration-free gaze tracking technique. In *Proc. International Conference on Pattern Recognition (ICPR)*, pages 201–204, 2000.

566. M. Shulze. Circular hough transform demonstration. http://markschulze.net/java/hough.

567. J. L. Sibert, M. Gokturk, and R. A. Lavine. The reading assistant: eye gaze triggered auditory prompting for reading remediation. In *UIST '00: Proceedings of the 13th annual ACM symposium on User interface software and technology*, pages 101–107, New York, NY, USA, 2000. ACM Press.

568. E. J. Sirevaag and J. A. Stern. *Engineering psychophysiology: Issues and applications.*, chapter Ocular measures of fatigue and cognitive factors. Lawrence Erlbaum and Associates, Inc., Mahwah, NJ, r. w. backs and w. boucsein (eds) edition, 2000.

569. L. Sirovich and M. Kirby. Low-dimensional procedure for the characterization of human faces. *Journal of the Optical Society of America*, 4:519–524, 1987.

570. H. S. Smallman, M. St. John, H. M. Oonk, and M. B. Cowen. Symbicons: A hybrid symbology that combines the best elements of symbols and icons. In *Proceedings of the Human Factors and Ergonomics Society 45th Annual Meeting*, pages 110–114, Santa Monica, CA, 2001. Human Factors and Ergonomics Society.

571. SMI. iview x: Eye and gaze tracker. http://www.smi.de/iv/.

572. C. D. Smith, T. M. Farrell, S. S. McNatt, and R. E. Metreveli. Assessing laparoscopic manipulative skills. *Am J Surg*, 181:547–550, 2001.

573. P. Smith, M. Shah, and N. daVitoria. Monitoring head-eye motion for driver alertness with one camera. *Pattern Recognition*, 2000.

574. L. Sobrado and J. C. Birget. Shoulder surfing resistant graphical passwords. Retrieved on September 29, 2005 at http://clam.rutgers.edu/ birget/grPssw/srgp.pdf.

575. P. D. Sozou, T. F. Cootes, C. J. Taylor, and E. C. di Mauro. Non-linear generalization of point distribution models using polynomial regression. *Image Vision Computing*, 13(5):451–457, 1995.

576. A. Spink. A user-centered approach to evaluating human interaction with web search engines: an exploratory study. *Information Processing and Management*, 38:401–426, 2002.

577. A. N. Srivastava. Mixture density mercer kernels: A method to learn kernels directly from data. In *Proc. SIAM International Conference on Data Mining (SDM)*, 2004.

578. A. N. Srivastava, J. Schumann, and B. Fischer. An ensemble approach to building mercer kernels with prior information. In *Proc. IEEE International Conference on Systems, Man and Cybernetics (SMC)*, pages 2352–2359, 2005.

579. Stanford Poynter Project. Front page entry points. http://www.poynterextra.org/et/i.htm.

580. H. Stanislaw and N. Todorov. Calculation of signal detection theory measures. *Behavior Research Methods, Instruments, & Computers*, 31(1):137–149, 1999.

581. L. W. Stark. *Eye Movements in Reading*, chapter Sequence of Fixations and Saccades in Reading, pages 151–163. Elsevier Science, Tarrytown, 1994. Ygge, J. and Lennerstrand, G. (Eds.).

582. L. W. Stark and S. R. Ellis. *Cognition and Visual Perception*, chapter Scanpaths Revisited: Cognitive Models Direct Active Looking. In Eye Movements, pages 193–226. Lawrence Erlbaum Associates, Hillsdale, 1981. Fisher D. F., Monty R. A. and Senders, J. W. (Eds.).

583. I. Starker and R. A. Bolt. A gaze-responsive self-disclosing display. In *CHI '90: Proceedings of the SIGCHI conference on Human factors in computing systems*, pages 3–10, New York, NY, USA, 1990. ACM Press.

584. C. P. Stein. Accurate Internal Camera Calibration Using Rotation, with Analysis of Sources of Errors. In *ICCV*, pages 230–236, June 1995.

585. L. B. Stelmach, W. J. Tam, and P. J. Hearty. Static and dynamic spatial resolution in image coding: An investigation of eye movements. In *Proc SPIE Human Vision, Visual Processing and Digital Display II*, volume 1453, pages 147–152, 1991.

586. M. Stephens. Bayesian Analysis of mixture Models with an Unknown Number of Components: An Alternative to reversible Jump Methods. *Annals of Statistics*, 28:40–74, 2000.

587. S. Stevenson, G. Kumar, and A. Roorda. Psychophysical and oculomotor reference points for visual direction measured with the adaptive optics scanning laser ophthalmoscope. *J. Vis.*, 7(9):137–137, 2007.

588. R. Stiefelhagen, J. Yang, and A. Waibel. A Model-Based Gaze-Tracking System. *International Journal of Artificial Intelligence Tools*, 6(2):193–209, 1997.

589. H. Strasburger, L. O. Harvey Jr., and I. Rentschler. Contrast thresholds for identification of numeric characters in direct and eccentric view. *Perception & Psychophysics*, 49:495–508, 1991.

590. D. L. Strayer, F. A. Drews, and W. A. Johnston. Cell phone-induced failures of visual attention during simulated driving. *Journal of Experimental Psychology: Applied*, 9(1):23–32, 2003.

591. D. L. Strayer and W. A. Johnston. Driven to distraction: Dual-task studies of simulated driving and conversing on a cellular telephone. *Psychological Science*, 12(4):462–466, 2001.

592. J. C. Stutts, J. W. Wilkins, and B. V. Vaughn. Why do people have drowsy driving crashes? input from drivers who just did. Technical report, AAA Foundation for Traffic Safety, 1999.

593. K. Subramonian, S. DeSylva, P. Bishai, P. Thompson, and G. Muir. Acquiring surgical skills: a comparative study of open versus laparoscopic surgery. *Eur Urol*, 45:346–51; author, 2004.

594. R. Swaminathan, M. D. Grossberg, and S. K. Nayar. Caustics of Catadioptric Cameras. In *IEEE ICCV 01*, volume II, pages 2–9, 2001.

595. J. Sweller. Cognitive load during problem solving: Effects on learning. *Cognitive Science*, 12:257–285, 1988.

596. J. Sweller. Cognitive technology: Some procedures for facilitating learning and problem solving in mathematics and science. *Journal of Educational Psychology*, 81(4):457–466, 1989.

597. J. Sweller, P. Chandler, P. Tierney, and M. Cooper. Cognitive load as a factor in the structuring of technical material. *Journal of Experimental Psychology*, 119(2):176–192, 1990.

598. J. Sweller, J. J. G. van Merrienboer, and F. G. W. C. Paas. Cognitive architecture and instructional design. *Educational Psychology Review*, 10(3):251–296, 1998.

599. D. L. Swets and J. J. Weng. Using discriminant eigenfeatures for image retrieval. *IEEE Transactions on Pattern Analysis and Machine Intelligence*, 18(8):831–836, 1996.

600. H. Takagi. Development of an eye-movement enhanced translation support system. In *APCHI '98: Proceedings of the Third Asian Pacific Computer and Human Interaction*, page 114, Washington, DC, USA, 1998. IEEE Computer Society.

601. M. Takeda. *Current Oculomotor Research: Physiological and Psychological Aspects*, chapter Eye Movements while Viewing a Foreign Movie with Subtitles, pages 313–315. Plenum Press, New York, 1999. Ulm, W., Heubel, H. and Mergner, T. (Eds.).

602. K. Talmi and J. Liu. Eye and gaze tracking for visually controlled interactive stereoscopic displays. *Signal Processing: Image Communication*, 14:799–810, 1999.

603. K.-H. Tan, D. J. Kriegman, and N. Ahuja. Appearance-based Eye Gaze Estimation. In *WACV*, pages 191–195, 2002.

604. P.-N. Tan. *Introduction to data mining*. Pearson Addison Wesley: Boston, 2005.

605. S. Tan, J. L. Dale, and A. Johnston. Performance of three recursive algorithms for fast space-variant Gaussian filtering. *Real-Time Imaging*, 9(3):215–228, 2003.

606. H. Tao and T. S. Huang. Visual estimation and compression of facial motion parameters: Elements of a 3d model-based video coding system. *Int'l Journal of Computer Vision*, 50(2):111–125, 2002.

607. D. S. Taubman and M. W. Marcellin. *JPEG2000 – Image Compression Fundamentals, Standards and Practice*. Kluwer Academic Publishers, 2001.

608. J. Tchalenko, A. James, A. Darzi, and G.-Z. Yang. Visual search in parallel environments – eye hand coordination in minimal invasive access surgery. *ECEM*, 2003.

609. J. H. Ten Kate, E. E. E. Frietman, W. Willems, B. M. Ter Haar Romeny, and E. Tenkink. Eye-switch controlled communication aids. In *Proceedings of the 12th International Conference on Medical & Biological Engineering*, 1979.

610. The National Library of Medicine. The Visible Human Project [online]. http://www.nlm.nih.gov/research/visible/visible_human.html, 2003.

611. W. M. Theimer. Phase-based binocular vergence control and depth reconstruction using active vision. *CVGIP: Image Understanding*, 60(3):343–358, 1994.

612. S. Theodoridis and K. Koutroumbas. *Pattern Recognition*. Academic Press, 1999.

613. J. Thorpe and P. van Oorschot. Graphical dictionaries and the memorable space of graphical passwords. *In Proceedings of the 13th USENIX Security Symposium*, pages 135–150, August 2004.

614. Y. Tian, T. Kanade, and J. F. Cohn. Dual-state parametric eye tracking. In *Proc. IEEE International Conference on Automatic Face and Gesture Recognition (FGR)*, pages 110–115, 2000.

615. Y. Tian, T. Kanade, and J. F. Cohn. Evaluation of gabor-wavelet-based facial action unit recognition in image sequences of increasing complexity. *Proc. of FGR02*, pages 218–223, 2002.

616. Y. Tian, T. Kanade, and J. F. Cohn. Eye-state detection by local regional information. In *Proceedings of the International Conference on Multimodal User Interface*, pages 143–150, Oct 2000.

617. Y. Tian, T. Kanade, and J. F. Cohn. Recognizing action units for facial expression analysis. *IEEE Trans. Pattern Anal. Machine Intell.*, 23(2):97–115, 2001.

618. D. H. Tiley, C. W. Erwin, and D. T. Gianturco. Drowsiness and driving: Preliminary report of a population survey. Technical report, Society of Automotive Engineers, Detroit, MI, 1973.

619. M. E. Tipping. Sparse Bayesian Learning and the Relevance Vector Machine. *Journal of Machine Learning Research*, 1:211–244, 2001.

620. M. D. Tisdall and M. S. Atkins. Using human and model performance to compare mri reconstructions. *IEEE Transactions on Medical Imaging*, 25(11):1510–1517, 2006.

621. D. M. Titterington, A. F. M. Smith, and U. E. Markov. *Statistical Analysis of Finite Mixture Distributions*. New York: Wiley, 1985.

622. Tobii Technology. Clearview software.
http://www.tobii.com/products_-_services/
testing_and_research_products/tobii_clearview_analysis_software/.

623. Tobii Technology. Tobii x50. http://www.tobii.com/ products_-_services/
testing_and_research_products/ tobii_x50_eye_tracker/.

624. C. Tomasi and T. Kanade. Detection and tracking of point features. *Carnegie Mellon University Technical Report CMU-CS-91-132*, April 1991.

625. S. S. Tomkins. *Affect, imagery, consciousness*. Springer, New York, 1962.

626. H. M. Tong and R. A. Fisher. Progress report on an eye-slaved area-of-interest visual display (report no. afhrl-tr-84-36, air force human resources laboratory, brooks air force base, texas.). In *Proceedings of IMAGE III Conference*, 1984.

627. L. Torresani and C. Bregler. Space-time tracking. *Proc. of ECCV*, 2002.

628. G. M. Treece, R. W. Prager, and A. H. Gee. Regularised marching tetrahedra: Improved iso-surface extraction. *Computers and Graphics UK*, 23(4):583–598, 1999.

629. E. Trucco and A. Verri. *Introductory Techniques for 3-D Computer Vision*. Prentice-Hall, 1998.

630. L. Trujillo, G. Olague, R. I. Hammoud, B. Hernandez, and E. Romero. Automatic feature localization in thermal images for facial expression recognition. *Journal of Computer Vision and Image Understanding (CVIU)*, 106(2 and 3):258–230, 2007.

631. R. Y. Tsai. A versatile camera calibration technique for high-accuracy 3d machine vision metrology using off-the-shelf tv cameras and lenses. *IEEE Journal of Robotics and Automation*, RA-3(4):323–344, 1987.

632. S. Tseng. Comparison of holistic and feature based approaches to face recognition. Master's thesis, School of Computer Science and Information Technology, Faculty of Applied Science, Royal Melbourne Institute of Technology University, Melbourne, Victoria, Australia, July 2003.

633. N. Tsumura, M. N. Dang, T. Makino, and Y. Miyake. Estimating the Directions to Light Sources Using Images of Eye for Reconstructing 3D Human Face. In *Proc. of IS&T/SID's Eleventh Color Imaging Conference*, pages 77–81, 2003.

634. M. Turk and A. Pentland. Eigenfaces for recognition. *Journal of Cognitive Neuroscience*, 3(1):71–86, 1991.

635. M. A. Turk and A. P. Pentland. Face recognition using eigenfaces. In *Proc. IEEE Computer Vision and Pattern Recognition '91*, pages 586–591, 1991.

636. H. Ueno, M. Kaneda, and M. Tsukino. Development of drowsiness detection system. In *Proc. Conf. Vehicle navigation & information Systems Conference*, 1994.

637. G. Underwood, P. Chapman, N. Brocklehurst, J. Underwood, and D. Crundall. Visual attention while driving: Sequences of eye fixations made by experienced and novice drivers. *Ergonomics*, 46(6):629–646, 2003.

638. M. Unser. Sum and Difference Histograms for Texture Classification. *IEEE Transactions on Pattern Analysis and Machine Intelligence*, 8(1):118–125, 1986.

639. H. Uwano, M. Nakamura, A. Monden, and K. Matsumoto. Analyzing individual performance of source code review using reviewers' eye movement. In *ETRA'06: Proceedings of the 2006 symposium on Eye tracking research & applications*, pages 133–140, New York, NY, USA, 2006. ACM Press.

640. W. Vanlaar, H. Simpson, D. Mayhew, and R. Robertson. Fatigued and drowsy driving: Attitudes, concern and practices of ontario drivers. Technical report, Traffic Injury Research Foundation, July 2007.

641. V. N. Vapnik. *Statistical Learning Theory*. New York: John Wiley and Sons, 1998.

642. V. N. Vapnik. *The nature of statistical learning theory*. Springer, New York, 1995.

643. B. Velichkovsky, A. Sprenger, and P. Unema. Towards gaze-mediated interaction: Collecting solutions of the "midas touch problem". In *INTERACT '97: Proceedings of the IFIP TC13 Interantional Conference on Human–Computer Interaction*, pages 509–516, London, UK, UK, 1997. Chapman & Hall, Ltd.

644. R. Vertegaal. Attentive user interfaces. *Commun. ACM*, 46(3):30–33, 2003.

645. R. Vertegaal. The gaze groupware system: mediating joint attention in multiparty communication and collaboration. In *CHI '99: Proceedings of the SIGCHI conference on Human factors in computing systems*, pages 294–301, New York, NY, USA, 1999. ACM Press.

646. R. Vertegaal, A. Mamuji, C. Sohn, and D. Cheng. Media eyepliances: using eye tracking for remote control focus selection of appliances. In *CHI '05: Extended abstracts on Human factors in computing systems*, pages 1861–1864, New York, NY, USA, 2005. ACM Press.

647. R. Vertegaal, I. Weevers, and C. Sohn. Gaze-2: an attentive video conferencing system. In *CHI '02: Extended abstracts on Human factors in computing systems*, pages 736–737, New York, NY, USA, 2002. ACM Press.

648. V. Vezhnevets and A. Degtiareva. Robust and accurate eye contour extraction. In *Proc. Graphicon*, pages 81–84, Moscow, Russia, September 2003 2003.

649. T. W. Victor, J. L. Harbluk, and J. A. Engstrom. Sensitivity of eye-movement measures to in-vehicle task difficulty. *Transportation Reseach Part F*, 8:167–190, 2005.

650. I. Viola, M. E. Gröller, and A. Kanitsar. Importance-driven volume rendering. In *Proceedings of IEEE Visualization'04*, pages 139–145, 2004.

651. I. Viola and M. Sbert. Importance-driven focus of attention. *IEEE Trans. Vis. and Comp. Graph.*, 12(5):933–940, 2006.

652. P. Viola and M. Jones. Robust real-time object detection. *Int'l Journal of Computer Vision*, 57(2):137–154, 2004.

653. P. Viviani. *Eye Movements and Their Role in Visual and Cognitive Processes*, chapter 8. Elsevier Science, Amsterdam, 1990.

654. C. S. Wallace. *Statistical and Inductive Inference by Minimum Message Length*. Springer, 2005.

655. E. L. Waltz. Information understanding: integrating data fusion and data mining processes. In *Proceedings of Proceedings of the 1998 IEEE International Symposium on Circuits and Systems, ISCAS '98, Monterey, CA*, volume 6, pages 553–556, 1998.

656. J. G. Wang and E. Sung. Study on eye gaze estimation. *IEEE Trans. on Systems, Man and Cybernetics, Part B*, 32(3):332–350, 2002.

657. J.-G. Wang, E. Sung, and R. Venkateswarlu. Eye Gaze Estimation from a Single Image of One Eye. In *IEEE ICCV 03*, pages 136–143, 2003.

658. J. S. Wang, R. R. Knipling, and M. J. Goodman. The role of driver inattention in crashes; new statistics from the 1995 crashworthiness data system. In *40th Annual Proceedings of Association for the Advancement of Automotive Medicine.*, Vancouver, British Columbia, 1996.

659. L. Wang, Y. Zhao, K. Mueller, and A. Kaufman. The magic volume lens: An interactive focus+context technique for volume rendering. In *Visualization 2005, IEEE VIS'05*, pages 367–374, 1995.

660. P. Wang, M. B. Green, Q. Ji, and J. Wayman. Automatic eye detection and its validation. *IEEE Workshop on Face Recognition Grand Challenge Experiments (with CVPR)*, 3, 2005.

661. P. Wang and Q. Ji. Learning discriminant features for multi-view face and eye detection. *Proc. of CVPR05*, 1:373–379, 2005.

662. Y. Wang, T. Tan, and K.-F. Loe. Joint region tracking with switching hypothesized measurements. *Proc. of ICCV03*, 1:75–82, 2003.

663. Z. Wang, L. Lu, and A. C. Bovik. Foveation scalable video coding with automatic fixation selection. *IEEE Transactions on Image Processing*, 12(2):243–254, 2003.

664. D. J. Ward and D. J. C. MacKay. Fast hands-free writing by gaze direction. *Nature*, 418(6900):838, 2002.

665. P. R. Warshaw. Application of selective attention theory to television advertising displays. *Journal of Applied Psychology*, 63(3):366–372, 1978.

666. S. Wasserman and K. Faust. *Social Network Analysis: Methods and Applications*. Cambridge University Press, 1994.

667. A. Webb. *Statistical Pattern Recognition*. Oxford University Press Inc., New York, 1999.

668. R. H. Webb and G. W. Hughes. Scanning laser ophthalmoscope. *IEEE Trans. Biomed. Eng.*, 28:488–492, 1981.

669. R. H. Webb, G. W. Hughes, and F. C. Delori. Confocal scanning laser ophthalmoscope. *Appl. Opt.*, 26:1492–1499, 1987.

670. R. H. Webb, G. W. Hughes, and O. Pomerantzeff. Flying spot tv ophthalmoscope. *Appl. Opt.*, 19:2991–2997, 1980.

671. W. W. Wierwille and F. L. Eggemeier. Recommendations for mental workload measurement in a test and evaluation environment. *Human Factors*, pages 263–281, 1993.

672. J. West, A. Haake, E. Rozanski, and K. Karn. eyepatterns: software for identifying patterns and similarities across fixation sequences. In *ETRA '06: Proceedings of the 2006 symposium on Eye tracking research & applications*, pages 149–154, New York, NY, USA, 2006. ACM Press.

673. C. Wester, P. F. Judy, M. Polger, R. G. Swensson, U. Feldman, and S. E. Selzer. Influence of visual distracters on detectability of liver nodules on contrast-enhanced spiral computed tomography scans. *Academic Radiology*, 4(5):335–342, 1997.

674. R. Westermann and B. Sevenich. Accelerated volume ray-casting using texture mapping. In *Proceedings IEEE Visualization 2001*, pages 271–278, 2001.

675. G. Westheimer. *Medical Physiology*, volume 1, chapter 16 The eye, pages 481–503. The C.V. Mosby Company, 1980. V. B. Mountcastle (Ed.).

676. C. D. Wickens. Multiple resources and performance prediction. *Theoretical Issues in Ergonomics Science*, 3(2):159–177, 2002.

677. S. Wiedenbeck, J. Waters, J. C. Birget, A. Brodskiy, and N. Memon. Authentication using graphical passwords: Basic results. *In Human–Computer Interaction International (HCII 2005)*, July 2005.

678. S. Wiedenbeck, J. Waters, J. C. Birget, A. Brodskiy, and N. Memon. Authentication using graphical passwords: Effects of tolerance and image choice. *In*

Proceedings of the Symposium on Usable Privacy and Security (SOUPS), July 2005.

679. S. Wiedenbeck, J. Waters, L. Sobrado, and J. C. Birget. Design and evaluation of a shoulder-surfing resistant graphical password scheme. *In Proceedings of Advanced Visual Interfaces (AVI2006)*, May 2006.

680. W. W. Wierwille, L. A. Ellsworth, S. S. Wreggit, R. J. Fairbanks, and C. L. Kirn. Research on vehicle-based driver status/performance monitoring; development, validation, and refinement of algorithms for detection of driver drowsiness. Technical report, Virginia Tech Transportation Institute, 1994.

681. L. Wilkinson and the Task Force on Statistical Inference. Statistical methods in psychology journals: Guidelines and explanations [Electronic version]. *American Psychologist*, 54:594–604, 1999.

682. L. Wiskott, J. M. Fellous, N. Krüger, and C. Von der Malsburg. Face recognition by elastic bunch graph matching. *IEEE Trans. on PAMI*, 19(7):775–779, 1997.

683. C. M. Wittenbrink, T. Malzbender, and M. E. Goss. Opacity-weighted color interpolation for volume sampling. In *Proc. 1998 Symposium on Volume Visulization*, pages 135–142, Oct 1998.

684. M. Woo, J. Neider, T. Davis, D. Shreiner, and OpenGL Architecture Review Board. *OpenGL Programming Guide*. Addison Wesley, Reading, Massachusetts, 3rd edition, 1999.

685. D. Wooding. Eye movements of large populations: Ii. deriving regions of interest, coverage, and similarity using fixation maps. *Behavior Research Methods, Instruments, & Computers*, 34(4):518–52, 2002.

686. C. Wu, C. Liu, H. Shum, Y. Xy, and Z. Zhang. Automatic eyeglasses removal from face images. *IEEE Transactions on Pattern Analysis and Machine Intelligence*, 26(3):322–336, 2004.

687. G. Wyszecki and W. S. Stiles. Color science: Concepts and methods, quantitative data and formulae. John Wiley & Sons, 1982.

688. J. Xiao, S. Baker, I. Matthews, and T. Kanade. Real-time combined 2d+3d active appearance models. *Proc. of CVPR04*, 2:535–542, 2004.

689. J. Xiao, J. Chai, and T. Kanade. A closed-form solution to non-rigid shape and motion recovery. *Int'l Journal of Computer Vision*, 67(2):233–246, 2006.

690. J. Xiao, T. Moriyama, T. Kanade, and J. F. Cohn. Robust full-motion recovery of head by dynamic templates and re-registration techniques. *International Journal of Imaging Systems and Technology*, 13:85–94, 2003.

691. K. Xie, J. Yang, and Y. M. Zhu. Multi-resoution LOD volume rendering in medicine. *Lecture notes in computer science*, 3516:925–933, 2005.

692. X. Xie, R. Sudhakar, and H. Zhuang. On improving eye feature extraction using deformable templates. *PR*, 27(6):791–799, 1994.

693. L-Q. Xu, D. Machin, and P. Sheppard. A Novel Approach to Real-time Nonintrusive Gaze Finding. In *BMVC*, pages 58–67, 1998.

694. M. Yamato, K. Inoue, A. Monden, K. Torii, and K.-I. Matsumoto. Button selection for general guis using eye and hand together. In *AVI '00: Proceedings of the working conference on Advanced visual interfaces*, pages 270–273, New York, NY, USA, 2000. ACM Press.

695. S. Yan, X. Hou, S. Z. Li, H. Zhang, and Q. Cheng. Face alignment using view-based direct appearance models. *Special Issue on Facial Image Processing, Analysis and Synthesis, Int'l Journal of Imaging Systems and Technology*, 13(1):106–112, 2003.

696. G.-Z. Yang, L. Dempere-Marco, X.-P. Hu, and A. Rowe. Visual search: psychophysical models and practical applications. *Image and Vision Computing*, 20:273–287, 2002.

697. M. Yang, D. J. Kriegman, and N. Ahuja. Detecting Faces in Images: A Survey. *IEEE Transactions on Pattern Analysis and Machine Intelligence*, 24(1):34–58, 2002.

698. K. Yano, K. Ishihara, M. Makikawa, and H. Kusuola. Detection of eye blinking from video camera with dynamic roi fixation. *IEEE Trans. SMC*, 6:335–339, 1999.

699. A. L. Yarbus. *Eye Movement and Vision.* Plenum Press, New York, 1967.

700. A. Yilmaz, K. Shafique, and M. Shah. Estimation of rigid and non-rigid facial motion using anatomical face model. *Proc. of ICPR02*, 1:377–380, 2002.

701. L. Yin and A. Basu. Realistic animation using extended adaptive mesh for model based coding. In *Proceedings of Energy Minimization methods in Computer Vision and Pattern Recognition*, pages 269–284, 1999.

702. D. H. Yoo and M. J. Chung. A novel non-intrusive eye gaze estimation using cross-ratio under large head motion. *Computer Vision and Image Understanding*, 98(1):25–51, 2005.

703. D. H. Yoo, B. R. Lee, and M. J. Chung. Non-contact eye gaze tracking system by mapping of corneal reflections. In *FGR '02: Fifth IEEE International Conference on Automatic Face and Gesture Recognition*, pages 101–106, Washington, DC, USA, 2002.

704. A. Yoshida, J. Rolland, and J. H. Reif. Design and applications of a high resolution insert head-mounted display. In *IEEE Virtual Reality Annual International Symposium (VRAIS'95)*, pages 84–93, 1995.

705. K. Young. Boffins use play-doh to fool biometrics. *Posted on vnunet.com*, December 2005. Retrieved on September 12, 2006 at http://www.itweek.co.uk/vnunet/news/2147491/hackers-play-doh-fool.

706. L. Young and D. Sheena. Methods & designs: Survey of eye movement recording methods. *Behavioral Research Methods & Instrumentation*, 7(5):397–429, 1975.

707. A. Yuille et al. *Active Vision*, chapter 2, pages 21–38. MIT Press, 1992.

708. A. Yuille, P. Haallinan, and D. S. Cohen. Feature extraction from faces using deformable templates. *Int'l Journal of Computer Vision*, 8(2):99–111, 1992.

709. A. L. Yuille, D. S. Cohen, and P. W. Hallinan. Feature extraction from faces using deformable templates. *IJCV*, 8(2):99–111, 1992.

710. W. H. Zangemeister, U. Oechsner, and C. Freksa. Short-term adaptation of eye movements in patients with visual hemifield defects indicates high level control of human scan path. *Optom Vision Science*, 72(7):467–477, 1995.

711. S. Zhai, C. Morimoto, and S. Ihde. Manual and gaze input cascaded (magic) pointing. In *CHI '99: Proceedings of the SIGCHI conference on Human factors in computing systems*, pages 246–253, New York, NY, USA, 1999. ACM Press.

712. H. Zhang, M. R. H. Smith, and G. J. Witt. Driver state assessment and driver support systems. In *SAE Convergence 2006 Proceedings (CD)*, Detroit, MI, 2006.

713. H. Zhang, M. R. H. Smith, and G. J. Witt. Identification of real-time diagnostic measures of visual distraction with an automatic eye-tracking system. *Human Factors*, 48(4):805–822, 2006.

714. Y. Zhang, Y. Owechko, and J. Zhang. Driver cognitive workload estimation: A data-driven perspective. In *Proceedings of IEEE Intelligent Transportation Systems Conference, Washington, D.C.*, pages 642–647, 2004.

715. Z. Zhang, M. Lyons, M. Schuster, and S. Akamatsu. Comparison between geometry-based and gabor-wavelets-based facial expression recognition using multi-layer perceptron. *Proc. of FGR98*, pages 454–459, 1998.

716. B. Zheng, Z. Janmohamed, and C. L. MacKenzie. Reaction times and the decision-making process in endoscopic surgery. *Surg Endosc*, 17:1475–1480, 2003.

717. J. Zhu and J. Yang. Subpixel eye gaze tracking. In *Proc. IEEE International Conference on Automatic Face and Gesture Recognition*, pages 131–136, May 2002.

718. Z. Zhu, K. Fujimura, and Q. Ji. Real-Time Eye Detection and Tracking Under Various Light Conditions. In *Proc. the 2002 Symposium on Eye Tracking Research & Applications*, pages 139–144, 2002.

719. Z. Zhu and Q. Ji. Robust Real-Time Eye Detection and Tracking under Variable Lighting Conditions and Various Face Orientations. *Computer Vision and Image Understanding (CVIU)*, 98:124–154, 2005.

720. Z. Zhu, Q. Ji, K. Fujimura, and K. Lee. Combining kalman filtering and mean shift for real time eye tracking under active ir illumination. *Proc. of ICPR02*, 4:318–321, 2002.

Index

Printing: Krips bv, Meppel, The Netherlands
Binding: Stürtz, Würzburg, Germany

$\frac{14}{83 \ 16}$